A First Course in Fun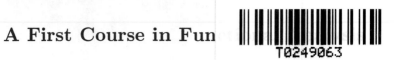

T0249063

A First Course in Functional Analysis

Theory and Applications

Rabindranath Sen

ANTHEM PRESS
LONDON · NEW YORK · DELHI

Anthem Press
An imprint of Wimbledon Publishing Company
www.anthempress.com

This edition first published in UK and USA 2014
by ANTHEM PRESS
75–76 Blackfriars Road, London SE1 8HA, UK
or PO Box 9779, London SW19 7ZG, UK
and
244 Madison Ave. #116, New York, NY 10016, USA

First published in hardback by Anthem Press in 2013

British Library Cataloguing-in-Publication Data
A catalogue record for this book is available from the British Library.

Library of Congress Cataloging-in-Publication Data
A catalogue record for this book has been requested.

ISBN-13: 978 1 78308 324 4 (Pbk)
ISBN-10: 1 78308 324 7 (Pbk)

This title is also available as an ebook.

In memory

of my parents

Preface

This book is the outgrowth of the lectures delivered on functional analysis and allied topics to the postgraduate classes in the Department of Applied Mathematics, Calcutta University, India. I feel I owe an explanation as to why I should write a new book, when a large number of books on functional analysis at the elementary level are available. Behind every abstract thought there is a concrete structure. I have tried to unveil the motivation behind every important development of the subject matter. I have endeavoured to make the presentation lucid and simple so that the learner can read without outside help.

The first chapter, entitled 'Preliminaries', contains discussions on topics of which knowledge will be necessary for reading the later chapters. The first concepts introduced are those of a set, the cardinal number, the different operations on a set and a partially ordered set respectively. Important notions like Zorn's lemma, Zermelo's axiom of choice are stated next. The concepts of a function and mappings of different types are introduced and exhibited with examples. Next comes the notion of a linear space and examples of different types of linear spaces. The definition of subspace and the notion of linear dependence or independence of members of a subspace are introduced. Ideas of partition of a space as a direct sum of subspaces and quotient space are explained. 'Metric space' as an abstraction of real line $\mathbb{R}$ is introduced. A broad overview of a metric space including the notions of convergence of a sequence, completeness, compactness and criterion for compactness in a metric space is provided in the first chapter. Examples of a non-metrizable space and an incomplete metric space are also given. The contraction mapping principle and its application in solving different types of equations are demonstrated. The concepts of an open set, a closed set and an neighbourhood in a metric space are also explained in this chapter. The necessity for the introduction of 'topology' is explained first. Next, the axioms of a topological space are stated. It is pointed out that the conclusions of the Heine-Borel theorem in a real space are taken as the axioms of an abstract topological space. Next the ideas of openness and closedness of a set, the neighbourhood of a point in a set, the continuity of a mapping, compactness, criterion for compactness and separability of a space naturally follow.

Chapter 2 is entitled 'Normed Linear Space'. If a linear space admits a metric structure it is called a metric linear space. A normed linear space is a type of metric linear space, and for every element x of the space there exists a positive number called norm x or $\|x\|$ fulfilling certain axioms. A normed linear space can always be reduced to a metric space by the choice of a suitable metric. Ideas of convergence in norm and completeness of a normed linear space are introduced with examples of several normed linear spaces, Banach spaces (complete normed linear spaces) and incomplete normed linear spaces.

Continuity of a norm and equivalence of norms in a finite dimensional normed linear space are established. The definition of a subspace and its various properties as induced by the normed linear space of which this is a subspace are discussed. The notion of a quotient space and its role in generating new Banach spaces are explained. Riesz's lemma is also discussed.

Chapter 3 dwells on Hilbert space. The concepts of inner product space, complete inner product or Hilbert space are introduced. Parallelogram law, orthogonality of vectors, the Cauchy-Bunyakovsky-Schwartz inequality, and continuity of scalar (inner) product in a Hilbert space are discussed. The notions of a subspace, orthogonal complement and direct sum in the setting of a Hilbert space are introduced. The orthogonal projection theorem takes a special place.

Orthogonality, various orthonormal polynomials and Fourier series are discussed elaborately. Isomorphism between separable Hilbert spaces is also addressed. Linear operators and their elementary properties, space of linear operators, linear operators in normed linear spaces and the norm of an operator are discussed in Chapter 4. Linear functionals, space of bounded linear operators and the uniform boundedness principle and its applications, uniform and pointwise convergence of operators and inverse operators and the related theories are presented in this chapter. Various types of linear operators are illustrated. In the next chapter, the theory of linear functionals is discussed. In this chapter I introduce the notions of a linear functional, a bounded linear functional and the limiting process, and assert continuity in the case of boundedness of the linear functional and vice-versa. In the case of linear functionals apart from different examples of linear functionals, representation of functionals in different Banach and Hilbert spaces are studied. The famous Hahn-Banach theorem on the extension on a functional from a subspace to the entire space with preservation of norm is explained and the consequences of the theorem are presented in a separate chapter. The notions of adjoint operators and conjugate space are also discussed. Chapter 6 is entitled 'Space of Bounded Linear Functionals'. The chapter dwells on the duality between a normed linear space and the space of all bounded linear functionals on it. Initially the notions of dual of a normed linear space and the transpose of a bounded linear operator on it are introduced. The zero spaces and range spaces of a bounded linear operator and of its duals are related. The duals of $L_p([a,b])$ and $C([a,b])$ are described. Weak convergence in a normed linear space and its dual is also discussed. A reflexive normed linear space is one for which the canonical embedding in the second dual is surjective (one-to-one). An elementary proof of Eberlein's theorem is presented. Chapter 7 is entitled 'Closed Graph Theorem and its Consequences'. At the outset the definitions of a closed operator and the graph of an operator are given. The closed graph theorem, which establishes the conditions under which a closed linear operator is bounded, is provided. After introducing the concept of an open mapping, the open mapping theorem and the bounded inverse theorem are proved. Application of the open mapping theorem is also provided. The next chapter bears the title 'Compact Operators on Normed Linear Spaces'. Compact linear operators are very important in applications. They play a crucial role in the theory of integral equations and in various problems of mathematical physics. Starting from the definition of compact operators, the criterion for compactness of a linear operator with a finite dimensional domain or range in a normed linear space and other results regarding compact linear operators are established. The spectral properties of a compact linear operator are studied. The notion of the Fredholm alternative is discussed and the relevant theorems are provided. Methods of finding an approximate solution of certain equations involving compact operators in a normed linear space are explored. Chapter 9 bears the title 'Elements of Spectral Theory on Self-adjoint Operators in Hilbert Spaces'. Starting from the definition of adjoint operators, self-adjoint operators and their various properties are elaborated upon the context of a Hilbert space. Quadratic forms and quadratic Hermitian forms are introduced in a Hilbert space and their bounds are discovered. I define a unitary operator in a Hilbert space and the situation when two operators are said to be unitarily equivalent, is explained. The notion of a projection operator in a Hilbert space is introduced and its various properties are investigated. Positive operators and the square root of operators in a Hilbert space are introduced and their properties are studied. The spectrum of a self-adjoint operator in a Hilbert space is studied and the point spectrum and continuous spectrum are explained. The notion of invariant subspaces in a Hilbert space is also brought within the purview of the discussion. Chapter 10 is entitled 'Measure and Integration in Spaces'. In this chapter I discuss the theory of Lebesgue integration and p-integrable functions on $\mathbb{R}$. Spaces of these functions provide very useful examples of many theorems in functional analysis. It is pointed out that the concept of the Lebesgue measure is a generalization of the idea of subintervals of given length in $\mathbb{R}$ to a class of subsets in $\mathbb{R}$. The ideas of the Lebesgue outer measure of a set $E \subset \mathbb{R}$, Lebesgue measurable set E and the Lebesgue measure of E are introduced. The notions of measurable functions and integrable functions in the sense of Lebesgue are explained. Fundamental theorems of Riemann integration and Lebesgue integration, Fubini and Toneli's theorem, are stated and explained. L_p spaces (the space of functions p-integrable on a measure subset E of $\mathbb{R}$) are introduced, that (E) is complete and related properties discussed. Fourier series and then Fourier integral for functions are investigated. In the next chapter, entitled 'Unbounded Linear Operators', I first give some examples of differential operators that are not bounded. But these are closed operators, or at least have closed linear extensions. It is indicated in this chapter that many of the important theorems that hold for continuous linear operators on a Banach space also hold for closed linear operators. I define the different states of an operator depending on whether

the range of the operator is the whole of a Banach space or the closure of
the range is the whole space or the closure of the range is not equal to
the space. Next the characterization of states of operators is presented.
Strictly singular operators are then defined and accompanied by examples.
Operators that appear in connection with the study of quantum mechanics
also come within the purview of the discussion. The relationship between
strictly singular and compact operators is explored. Next comes the study
of perturbation theory. The reader is given an operator 'A', the certain
properties of which need be found out. If 'A' is a complicated operator, we
sometimes express 'A = T + B' where 'T' is a relatively simple operator and
'B' is related to 'T' in such a manner that knowledge about the properties
of 'T' is sufficient to gain information about the corresponding properties
of 'A'. In that case, for knowing the specific properties of 'A', we can
replace 'A' with 'T', or in other words we can perturb 'A' by 'T'. Here
we study perturbation by a bounded linear operator and perturbation by
strictly singular operator. Chapter 12 bears the title 'The Hahn-Banach
Theorem and the Optimization Problems'. I first explain an optimization
problem. I define a hyperplane and describe what is meant by separating
a set into two parts by a hyperplane. Next the separation theorems for
a convex set are proved with the help of the Hahn-Banach theorem. A
minimum Norm problem is posed and the Hahn-Banach theorem is applied
to the proving of various duality theorems. Said theorem is applied to prove
Chebyshev approximation theorems. The optimal control problem is posed
and the Pontryagin's problem is mentioned. Theorems on optimal control of
rockets are proved using the Hahn-Banach theorem. Chapter 13 is entitled
'Variational Problems' and begins by introducing a variational problem.
The aim is to investigate under which conditions a given functional in a
normed linear space admits of an optimum. Many differential equations are
often difficult to solve. In such cases a functional is built out of the given
equation and minimized. One needs to show that such a minimum solves
the given equation. To study those problems, a Gâteaux derivative and a
Fréchet derivative are defined as a prerequisite. The equivalence of solving
a variational problem and solving a variational inequality is established.
I then introduce the Sobolev space to study the solvability of differential
equations. In Chapter 14, entitled 'The Wavelet Analysis', I provide a
brief introduction to the origin of wavelet analysis. It is the outcome of
the confluence of mathematics, engineering and computer science. Wavelet
analysis has begun to play a serious role in a broad range of applications
including signal processing, data and image compression, the solving of
partial differential equations, the modeling of multiscale phenomena and
statistics. Starting from the notion of information, we discuss the scalable
structure of information. Next we discuss the algebra and geometry of
wavelet matrices like Haar matrices and Daubechies's matrices of different
ranks. Thereafter come the one-dimensional wavelet systems where the
scaling equation associated with a wavelet matrix, the expansion of a

function in terms of wavelet system associated with a matrix and other results are presented. The final chapter is concerned with dynamical systems. The theory of dynamical systems has its roots in the theory of ordinary differential equations. Henry Poincaré and later Ivar Benedixon studied the topological properties of the solutions of autonomous ordinary differential equations (ODEs) in the plane. They did so with a view of studying the basic properties of autonomous ODEs without trying to find out the solutions of the equations. The discussion is confined to one-dimensional flow only.

Prerequisites The reader of the book is expected to have a knowledge of set theory, elements of linear algebra as well as having been exposed to metric spaces.

Courses The book can be used to teach two semester courses at the M.Sc. level in universities (MS level in Engineering Institutes):

(i) Basic course on functional analysis. For this Chapters 2–9 may be consulted.

(ii) Another course may be developed on linear operator theory. For this Chapters 2, 3–5, 7–9 and 11 may be consulted. The Lebesgue measure is discussed at an elementary level in Chapter 10; Chapters 2–9 can, however, be read without any knowledge of the Lebesgue measure.

Those who are interested in applications of functional analysis may look into Chapters 12 and 13.

Acknowledgements I wish to express my profound gratitude to my advisor, the late Professor Parimal Kanti Ghosh, former Ghose professor in the Department of Applied Mathematics, Calcutta University, who introduced me to this subject. My indebtedness to colleagues and teachers like Professor J. G. Chakraborty, Professor S. C. Basu is duly acknowledged. Special mention must be made of my colleague and friend Professor A. Roy who constantly encouraged me to write this book. My wife Mrs. M. Sen offered all possible help and support to make this project a success, and thanks are duly accorded. I am also indebted to my sons Dr. Sugata Sen and Professor Shamik Sen for providing editorial support. Finally I express my gratitude to the inhouse editors and the external reviewer. Several improvements in form and content were made at their suggestion.

Contents

Introduction

Functional analysis is an abstract branch of mathematics that grew out of classical analysis. It represents one of the most important branches of the mathematical sciences. Together with abstract algebra and mathematical analysis, it serves as a foundation of many other branches of mathematics. Functional analysis is in particular widely used in probability and random function theory, numerical analysis, mathematical physics and their numerous applications. It serves as a powerful tool in modern control and information sciences.

The development of the subject started from the beginning of the twentieth century, mainly through the initiative of the Russian school of mathematicians. The impetus came from the developments of linear algebra, linear ordinary and partial differential equations, calculus of variation, approximation theory and, in particular, those of linear integral equations, the theory of which had the greatest impact on the development and promotion of modern ideas. Mathematicians observed that problems from different fields often possess related features and properties. This allowed for an effective unifying approach towards the problems, the unification being obtained by the omission of inessential details. Hence the advantage of such an abstract approach is that it concentrates on the essential facts, so that they become clearly visible.

Since any such abstract system will in general have concrete realisations (concrete models), we see that the abstract method is quite versatile in its applications to concrete situations. In the abstract approach, one usually starts from a set of elements satisfying certain axioms. The nature of the elements is left unspecified. The theory then consists of logical consequences, which result from the axioms and are derived as theorems once and for all. This means that in the axiomatic fashion one obtains a mathematical structure with a theory that is developed in an abstract way.

For example, in algebra this approach is used in connection with fields, rings and groups. In functional analysis, we use it in connection with 'abstract' spaces; these are all of basic importance.

In functional analysis, the concept of space is used in a very wide and surprisingly general sense. An abstract space will be a set of (unspecified) elements satisfying certain axioms, and by choosing different sets of axioms, we obtain different types of abstract spaces.

The idea of using abstract spaces in a systematic fashion goes back to M. Fréchet (1906) and is justified by its great success. With the introduction of abstract space in functional analysis, the language of geometry entered the arena of the problems of analysis. The result is that some problems of analysis were subjected to geometric interpretations. Furthermore many conjectures in mechanics and physics were suggested, keeping in mind the two-dimensional geometry. The geometric methods of proof of many theorems came into frequent use.

The generalisation of algebraic concepts took place side by side with that of geometric concepts. The classical analysis, fortified with geometric and algebraic concepts, became versatile and ready to cope with new problems not only of mathematics but also of mathematical physics. Thus functional analysis should form an indispensable part of the mathematics curricula at the college level.

CHAPTER 1

PRELIMINARIES

In this chapter we recapitulate the mathematical preliminaries that will be relevant to the development of functional analysis in later chapters. This chapter comprises six sections. We presume that the reader has been exposed to an elementary course in real analysis and linear algebra.

1.1 Set

The theory of sets is one of the principal tools of mathematics. One type of study of set theory addresses the realm of logic, philosophy and foundations of mathematics. The other study goes into the highlands of mathematics, where set theory is used as a medium of expression for various concepts in mathematics. We assume that the sets are 'not too big' to avoid any unnecessary contradiction. In this connection one can recall the famous 'Russell's Paradox' (Russell, 1959). A set is a collection of distinct and distinguishable objects. The objects that belong to a set are called elements, members or points of the set. If an object a belongs to a set A, then we write $a \in A$. On the other hand, if a does not belong to A, we write $a \notin A$. A set may be described by listing the elements and enclosing them in braces. For example, the set A formed out of the letters a, a, a, b, b, c can be expressed as $A = \{a, b, c\}$. A set can also be described by some defining properties. For example, the set of natural numbers can be written as $\mathbb{N} = \{x : x, \text{ a natural number}\}$ or $\{x|x, \text{ a natural number}\}$. Next we discuss *set inclusion*. If every element of a set A is an element of the set B, A is said to be a *subset* of the set B or B is said to be a superset of A, and this is denoted by $A \subseteq B$ or $B \supseteq A$. Two sets A and B are said to be *equal* if every element of A is an element of B and every element of B is an element of A–in other words if $A \subseteq B$ and $B \subseteq A$. If A is equal to B, then we write $A = B$. A set is generally completely determined by its elements, but there may be a set that has no element in it. Such a set is called an *empty (or void or null) set* and the empty set is denoted by Φ

(Phi). $\Phi \subset A$; in other words, the null set is included in any set A – this fact is vacuously satisfied. Furthermore, if A is a subset of B, $A \neq \Phi$ and $A \neq B$, then A is said to be a *proper subset* of B (or B is said to properly contain A). The fact that A is a *proper subset* of B is expressed as $A \subset B$. Let A be a set. Then the set of all subsets of A is called the *power set* of A and is denoted by $P(A)$. If A has three elements like letters p, q and r, then the set of all subsets of A has $8(= 2^3)$ elements. It may be noted that the null set is also a subset of A. A set is called a *finite set* if it is empty or it has n elements for some positive integer n; otherwise it is said to be *infinite*. It is clear that the empty set and the set A are members of $P(A)$. A set A is called *denumerable* or *enumerable* if it is in one-to-one correspondence with the set of natural numbers. A set is called *countable* if it is either finite or denumerable. A set that is not countable is called *uncountable*.

We now state without proof a few results which might be used in subsequent chapters:

(i) An infinite set is equivalent to a subset of itself.

(ii) A subset of a countable set is a countable set.

The following are examples of countable sets: a) the set **J** of all integers, b) the set $\mathbb{Q}$ of all rational numbers, c) the set **P** of all polynomials with rational coefficients, d) the set all straight lines in a plane each of which passes through (at least) two different points with rational coordinates and e) the set of all rational points in $\mathbb{R}^n$.

Examples of uncountable sets are as follows: (i) an open interval $]a, b[$, a closed interval $[a, b]$ where $a \neq b$, (ii) the set of all irrational numbers. (iii) the set of all real numbers. (iv) the family of all subsets of a denumerable set.

1.1.1   Cardinal numbers

Let all the sets be divided into two families such that two sets fall into one family if and only if they are equivalent. This is possible because the relation $\sim$ between the sets is an equivalence relation. To every such family of sets, we assign some arbitrary symbol and call it the *cardinal number* of each set of the given family. If the cardinal number of a set A is α, $\overline{\overline{A}} = \alpha$ or card $A = \alpha$. The cardinal number of the empty set is defined to be 0 (zero). We designate the number of elements of a nonempty finite set as the cardinal number of the finite set. We assign $\aleph_0$ to the class of all denumerable sets and as such $\aleph_0$ is the cardinal number of a denumerable set. c, the first letter of the word 'continuum' stands for the cardinal number of the set $[0, 1]$.

1.1.2   The algebra of sets

In the following section we discuss some operations that can be performed on sets. By universal set we mean a set that contains all the sets under reference. The universal set is denoted by U. For example, while discussing the set of real numbers we take $\mathbb{R}$ as the universal set. Once again for sets of complex numbers the universal set is the set $\mathbb{C}$ of complex numbers. Given two sets A and B, the *union* of A and B is denoted by $A \cup B$ and stands for a set whose every element is an element of either A or B (including elements of both A and B). $A \cup B$ is also called the sum of A and B and is written as $A + B$. The *intersection* of two sets A and B is denoted by $A \cap B$, and is a set, the elements of which are the elements common to both A and B. The intersection of two sets A and B is also called the product of A and B and is denoted by $A \cdot B$. The *difference* of two sets A and B is denoted by $A - B$ and is defined by the set of elements in A which are not elements of B. Two sets A and B are said to be *disjoint* if $A \cap B = \Phi$. If $A \subseteq B$, $B - A$ will be called the *complement* of A with reference to B. If B is the universal set, A^c will denote the *complement* of A and will be the set of all elements which are not in A.

Let A, B and C be three non-empty sets. Then the following laws hold true:

1. **Commutative laws**

   $A \cup B = B \cup A$ and $A \cap B = B \cap A$

2. **Associative laws**

   We have a finite number of sets

   $A \cup (B \cup C) = (A \cup B) \cup C$ and $(A \cap B) \cap C = A \cap (B \cap C)$

3. **Distributive laws**

   $A \cap (B \cup C) = (A \cap B) \cup (A \cap C)$

   $(A \cup B) \cap C = (A \cap C) \cup (A \cap B)$

4. **De Morgan's laws**

   $(A \cup B)^c = (A^c \cap B^c)$ and $(A \cap B)^c = (A^c \cup B^c)$

   Suppose we have a finite class of sets of the form $\{A_1, A_2, A_3, \ldots, A_n\}$, then we can form $A_1 \cup A_2 \cup A_3 \cup \ldots A_n$ and $A_1 \cap A_2 \cap A_3 \cap \ldots A_n$. We can shorten the above expression by using the index set $I = \{1, 2, 3, \ldots, n\}$. The above expressions for union and intersection can be expressed in short by $\cup_{i \in I} A_i$ and $\cap_{i \in I} A_i$ respectively.

1.1.3 *Partially ordered set*

Let A be a set of elements $a, b, c, d, \ldots$ of a certain nature. Let us introduce between certain pairs (a, b) of elements of A the relation $a \leq b$ with the properties:

   (i) If $a \leq b$ and $b \leq c$, then $a \leq c$ (*transitivity*)

   (ii) $a \leq a$ (*reflexivity*)

   (iii) If $a \leq b$ and $b \leq a$ then $a = b$

Such a set A is said to be *partially ordered* by $\leq$ and a and b, satisfying $a \leq b$ and $b \leq a$ are said to be *congruent*. A set A is said to be *totally ordered* if for each pair of its elements a, b, $a \leq b$ or $b \leq a$.

A subset B of a partially ordered set A is said to be *bounded above* if there is an element b such that $y \leq b$ for all $y \in B$, the element b is called an *upper bound* of B. The smallest of all *upper bounds* of B is called the *least upper bound (l.u.b.)* or *supremum* of B. The terms *bounded below* and *greatest lower bound (g.l.b.)* or *infimum* can be analogously defined. Finally, an element $x_0 \in A$ is said to be *maximal* if there exists in A no element $x \neq x_0$ satisfying the relation $x_0 \leq x$. The natural numbers are *totally ordered* but the branches of a tree are not. We next state a highly important lemma known as Zorn's lemma.

1.1.4 *Zorn's lemma*

Let X be a partially ordered set such that every totally ordered subset of X has an upper bound in X. Then X contains a maximal element.

Although the above statement is called a lemma it is actually an axiom.

1.1.5 *Zermelo's theorem*

Every set can be well ordered by introducing certain order relations.

The proof of Zermelo's theorem rests upon *Zermelo's axiom of arbitrary choice*, which is as follows:

If one system of nonempty, pair-wise disjoint sets is given, then there is a new set possessing exactly one element in common with each of the sets of the system.

Zorn's Lemma, Zermelo's Axiom of Choice and well ordering theorem are equivalent.

1.2   Function, Mapping

Given two nonempty sets X and Y, the *Cartesian product* of X and Y, denoted by $X \times Y$ is the set of all ordered pairs (x, y) such that $x \in X$ and $y \in Y$.

Thus $X \times Y = \{(x, y) : x \in X, y \in Y\}$.

1.2.1 *Example*

Let $X = \{a, b, c\}$ and let $Y = \{d, e\}$. Then, $X \times Y = \{(a, d), (b, d), (c, d), (a, e), (b, e), (c, e)\}$.

It may be noted that the Cartesian product of two countable sets is countable.

1.2.2 *Function*

Let X and Y be two nonempty sets. A function from X to Y is a subset of $X \times Y$ with the property that no two members of f have the same first coordinate. Thus $(x, y) \in f$ and $(x, z) \in f$ imply that $y = z$. The *domain* of a function f from X to Y is the subset of X that consists of all first coordinates of members of f. Thus x is in the domain of f if and only if $(x, y) \in f$ for some $y \in Y$.

The *range* of f is the subset of Y that consists of all second coordinates of members of f. Thus y is in the range of f if and only if $(x, y) \in f$ for some $x \in X$. If f is a function and x is a point in the domain of f then $f(x)$ is the second coordinate of the unique member of f whose first coordinate is x.

Thus $y = f(x)$ if and only if $(x, y) \in f$. This point $f(x)$ is called the image of x under f.

1.2.3 Mappings: into, onto (surjective), one-to-one (injective) and bijective

A function f is said to be a mapping of X into Y if the domain of f is X and the range of f is a subset of Y. A function f is said to be a mapping of X onto Y *(surjective)* if the domain of f is X and the range of f is Y.

The fact that f is a mapping of X onto Y is denoted by $f : X \xrightarrow{onto} Y$. A function f from X to Y is said to be *one-to-one (injective)* if distinct points in X have distinct images under f in Y. Thus f is one-to-one if and only if $(x_1, y) \in f$ and $(x_2, y) \in f$ imply that $x_1 = x_2$. A function from X to Y is said to be *bijective* if it is both *injective* and *surjective*.

1.2.4 Example

Let $X = \{a, b, c\}$ and let $Y = \{d, e\}$. Consider the following subsets of $X \times Y$:

$F = \{(a, b), (b, c), (c, d), (a, c)\}$, $G = \{(a, d), (b, d), (c, d)\}$,
$H = \{(a, d), (b, e), (c, e)\}$, $\phi = \{(a, c), (b, d)\}$

The set F is not a function from X to Y because (a, d) and (a, e) are distinct members of F that have the same first coordinate. The domain of both G and H is X and the domain of ϕ is (a, b).

1.3 Linear Space

A nonempty set is said to be a space if the set is closed with respect to certain operations defined on it. It is apparent that the elements of some sets (i.e., set of finite matrices, set of functions, set of number sequences) are closed with respect to addition and multiplication by a scalar. Such sets have given rise to a space called linear space.

Definition. Let E be a set of elements of a certain nature satisfying the following axioms:

(i) E is an additive abelian group. This means that if x and $y \in E$, then their sum $x + y$ also belongs to the same set E, where the operation of addition satisfies the following axioms:

   (a) $x + y = y + x$ (commutativity);
   (b) $x + (y + z) = (x + y) + z$ (associativity);
   (c) There exists a uniquely defined element 0, such that $x + \theta = x$ for any x in E;
   (d) For every element $x \in E$ there exists a unique element $(-x)$ of the same space, such that $x + (-x) = \theta$.
   (e) The element θ is said to be the null element or zero element of E and the element $-x$ is called the inverse element of x.

(ii) A scalar multiplication is said to be defined if for every $x \in E$, for any scalar λ (real or complex) the element $\lambda x \in E$ and the following conditions are satisfied:

   (a) $\lambda(\mu x) = \lambda \mu x$ (associativity)

   (b) $\left.\begin{array}{l} \lambda(x + y) = \lambda x + \lambda y \\ (\lambda + \mu)x = \lambda x + \mu x \end{array}\right\}$ (distributivity)

   (c) $1 \cdot x = x$

The set E satisfying the axioms (i) and (ii) is called a *linear* or *vector space*. This is said to be a *real* or *complex space* depending on whether the set of multipliers are *real* or *complex*.

1.3.1   Examples

(i) Real line $\mathbb{R}$

The set of all real numbers for which the ordinary additions and multiplications are taken as linear operations, is a real linear space $\mathbb{R}$.

(ii) The Euclidean space $\mathbb{R}^n$, unitary space C^n, and complex plane $\mathbb{C}$

Let X be the set of all ordered n-tuples of real numbers. If $x = (\xi_1, \xi_2, \ldots, \xi_n)$ and $y = (\eta_1, \eta_2, \ldots, \eta_n)$, we define the operations of addition and scalar multiplication as $x + y = (\xi_1 + \eta_1, \xi_2 + \eta_2, \ldots, \xi_n + \eta_n)$ and $\lambda x = (\lambda \xi_1, \lambda \xi_2, \ldots, \lambda \xi_n)$. In the above equations, λ is a real scalar. The above linear space is called the real n-dimensional space and denoted by $\mathbb{R}^n$. The set of all ordered n-tuples of complex numbers, $\mathbb{C}^n$, is a linear space with the operations of additions and scalar multiplication defined as above. The complex plane $\mathbb{C}$ is a linear space with addition and multiplication of complex numbers taken as the linear operations over $\mathbb{R}$ (or $\mathbb{C}$).

(iii) Space of $m \times n$ matrices, $\mathbb{R}^{m \times n}$

$\mathbb{R}^{m \times n}$ is the set of all matrices with real elements. Then $\mathbb{R}^{m \times n}$ is a *real linear space* with addition and scalar multiplication defined as follows:

Let $A = \{a_{ij}\}$ and $B = \{b_{ij}\}$ be two $m \times n$ matrices. Then $A + B = \{a_{ij} + b_{ij}\}$. $\alpha A = \{\alpha a_{ij}\}$, where α is a scalar. In this space $-A = \{-a_{ij}\}$ and the matrix with all its elements as zeroes is the zero element of the space $\mathbb{R}^{m \times n}$.

(iv) Sequence space l_∞

Let X be the set of all bounded sequences of complex numbers, i.e., every element of X is a complex sequence $x = \{\xi_i\}$ such that $|\xi_i| < C_i$, where C_i is a real number for each i. If $y = \{\eta_i\}$ then we define $x + y = \{\xi_i + \eta_i\}$ and $\lambda x = \{\lambda \xi_i\}$. Thus, l_∞ is a linear space, and is called a *sequence space*.

(v) $C([a, b])$

Let X be the set of all real-valued continuous functions x, y, etc, which are functions of an independent variable t defined on a given closed interval $J = [a, b]$. Then X is closed with respect to additions of two continuous functions and multiplication of a continuous function by a scalar, i.e., $(x + y)(t) = x(t) + y(t)$, $\alpha x(t) = (\alpha x(t))$ where α is a scalar.

(vi) Space l_p, Hilbert sequence space l_2

Let $p \geq 1$ be a fixed real number. By definition each element in the space l_p is a sequence $x = \{\xi_i\} = \{\xi_1, \xi_2, \ldots, \xi_n, \ldots\}$, such that $\sum_{i=1}^{\infty} |\xi_i|^p < \infty$, for p real and $p \geq 1$, if $y \in l_p$ and $y = \{\eta_i\}$, $x + y = \{\xi_i + \eta_i\}$ and $\alpha x = \{\alpha \xi_i\}$, $\alpha \in \mathbb{R}$. Since $|\xi_i|^p + |\eta_i|^p \leq 2^p \max(|\xi_i|^p, |\eta_i|^p) \leq 2^p(|\xi_i|^p + |\eta_i|^p)$, it follows that $\sum_{i=1}^{\infty} |\xi_i + \eta_i|^p < \infty$. Therefore, $x + y \in l_p$. Similarly, we can show that $\alpha x \in l_p$ where α is a scalar. Hence, l_p is a linear space with respect to the algebraic operations defined above. If $p = 2$, the space l_p becomes l_2, a square summable space which possesses some special properties to be revealed later.

(vii) Space $L_p([a, b])$ of all Lebesgue pth integrable functions

Let f be a Lebesgue measurable function defined on $[a, b]$ and $0 < p < \infty$. Since $f \in L_p([a, b])$, we have $\int_a^b |f(t)|^p dt < \infty$. Again, if $g \in L_p([a, b])$, i.e., $\int_a^b |g(t)|^p dt < \infty$. Since $f + g|^p \leq 2^p \max(|f|^p, |g|^p) \leq 2^p(|f|^p + |g|^p)$, $f \in L_p([a, b])$, $g \in L_p([a, b])$ imply that $(f + g) \in L_p([a, b])$ and $\alpha f \in L_p([a, b])$. This shows that $L_p([a, b])$ is a linear space. If $p = 2$, we get $L_2([a, b])$, which is known as the space of square integrable functions. The space possesses some special properties.

1.3.2 Subspace, linear combination, linear dependence, linear independence

A subset X of a linear space E is said to be a *subspace* if X is a linear space with respect to vector addition and scalar multiplication as defined in E. The vector of the form $x = \alpha_1 x_1 + \alpha_2 x_2 + \cdots + \alpha_n x_n$ is called a *linear combination* of vectors $x_1, x_2, \ldots, x_n$, in the linear space E, where $\alpha_1, \alpha_2, \ldots, \alpha_n$ are real or complex scalars. If X is any subset of E, then the set of linear combinations vectors in X forms a subspace of E. The subspace so obtained is called the subspace *spanned* by X and is denoted by *span* X. It is, in fact, the smallest subspace of E containing X. In other words it is the intersection of all subspaces of E containing X.

A finite set of vectors $\{x_1, x_2, \ldots, x_n\}$ in X is said to be *linearly dependent* if there exist scalars $\{\alpha_1, \alpha_2, \ldots, \alpha_n\}$, not all zeroes, such that $\alpha_1 x_1 + \alpha_2 x_2 + \cdots + \alpha_n x_n = 0$ where $\{\alpha_1, \alpha_2, \ldots, \alpha_n\}$ are scalars, real or complex. On the other hand, if for all scalars $\{\alpha_1, \alpha_2, \ldots, \alpha_n\}$, $\alpha_1 x_1 + \alpha_2 x_2 + \cdots + \alpha_n x_n = 0 \Rightarrow \alpha_1 = 0, \; \alpha_2 = 0, \; \ldots, \; \alpha_n = 0$, then the set of vectors is said to be *linearly independent*.

A subset X (finite or infinite) of E is linearly independent if every finite subset of X is linearly independent. As a convention we regard the empty set as linearly independent.

1.3.3 Hamel basis, dimension

A subset L of a *linear space* E is said to be a *basis* (or *Hamel basis*) for E if (i) L is a linearly independent set, and (ii) L spans the whole space.

In this case, any non-zero vector x of the space E can be expressed uniquely as a linear combination of finitely many vectors of L with the scalar coefficients that are not all zeroes. Clearly any maximal linearly independent set (to which no new non-zero vector can be added without destroying linear independence) is a *basis* for L and any minimal set spanning L is also a *basis* for L.

1.3.4 Theorem

Every linear space $X \neq \{\theta\}$ has a Hamel basis.

Let L be the set of all linearly independent subsets of X. Since $X \neq \{\theta\}$, it has an element $x \neq \theta$ and $\{x\} \in L$, therefore $L \neq \Phi$. Let the partial ordering in L be denoted by 'set inclusion'. We show that for every totally ordered subset $L_\alpha, \; \alpha \in A$ of L, the set $\overline{L} = \cup[\overline{L}_\alpha : \alpha \in A]$ is also in L. Otherwise, $\{\overline{L}\}$ would be generated by a proper subset $T \subset \overline{L}$. Therefore, for every $\alpha \in A$, $\{\overline{L}_\alpha\}$ is generated by $T_\alpha = T \cap \overline{L}_\alpha$. However, the linear independence of $\overline{L}_\alpha$ implies $T_\alpha = \overline{L}_\alpha$. Thus, $T = \cup[T \cap \overline{L}_\alpha : \alpha \in A] = \cup[T_\alpha : \alpha \in A] = \cup[\overline{L}_\alpha : \alpha \in A] = \overline{L}$, contradicting the assumption that T is a proper subset $\overline{L}_\alpha$. Thus, the conditions of *Zorn's lemma* having been satisfied, there is a maximal $M \in L$. Suppose $\{M\}$ is a proper subspace of X. Let $y \in X$ and $y \notin \{M\}$. The subspace Y of X generated by M and y then contains $\{M\}$ as a proper subspace. If, for any proper subset

$T \subset M$, T and also y generate Y, it follows that T also generates $\{M\}$, thus contradicting the concept that M is linearly independent. There is thus no $y \in X$, $y \notin \{M\}$. Hence M generates X.

A linear space X is said to be *finite dimensional* if it has a finite basis. Otherwise, X is said to be *infinite dimensional*.

1.3.5 Examples

(i) Trivial linear space

Let $X = \{\theta\}$ be a trivial linear space. We have assumed that Φ is a linearly independent set. The span of Φ is the intersection of all subspaces of X containing Φ. However, θ belongs to every subspace of X. Hence it follows that the *Span* $\Phi = \{\theta\}$. Therefore, Φ is a basis for X.

(ii) $\mathbb{R}^n$

Consider the real linear space $\mathbb{R}^n$ where every $x \in \mathbb{R}^n$ is an ordered n-tuple of real numbers. Let $e_1 = (1, 0, 0, \ldots, 0)$, $e_2 = (0, 1, 0, 0, \ldots, 0), \ldots, e_n = (0, 0, \ldots, 1)$. We may note that $\{e_i\}$, $i = 1, 2, \ldots, n$ is a linearly independent set and spans the whole space $\mathbb{R}^n$. Hence, $\{e_1, e_2, \ldots, e_n\}$ forms a basis of $\mathbb{R}^n$. For $n = 1$, we get $\mathbb{R}^1$ and any singleton set comprising a non-zero element forms a basis for $\mathbb{R}^1$.

(iii) $\mathbb{C}^n$

The complex linear space $\mathbb{C}^n$ is a linear space where every $x \in \mathbb{C}^n$ is an ordered n-tuple of complex numbers and the space is finite dimensional. The set $\{e_1, e_2, \ldots, e_n\}$, where e_i is the ith vector, is a basis for $\mathbb{C}^n$.

(iv) $C([a, b])$, $P_n([a, b])$

$C([a, b])$ is the space of continuous real functions in the closed interval $[a, b]$. Let $B = \{1, x, x^2, \ldots, x^n, \ldots\}$ be a set of functions in $C([a, b])$. It is apparent that B is a basis for $C([0, 1])$. $P_n([a, b])$ is the space of real polynomials of order n defined on $[a, b]$. The set $B_n = \{1, x, x^2, \ldots, x_n\}$ is a basis in $P_n([a, b])$.

(v) $\mathbb{R}^{m \times n}(\mathbb{C}^{m \times n})$

$\mathbb{R}^{m \times n}$ is the space of all matrices of order $m \times n$. For $i = 1, 2, \ldots, n$ let $E_{i \times j}$ be the $m \times n$ matrix with (i, j)th entry as 1 and all other entries as zero. Then, $\{E_{ij} : i = 1, 2, \ldots, m; j = 1, 2, \ldots, n\}$ is a basis for $(\mathbb{R})^{m \times n}(C^{m \times n})$.

1.3.6 Theorem

Let E be a finite dimensional linear space. Then all the bases of E have the same number of elements.

Let $\{e_1, e_2, \ldots, e_n\}$ and $\{f_1, f_2, \ldots, f_n, f_{n+1}\}$ be two different bases in E. Then, any element f_i can be expressed as a linear combination of

$e_1, e_2, \ldots, e_n$; i.e., $f_i = \sum_{i=1}^{n} a_{ij} e_j$. Since f_i, $i = 1, 2, \ldots, n$ are linearly independent, the matrix $[a_{ij}]$ has rank n. Therefore, we can express f_{n+1} as $f_{n+1} = \sum_{j=1}^{n} a_{n+1j} e_j$. Thus the elements $f_1, f_2, \ldots, f_{n+1}$ are not linearly independent. Since $\{f_1, f_2, \ldots, f_{n+1}\}$ forms a basis for the space it must contain a number of linearly independent elements, say $m (\leq n)$. On the other hand, since $\{f_i\}$, $i = 1, 2, \ldots, n+1$ forms a basis for E, e_i can be expressed as a linear combination of $\{f_j\}$, $j = 1, 2, \ldots, n+1$ such that $n \leq m$. Comparing $m \leq n$ and $n \leq m$ we conclude that $m = n$. Hence the number of elements of any two bases in a finite dimensional space E is the same.

The above theorem helps us to define the *dimension of a finite dimensional space.*

1.3.7 Dimension, examples

The *dimension* of a finite dimensional linear space E is defined as *the number of elements* of any basis of the space and is written as dim E.

(i) dim $\mathbb{R}$ = dim $\mathbb{C}$ = 1

(ii) dim $\mathbb{R}^n$ = dim $\mathbb{C}^n = n$

For an infinite dimensional space it can be shown that all bases are equivalent sets.

1.3.8 Theorem

If E is a linear space all the bases have the same cardinal number.

Let $S = \{e_i\}$ and $T = \{f_i\}$ be two bases. Suppose S is an infinite set and has cardinal number α. Let β be the cardinal number of T. Every $f_i \in T$ is a linear combination, with non-zero coefficients, of a finite number of elements $e_1, e_2, \ldots, e_n$ of S and only a finite number of elements of T are associated in this way with the same set $e_1, e_2, \ldots, e_n$ or some subset of it. Since the cardinal number of the set of finite subsets of S is the same as that of S itself, it follows that $\beta \leq \aleph_0$, $\beta \leq \alpha$. Similarly, we can show that $\alpha \leq \beta$. Hence, $\alpha = \beta$. Thus the common cardinal number of all bases in an infinite dimensional space E is defined as the *dimension* of E.

1.3.9 Direct sum

Here we consider the representation of a linear space E as a direct sum of two or more subspaces. Let E be a linear space and $X_1, X_2, \ldots, X_n$ be n subspaces of E. If $x \in E$ has an unique representation of the form $x = x_1 + x_2 + \cdots + x_n$, $x_i \in X_i$, $i = 1, 2, \ldots, n$, then E is said to be the direct sum of its subspaces $X_1, X_2, \ldots, X_n$. The above representation is called the decomposition of the element x into the elements of the subspaces $X_1, X_2, \ldots, X_n$. In that case we can write $E = X_1 \oplus X_2 \oplus \cdots X_n = \sum_{i=1}^{n} \oplus X_i$.

1.3.10 Quotient spaces

Let M be a subspace of a linear space E. The *coset* of an element $x \in E$ with respect to M, denoted by $x+M$ is defined as $x+M = \{x+m : m \in M\}$. This can be written as $E/M = \{x + M : x \in E\}$. One observes that $M = \theta + M$, $x_1 + M = x_2 + M$ if and only if $x_1 - x_2 \in M$ and as a result, for each pair $x_1, x_2 \in E$, either $(x_1 + M) \cap (x_2 + M) = \theta$ or $x_1 + M = x_2 + M$. Further, if $x_1, x_2, y_1, y_2 \in E$, it then follows that $x_1 - x_2 \in M$ and $y_1 - y_2 \in M$, $(x_1 + x_2) - (y_1 + y_2) \in M$ and for any scalar λ, $(\lambda x_1 - \lambda x_2) \in M$ because M is a linear subspace. We define the linear operations on E/M by

$$\begin{cases} (x + M) + (y + M) = (x + y) + M, \\ \lambda(x + M) = \lambda x + M \text{ where } x, y \in M, \ \lambda \text{ is a scalar (real or complex).} \end{cases}$$

It is clearly apparent that E/M under the linear operations defined above is a linear space over $\mathbb{R}$ (or $\mathbb{C}$). The linear space E/M is called a *quotient space* of E by M. The function $\phi : E \to E/M$ defined by $\phi(x) = x + M$ is called canonical mapping of E onto E/M. The dimension of E/M is called the codimension (*codim*) of M in E. Thus, $\text{codim } M = \dim(E/M)$. The *quotient space* has a simple geometrical interpretation. Let the linear space $E = R^2$ and the subspace M be given by the straight line as fig. 1.1.

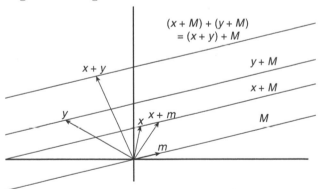

Fig. 1.1 Addition in quotient space

1.4 Metric Spaces

Limiting processes and continuity are two important concepts in classical analysis. Both these concepts in real analysis, specifically in $\mathbb{R}$ are based on distance. The concept of distance has been generalized in abstract spaces yielding what are known as metric spaces. For two points x, y in an abstract space let $d(x, y)$ be the distance between them in $\mathbb{R}$, i.e., $d(x, y) = |x - y|$. The concept of distance gives rise to the concept of limit, i.e., $\{x_n\}$ is said to tend to x as $n \to \infty$ if $d(x_n, x) \to 0$ as $n \to \infty$. The concept of continuity can be introduced through the limiting process. We replace the set of real numbers underlying $\mathbb{R}$ by an abstract set X of elements (all the attributes of which are known, but the concrete forms are not spelled out) and introduce on X a distance function. This will help us in studying different classes of problems within a single umbrella and drawing some conclusions that are universally valid for such different sets of elements.

1.4.1   Definition: metric space, metric

A metric space is a pair (X, ρ) where X is a set and ρ is a metric on X (or a distance function on X) that is a function defined on $X \times X$ such that for all $x, y, z \in X$ the following axioms hold:

1. ρ is real-valued, finite and non-negative

2. $\rho(x, y) = 0 \Leftrightarrow x = y$

3. $\rho(x, y) = \rho(y, x)$ (Symmetry)

4. $\rho(x, y) \leq \rho(x, z) + \rho(z, y)$ (Triangle Inequality)

These axioms obviously express the fundamental properties of the distance between the points of the three-dimensional Euclidean space.

A subspace $(Y, \tilde{\rho})$ of (X, ρ) is obtained by taking a subset $Y \subset X$ and restricting ρ to $Y \times Y$. Thus the metric on Y is the restriction $\tilde{\rho} = \rho|_{Y \times Y}$. $\tilde{\rho}$ is called the metric induced on Y by ρ.

In the above, X denotes the Cartesian product of sets. $A \times B$ is the set of ordered pairs (a, b), where $a \in A$ and $b \in B$. Hence, $X \times X$ is the set of all ordered pairs of elements of X.

1.4.2   Examples

(i) Real line $\mathbb{R}$

This is the set of all real numbers for which the metric is taken as $\rho(x, y) = |x - y|$. This is known as the usual metric in $\mathbb{R}$.

(ii) Euclidean space $\mathbb{R}^n$, unitary space $\mathbb{C}^n$, and complex plane $\mathbb{C}$

Let X be the set of all ordered n-tuples of real numbers. If $(\xi_1, \xi_2, \ldots, \xi_n)$ and $y = (\eta_1, \eta_2, \ldots, \eta_n)$ then we set

$$\rho(x, y) = \sqrt{\sum_{i=1}^{n} (\xi_i - \eta_i)^2} \tag{1.1}$$

It is easily seen that $\rho(x, y) \geq 0$. Furthermore, $\rho(x, y) = \rho(y, x)$.

Let, $z = (\zeta_1, \zeta_2, \ldots, \zeta_n)$. Then, $\rho^2(x, z) = \sum_{j=1}^{n}(\xi_i - \zeta_i)^2$

$$= \sum_{i=1}^{n}(\xi_i - \eta_i)^2 + \sum_{i=1}^{n}(\eta_i - \zeta_i)^2 + 2\sum_{i=1}^{n}(\xi_i - \eta_i)(\eta_i - \zeta_i)$$

Now by the Cauchy-Bunyakovsky-Schwartz inequality [see 1.4.3]

$$\sum_{i=1}^{n}(\xi_i - \eta_i)(\eta_i - \zeta_i) \leq \left(\sum_{i=1}^{n}(\xi_i - \eta_i)^2\right)^{1/2} \left(\sum_{i=1}^{n}(\eta_i - \zeta_i)^2\right)^{1/2}$$

$$\leq \rho(x, y)\rho(y, z)$$

Thus, $\rho(x, z) \leq \rho(x, y) + \rho(y, z)$.

Hence, all the axioms of a metric space are fulfilled. Therefore, $\mathbb{R}^n$ under the metric defined by (1.1) is a metric space and is known as the n-dimensional Euclidean space. If x, y, z denote three distinct points in $\mathbb{R}^2$ then the inequality $\rho(x, z) \leq \rho(x, y) + \rho(y, z)$ implies that the length of any side of a triangle is always less than the sum of the lengths of the other two sides of the triangle obtained by joining x, y, z. Hence, axiom 4) of the set of metric axioms is known as the triangle inequality. n-dimensional unitary space $\mathbb{C}^n$ is the space of all ordered n-tuples of complex numbers with metric defined by $\rho(x, y) = \sqrt{\sum_{i=1}^{n}(\xi_i - \eta_i)^2}$. When $n = 1$ this is the complex plane $\mathbb{C}$ with the usual metric defined by $\rho(x, y) = |x - y|$.

(iii) Sequence space l_∞

Let X be the set of all bounded sequences of complex numbers, i.e., every element of X is a complex sequence $x = (\xi_1, \xi_2, \ldots, \xi_n)$ or $x = \{\xi_i\}$ such that $|\xi_i| \leq C_i$, where C_i for each i is a real number. We define the metric as $\rho(x, y) = \sup_{i \in N} |\xi_i - \eta_i|$, where $y = \{\eta_i\}$, $N = \{1, 2, 3, \ldots\}$, and 'sup' denotes the least upper bound (l.u.b.). l_∞ is called a sequence space because each element of X (each point in X) is a sequence.

(iv) $C([a, b])$

Let X be the set of all real-valued continuous functions x, y, etc, that are functions of an independent variable t defined on a given closed interval $J = [a, b]$.

We choose the metric defined by $\rho(x, y) = \max_{t \in J} |x(t) - y(t)|$ where max denotes the maximum. We may note that $\rho(x, y) \geq 0$ and $\rho(x, y) = 0$ if and only if $x(t) = y(t)$. Moreover, $\rho(x, y) = \rho(x, y)$. To verify the triangular inequality, we note that

$$|x(t) - z(t)| \leq |x(t) - y(t)| + |y(t) - z(t)|$$

$$\leq \max_t |x(t) - y(t)| + \max_t |y(t) - y(t)$$
$$\leq \rho(x, y) + \rho(y, z), \text{ for every } t \in [0, 1]$$

Hence, $\rho(x, z) \leq \rho(x, y) + \rho(y, z)$. Thus, all the axioms of a metric space are satisfied.

The set of all continuous functions defined on the interval $[a, b]$ with the above metric is called the space of continuous functions and is defined on J and denoted by $C([a, b])$. This is a function space because every point of $C([a, b])$ is a function.

(v) Discrete metric space

Let X be a set and let ρ be defined by, $\rho(x, y) = \begin{cases} 0, & \text{if } x = y \\ 1, & \text{if } x \neq y \end{cases}$.

The above is called a discrete metric and the set X endowed with the above metric is a discrete metric space.

(vi) The space $M([a, b])$ of bounded real functions

Consider a set of all bounded functions $x(t)$ of a real variable t, defined on the segment $[a, b]$. Let the metric be defined by $\rho(x, y) = \sup_t |x(t) - y(t)|$.

All the metric axioms are fulfilled with the above metric. The set of real bounded functions with the above metric is designated as the space $M[a, b]$. It may be noted that $C[a, b] \subseteq M([a, b])$.

(vii) The space $BV([a, b])$ of functions of bounded variation

Let $BV([a, b])$ denote the class of all functions of bounded variation on $[a, b]$, i.e., all f for which the total variation $V(f) = \sup \sum_{i=1}^{n} |f(x_i) - f(x_{i+1})|$ is finite, where the supremum is taken over all partitions, $a = x_0 < x_1 < x_2 < \cdots < x_n = b$. Let us take $\rho(f, g) = V(f - g)$. If $f = g$, $v(f - g) = 0$. Else, $V(f - g) = 0$ if and only if f and g differ by a constant.

$$\rho(f, g) = \rho(g, f) \quad \text{since} \quad V(f - g) = V(g - f).$$

If h is a function of bounded variation,

$$\rho(f, h) = V(f - h) = \sup \sum_{i=1}^{n} |(f(t_i) - h(t_i)) - (f(t_{i-1}) - h(t_{i-1}))|$$

$$= \sup \sum_{i=1}^{n} |(f(t_i) - g(t_i) + g(t_i) - h(t_i))$$

$$- (f(t_{i-1}) - g(t_{i-1}) + g(t_{i-1}) - h(t_{i-1}))|$$

$$\leq \sup \sum_{i=1}^{n} |(f(t_i) - g(t_i)) - (f(t_{i-1}) - g(t_{i-1}))|$$

$$+ \sup \sum_{i=1}^{n} |(g(t_i) - h(t_i)) - (g(t_{i-1}) - h(t_{i-1}))|$$

$$\leq V(f - g) + V(g - h) = \rho(f, g) + \rho(g, h)$$

Thus all the axioms of a metric space are fulfilled.

If $BV([a, b])$ is decomposed into equivalent classes according to the equivalence relation defined by $f \cong g$, and if $f(t) - g(t)$ is constant on $[a, b]$, then this $\rho(f, g)$ determines a metric $\widetilde{\rho}$ on the space $\overline{BV}([a, b])$ of such equivalent classes in an obvious way. Alternatively we may modify the definition of ρ so as to obtain a metric on the original class $BV([a, b])$. For example, $\rho(f, g) = |f(a) - g(a)| + V(f - g)$ is a metric of $BV([a, b])$. The subspace of the metric space, consisting of all $f \in BV([a, b])$ for which $f(a) = 0$, can naturally be identified with the space $\overline{BV}([a, b])$.

(viii) The space c of convergent numerical sequences

Let X be the set of convergent numerical sequences $x = \{\xi_1, \xi_2, \xi_3, \ldots, \xi_n, \ldots\}$, where $\lim_i \xi_i = \xi$. Let $x = \{\xi_1, \xi_2, \xi_n, \ldots\}$ and $y = \{\eta_1, \eta_2, \eta_n, \ldots\}$. Set $\rho(x, y) = \sup_i |\xi_i - \eta_i|$.

(ix) The space m of bounded numerical sequences

Let X be the sequence of bounded numerical sequences $x = \{\xi_1, \xi_2, \ldots, \xi_n, \ldots\}$, implying that for every x there is a constant $K(X)$ such that $|\xi_i| \leq K(X)$ for all i. Let $x = \{\xi_i\}$, $y = \{\eta_i\}$ belong to X. Introduce the metric $\rho(x, y) = \sup_i |\xi_i - \eta_i|$.

It may be noted that the space c of convergent numerical sequences is a subspace of the space m of bounded numerical sequences.

(x) Sequence space s

This space consists of the set of all (not necessarily bounded) sequences of complex numbers and the metric ρ is defined by

$$\rho(x, y) = \sum_{i=1}^{n} \frac{1}{2^i} \frac{|\xi_i - \eta_i|}{1 + |\xi_i - \eta_i|} \quad \text{where} \quad x = \{\xi_i\} \quad \text{and} \quad y = \{\eta_i\}.$$

Axioms **1-3** of a metric space are satisfied. To see that $\rho(x, y)$ also satisfies axiom **4** of a metric space, we proceed as follows:

Let $\quad f(t) = \dfrac{t}{1 + t}$, $t \in R$. Since $f'(t) = \dfrac{1}{(1 + t)^2} > 0$,

$f(t)$ is monotonically increasing.

Hence $\quad |a + b| \leq |a| + |b| \implies f(|a + b|) \leq f(|a|) + f(|b|)$.

Thus, $\quad \dfrac{|a + b|}{1 + |a + b|} \leq \dfrac{|a| + |b|}{1 + |a| + |b|} \leq \dfrac{|a|}{1 + |a|} + \dfrac{|b|}{1 + |b|}$.

Let $a = \xi_i - \zeta_i$, $b = \zeta_i - \eta_i$, where $z = \{\zeta_i\}$.

Thus, $\dfrac{|\xi_i - \eta_i|}{1 + |\xi_i - \eta_i|} \leq \dfrac{|\xi_i - \zeta_i|}{1 + |\xi_i - \zeta_i|} + \dfrac{|\zeta_i - \eta_i|}{1 + |zeta_i - \eta_i|}$

Hence, $\rho(x, y) \leq \rho(x, z) + \rho(z, y)$, indicating the axiom on 'triangle inequality' has been satisfied. Thus s is a metric space.

Problems

1. Show that $\rho(x, y) = \sqrt{x - y}$ defines a metric on the set of all real numbers.

2. Show that the set of all n-tuples of real numbers becomes a metric space under the metric $\rho(x, y) = \max\{|x_1 - y_1|, \ldots, |x_n - y_n|\}$ where $x = \{x_i\}$, $y = \{y_i\}$.

3. Let $\mathbb{R}$ be the space of real or complex numbers. The distance ρ of two elements f, g shall be defined as $\rho(f, g) = \varphi(|f - g|)$ where $\varphi(x)$ is a function defined for $x \geq 0$, $\varphi(x)$ is twice continuously differentiable and strictly monotonic increasing (that is, $\varphi'(x) > 0$), and $\varphi(0) = 0$. Then show that $\rho(f, g) = 0$ if and only if $f = g$.

4. Let $C([B])$ be the space of continuous (real or complex) functions f, defined on a closed bounded domain B on $\mathbb{R}^n$. Define $\rho(f, g) = \varphi(r)$ where $r = \max\limits_B |f - g|$. For $\varphi(r)$ we make the same assumptions as in example 3. When $\varphi''(r) < 0$, show that the function space $\mathbb{R}$ is metric, but, when $\varphi''(r) > 0$ the space $\mathbb{R}$ is no more metric.

1.4.3   Theorem (Hölder's inequality)

If $p > 1$ and q is defined by $\dfrac{1}{p} + \dfrac{1}{q} = 1$

(H1) $\displaystyle\sum_{i=1}^{n} |x_i y_i| \leq \left[\sum_{i=1}^{n} |x_i|^p\right]^{1/p} \left[\sum_{i=1}^{n} |y_i|^q\right]^{1/q}$

for any complex numbers $x_1, x_2, x_3, \ldots, x_n, y_1, \ldots, y_n$.

(H2)   If $x \in \ell_p$ i.e., p^{th} power summable, $y \in \ell_q$ where p, q are defined as above, $x = \{x_i\}, y = \{y_i\}$.

We have $\displaystyle\sum_{i=1}^{\infty} |x_i y_i| \leq \left[\sum_{i=1}^{\infty} |x_i|^p\right]^{1/p} \left[\sum_{i=1}^{\infty} |y_i|^q\right]^{1/q}$. The inequality is known as Hölder's inequality for sum.

(H3)   If $x(t) \in L_p(0, 1)$ i.e. p^{th} power integrable, $y(t) \in L_q(0, 1)$ i.e. q^{th} power integrable, where p and q are defined as above, then

$$\int_0^1 |x(t)y(t)|dt \leq \left(\int_0^1 |x(t)^p dt\right)^{1/p} \left(\int_0^1 |y(t)^q dt\right)^{1/q}.$$

The above inequality is known as Hölder's inequality for integrals. Here p and q are said to be conjugate to each other.

Proof: We first prove the inequality

$$a^{1/p}b^{1/q} \le \frac{a}{p} + \frac{b}{q}, \quad a \ge 0,\ b \ge 0. \tag{1.2}$$

In order to prove the inequality we consider the function

$$f(t) = t^{\alpha} - \alpha t + \alpha - 1 \text{ defined for } 0 < \alpha < 1,\ t \ge 0.$$

Then, $\quad f'(t) = \alpha(t^{\alpha-1} - 1)$

so that $\quad f(1) = f'(1) = 0$

$$f'(t) > 0 \quad \text{for } 0 < t < 1 \quad \text{and} \quad f'(t) < 0 \quad \text{for } t > 1.$$

It follows that $f(t) \le 0$ for $t \ge 0$. The inequality is true for $b = 0$ since $p > 1$. Suppose $b > 0$ and let $t = a/b$ and $\alpha = 1/p$. Then

$$f\left(\frac{a}{b}\right) = \left(\frac{a}{b}\right)^{\alpha} - \frac{1}{p} \cdot \frac{a}{b} + \frac{1}{p} - 1 \le 0.$$

Multiplying by b, we obtain,

$$a^{1/p}b^{1-\frac{1}{p}} \le \frac{a}{p} + b\left(1 - \frac{1}{p}\right) = \frac{a}{p} + \frac{b}{q} \quad \text{since } 1 - \frac{1}{p} = \frac{1}{q}.$$

Applying this to the numbers

$$a_j = \frac{|x_j|^p}{\displaystyle\sum_{i=1}^{\infty} |x_i|^p}, \quad b_j = \frac{|y_j|^q}{\displaystyle\sum_{i=1}^{n} |y_i|^q}$$

for $j = 1, 2, \ldots n$, we get

$$\frac{|x_j y_j|}{\left[\displaystyle\sum_{j=1}^{n} |x_j|^p\right]^{1/p} \left[\displaystyle\sum_{j=1}^{n} |y_j|^q\right]^{1/q}} \le \frac{a_j}{p} + \frac{b_j}{q}, \quad j = 1, 2, \ldots, n.$$

By adding these inequalities the RHS takes the form

$$\frac{\displaystyle\sum_{j=1}^{n} a_j}{p} + \frac{\displaystyle\sum_{j=1}^{n} b_j}{q} = \frac{1}{p} + \frac{1}{q} = 1.$$

LHS gives the Hölder's inequality (H1)

$$\sum_{j=1}^{n} |x_j y_j| \le \left(\sum_{j=1}^{n} |x_j|^p\right)^{1/p} \left(\sum_{j=1}^{n} |y_j|^q\right)^{1/q} \tag{1.3}$$

which proves (1).

To prove (H2), we note that

$$x \in \ell_p \Rightarrow \sum_{j=1}^{\infty} |x_j|^p < \infty \qquad [\text{see } \mathbf{H2}],$$

$$y \in \ell_q \Rightarrow \sum_{j=1}^{\infty} |y_j|^q < \infty \qquad [\text{see } \mathbf{H2}].$$

Taking $\quad a_j = \dfrac{|x_j|^p}{\sum\limits_{i=1}^{\infty} |x_i|^p}, \; b_j = \dfrac{|y_j|^q}{\sum\limits_{i=1}^{\infty} |y_i|^q} \quad$ for $j = 1, 2, \ldots \infty$

we obtain as in above,

$$\frac{|x_j y_j|^p}{\left[\sum\limits_{i=1}^{\infty} |x_i|^p\right]^{1/p} \left[\sum\limits_{i=1}^{\infty} |y_i|^q\right]^{1/q}} \leq \frac{a_j}{p} + \frac{b_j}{q}.$$

Summing over both sides for $j = 1, 2, \ldots, \infty$ we obtain the Hölder's inequality for sums.

In case $p = 2$, then $q = 2$, the above inequality reduces to the **Cauchy-Bunyakovsky-Schwartz inequality**, namely

$$\sum_{i=1}^{\infty} |x_i y_i| \leq \left(\sum_{i=1}^{\infty} |x_i|^p\right)^{1/p} \left(\sum_{i=1}^{\infty} |y_i|^q\right)^{1/q}.$$

The Cauchy-Bunyakovsky-Schwartz has numerous applications in a variety of mathematical investigations and will find important applications in some of the later chapters.

To prove (H3) we note that

$$x(t) \in L_p(0,1) \Rightarrow \int_0^1 |x(t)^p dt < \infty \qquad [\text{see } \mathbf{H3}],$$

$$y(t) \in L_p(0,1) \Rightarrow \int_0^1 |y(t)|^q dt < \infty \qquad [\text{see } \mathbf{H3}].$$

Taking $\quad a = \dfrac{|x(t)|^p}{\int_0^1 |x(t)|^p dt}$ and $b = \dfrac{|y(t)|^q}{\int_0^1 |y(t)|^q dt}$ in the inequality

$a^{1/p} b^{1-\frac{1}{p}} \leq \dfrac{a}{p} + \dfrac{b}{q}$, and integrating from 0 to 1, we have

$$\frac{\int_0^1 |x(t)y(t)| dt}{(\int_0^1 |x(t)|^p dt)^{1/p} (\int_0^1 |y(t)|^q dt)^{1/q}} \leq 1 \qquad (1.4)$$

which yields the Hölder's inequality for integrals.

1.4.4 Theorem (Minkowski's inequality)

(M1) If $p \geq 1$, then

$$\left[\sum_{i=1}^{n} |x_i + y_i|^p\right]^{1/p} \leq \left[\sum_{i=1}^{n} |x_i|^p\right]^{1/p} + \left[\sum_{i=1}^{n} |y_i|^p\right]^{1/p}$$

for any complex numbers $x_1, \ldots x_n, y_1, y_2, \ldots, y_n$.

(M2) If $p \geq 1$, $\{x_i\} \in \ell_p$, i.e. pth power summable, $y = \{y_i\} \in \ell_q$, i.e., qth power summable, where p and q are conjugate to each other, then

$$\left(\sum_{i=1}^{\infty} |x_i + y_i|^p\right)^{1/p} \leq \left(\sum_{i=1}^{\infty} |x_i|^p\right)^{1/p} + \left(\sum_{i=1}^{\infty} |y_i|^p\right)^{1/p}.$$

(M3) If $x(t)$ and $y(t)$ belong to $L_p(0,1)$, then

$$\left(\int_0^1 |x(t) + y(t)^p dt\right)^{1/p} \leq \left(\int_0^1 |x(t)^p dt\right)^{1/p} + \left(\int_0^1 |y(t)|^p dt\right)^{1/p}.$$

Proof: If $p = 1$ and $p = \infty$ the (M1) is easily seen to be true. Suppose $1 < p < \infty$, then

$$\left[\sum_{i=1}^{n} |x_i + y_i|^p\right]^{1/p} \leq \left[\sum_{i=1}^{n} (|x_i| + |y_i|)^p\right]^{1/p}. \tag{1.5}$$

Moreover, $(|x_i| + |y_i|)^p = (x_i| + |y_i|)^{p-1}|x_i| + (|x_i| + |y_i|)^{p-1}|y_i|$.

Summing these identities for $i = 1, 2, \ldots, n$,

$$\sum_{i=1}^{n} (|x_i + y_i|^{p-1})|x_i| \leq \left[\sum_{i=1}^{n} |x_i|^p\right]^{1/p} \left[\sum_{i=1}^{n} ((|x_i| + |y_i|)^{p-1})^q\right]^{1/q}$$

$$= \left[\sum_{i=1}^{p} |x_i|^p\right]^{1/p} \left[\sum_{i=1}^{n} (|x_i| + |y_i|)^p\right]^{1/q}.$$

Similarly we have,

$$\sum_{i=1}^{n} (|x_i| + |y_i|)^{p-1}|y_i| \leq \left[\sum_{i=1}^{n} |y_i|^p\right]^{1/p} \left[\left(\sum_{i=1}^{n} (|x_i| + |y_i|^p)\right)\right]^{1/q}$$

From the above two inequalities,

$$\sum_{i=1}^{n} (|x_i| + |y_i|)^p \leq \left[\left(\sum_{i=1}^{n} |x_i|^p\right)^{1/p} + \left(\sum_{i=1}^{n} |y_i|^p\right)^{1/p}\right].$$

$$\left[\left(\sum_{i=1}^{n}(|x_i|+|y_i|)^p\right)\right]^{1/q} \quad (1.6)$$

or, $$\left[\left(\sum_{i=1}^{n}(|x_i|+|y_i|)^p\right)\right]^{1/p} \leq \left[\sum_{i=1}^{n}|x_i|^p\right]^{1/p} + \left[\sum_{i=1}^{n}|y_i|^p\right]^{1/p}$$

assuming that $\sum_{i=1}^{n}(|x_i|+|y_i|)^p \neq 0$.

From (1.5) and (1.6) we have

$$\left[\sum_{i=1}^{n}|x_i+y_i|^p\right]^{1/p} \leq \left[\sum_{i=1}^{n}|x_i|^p\right]^{1/p} + \left[\sum_{i=1}^{n}|y_i|^p\right]^{1/p} \quad (1.7)$$

(**M2**) is true for $p=1$ and $p=\infty$.

To, prove (**M2**) for $1 < p < \infty$, we note that

$$x = \{x_i\} \in \ell_p \Rightarrow \sum_{i=1}^{n}|x_i|^p < \infty \quad \text{also } y = \{y_i\} \in \ell_p \Rightarrow \sum_{i=1}^{n}|y_i|^p < \infty.$$

We examine $\sum_{i=1}^{n}|x_i+y_i|^p$.

Let us note that $z = \{z_i\} \in \ell_p \Rightarrow z' = \{|z_i|^{p-1}\} \in \ell_q$.

On applying twice Hölder's inequality to the sequences $\{x_i\} \in \ell_p$ and $\{|x_i+y_i|^{p-1}\} \in \ell_q$, corresponding to $\{y_i\} \in \ell_p$ we get,

$$\sum_{i=1}^{\infty}|x_i+y_i|^p \leq \sum_{i=1}^{\infty}|x_i+y_i|^{p-1}|x_i| + \sum_{i-1}^{n}|x_i+y_i|^{p-1}|y_i|$$

$$\leq \left(\sum_{i=1}^{\infty}|x_i+y_i|^{(p-1)q}\right)^{1/q}\left[\left(\sum_{i=1}^{\infty}|x_i|^p\right)^{1/p} + \left(\sum_{i=1}^{\infty}|y_i|^q\right)^{1/q}\right]$$

$$= \left(\sum_{i=1}^{\infty}|x_i+y_i|^p\right)^{1/q}\left[\left(\sum_{i=1}^{\infty}|x_i|^p\right)^{1/p} + \left(\sum_{i=1}^{\infty}|y_i|^q\right)^{1/q}\right].$$

Assuming $\left(\sum_{i=1}^{\infty}|x_i+y_i|^p\right) \neq 0$, the above inequality yields on division

by $\left(\sum_{i=1}^{\infty}|x_i+y_i|^p\right)^{1/q}$,

$$\left(\sum_{i=1}^{\infty}|x_i+y_i|^p\right)^{1/p} \leq \left[\left(\sum_{i=1}^{\infty}|x_i|^p\right)^{1/p} + \left(\sum_{i=1}^{\infty}|y_i|^q\right)^{1/p}\right] \quad (1.8)$$

It is easily seen that (**M3**) is true for $p = 1$ and $p = \infty$. To prove (M3) for $1 < p < \infty$ we proceed as follows.

Let $\quad x(t) \in L_p(0,1) \quad$ i.e. $\quad \displaystyle\int_0^1 |x(t)|^p dt < \infty$

$\qquad y(t) \in L_p(0,1) \quad$ i.e. $\quad \displaystyle\int_0^1 |y(t)|^p dt < \infty$

If $\quad z(t) \in L_p(0,1) \quad$ i.e. $\quad \displaystyle\int_0^1 |z(t)|^p dt < \infty$

i.e. $\displaystyle\int_0^1 (|z(t)|^{p-1})^{\frac{p}{p-1}} dt < \infty \quad$ i.e. $\quad |z(t)|^{p-1} \in L_q(0,1)$.

Let us consider the integral $\displaystyle\int_0^1 |x(t) + y(t)|^p dt$

for $\quad 1 < p < \infty, |x(t) + y(t)|^p \le |x(t)|^p + |y(t)|^p$

$$\le 2^p(|x(t)|^p + |y(t)|^p)$$

Hence, $\qquad \displaystyle\int_0^1 |x(t) + y(t)|^p dt \le 2^p \left(\int_0^1 |x(t)|^p dt + \int_0^1 |y(t)|^p dt \right)$

$$< \infty \quad \text{since } x(t), y(t) \in L_p(0,1)$$

Furthermore, $\displaystyle\int_0^1 |x(t) + y(t)|^p dt < \infty \Rightarrow \int_0^1 (|x(t) + y(t)|)^{\frac{p}{p-1}} dt < \infty$

$$\Rightarrow \qquad \int_0^1 (|x(t) + y(t)|)^{p-1} dt \in L_q(0,1)$$

where p and q are conjugate to each other.

Using Hölder's inequality we conclude

$$\int_0^1 |x(t) + y(t)|^p dt \le \int_0^1 |x(t) + y(t)|^{p-1} |x(t)| dt$$

$$+ \int_0^1 |x(t) + y(t)|^{p-1} |y(t)| dt$$

$$\le \left(\int_0^1 |x(t) + y(t)|^{(p-1)/q} dt \right)^{1/q} \left(\int_0^1 |x(t)|^p dt \right)^{1/p}$$

$$+ \left(\int_0^1 |x(t) + y(t)|^{(p-1)q} dt \right)^{1/q} \left(\int_0^1 |y(t)|^p dt \right)^{1/p}$$

$$= \left(\int_0^1 |x(t) + y(t)|^p dt \right)^{1/q} \left[\left(\int_0^1 |x(t)|^p dt \right)^{1/p} \right.$$

$$\left. + \left(\int_0^1 |y(t)|^p dt \right)^{1/p} \right].$$

Assuming that $\int_0^1 |x(t) + y(t)|^p dt \neq 0$ and dividing both sides of the above inequality by $\left(\int_0^1 |x(t) + y(t)|^p dt \right)^{1/q}$, we get

$$\left(\int_0^1 |x(t) + y(t)|^p dt \right)^{1/p}$$
$$\leq \left[\left(\int_0^1 |x(t)|^p dt \right)^{1/p} + \left(\int_0^1 |y(t)|^p dt \right)^{1/p} \right] \qquad (1.9)$$

Problems

1. Show that the Cauchy-Bunyakovsky-Schwartz inequality implies that

$$(|\xi_1| + |\xi_2| + \cdots + |\xi_n|)^2 \leq n(|\xi_1|^2 + \cdots + |\xi_n|^2).$$

2. In the plane of complex numbers show that the points z on the open unit disk $|z| < 1$ form a metric space if the metric is defined as

$$\rho(z_1, z_2) = \frac{1}{2} \log \frac{1 + u}{1 - u}, \quad \text{where} \quad u = \left| \frac{z_1 - z_2}{1 - \bar{z}_1 z_2} \right|.$$

1.4.5   The spaces $l_p^{(n)}, l_\infty^{(n)}, l_p, p \geq 1, l_\infty$

(i) The spaces $l_p^{(n)}, l_\infty^{(n)}$

Let X be an n-dimensional arithmetic space, i.e., the set of all possible n-tuples of real numbers and let $x = \{x_1, x_2, \ldots, x_n\}$, $y = \{y_1, y_2, \ldots, y_n\}$, and $p \geq 1$.

We define $\rho_p(x, y) = \left(\sum_{i=1}^n |x_i - y_i|^p \right)^{1/p}$. Let $\max_{1 \leq i \leq n} |x_i - y_i| = |x_k - y_k|$. Then,

$$\rho_p(x, y) = |x_k - y_k| \left(1 + \sum_{\substack{i=1 \\ i \neq k}}^n \left(\left| \frac{x_i - y_i}{x_k - y_k} \right| \right)^p \right)^{1/p}.$$

Making $p \to \infty$, we get $\rho_\infty(x, y) = \max_{1 \leq i \leq n} |x_i - y_i|$. It may be noted that $\rho_p(x, y) \ \forall \ x, y \in X$ satisfies the axioms **1-3** of a metric space. Since by Minkowski's inequality,

$$\left(\sum_{i=1}^n |x_i - z_i|^p \right)^{1/p} \leq \left(\sum_{i=1}^n |x_i - y_i|^p \right)^{1/p} + \left(\sum_{i=1}^n |y_i - z_i|^p \right)^{1/p}$$

axiom **4** of a metric space is satisfied. Hence the set X with the metric $\rho_p(x, y)$ is a metric space and is called $l_\infty^{(n)}$.

(ii) The spaces l_p, $p \geq 1$, l_∞

Let X be the set of sequences $x = \{x_1, x_2, \ldots, x_n\}$ of real numbers. x is said to belong to the space l_p if $\sum\limits_{i=1}^{\infty} |x_i|^p < \infty$ ($p \geq 1$, p fixed). In l_p we introduce the metric $\rho(x, y)$ for $x = \{x_i\}$ and $y = \{y_i\}$ as

$\rho_p(x, y) = \left(\sum\limits_{i=1}^{\infty} |x_i - y_i|^p \right)^{1/p}$. The metric is a natural extension of the

metric in $l_p^{(n)}$ when $n \to \infty$. To see that the series for ρ_p converges for $x, y \in l_p$ we use Minkowski's inequality (**M2**). It may be noted that the above metric satisfies axioms **1-3** of a metric space. If $z = \{z_i\} \in l_p$, then

the Minkowski's inequality (**M2**) yields $\rho(x, y) = \left(\sum\limits_{i=1}^{\infty} |x_i - z_i|^p \right)^{1/p} =$

$\left[\sum\limits_{i=1}^{\infty} |(x_i - y_i) + (y_i - z_i)|^p \right]^{1/p} \leq \rho(x, y) + \rho(y, z)$ Thus l_n is a metric space.

If $p = 2$, we have the space l_2 with the metric $\rho(x, y) = \left(\sum\limits_{i=1}^{\infty} |x_i - y_i|^2 \right)^{1/2}$.

Later chapters will reveal that l_2 possesses a special property in that it admits of a scalar product and hence becomes a Hillbert space.

l_∞ is the space of all bounded sequences, i.e., all $x = \{x_1, x_2, \ldots, .\}$ for which $\sup\limits_{1 \leq i \leq \infty} |x_i| < \infty$, with metric $\rho(x, y) = \sup\limits_{1 \leq i \leq \infty} |x_i - y_i|$ where $y = \{y_1, y_2, \ldots, .\}$.

1.4.6 The complete spaces, non-metrizable spaces

(i) Complex spaces

Together with the real spaces $C([0, 1]), m, l_p$ it is possible to consider the complex space $\mathbb{C}([0, 1]), m, l_p$ corresponding to the real spaces. The elements of complex space $\mathbb{C}([0, 1])$ are complex-valued continuous functions of a real variable. Similarly, the elements of complex space m are elements that are bounded, as in the case of complex l_p spaces whose series of p-power of moduli converges.

(ii) Non-metrizable spaces

Let us consider the set $F([0, 1])$ of all real functions defined on the interval [0,1]. A sequence $\{x_n(t)\} \subset F([0, 1])$ will converge to $x(t) \in F([0, 1])$, if for any fixed t, we have $x_n(t) \to x(t)$. Thus the convergence of a sequence of functions in $F([0, 1])$ is a pointwise convergence. We will show that $F([0, 1])$ is not a metric space. Let M be the set of continuous functions in the metric space $F([0, 1])$. Using the properties of closure in the metric space, $\overline{\overline{M}} = \overline{M}$ [see 1.4.6]. Since $\overline{M}$ is a set of continuous functions, the limits are in the sense of uniform convergence. However, $F([0,1])$ admits of only pointwise convergence. This means int $\overline{M} = \Phi$, i.e., $\overline{M}$ is nowhere dense, that is therefore a first category set of functions [sec 1.4.9.]. Thus, $\overline{M}$ is the set of real functions and their limits are in the sense of pointwise convergence. Therefore, $\overline{\overline{M}}$ is a set of the second category [sec. 1.4.9] and the pointwise convergence is non-metrizable.

Problems

1. Find a sequence which converges to 0, but is not in any space l_p where $1 \leq p < \infty$.

2. Show that the real line with $\rho(x,y) = \dfrac{|x-y|}{1+|x-y|}$ is a metric space.

3. If (X, ρ) is any metric space, show that another metric of X is defined by
$$\widetilde{\rho}(x,y) = \frac{\rho(x,y)}{1+\rho(x,y)}.$$

4. Find a sequence $\{x\}$ which is in l^p with $p > 1$ but $x \notin l^1$.

5. Show that the set of continuous functions on $(-\infty, \infty)$ with
$$\rho(x,y) = \sum_{n=1}^{\infty} \frac{1}{2^n} \frac{\max[|x(t)-y(t)| : |t| \leq n]}{1-\max[|x(t)-y(t)| : |t| \leq n]} \text{ is a metric space.}$$

6. **Diameter, bounded set:** The diameter $D(A)$ of a non-empty set A in a metric space (x, ρ) is defined to be $D(A) = \sup\limits_{x,y \in A} \rho(x,y)$. A is said to be bounded if $D(A) < \infty$. Show that $A \subseteq B$ implies that $D(A) \leq D(B)$.

7. **Distance between sets:** The distance $D(A,B)$ between two non-empty sets A and B of a metric space (X, ρ) is defined to be $D(A,B) = \inf\limits_{\substack{x \in A \\ y \in B}} \rho(x,y)$. Show that D does not define a metric on the power set of X.

8. **Distance of a point from a set:** The distance $D(x, A)$ from a point x to a non-empty subset A of (X, ρ) is defined to be $D(x, A) = \inf\limits_{a \in A} \rho(x,a)$. Show that for any $x, y \in X$, $|D(x,A) - D(y,A)| \leq d(x,y)$.

1.4.7 Definition: ball and sphere

In this section we introduce certain concepts which are quite important in metric spaces. When applied to Euclidean spaces these concepts can be visualised as an extension of objects in classical geometry to higher dimensions. Given a point $x_0 \in X$ and a real number $r > 0$, we define three types of sets:

(a) $B(x_0, r) = \{x \in X | \rho(x, x_0) < r\}$ (**open ball**)

(b) $\overline{B}(x_0, r) = \{x \in X | \rho(x, x_0) \leq r\}$ (**closed ball**)

(c) $S(x_0, r) = \{x \in X | \rho(x, x_0) = r\}$ (**sphere**).

In all these cases x_0 is called the centre and r the ball radius. An open ball in the set X is a set of all points of X the distance of which from the centre x_0 is always less than the radius r.

Note 1.4.1. In working with metric spaces we borrow some terminology from Euclidean geometry. But we should remember that balls and spheres in an arbitrary metric space do not possess the same properties as balls and spheres in $\mathbb{R}^3$. An unusual property of a sphere is that it may be empty. For example a sphere in a discrete metric space is null, i.e., $S(x_0, r) = \Phi$ if $r \neq 1$. We next consider two related concepts.

1.4.8 Definition: open set, closed set, neighbourhood, interior point, limit point, closure

A subset M of a metric space X is said to be open if it contains a ball about each of its points. A subset K of X is said to be closed if its complement (in X) is open- that is, $K^C = X - K$ is open.

An open ball $B(x_0, \epsilon)$ of radius ϵ is often called an ϵ-neighbourhood of x_0. By a neighbourhood of x_0, we mean any subset of X which contains an ϵ-neighbourhood of x_0. We see that every neighbourhood of x_0 contains x_0. In other words, x_0 is a point in each of its neighbourhoods. If N is a neighbourhood of x_0 and $N \subseteq M$, then M is also a neighbourhood of x_0.

We call x_0 an interior point of a set $M \subseteq X$ if M is a neighbourhood of x_0. The interior of M is the set of all interior points of M and is denoted by M^0 or $\text{int}(M)$. $\text{int}(M)$ is open and is the largest open set contained in M. Symbolically, $\text{int}(M) = (x : x \in M$ and $B(x_0, \epsilon) \subseteq M)$ for some $\epsilon > 0$.

If $A \subseteq X$ and $x_0 \in M$, then x_0 is called the limit point of A if every neighbourhood of x_0 contains at least one point of A other than x_0. That is, x_0 is a limit point of A if and only if N_{x_0} is a neighbourhood of x_0 and implies that $(N_{x_0} - (x_0)) \cap A \neq \Phi$.

If for all neighbourhoods N_x of x, $N_x \cap A \neq \Phi$, then x is called a contact point. For each $A \subset X$, the set $\overline{A}$, consisting of all points which are either points of A or its limiting points, is called the closure of A. The closure of a set is a closed set and is the smallest closed set containing A.

Note 1.4.2. In what follows we show how different metrics yield different types of open balls. Let $X = \mathbb{R}^2$ be the Euclidean space. Then the unit open ball $B(0, 1)$ is given in Figure 1.2(a). If the l_∞ norm is used the unit open ball $B(0, 1)$ is the unit square as given in Figure 1.2(b). If the l_1 norm is used, the unit open ball $B(0, 1)$ becomes the 'diamond shaped' region shown in Figure 1.2(c). If we select $p > 2$, $B(0, 1)$ becomes a figure with curved sides, shown in Figure 1.2(d). The unit ball in $C[0, 1]$ is given in

Figure 1.2(e).

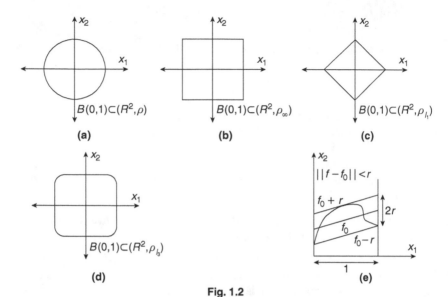

Fig. 1.2

Note 1.4.3. It may be noted that the closed sets of a metric space have the same basic properties as the closed numerical point sets, namely:

(i) $\overline{M \cup N} = \overline{M} \cup \overline{N}$;

(ii) $M \subset \overline{M}$;

(iii) $\overline{\overline{M}} = \overline{M} = M$;

(iv) The closure of an empty set is empty.

1.4.9   Theorem

In any metric space X, the empty set Φ and the full space X are open. To show that Φ is open, we must show that each point in Φ is the centre of an open ball contained in Φ; but since there are no points in Φ, this requirement is automatically satisfied. X is clearly open since every open ball centered on each of its points is contained in X. This is because X is the entire space.

Note 1.4.4. It may be noted that an open ball $B(0,r)$ on the real line is the bounded open interval $]-r,r[$ with its centre as the origin and a total length of $2r$. We may note that $[0,1[$ on the real line is not an open set since the interval being closed on the left, every bounded open interval with the origin as centre or in other words every open ball $B(0,r)$ contains points of $\mathbb{R}$ not belonging to $[0,1[$. On the other hand, if we consider $X = [0,1[$ as a space itself, then the set X is open. There is no inconsistency in the above statement, if we note that when $X = [0,1[$ is considered as a space, there are no points of the space outside $[0,1[$. However, when $X = [0,1[$ is considered as a subspace of $\mathbb{R}$, there are points of $\mathbb{R}$ outside X. One should take note of the fact that whether or not a set is open is relevant only with respect to a specific metric space containing it, but never on its own.

1.4.10 Theorem

In a metric space X each open ball is an open set.

Let $B(x_0, r)$ be a given ball in a metric space X. Let x be any point in $B(x_0, r)$. Now $\rho(x_0, x) < r$. Let $r_1 = r - \rho(x_0, x)$. Hence $B(x, r_1)$ is an open ball with centre x and radius r_1. We want to show that $B(x, r_1) \subseteq B(x_0, r)$. For if $y \in B(x, r_1)$, then

$$\rho(y, x_0 \le \rho(y, x) + \rho(x, x_0) < r_1 + \rho(x, x_0) = r, \quad \text{i.e.,} \quad y \in B(x_0, r).$$

Thus $B(x, r_1)$ is an open ball contained in $B(x_0, r)$. Since x is arbitrary, it follows that $B(x_0, r)$ is an open set. In what follows we state some results that will be used later on:

Let X be a metric space.

(i) A subset G of X is open $\Leftrightarrow$ it is a union of open balls.

(ii) (a) every union of open sets in X is open and (b) any finite intersection of open sets in X is open.

Note 1.4.5. The two properties mentioned in (ii) are vital properties of a metric space and these properties are established by using only the 'openness' of a set in a metric space. No use of distance or metric is required in the proof of the above theorem. These properties are germane to the development of 'topology' and 'topological spaces'. We discuss them in the next section.

We will next mention some properties of closed sets in a metric space. We should recall that a subset K of X is said to be closed if its complement $K^C = X - K$ is open.

(i) The null set Φ and the entire set X in a metric space are closed.

(ii) In a metric space a closed ball is a closed set.

(iii) In a metric space, (a) the intersection of closed sets is closed and (b) the union of a finite number of closed sets is a closed set.

Note 1.4.6. This leads to what is known as **closed set topology**.

1.4.11 Convergence, Cauchy sequence, completeness

In real analysis we know that a sequence of real numbers $\{\xi_i\}$ is said to tend to a limit l if the distance of $\{\xi_i\}$ from l is arbitrarily small $\forall i$ excepting a finite number of terms. In other words, it is the metric on $\mathbb{R}$ that helps us introduce the concept of convergence. This idea has been generalized in any metric space where convergence of a sequence has been defined with the help of the relevant metric.

Definition: convergence of a sequence, limit

A sequence $\{x_n\}$ in a metric space $X = (X, \rho)$ is said to converge or to be convergent if there is an $x \in X$ such that $\lim\limits_{n \to \infty} \rho(x, x_n) = 0$. x is called limit of $\{x_n\}$ and we write it as $x = \lim\limits_{n \to \infty} x_n$. In other words, given $\epsilon > 0$, $\exists\, n_0(\epsilon)$ s.t. $\rho(x, x_n) < \epsilon \,\forall\, n > n_0(\epsilon)$.

Note 1.4.7. The limit of a convergent sequence must be a point of the space X.

For example, let X be the open interval $[0,1[$ in $\mathbb{R}$ with the usual metric $\rho(x, y) = |x - y|$. Then the sequence $(1/2, 1/3, \ldots, 1/n, \ldots)$ is not convergent since '0', the point to which the sequence is supposed to converge, does not belong to the space X.

1.4.12   Theorem

A sequence $\{x_n\}$ of points of a metric space X can converge to one limit at most.

If the limit is not unique, let $x_n \to x$ and $x_n \to y$ as $n \to \infty, x \neq y$.

Then $\rho(x, y) \leq \rho(x_n, x) + \rho(x_n, y) < \epsilon + \epsilon$, for $n \geq n_0(\epsilon)$. Since ϵ is an arbitrary positive number, it follows that $x = y$.

1.4.13   Theorem

If a sequence $\{x_n\}$ of points of X converges to a point $x \in X$, then the set of numbers $\rho(x_n, \theta)$ is bounded for every fixed point θ of the space X.

Note 1.4.8. In some spaces the limit of a sequence of elements is directly defined. If we can introduce in this space a metric such that the limit induced by the metric coincides with the initial limit, the given space is called **metrizable**.

Note 1.4.9. It is known that in $\mathbb{R}$ the Cauchy convergence criterion ensures the existence of the limit. Yet in any metric space the fulfillment of the Cauchy convergence criterion does not ensure the existence of the limit. This needs the introduction of the notion of completeness.

1.4.14   Definition: Cauchy sequence, completeness

A sequence $\{x_n\}$ in a metric space $X = (X, \rho)$ is said to be a Cauchy sequence or convergent sequence if given $\epsilon > 0$, $\exists\, n_0(\epsilon)$, a positive integer such that $\rho(x_n, x_m) < \epsilon$ for $n, m > n_0(\epsilon)$.

Note 1.4.10. The converse of the theorem is not true for an arbitrary metric space, since there exist metric spaces that contain a Cauchy sequence but have no element that will be the limit.

1.4.15 Examples

(i) The space of rational numbers

Let X be the set of rational numbers, in which the metric is taken as $\rho(r_1, r_2) = |r_1 - r_2|$. Thus, X is a metric space. Let us take $r_1 = 1$, $r_2 = \frac{1}{2}, \cdots, r_n = \frac{1}{n}$. $\{r_n\}$ is a Cauchy sequence and $r_n \to 0$ as $n \to \infty$. On the other hand let us take $r_n = \left(1 + \frac{1}{n}\right)^n$ where n is an integer. $\{r_n\}$ is a Cauchy sequence. However, $\lim\limits_{n \to \infty} \left(1 + \frac{1}{n}\right)^n = e$, which is not a rational number.

(ii) The space of polynomials $P(t)(0 \le t \le 1)$

Let X be the set of polynomials $\boldsymbol{P(t)}$ $(0 \le t \le 1)$ and let the metric be defined by $\rho(P, Q) = \max\limits_t |P(t) - Q(t)|$. It can be seen that with the above metric, the space X is a metric space. Let $\{P_n(t)\}$ be the sequence of nth degree polynomials converging uniformly to a continuous function that is not a polynomial. Thus the above sequence of polynomials is a Cauchy sequence with no limit in the space (X, ρ). In what follows, we give some examples of complete metric spaces.

(iii) Completeness of $\mathbb{R}^n$ and $\mathbb{C}^n$

Let us consider $x_p \subseteq \mathbb{R}^n$. Then we can write $x_p = \{\xi_1^{(p)}, \xi_2^{(p)}, \ldots, \xi_n^{(p)}\}$. Similarly, $x_q = \{\xi_1^{(q)}, \ldots \xi_2^{(q)}, \ldots, \xi_n^{(q)}\}$. Then,

$$\rho(x_p, x_q) = \left(\sum_{i=1}^n |\xi_i^{(p)} - \xi_i^{(q)}|^2\right)^{1/2}.$$ Now, if $\{x_m\}$ is a Cauchy sequence, for every $\epsilon > 0$, $\exists\, n_0(\epsilon)$ s.t.

$$\rho(x_p, x_q) = \left(\sum_{i=1}^n |\xi_i^{(p)} - \xi_i^{(q)}|^2\right)^{1/2} < \epsilon \text{ for } p, q > n_0(\epsilon). \tag{1.10}$$

Squaring, we have for $p, q > n_0(\epsilon)$, $i = 1, 2, \ldots, n$, $|\xi_i^{(p)} - \xi_i^{(q)}|^2 < \epsilon^2 \Rightarrow$ $|\xi_i^{(p)} - \xi_i^{(q)}| < \epsilon$. This shows that for each fixed i, $(1 \le i \le n)$, the sequence $\{\xi_1^{(i)}, \xi_2^{(i)}, \ldots, \xi_n^{(i)}\}$ is a Cauchy sequence of real numbers. Therefore, $\xi_i^{(m)} = \xi_i \in \mathbb{R}$ as $m \to \infty$. Let us denote by x the vector, $x = (\xi_1, \xi_2, \ldots, \xi_n)$. Clearly, $x \in \mathbb{R}^n$. It follows from (1.1) that $\rho(x_m, x) \le \epsilon$ for $m \ge n_0(\epsilon)$. This shows that x is the limit of $\{x_m\}$ and this proves completeness because $\{x_m\}$ is an arbitrary Cauchy sequence. Completeness of $\mathbb{C}^n$ can be proven in a similar fashion.

(iv) Completeness of $C([a, b])$ and incompleteness of $S([a, b])$

Let $\{x_n(t)\} \subset C([a, b])$ be a Cauchy sequence. Hence, $\rho(x_n(t), x_m(t)) \to 0$ as $n, m \to \infty$ since $x_n(t), x_m(t) \in C([a, b])$. Thus, given $\epsilon > 0$, $\exists\, n_0(\epsilon)$ such that $\max\limits_{t \in [a,b]} |x_n(t) - x_m(t)| < \epsilon$ for $n, m \ge x_0(t)$. Hence, for every fixed $t = t_0 \in J = [a, b]$, $|x_n(t_0) - x_m(t_0)| < \epsilon$ for $m, n > n_0(\epsilon)$. Thus, $\{x_n(t_0)\}$ is a convergent sequence of real numbers. Since $\mathbb{R}$ is complete, $\{x_n(t_0)\} \to x(t_0) \in \mathbb{R}$. In this way we can associate with each $t \in J$ a unique real number $x(t)$ as limit of the sequence $\{x_n(t)\}$. This defines a (pointwise) function x on J and thus $x(t) \in C([a, b])$. Thus, it follows from (6.2), making $n \to \infty$, $\max |x_m(t) - x(t)| < \epsilon$ for $m \geq n_0(\epsilon)$ for every $t \in J$. Therefore, $\{x_m(t)\}$ converges uniformly to $x(t)$ on J. Since $x_m(t)$'s are continuous functions of t, and the convergence is uniform, the limit $x(t)$ is continuous on J. Hence, $x(t) \in C([a, b])$, i.e., $C([a, b])$, is complete.

Note 1.4.11. We would call $C([a, b])$ as real $C([a, b])$ if each member of $C([a, b])$ is real-valued. On the other hand, if each member of $C([a, b])$ is complex-valued then we call the space as complex $\mathbb{C}([a, b])$.

By arguing analogously as above we can show that complex $\mathbb{C}([a, b])$ is complete. We next consider the set X of all continuous real-valued functions on $J = [a, b]$. Let us define the metric $\rho(x(t), y(t))$ for $x(t), y(t) \in X$ as

$$\rho(x, y) = \int_a^b |x(t) - y(t)| dt.$$

We can easily see that the set X with the metric defined above is a metric space $S[a, b] = (X, \rho)$. We next show that $S[a, b]$ is not complete. Let us construct a $\{x_n\}$ as follows: If $a < c < b$ and for every n so large that $a < c - \frac{1}{n}$

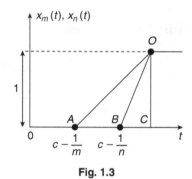

Fig. 1.3

We define

$$x_n(t) = \begin{cases} 0 & \text{if } a \leq t \leq c - \dfrac{1}{n} \\[2mm] nt - nc + 1 & \text{if } c - \dfrac{1}{n} \leq t \leq c \\[2mm] 1 & \text{if } c \leq t \leq b \end{cases}$$

For $n > m$

$$\int_a^b |x_n(t) - x_m(t)| dt = \Delta AOB = \frac{1}{2} \cdot 1 \cdot \left(c - \frac{1}{n} - c + \frac{1}{m}\right) < \frac{1}{2}\left(\frac{1}{n} + \frac{1}{m}\right)$$

Thus $\rho(x_n, x_m) \to 0$ as $n, m \to \infty$. Hence (x_n) is a Cauchy sequence. Now, let $x \in S[a, b]$, then $\lim\limits_{n \to i} \rho(x_n, x) = 0 \Rightarrow \begin{cases} x(t) = 0, & t \in [a, c) \\ x(t) = 1, & t \in (c, b] \end{cases}$

Since it is impossible for a continuous function to have the property, (x_n) does not have a limit.

(v) m, the space of bounded number sequences, is complete.

(vi) Completeness of l_p, $1 \leq p < \infty$.

(a) Let (x_n) be any Cauchy sequence in the space l_p where $x_n = \{\xi_1^{(n)}, \xi_2^{(n)}, \ldots, \xi_i^{(n)}, \ldots\}$. Then given $\epsilon > 0$, $\exists \, n_0(\epsilon)$ such that $\rho(x_n, x_m) < \epsilon$ for $n, m \geq n_0(\epsilon)$. Or, $\left(\sum_{i=1}^{\infty} |\xi_i^{(n)} - \xi_i^{(m)}|^p \right)^{i/p} < \epsilon$. It follows that for every $i = 1, 2, \ldots, |\xi_i^{(n)} - \xi_i^{(m)}| < \epsilon \, (n, m \geq n_0(\epsilon))$. We choose a fixed i. The above inequality yields $\{\xi_1^{(n)}, \xi_2^{(n)}, \ldots\}$ as a Cauchy sequence of numbers. The space $\mathbb{R}$ being complete $\{\xi_i^{(n)}\} \to \zeta_i \in \mathbb{R}$ as $n \to \infty$. Using these limits, we define $x = \{\xi_1, \xi_2, \ldots\}$ and show that $x \in l_p$ and $x_m \to x$ as $m \to \infty$. Since ϵ is an arbitrary small positive number,

$$\rho(x_n, x_m) < \epsilon \Rightarrow \sum_{i=1}^{k} |\xi_i^{(n)} - \xi_i^{(m)}|^p < \epsilon^p (k = 1, 2, \ldots). \text{ Making } n \to \infty,$$

we obtain for $m > n_0(\epsilon)$ $\sum_{i=1}^{k} |\xi_i^{(m)} - \xi_i|^p < \epsilon^p$. We may now let $k \to \infty$, then for $m > n_0(\epsilon)$, $\sum_{i=1}^{\infty} |\xi_i^{(m)} - \sum_i|^p < \epsilon^p$. This shows that $x_m - x = \{\xi_i^{(m)} - \xi_i\} \in l_p$. Since $x_m \in l_p$, it follows by the Minkowski inequality that $x = (x - x_m) + x_m \in l_p$. It also follows from the above inequality that $(\rho(x_n, x_m))^p \leq \epsilon^p$. Further, since ϵ is a small positive number, $x_m \to x$ as $m \to \infty$. Since $\{x_m\}$ is an arbitrary Cauchy sequence in l_p, this proves the completeness of $l_p, 1 \leq p \leq \infty$.

(b) Let $\{x_n\}$ be a Cauchy sequence in l_∞ where $x_n = \{\xi_1^{(n)}, \xi_2^{(n)}, \ldots, \xi_i^{(n)}, \ldots\}$. Thus for each $\epsilon > 0$, there is an N such that for $m, n > N$, we have $\sup_i |\xi_i^{(n)} - \xi_i^{(m)}| < \epsilon$. It follows that for each $i, \{\xi_i^{(n)}\}$ is a Cauchy sequence. Let $\xi_i = \lim_{n \to \infty} \xi_i^{(n)}$ and let $x = \{\xi_1, \xi_2, \ldots\}$. Now for each i and $n > N$, it follows that $|\xi_i - \xi_i^{(n)}| < \epsilon$. Therefore, $|\xi_i| \leq |\xi_i^{(n)}| + |\xi_i - \xi_i^{(n)}| \leq |\xi_i^{(n)}| + \epsilon$ for $n > N$. Hence, ξ_i is bounded for each i- i.e., $x \in l_\infty$ and $\{\xi_i^{(n)}\}$ converges to x in the l_∞ norm. Hence, l_∞ is complete under the metric defined for l_∞.

Problems

1. Show that in a metric space an 'open ball' is an open set and a 'closed ball' is a closed set.

2. What is an open ball $B(x_0; 1)$ on $\mathbb{R}$? In $\mathbb{C}$? In l_1? In $C([0, 1])$? In l_2?

3. Let X be a metric space. If $\{x\}$ is a subset of X consisting of a single point, show that its complement $\{x\}^c$ is open. More generally show that A^C is open if A is any finite subset of X.

4. Let X be a metric space and $B(x, r)$ the open ball in X with centre x and radius r. Let A be a subset of X with diameter less than r that intersects $B(x, r)$. Prove that $A \subseteq B(x, 2r)$.

5. Show that the closure $\overline{B(x_0, r)}$ of an open ball $B(x_0, r)$ in a metric space can differ from the closed ball $\overline{B}(x_0, r)$.

6. Describe the interior of each of the following subsets of the real line: the set of all integers; the set of rationals; the set of all irrationals; $]0, 1]$; $[0, 1]$; and $[0, 1[\cup \{1, 2\}$.

7. Give an example of an infinite class of closed sets, the union of which is not closed. Give an example of a set that (a) is both open and closed, (b) is neither open nor closed, (c) contains a point that is not a limit point of the set, and (d) contains no points that are not limit points of the set.

8. Describe the closure of each of the following subsets of the real line; the integers; the rationals; $]0, +\infty[;] - 1, 0[\cup]0, 1[$.

9. Show that the set of all real numbers constitutes an incomplete metric space if we choose $\rho(x, y) = |\arctan x - \arctan y|$.

10. Show that the set of continuous real-valued functions on $J = [0, 1]$ do not constitute a complete metric space with the metric $\rho(x, y) = \int_0^1 |x(t) - y(t)| dt$.

11. Let X be the metric space of all real sequences $x = \{\xi_i\}$ each of which has only finitely many nonzero terms, and $\rho(x, y) = \sum |\xi_i - \eta_i|$, where $y = \{\eta_i^{(n)}\}$. Show that $\{x_n\}$ with $x_n = \{\xi_j^{(n)}\}, \xi_j^{(n)} = j^{-2}$ for $j = 1, 2, \ldots, n$, and $\xi_j^{(n)} = 0$ for $j > n$ is a Cauchy sequence but does not converge.

12. Show that $\{x_n\}$ is a Cauchy sequence if and only if $\rho(x_{n+k}, x_n)$ converges to zero uniformly in k.

13. Prove that the sequence 0.1, 0.101, 0.101001, 0.1010010001,...is a Cauchy sequence of rational numbers that does not converge in the space of rational numbers.

14. In the space l_2, let $A = \{x = (x_1, x_2, \ldots) : |x_n| \leq \frac{1}{n}, n = 1, 2, \ldots\}$. Prove that A is closed.

1.4.16   Criterion for completeness

Definition (dense set, everywhere dense set and nowhere dense set)

Given A and B subsets of X, A is said to be dense in B if $B \subseteq \overline{A}$. A is said to be everywhere dense in X if $\overline{A} = X$. A is said to be *nowhere dense* in X, if $\overline{A}^C = X$ or $X - \overline{A} = X$ or $\overline{A} = \Phi$ or $\text{int}\overline{A} = \Phi$. The set of rational numbers is dense in $\mathbb{R}$.

As an example of a nowhere dense set in two-dimensional Euclidean space (the plane) is any set of points whose coordinates are both rational is an example of the first category [see 1.4.18]. It is the union of countable sets of 'one-point' sets. Although this set is of first category, it is nevertheless dense in $\mathbb{R}^2$.

We state without proof the famous *Cantor's intersection theorem*.

1.4.17 Theorem

Let a nested sequence of closed balls [i.e., each of which contains all that follows: $\overline{B}_1 \supseteq \overline{B}_2 \supseteq \ldots \overline{B}_n \ldots$] be given in a complete metric space X. If the radii of the balls tend to zero, then these balls have a unique common point.

1.4.18 Definition: first category, second category

A set M is said to be of the first category if it can be written as a countable union of nowhere dense sets. Otherwise, it is said to be of the second category. The set of rational points of a straight line is of the first category while that of the irrational points is of the second category as borne out by the following.

1.4.19 Theorem

A nonempty complete metric space is a set of the second category.

As an application of theorem 1.4.16, we prove the existence of nowhere differentiable functions on $[0, 1]$ that are continuous in the said interval. Let us consider the metric space $C_0([0,1])$ of continuous functions f for which $f(0) = f(1)$ with $\rho(f,g) = \max\{|f(x) - g(x)|, x \in [0,1]\}$. Then $C_0([0,1])$ is a complete metric space. We would like to show that those functions in $C_0([0,1])$ that are somewhere differentiable form a subset of the first category. $C_0([0,1])$ being complete, is of the second category. $C_0([0,1])$ can contain functions which are somewhere differentiable. Therefore, $C_0([0,1])$ can contain functions that are nowhere differentiable. For convenience we extend the functions of $C_0([0,1])$ to the entire axis by periodicity and to treat the space Γ of such extensions with the metric ρ defined above.

Let $K \subset \Gamma$ be the set of functions such that for some ξ, the set of numbers $\left[\dfrac{f(\xi + h) - f(\xi)}{h} : h > 0\right]$ is bounded. K contains the set of functions that are somewhere differentiable. We want to show that K is of the first category in Γ.

Let $K_n = \left[f \in \Gamma : \text{ for some } \xi, \left[\dfrac{f(\xi + h) - f(\xi)}{h}\right] \leq n, \text{ for all } h > 0\right]$.

Then $K = \bigcup_{n=1}^{\infty} K_n$. We shall show that for every $n = 1, 2, \ldots,$ (i) K_n is closed, and (ii) $\Gamma \sim K_n$ is everywhere dense in Γ. If (i) and (ii) are both true, $\Gamma \sim K_n = \Gamma$ or $K_n = \Phi$. Since K_n is a closed set, it follows that K_n is nowhere dense in Γ. Hence K will become nowhere dense in Γ.

For (i) let f be a limit point of K_n and let $\{f_n\}$ be a sequence in K_n converging to f. For each $k = 1, 2, \ldots$ let ξ_k be in $[0,1]$ such that $\left[\dfrac{f_k(\xi + h) - f_k(\xi_k)}{h}\right] \le n$ for all $h > 0$. Let ξ be a limit point of $\{\xi_k\}$ and $\{\xi_{k_j}\}$ converge to ξ. For $h > 0$ and $\epsilon > 0$,

$$\left|\frac{f(\xi + h) - f(\xi)}{h}\right| \le \left[\frac{f_{k_j}(\xi_{k_j} + h) - f_{k_j}(\xi_{k_j})}{h}\right] + \frac{1}{h}\{|f(\xi + h)$$
$$- f_{k_j}(\xi_{k_j} + h)| + |f_{k_j}(\xi_{k_j} + h) - f_{k_j}(\xi_{k_j})| + |f_{k_j}(\xi_k) - f(\xi_k)|$$
$$+ |f(\xi_k) - f(\xi)|\}$$

There is an $N = N(\epsilon h)$ such that $k > N$ implies that $\sup_t |f_k(t) - f(t)| \le \dfrac{\epsilon h}{4}$. Since f is continuous, there is an $M > N$ such that for $k_j > M$, we have, $|f(\xi + h) - f(\xi_{k_j} + h)| < \dfrac{\epsilon h}{4}$ and $|f(\xi_{k_j}) - f(\xi)| < \dfrac{\epsilon h}{4}$. Since $\lim\limits_{j\to\infty} \xi_{k_j} = \xi$, if $k_j > M$, we have $\left|\dfrac{f(\xi + h) - f(\xi)}{h}\right| < \left|\dfrac{f_{k_j}(\xi_{k_j} + h) - f_{k_j}(\xi_{k_j})}{h}\right| + \epsilon \le n + \epsilon$. It follows that $\left|\dfrac{f(|\xi + h) - f(\xi)}{h}\right| \le n$ for all $h > 0$. Thus $f \in K_n$ and K_n is closed.

For (ii) Let us suppose that $g \in \Gamma$. Let $\epsilon > 0$ and let us partition $[0, 1]$ into k equal intervals such that if x, x' are in the same interval of the partitioning, $|g(x) - g(x')| < \epsilon/2$ holds. Let us consider the ith subinterval $\frac{i-1}{k} \le x \le \frac{i}{k}$ and consider the rectangle with sides $g\left(\frac{i-1}{k}\right)$ and $g\left(\frac{i}{k}\right)$. For all points within the rectangle the ordinates satisfy $g\left(\frac{i-1}{k}\right) - \frac{\epsilon}{2} \le y \le g\left(\frac{i}{k}\right) + \frac{\epsilon}{2}$. Thus $\left(\frac{i}{k}, g\left(\frac{i}{k}\right)\right)$ is on the right-hand side of the rectangle and $\left(\frac{i-1}{k}, g\left(\frac{i-1}{k}\right)\right)$ is on the left-hand side of the rectangle. By joining these two points by a polygonal graph that remains within the rectangle and the line segments of which have slopes exceeding n in absolute value, we thus obtain a continuous function that is within ϵ of g and as because its slope exceeds n, it belongs to $\Gamma \sim K_n$. Thus $\Gamma - K_n$ is dense in Γ. Combining (i) and (ii) we can say that K_n and hence K is nowhere dense in Γ.

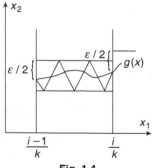

Fig. 1.4

1.4.20   *Isometric mapping, isometric spaces, metric completion, necessary and sufficient conditions*

Definition: isometric mapping, isometric spaces

Let $X_1 = (X_1, \rho_1)$ and $X_2 = (X_2, \rho_2)$ be two metric spaces. Then

(a) A mapping f of x_1 into X_2 is said to be *isometric* or an *isometry* if f preserves distance-i.e., for all $x, y \in X_1$, $\rho_2(fx, fy) = \rho_1(x, y)$ where fx and fy are images of x and y respectively.

(b) The space X_1 is said to be *isometric* to X_2 if there exists an one-to-one and onto (bijective) isometry of X_1 onto X_2. The spaces X_1 and X_2 are called *isomtric spaces*.

In what follows we aim to show that every metric space can be embedded in a complete metric space in which it is dense. If the metric space X is dense in $\widetilde{X}$, then $\widetilde{X}$ is called the metric completion of X. For example, the space of real numbers is the completion of the X of rational numbers corresponding to the metric $\rho(x, y) = |x - y|, x, y \in X$.

1.4.21   *Theorem*

Any metric space admits of a completion.

1.4.22   *Theorem*

If $X \subseteq \hat{X}, X$ is dense in $\hat{X}$, and any fundamental sequence of points of X has a limit in $\hat{X}$, then $\hat{X}$ is a completion of X.

1.4.23   *Theorem*

A subspace of a complete metric space is complete if and only if it is closed.

1.4.24   *Theorem*

Given a metric space X_1, assume that this space is incomplete, i.e., there exists in this space a Cauchy sequence that has no limit in X_1. Then there exists a complete space X_2 such that it has a subset X_2' everywhere dense in X_2 and isometric to X_1.

Problems

1. Let X be a metric space. If (x_n) and (y_n) are sequences in X such that $x_n \to x$ and $y_n \to y$, show that $\rho(x_n, y_n) \to \rho(x, y)$.

2. Show that a Cauchy sequence is convergent $\Leftrightarrow$ it has a convergent subsequence.

3. Exhibit a non-convergent Cauchy sequence in the space of polynomials on $[0,1]$ with uniform metric.

4. If ρ_1 and ρ_2 are metrics on the same set X and there are positive numbers a and b such that for all $x, y \in X, a\rho_1(x, y) \leq \rho_2(x, y) \leq b\rho_1(x, y)$, show that the Cauchy sequences in (X, ρ_1) and (X, ρ_2) are the same.

5. Using completeness in $\mathbb{R}$, prove completeness of C.

6. Show that the set of real numbers constitute an incomplete metric space, if we choose $\rho(x, y) = |\arctan x - \arctan y|$.

7. Show that a discrete metric space is complete.

8. Show that the space C of convergent numerical sequence is complete with respect to the metric you are to specify.

9. Show that convergence in C implies coordinate-wise convergence.

10. Show that the set of rational numbers is dense in $\mathbb{R}$.

11. Let X be a metric space and A a subset of X. Prove that A is everywhere dense in X $\Leftrightarrow$. The only closed superset of A is X $\Leftrightarrow$ the only open set disjoint from A is Φ.

12. Prove that a closed set F is nowhere dense if and only if it contains no open set.

13. Prove that if E is of the first category and $A \subseteq E$, then A is also of the first category.

14. Show that a closed set is nowhere dense $\Leftrightarrow$ its complement is everywhere dense.

15. Show that the notion of being nowhere dense is not the opposite of being everywhere dense. [Hint: Let $\mathbb{R}$ be a metric space with the usual metric and consider the subset consisting of the open interval $]1,2[$. The interior of the closure of this set is non-empty whereas the closure of $]1,2[$ is certainly not all or $\mathbb{R}$.]

1.4.25 *Contraction mapping principle*

1.4.26 *Theorem*

In a complete metric space X, let A be a mapping that maps the elements of the space X again into the elements of this space. Further for all x and y in X, let $\rho(A(x), A(y)) \leq \alpha\rho(x, y)$ with $0 \leq \alpha < 1$ independent of x and y. Then, there exists a unique point x^* such that $A(x^*) = x^*$. The point x^* is called a fixed point of A.

Proof: Starting from an arbitrary element $x_0 \in X$, we build up the sequence $\{x_n\}$ such that $x_1 = A(x_0), x_2 = A(x_1), \ldots, x_n = A(x_{n-1}), \ldots$. It is to be shown that $\{x_n\}$ is a Cauchy or fundamental sequence. For this we note that,

$$\rho(x_1, x_2) = \rho(A(x_0), A(x_1)) \leq \alpha\rho(x_0, x_1) = \alpha\rho(x_0, A(x_0)),$$
$$\rho(x_2, x_3) = \rho(A(x_1), A(x_2)) \leq \alpha\rho(x_1, x_2) \leq \alpha^2\rho(x_0, A(x_0)),$$
$$\cdots \quad \cdots \quad \cdots \quad \cdots \quad \cdots \quad \cdots \quad \cdots \quad \cdots$$
$$\rho(x_n, x_{n+1}) \leq \alpha^n\rho(x_0, A(x_0))$$

Further, $\rho(x_n, x_{n+p}) \leq \rho(x_n, x_{n+1}) + \cdots + \rho(x_n, x_{n+1})$

$$+ \cdots + \rho(x_{n+p-1}, x_{n+p})$$
$$\leq (\alpha^n + \alpha^{n+1} + \cdots + \alpha^{n+p-1})\rho(x_0, A(x_0))$$
$$= \frac{\alpha^n - \alpha^{n+p}}{1 - \alpha} \rho(x_0, A(x_0)).$$

Since, by hypothesis, $\rho(x_n, x_{n+p}) \leq \frac{\alpha^n}{1-\alpha}\rho(x_0, A(x_0))$, therefore $\rho(x_n, x_{n+p}) \to 0$ as $n \to \infty$, $p > 0$. Thus, (x_n) is a Cauchy sequence. Since the space is complete, there is an element $x^* \in X$, the limit of the sequence, $x^* = \lim_{n \to \infty} x_n$.

We shall show that $A(x^*) = x^*$.

$$\rho(x^*, A(x^*)) \leq \rho(x^*, x_n) + \rho(x_n, A(x^*))$$
$$\leq \rho(x^*, x_n) + \rho(A(x_{n-1}), A(x^*))$$
$$\leq \rho(x^*, x_n) + \alpha\rho(x_{n-1}, x^*)$$

For n sufficiently large, we can write, $\rho(x^*, x_n) < \epsilon/2, \rho(x^*, x_{n-1}) < \epsilon/2\alpha$, for any given ϵ. Hence $\rho(x^*, A(x^*)) < \epsilon$. Since $\epsilon > 0$ is arbitrary, $\rho(x^*, A(x^*)) = 0$, i.e., $A(x^*) = x^*$.

Let us assume that there exists two elements, $x^* \in X, y^* \in Y, x^* \neq y^*$ satisfying $A(x^*) = x^*$ and $A(y^*) = y$. Then, $\rho(x^*, y^*) = \rho(A(x^*), A(y^*)) \leq \alpha\rho(x^*, y^*)$. Since $x^* \neq y^*$, and $\alpha < 1$, the above inequality is impossible unless $\rho(x^*, y^*) = 0$, i.e, $x^* = y^*$. Making $p \to \infty$, in the inequality $\rho(x_n, x_{n+p}) \leq \frac{\alpha^n - \alpha^{n+p}}{1-\alpha}\rho(x_0, A(x_0))$, we obtain $\rho(x_n, x^*) \leq \frac{\alpha^n}{1-\alpha}\rho(x_0, A(x_0))$.

Note 1.4.12. Given an equation $F(x) = 0$, where $F : \mathbb{R}^n \to \mathbb{R}^n$, we can write the equation $F(x) = 0$ in the form $x = x - F(x)$. Denoting $x - F(x)$ by $A(x)$, we can see that the problem of finding the solution of $F(x) = 0$ is equivalent to finding the fixed point of $A(x)$ and vice versa.

1.4.27 Applications

(i) **Solution of a system of linear equations by the iterative method**

Let us consider the real n-dimensional space. If $x = (\xi_1, \xi_2, \ldots, \xi_n)$ and $y = (\eta_1, \eta_2, \ldots, \eta_n)$, let us define the metric as $\rho(x, y) = \max_i |\xi_i - \eta_i|$. Let us consider $y = Ax$, where A is an $n \times n$ matrix, i.e., $A = (a_{ij})$. The system of linear equations is given by $\eta_i = \sum_{j=1}^{n} a_{ij}\xi_j$, $i = 1, 2, \ldots, n$. Then

$$\rho(y_1, y_2) = \rho(A\xi_1, A\xi_2) \text{ yields } \max_i |\eta_i^{(1)} - \eta_i^{(2)}| = \max_i \left| \sum_j a_{ij}(\xi_j^{(1)} - \xi_j^{(2)}) \right| \leq$$

$\max_i \sum_{j=1}^{n} |a_{ij}|\rho(x_1, x_2)$. Now if it is assumed that $\sum_j |a_{ij}| < 1$, for all i, then the contraction mapping principle becomes applicable and consequently the matrix A has a unique fixed point.

(ii) Existence and uniqueness of the solution of an integral equation

1.4.28   Theorem

Let $k(t, s)$ be a real valued function defined in the square $a \leq t, s \leq b$ such that $\int_a^b \int_a^b k^2(t, s) dt \, ds < \infty$. Let $f(t) \in L_2([a, b])$ i.e., $\int_a^b |f(t)|^2 dt < \infty$. Then the integral equation $x(t) = f(t) + \lambda \int_a^b k(t, s) x(s) ds$ has a unique solution $x(t) \in L_2([a, b])$ for every sufficiently small value of the parameter λ.

Proof: Consider the operator $Ax(t) = f(t) + \lambda \int_a^b k(t, s) x(s) ds$. Let $x(t) \in L_2([a, b])$, i.e., $\int_a^b x^2(t) dt < \infty$. We first show that for $x(t) \in L_2([a, b])$, $Ax \in L_2([a, b])$.

$$\int_a^b (Ax)^2 dt = \int_a^b f^2(t) dt + 2\lambda \int_a^b f(t) \left(\int_a^b k(t, s) x(s) ds \right) dt$$
$$+ \lambda^2 \int_a^b \left(\int_a^b k(t, s) x(s) ds \right)^2 dt$$

Using Fubini's theorem th. 10.5 and the square integrability of $k(t, s)$ we can show that

$$\int_a^b f(t) \left(\int_a^b k(t, s) x(s) ds \right) dt = \int_a^b \int_a^b k(t, s) x(s) f(t) dt \, ds$$
$$\leq \left(\int_a^b \int_a^b k^2(t, s) dt \, ds \right)^{1/2} \left(\int_a^b f^2(t) dt \right)^{1/2} \cdot \left(\int_a^b x^2(s) ds \right)^{1/2}$$
$$< +\infty$$

Similarly we have $\int_a^b \left(\int_a^b k(t, s) x(s) ds \right)^2 dt < \infty$. Thus, $A(x) \in L_2([a, b])$. Therefore, $A : L_2([a, b]) \to L_2([a, b])$. Using the metric in $L_2([a, b])$, i.e., given $x(t), y(y) \in L_2([a, b])$,

$$\rho(Ax, Ay) = \left(\int_a^b |Ax - Ay|^2 dt \right)^{1/2}$$

$$= |\lambda| \left[\int_a^b \left(\int_a^b k(t,s)(x(s) - y(s))ds \right)^2 dt \right]^{1/2}$$

$$\leq |\lambda| \left(\int_a^b \int_a^b |k(t,s)|^2 dt\, ds \right)^2 \left(\int_a^b |x(s) - y(s)|^2 ds \right)^{1/2}$$

$$= |\lambda| \left(\int_a^b \int_a^b |k(t,s)|^2 dt\, ds \right)^{1/2} \rho(x,y) < \alpha \rho(x,y)$$

where $\alpha = |\lambda| \left(\int_a^b \int_a^b |k(t,s)|^2 dt\, ds \right)^{1/2}$ and $\alpha < 1$ if $|\lambda| <$ $\left(\int_a^b \int_a^b |k(t,s)^2|dt\, ds \right)^{-1/2}$. Thus the contraction mapping principle holds, proving the existence and uniqueness of the solution of the given integral equation for values of λ satisfying the above inequality.

(iii) Existence and uniqueness of solution for ordinary differential equations

Definition: Lipschitz condition

Let E be a connected open set in the plane $\mathbb{R}^2$ of the form $E = \,]s_0 - a, s_0 + a[\times]t_0 - b, t_0 + b[$, where $a > 0, b > 0, (s_0, t_0) \in E$. Let f be a real function defined on E. We shall say that f satisfies a Lipschitz condition in t on E, with Lipschitz condition M if for every (s, t_1) and (s, t_0) in E, and $s \in]s_0 - a, s_0 + a[$, we have $|f(s, t_1) - f(s, t_0)| \leq M|t_1 - t_0|$.

Let $(s_0, t_0) \in E$. By a local solution passing through (s_0, t_0) we mean a function φ defined on s_0, $\varphi(s_0) = t_0$, s, $\varphi(s) \in E$ for every $s \in]s_0 - a, s_0 + a[$ and $\varphi'(s) = f(s, \varphi(s))$ for every $s \in]s_0 - a, s_0 + a[$.

1.4.29   Theorem

If f is continuous on the open connected set $E =]s_0 - a, s_0 + a[$ and satisfies a Lipschitz condition in t on E, then for every $(s_0, t_0) \in E$, the differential equation $\frac{dt}{ds} = f(s, t)$ has a unique local solution passing through (s_0, t_0).

Proof: We first show that the function φ defined on the interval $]s_0 - a, s_0 + a[$ such that $\varphi(s_0) = t_0$ and $\varphi'(s) = f(s, \varphi(s))$ for every s in the said interval is of the form $\varphi(s) = t_0 + \int_{s_0}^{s} f(t', \varphi(t'))dt'$. It may be observed from the above form that $\varphi(s_0) = t_0, \varphi(s)$, is differentiable and $\varphi'(s) = f(s, \varphi(s))$.

Let $E_1 \in E =]s_0 - a, s_0 + a[\times]t_0 - a, t_0 + b[$ $a > 0$, $b > 0$ be an open connected set containing (s_0, t_0). Let f be bounded on E_1 and let $|f(s, t)| \leq A$ for all $(s, t) \in E_1$. Let $d > 0$ be such that (a) the rectangle

$R \subseteq E_1$ where $R =]s_0 - d, s_0 + d[\times]t_0 - dA, t_0 + dA[$ and (b) $Md < 1$, where M is a Lipschitz constant for f in E.

Let $J =]s_0 - d, s_0 + d[$. The set B of continuous functions ψ on J such that $\psi(s_0) = t_0$ and $|\psi(s) - t_0| \leq dA$ for every $s \in J$ is a complete metric space under the uniform metric ρ.

Consider the mapping T defined by $(T\psi)(s) = t_0 + \int_{s_0}^{s} f(t, \psi(t))dt$ for $\psi \in B$ and $s \in J$.

Now $(T\psi)(s_0) = t_0$, $T\psi$ is continuous and for every $s \in J$, $|T\psi(s) - t_0| = \left| \int_{s_0}^{s} f(t, \psi(t))dt \right| \leq \int_{s_0}^{s} |f(t, \psi(t))|dt \leq dA$. Hence $T\psi \in B$. Thus T maps B into B.

We now show that T is a contraction. Let $\psi_1, \psi_2 \in B$. Then for every $s \in J$,

$$|T\psi_1(s) - T\psi_2(s)| = \left| \int_{s_0}^{s} (f(t', \psi_1(t'))) - f(t', \psi_2(t'))dt' \right|$$
$$\leq Md \max[|\psi_1(t') - \psi_2(t')| : t' \in J]$$

so that $\rho(T\psi_1, T\psi_2) \leq Md\rho(\psi_1, \psi_2)$. Hence, T is a contraction.

We next show that the local solution can be extended across E_1. Let $J = J_1, d = d_1, s_0 + d = s_1$ and $\varphi(s_1) = t_1$. By theorem 1.4.29 applied to (s_1, t_1) we obtain J_2, d_2 and (s_2, t_2). The solution functions $\varphi = \phi_1$ on J_1 and ϕ_2 on J_2 agree on an interval and so yields a solution on $J_1 \cup J_2$. In this way, we obtain a sequence $\{(s_n, t_n)\}$ with $s_{n+1} > s_n, n = 1, 2, \ldots$. We assume that E_1 is bounded and show that the distance of (s_n, t_n) from the boundary of E_1 converges to zero. If $(s_n, t_n) \in E_1$, we denote by δ_n the distance of (s_n, t_n) from the boundary. We take $d_n = Min \left[\dfrac{\delta_n}{\sqrt{A^2 + 1}}, \dfrac{1}{2M} \right]$, so that $Md_n < 1$ and $d_n \leq \dfrac{\delta_n}{\sqrt{A^2 + 1}}$. Thus $s_{n+1} = s_n + d_n$ and $\varphi(s_{n+1}) = t_{n+1}$.

Hence $(s_{n+1}, t_{n+1}) \in E_1$. Since $d_n > 0$ for all n, $\displaystyle\sum_{n=1}^{\infty} d_n < \infty$. If $\dfrac{\delta_n}{\sqrt{A^2 + 1}}$ is smaller than $\dfrac{1}{2}M$, $d_n = \dfrac{\delta_n}{\sqrt{A^2 + 1}}$. Since $\displaystyle\sum_{n=1}^{\infty} d_n < \infty$, $\displaystyle\sum_{n=1}^{\infty} \delta_n = \sqrt{(A^2 + 1)} \sum_{n=1}^{\infty} d_n < \infty$. On the other hand, if $\dfrac{1}{2}M$ is smaller than $\dfrac{\delta_n}{\sqrt{A^2 + 1}}$, then $d_n = \dfrac{1}{2}M$. Since $\displaystyle\sum_{n=1}^{\infty} d_n < \infty$, $\displaystyle\lim_{n \to \infty} \dfrac{n}{2M} < \infty$. Hence M must be of the order Kn, where K is finite. Now, the Lipschitz constant M cannot be arbitrarily large. Hence $\dfrac{\delta_n}{\sqrt{A^2 + 1}} < \dfrac{1}{2}M$ and $\displaystyle\sum_{n=1}^{\infty} d_n < \infty$.

Therefore $\delta_n \to 0$ as $n \to \infty$. Keeping in mind that D is the union of an increasing sequence of sets, each having the above properties of E_1, we have the following theorem.

1.4.30 Theorem

If f is continuous on an open connected set E and satisfies the Lipschitz condition in t on E, then for every $(s_0, t_0) \in E$ the differential equation $\dfrac{dt}{ds} = f(s, t)$ has a unique solution $t = \varphi(s)$ such that $t_0 = \varphi(s_0)$ and such that the curve given by the solution passes through E from boundary to boundary.

1.4.31 Quasimetric space

If we relax the condition $\rho(x, y) = 0 \Leftrightarrow x = y$, we get what is known as a quasimetric space. Formally, a quasimetric space is a pair of (X, q) where X is a set and q (quasidistance) is a real function defined on $X \times X$ such that for all $x, y, z \in X$ we have $q(x, x) = 0$ and $q(x, z) \leq q(x, y) + q(z, y)$. We next aim to show that the quasidistance is symmetric and non-negative.

If we take $x = y$, then $q(x, z) \leq q(x, x) + q(z, x)$. Since $q(x, x) = 0$, $q(x, z) \leq q(z, x)$. Similarly, we can show $q(z, x) \leq q(x, z)$. This is only possible if $q(x, z) = q(z, x)$, which proves symmetry.

Taking $x = z$ in the inequality, we have $0 \leq 2q(z, y)$ or $q(z, y) \geq 0$, which shows non-negativity.

Combining $q(x, y) \geq q(x, z) - q(y, z)$ and $q(x, y) \geq q(y, z) - q(x, z)$, we can write $|q(x, z) - q(y, z)| \leq q(x, y)$.

1.4.32 Example

Let $\mathbb{R}^2$ be the two-dimensional plane, $x = (\xi_1, \eta_1), y = (\xi_2, \eta_2)$, where $x, y \in \mathbb{R}^2$. The quasidistance between x, y is given by $q(x_1, x_2) = |\xi_1 - \xi_2|$. We will show that $\mathbb{R}^2$ with the above quasidistance between two points is a quasi-metric space.

Firstly, $q(x_1, x_1) = 0$. If $x_3 = (\xi_3, \eta_3)$,

$$q(x_1, x_2) = |\xi_1 - \xi_3| \leq |\xi_1 - \xi_2| + |\xi_2 - \xi_3| = q(x_1, x_2) + q(x_3, x_2)$$

Hence $\mathbb{R}^2$ with the above quasi-distance is a quasimetric space.

Note 1.4.13. $q(x_1, x_2) = 0 \Rightarrow x_1 = x_2$. Thus a quasimetric space is not necessarily a metric space.

Problems

1. Show that theorem 1.4.26 fails to hold if T has only the property $\rho(Tx, Ty) < \rho(x, y)$.

2. If $T : X \to X$ satisfies $\rho(Tx, Ty) < \rho(x, y)$ when $x \neq y$ and T has a fixed point, show that the fixed point is unique; here (X, ρ) is a metric space.

3. Prove that if T is a contraction in a complete metric space and $x \in X$, then

$$T \lim_{n \to \infty} T^n x = \lim_{n \to \infty} T^{n+1} x.$$

4. If T is a contraction, show that $T^n (n \in N)$ is a contraction. If T^n is a contraction for $n > 1$, show that T need not be a contraction.

5. Show that f defined by $f(t, x) = |\sin(x)| + t$ satisfies a Lipschitz condition on the whole tx-plane with respect to its second argument, but that $\dfrac{\partial f}{\partial x}$ does not exist when $x = 0$. What fact does this illustrate?

6. Does f defined by $f(t, x) = |x|^{1/2}$ satisfy a Lipschitz condition?

7. Show that the differential equation $\dfrac{d^2 u}{dx^2} = -f(x)$ where $u \in C^2(0, 1)$, $f(x) \in C(0, 1)$ and $u(0) = u(1) = 0$, is equivalent to the integral equation $u(x) = \displaystyle\int_0^1 G(x, t) f(t) dt$ where $G(x, t)$ is defined as

$$G(x, t) = \begin{cases} x(1 - t) & x \le t \\ t(1 - x) & t \le x \end{cases}.$$

8. For the vector iteration $\begin{pmatrix} x_{n+1} \\ y_{n+1} \end{pmatrix} = \begin{pmatrix} 2x_n \\ \frac{1}{2} x_n \end{pmatrix}$ show that $x = y = 0$ is a fixed point.

9. Let $X = \{x \in \mathbb{R} : x \ge 1\} \subset \mathbb{R}$ and let the mapping $T : X \to X$ be defined by $Tx = x/2 + x^{-1}$. Show that T is a contraction.

10. Let the mapping $T : [a, b] \to [a, b]$ satisfy the condition $|Tx - Ty| \le k|x - y|$, for all $x, y \in [a, b]$. (a) Is T a contraction? (b) If T is continuously differentiable, show that T satisfies a Lipschitz condition. (c) Does the converse of (b) hold?

11. Apply the Banach fixed theorem to prove that the following system of equations has a unique solution:

$$2\xi_1 + \xi_2 + \xi_3 = 4$$

$$\xi_1 + 2\xi_2 + \xi_3 = 4$$

$$\xi_1 + \xi_2 + 2\xi_3 = 4$$

12. Show that $x' = 3x^{2/3}$, $x(0) = 0$ has infinitely many solutions, x, given by $x(t) = 0$ if $t < c$ and $x(t) = (t - c)^3$ if $t \ge c$, where $c > 0$ is any constant. Does $3x^{2/3}$ on the right-hand side satisfy a Lipschitz condition?

13. **Pseudometric:** A finite pseudometric on a set X is a function $\rho : X \times X \to \mathbb{R}$ satisfying for all $x \in X$ conditions (1), (3) and (4) of Section 1.4.1 and 2 (i.e. $\rho(x, x) = 0$, for all $x \in X$).

What is the difference between a metric space and a pseudometric space? Show that $\rho(x,y) = |\xi_i - \eta_i|$ defines a pseudometric on the set of all ordered pairs of real numbers, where $x = (\xi_1, \xi_2)$, $y = (\eta_1, \eta_2)$.

14. Show that the (real or complex) vector space $\mathbb{R}^n$ of vectors $x = \{x_1, \ldots, x_n\}$, $y = \{y_1, \ldots, y_n\}$ becomes a pseudometric space by introducing the distance as a vector: $\rho(x,y) = (p_1|x_1 - y_1|, p_2|x_2 - y_2|, \ldots, p_n|x_n - y_n|)$. The $p_j (j = 1, 2, \ldots, n)$ are fixed positive constants. The order is introduced as follows: $x, y \in \mathbb{R}^n, x \leq y \Leftrightarrow x_i \leq y_i, i = 1, 2, \ldots, n$.

15. Show that the space $\mathbb{R}$ of real or complex valued functions $f(x_1, x_2, \ldots, x_n)$ that are continuous on the closure of B (i.e., $\overline{B}$ of $\mathbb{R}$, is pseudometric when the distance is the function $\rho(f(x), g(x)) = p(x)|f(x) - g(x)|$, and $p(x)$ is a given positive function in $\mathbb{R}$).

1.4.33 *Separable space*

Definition: separable space A space X is said to be separable if it contains a countable everywhere dense set; in other words, if there is in X as sequence $(x_1, x_2, \ldots, x_n)$ such that for each $x \in X$ we find a subsequence $\{x_{n_1}, x_{n_2}, \ldots, x_{n_k}, \ldots\}$ of the above sequence, which converges to x. If X is a metric space, then separability can be defined as follows: There exists a sequence $\{x_1, x_2, \ldots, x_n, \ldots\}$ in X such that we find an element x_{n_0} of it for every $x \in X$ and every $\epsilon > 0$ satisfying $\rho(x, x_{n_0}) < \epsilon$.

The separability of the n-dimensional Euclidean space $\mathbb{R}^n$

The set $\mathbb{R}^{n_0}$, which consists of all points in the space $\mathbb{R}^n$ with rational coordinates, is countable and everywhere dense in $\mathbb{R}^n$.

The separability of the space $C([0, 1])$

In the space $C([0, 1])$, the set C_0, consisting of all polynomials with rational coefficients, is countable. Take any function $x(t) \in C([0, 1])$. By the Weierstrass approximation theorem [Theorem 1.4.34] there is a polynomial $p(t)$ s.t.

$$\max_t |x(t) - p(t)| < \epsilon/2,$$

$\epsilon > 0$ being any preassigned number. On the other hand, there exists another polynomial $p_0(t)$ with rational coefficients, s.t.,

$$\max_t |p(t) - p_0(t)| < \epsilon/2.$$

Hence $\rho(x, p_0) = \max_t |x(t) - p_0(t)| < \epsilon$. Hence $C([0, 1])$ is separable.

1.4.34 *The Weierstrass approximation theorem*

If $[a, b]$ is a closed interval on the real-line, then the polynomials with real coefficients are dense in $C([a, b])$.

In other words, every continuous function on $[a, b]$ is the limit of a uniformly convergent sequence of polynomials.

$$B_n(x) = \sum_{k=0}^{x} \binom{n}{k} x^k (1-x)^{n-k} f\left(\frac{k}{n}\right)$$

are called Bernstein polynomials associated with f. We can prove our theorem by finding a Bernstein polynomial with the required property.

Note 1.4.14. The Weierstrass theorem for $C([0,1])$ says in effect that all real linear combinations of functions $1, x, x^2, \ldots, x^n$ are dense in $[0,1]$.

The Separability of the Space $l_p (1 < p < \infty)$

Let E_0 be the set of all elements x of the form $(r_1, r_2, \ldots, r_n)$ where r_i are rational numbers and n is an arbitrary natural number. E_0 is countable. We would like to show that E_0 is everywhere dense in l_p. Let us take an element $x = \{x_i\} \in l_p$ and let an arbitrary $\epsilon > 0$ be given. We find a natural number n_0 such that for $n > n_0$ $\sum_{k=n+1}^{\infty} |\xi_k|^p < \frac{\epsilon^p}{2}$. Next, take an element $x_0 = (r_1, r_2, \ldots, r_{n}, 0, 0 \ldots)$ such that $\sum_{k=1}^{\infty} |\xi_k - r_k|^p < \frac{\epsilon^p}{2}$. Then,

$$[\rho(x, x)]^p = \sum_{k=1}^{\infty} |\xi_k - r_k|^p + \sum_{k=n+1}^{\infty} |\xi_k|^p < \frac{\epsilon^p}{2} + \frac{\epsilon^p}{2} = \epsilon^p \text{ where } \rho(x, x_0) < \epsilon.$$

The space s is separable.

The space m of bounded numerical sequences is inseparable.

Problems

1. Which of the spaces $\mathbb{R}^n, \mathbb{C}, l_\infty$ are separable?

2. Using the separability property of $C([a, b])$, show that $L_p([a, b]), a < b$ (the space of pth power integrable functions) is separable.

1.5   Topological Spaces

The predominant feature of a metric space is its metric or distance. We have defined open sets and closed sets in a metric space in terms of a metric or distance. We have proved certain results for open sets (see results (i) and (ii) stated after theorem 1.4.11.) in a metric space. The assertions of the above results are taken as axioms in a topological space and are used to define an open set. No metric is used in a topological space. Thus a metric space is a topological space but a topological space is not always a metric space.

Open sets play a crucial role in a topological space.

Many important concepts such as limit points, continuity and compactness (to be discussed in later sections) can be characterised in terms of open sets. It will be shown in this chapter that a continuous mapping sends an open set back into an open set.

We think of deformation as stretching and bending without tearing. This last condition implies that points that are neighbours in one configuration are neighbours in another configuration, a fact that we should recognize as a description of continuity of mapping. The notion of 'stretching and bending' can be mathematically expressed in terms of functions. The notion of 'without tearing' can be expressed in terms of continuity. Let us transform a figure A into a figure A' subject to the following conditions:

(1) To each distinct point p of A corresponds one point p' of A' and vice versa.

(2) If we take any two points p, q of A and move p so that the distance between it and q approaches zero, then the distance between the corresponding points p' and q' of A' will also approach zero, and vice versa. If we take a circle made out of a rubber sheet and deform it subject to the above two conditions, then we get an ellipse, a triangle or a square but not a figure eight, a horseshoe or a single point.

These types of transformations are called topological transformations and are different from the transformations of elementary geometry or of projective geometry. A topological property is therefore a property that remains invariant under such a transformation or in particular deformation. In a more sophisticated fashion one may say that a topological property of a topological space X is a property that is possessed by another topological space Y homeomorphic to X (homeomorphism will be explained later in this chapter). In this section we mention some elementary ideas of a topological space t. The notions of neighbourhood, limiting point and interior of a set amongst others that will be discussed in this section.

1.5.1   Topological space, topology

Let X be a non-empty set. A class $\mathfrak{I}$ of subsets of X is called a *topology* on X if it satisfies the following conditions:

(i) $\phi, X \in \mathfrak{I}$.

(ii) The union of every class of sets in $\mathfrak{I}$ is a set in $\mathfrak{I}$.

(iii) The intersection of finitely many members of $\mathfrak{I}$ is a member of $\mathfrak{I}$.

Accordingly, one defines a *topological space* $(X, \mathfrak{I})$ as a set X and a class $\mathfrak{I}$ of subsets of X such that $\mathfrak{I}$ satisfies the axioms (i) to (iii). The member of $\mathfrak{I}$ are called *open sets*.

1.5.2 Example

Let $X = (\alpha_1, \alpha_2, \alpha_3)$. Consider $\Im_1 = \{\phi, X, \{\alpha_1\}, \{\alpha_1, \alpha_2\}\}$, $\Im_2 = \{\phi, X, \{\alpha_1\}, \{\alpha_2\}\}$, $\Im_3 = \{\phi, x, \{\phi_1\}, \{\alpha_2\}, \{\alpha_3\}, \{\alpha_1, \alpha_2\}, \{\alpha_2, \alpha_3\}\}$ and $\Im_4 = \{\phi, X\}$.

Here, $\Im_1$, $\Im_3$, and $\Im_4$ are topologies, but $\Im_2$ is not a topology due to the fact that $\{\alpha_1\} \cup \{\alpha_2\} = \{\alpha_1, \alpha_2\} \notin \Im_2$.

1.5.3 Definition: indiscrete topology, discrete topology

An indiscrete topology denoted by '$\mathbb{J}$', has only two members, Φ and X. The topological space $(X, \mathbb{J})$ is called an indiscrete topological space. Another trivial topology for a non-empty set X is the discrete topology denoted by '$\mathcal{D}$'. The discrete topology for X consists of all subsets of X. A topological space $(X, \mathcal{D})$ is called a discrete topological space.

1.5.4 Example

Let $X = \mathbb{R}$. Consider the topology 'S' where $\Phi \in S$. If $G \subseteq \mathbb{R}$ and $G \neq \Phi$, then $G \in S$ if for each $p \in G$ there is a set H of the form $\{x \in \mathbb{R} : a \leq x < b\}, a < b$, such that $p \in H$ and $H \subseteq G$. The set $H = \{x \in R : a \leq x < b\}$ is called a right-half open interval. Thus a nonempty set G is S-open if for each $p \in G$, there is a right-half open interval H such that $p \in H \subseteq G$. The topology defined above is called a limit topology.

1.5.5 Definition: usual topology, upper limit topology, lower limit topology

Let $X = \mathbb{R}$ {real}. Let us consider a topology $\Im = \{\Phi, \mathbb{R}, \{]a, b[\}, a < b$ and all unions of open intervals} on $X = \mathbb{R}$. This type of topology is called the usual topology.

Let $X = \mathbb{R}$ be the non-empty set and $\Im = \{\phi, \mathbb{R}, \{]a, b]\}, a < b$ and union of left-open right closed intervals}. This type of topology is called the upper limit topology.

Let $X = \mathbb{R}$ be the non-empty set and $\Im = \{\Phi, X = \mathbb{R}, \{[a, b[\}, a < b$ and union of left-closed right-open intervals}. Then this type of topology is called lower limit topology.

1.5.6 Examples

(i) (Finite Complement Topology)

Let us consider an infinite set X and let $\Im = \{\Phi, X, A \subset X | A^C$ be a finite subset of $X\}$. Then we see that $\Im$ is a topology and we call it a finite complement topology.

(ii) (Countable complement topology)

Let X be a non-enumerable set and $\Im = \{\Phi, X, A \subset X | A^C$ be a countable complement}. Then $\Im$ is a topology and will be known as a countable complement topology.

(iii) In the usual topology in the real line, a single point set is closed.

1.5.7 Definition: T_1-space, closure of a set

A topological space is called a T_1-space if each set consisting of a single point is closed.

1.5.8 Theorem

Let X be a topological space. Then (i) any intersection of closed sets in X is a closed set and (ii) finite union of closed sets in X is closed.

Closure of a set

If A is a subset of a topological space, then the closure of A (denoted by $\overline{A}$) is the intersection of all closed supersets of A. It is easy to see that the closure of A is a closed superset of A that is contained in every closed superset of A and that A is closed $\Leftrightarrow A = \overline{A}$.

A subset of a topological space X is said to *dense* or *everywhere dense* if $\overline{A} = X$.

1.5.9 Definition: neighbourhood

Let $(X, \Im)$ be a topological space and $x \in X$ be an arbitrary point. A subset $N_x \subseteq X$ containing the point $x \in X$ is said to be a neighbourhood of x if $\exists$ a $\Im$-open set G_x such that $x \in G_x \subseteq N_x$.

Clearly, every open set G_x containing x is a neighbourhood of x.

1.5.10 Examples

(i)$]x - \epsilon_1, \ x + \epsilon_2[$ is an open neighbourhood in $(\mathbb{R}, U)$.

(ii)$]x, x + \epsilon]$ is an open neighbourhood in $(\mathbb{R}, U_L)$, where U_L is a topology defined above $\mathbb{R}$.

(iii)$[x - \epsilon, \ x[$ is an open neighbourhood in $(\mathbb{R}, U_\mathbb{R})$.

(iv)Let $\{X, \Im\}$ be a topological space and N_1, N_2 be any two neighbourhoods of the element $x \in X$. Then $N_1 \cup N_2$ is also a neighbourhood of x.

Note 1.5.1. A $\Im$-neighbourhood of a point need not be a $\Im$-open set but a $\Im$-open set is a $\Im$-neighbourhood of each of its points.

1.5.11 Theorem

A necessary and sufficient condition that a set G in a topological space $(X, \Im)$ be open is that G contains a neighbourhood of each $x \in G$.

Problems

1. Show that the indiscrete topology $\mathbb{J}$ satisfies all the conditions of 1.5.1.

2. Show that the discrete topology $\mathcal{D}$ satisfies all the conditions of 1.5.1.

3. If $\mathcal{D}$ represents the discrete topologies for X, show that every subset of X is both open and closed.

4. If $\{I_i, i = 1, 2, \ldots, n\}$ is a finite collection of open intervals such that $\cap\{I_i, i = 1, 2, \ldots, n\} \neq \Phi$, show that $\cap\{I_i, i = 1, 2, \ldots, n\}$ is an open interval.

5. Show that any finite set of real numbers is closed in the usual topology for $\mathbb{R}$.

6. Which of the following subsets of $\mathbb{R}$ are U-neighbourhoods of 2?

    (i) $]1, 3[$ (ii) $[1, 3]$ (iii) $]1, 3]$ (iv) $[1, 3[$ (v) $[2, 3[$.

1.5.12   Bases for a topology

In 1.5.1 the topologies U and S for $\mathbb{R}$ and $\mathcal{T}$ for J were introduced. These neighbourhoods for each point were specified and then a set was declared to be a member of the topology if and only if the set contains a neighbourhood of each of its points. This is an extremely useful way of defining a topology. It should be clear, however, that neighbourhoods must have certain properties. In what follows we present a characterization of the neighbourhoods in a topological space.

1.5.13   Theorem

Let $(X, \mathfrak{S})$ be a topological space, and for each $p \in X$ let u_p be the family of $\mathfrak{S}$-neighbourhoods of p. Then:

(i) If $U \in u_p$ then $p \in U$.

(ii) If $U \in u_p$ and $V \in u_p$, then, by the definition of neighbourhood, there are $\mathfrak{S}$-open sets G_1 and G_2 such that $p \in G_1 \subseteq U$ and $p \in G_2 \subseteq V$. Now, $p \in G_1 \cap G_2$ where $G_1 \cap G_2$ is a $\mathfrak{S}$-open set. Since $p \in G_1 \cap G_2 \subseteq U \cap V$ it follows that $U \cap V$ is a $\mathfrak{S}$-neighbourhoods of p. Hence $U \cap V \in u_p$.

(iii) If $U \in u_p$, the family of $\mathfrak{S}$-neighbourhoods of p, $\exists$ an open set G such that $p \in G \subseteq U$. Therefore, $p \in G \subseteq U \subseteq V$ and V is a $\mathfrak{S}$-neighbourhood of p. Hence $V \in u_p$.

(iv) If $U \in u_p$, then there is a $\mathfrak{S}$-open set V such that $p \in V \subseteq U$. Since V is a $\mathfrak{S}$-open set, V is a neighbourhood of each of its points. Therefore $V \in u_q$ for each $q \in V$.

1.5.14   Theorem

Let X be a non-empty set and for each $p \in X$, let $\mathcal{B}_p$ be a non-empty collection of subsets of X such that

(i) If $B \in \mathcal{B}_p$, then $p \in B$.

(ii) If $B \in \mathcal{B}_p$, and $C \in \mathcal{B}_p$, then $B \cap C \in \mathcal{B}_p$.

If $\mathfrak{S}$ consists of the empty set together with all non-empty subsets G of X having the property that $p \in G$ implies that there is a $B \in \mathcal{B}_p$ such that $B \subseteq G$, then $\mathfrak{S}$ is a topology for X.

1.5.15 Definition: base at a point, base of a topology

Base at a point

Let $(X, \Im)$ be a topological space and for each $x \in X$, let $\mathcal{B}_x$ be a nonempty collection of $\Im$-neighbourhoods of x. We shall say that $\mathcal{B}_x$ is a base for the $\Im$-neighbourhood system of x if for each $\Im$-neighbourhood N_x of x there is a $\mathcal{B} \in \mathcal{B}_x$ such that $x \in \mathcal{B}_x \subseteq N_x$.

If $\mathcal{B}_x$ is a base for the $\Im$-neighbourhood system of x, then the members of $\mathcal{B}_x$ will be called **basic $\Im$-neighbourhoods** of x.

Base for a topology

Let $(X, \Im)$ be a topological space and $\mathcal{B}$ be a non-empty collection of subsets of X such that

(i) $\mathcal{B} \in \Im$

(ii) $\forall \ x \ \in \ X$, $\forall$ neighbourhoods N_x of x, $\exists \ \mathcal{B}_x \ \in \ \mathcal{B}$ such that $x \in \mathcal{B}_x \subseteq N_x$.

Thus $\mathcal{B}$ is called the base of the topology and the sets belonging to $\mathcal{B}$ are called **basic open sets**.

1.5.16 Examples

(i) Consider the usual topology U for the set of real numbers $\mathbb{R}$. The set of all open intervals of lengths $2/n(n = 1, 2, \ldots)$ is a base for $(\mathbb{R}, U)$. The open intervals $\left\{ x : |x - x_0| < \dfrac{1}{n}, (n = 1, 2, \ldots) \right\}$ for $(\mathbb{R}, U)$ form a base at x_0.

(ii) In the case of a point in a metric space, an open ball centered on the point is a neighbourhood of the point, and the class of all such open balls is a base for the point. In the theorem below we can show a characterization of 'openness' of a set in terms of members of the base $\mathcal{B}$ for a topological space $(X, \Im)$.

1.5.17 Theorem

Let $(X, \Im)$ be a topological space and $\mathcal{B}$, a base of the topology, then, a necessary condition that a set $G \subseteq X$ be open is that G can be expressed as union of members of $\mathcal{B}$.

1.5.18 Definition: first countable, second countable

A topological space $(X, \Im)$ that has a countable local base at each $x \in X$ is called **first countable**. A topological space $(X, \Im)$ is said to be **second countable** if $\exists$ a countable base for the topology.

1.5.19 Lindelöf's theorem

Let X be a second countable space. If a non-empty open set G in X is represented as the union of a class $\{G_i\}$ of open sets, then G can be represented as a countable union of G_i's.

1.5.20   Example

$(\mathbb{R}, U)$ is first countable. $(\mathbb{R}, U_L)$ is second countable. It is to be noted that a second countable space is also first countable.

Problems

1. For each $p \in \mathbb{R}$ find a collection $\mathcal{B}_p$ such that $\mathcal{B}_p$ is a base for the $\mathcal{D}$-neighbourhood system of p.

2. Let $X = \{a, b, c\}$ and let $\Im = \{X, \Phi, X, \{a\}, \{b, c\}\}$. Show that $\Im$ is a topology for X.

3. For each $p \in X$ find a collection $\mathcal{B}_p$ of basic $\Im$-neighbourhoods of p.

4. Prove that open rectangles in the Euclidean plane form an open base.

1.5.21   Limit points, closure and interior

We have characterized $\Im$-open sets of a topological space $(X, \Im)$ in terms of $\Im$-neighbourhoods. We now introduce and examine another concept that is conveniently described in terms of neighbourhoods.

Definition: limit point, contact point, isolated point, derived set, closure

Let $(X, \Im)$ be a topological space and let A be a subset of X. The point $x \in X$ is said to be a **limit point** of A if every $\Im$-neighbourhood of x contains at least one point of A other than x. That is, x is a limit point of A if and only if N_x a $\Im$-*neighbourhood of* x satisfies the condition $N_x \cap (A - \{x\}) \neq \Phi$.

If $\forall$ neighbourhoods N_x of x, s.t. $N_x \cap A \neq \Phi$, then x is called a *contact point* of A. $D(A) = \{x : x \text{ is a limit point of } A\}$ is called the *derived* set of A.

$\overline{A \cup D(A)}$ is called the *closure* of A denoted by $\overline{A}$.

Problems

Let $X = \mathbb{R}$ and $A =]0, 1[$. Then find $D(A)$ for the following cases:

(i) $\Im = U$, the usual topology on $\mathbb{R}$

(ii) $\Im = U_L$, the lower limit topology on $\mathbb{R}$

(iii) $\Im = U_{\mathbb{R}}$, the upper limit topology on $\mathbb{R}$.

1.5.22   Theorem

Let $(X, \Im)$ be a topological space. Let A be a subset of X and $D(A)$ the set of all limit points of A. Then $A \cup D(A)$ is $\Im$-closed.

1.5.23   Definition: $\Im$-closure

Let $(X, \Im)$ be the topological space and A be a subset of X. The $\Im$-closure of A denoted by $\overline{A}$ is the smallest $\Im$-closed subset of X that contains A.

1.5.24 Theorem

Let $(X, \Im)$ be a topological space and let A be a subset of X. Then $\overline{A} = A \cup D(A)$ where $D(A)$ is the set of limit points of A. It follows from the previous theorem that $A \cup D(A)$ is a closed set.

1.5.25 Definition: interior, exterior, boundary

Let $(X, \Im)$ be a topological space and let A be a subset of X. A point x is a $\Im$-interior point of A if A is a $\Im$-neighbourhood of x. The $\Im$-**interior** of A denoted by **Int** A is the set of all interior points of A.

$x \in X$ is said to be an **exterior point** of A if x is an interior point of $A^C = X \sim A$.

A point in X is called a **boundary point** of A if each $\Im$-neighbourhood of x contains points both of A and of A^C. The $\Im$-boundary of A is the set of all boundary points of A.

1.5.26 Example

Consider the space $(\mathbb{R}, U)$. The point $\frac{1}{2}$ is an interior point of $[0, 1]$, but neither 0 nor 1 is an interior point of $[0, 1]$. The U-interior of $[0, 1]$ is $]0, 1[$. In the U_L-topology for $\mathbb{R}$, '0' is an interior point of $[0, 1]$ but 1 is not. The U_L-interior of $[0, 1]$ is $[0, 1[$.

1.5.27 Definition: separable space

Let $(X, \Im)$ be a topological space. If $\exists$ a denumerable (enumerable) subset A of $X, A \subseteq X$ such that $\overline{A} = X$, then X is called a **separable space**. Or, in other words, a topological space is said to be **separable** if it contains a denumerable everywhere dense subset.

1.5.28 Example

Let $X = \mathbb{R}$ and A be the set of all intervals. Now let $D(A)$ be the derived set of A. It is clear that $D(A) = \mathbb{R}$. Hence $\overline{A} = A \cup D(A) = A \cup \mathbb{R} = X$. Hence A is everywhere dense in $X = \mathbb{R}$. Similarly, the set $\mathbb{Q}$ of rational points, which is also a subset of $\mathbb{R}$, is everywhere dense in $\mathbb{R}$. Again, since $\mathbb{Q}$ is countable the topological space $\mathbb{R}$ is separable.

1.6 Continuity, Compactness

1.6.1 Definition: '$\epsilon - \delta$' continuity

Let $D \subset \mathbb{R}$ (or $\mathbb{C}$), $f : D \to \mathbb{R}$ (or $\mathbb{C}$) and $a \in D$. The function f is said to be continuous at a if $\lim\limits_{x \to a} f(x) = f(a)$. In other words, given $\epsilon > 0$, $\exists$ a $\delta = \delta(\epsilon)$ such that $|x - a| < \delta \Rightarrow |f(x) - f(a)| < \epsilon$.

Note 1.6.1. f is said to be continuous in D if f is continuous at every $a \in D$.

1.6.2   Definition: continuity in a metric space

Given two metric spaces (X, ρ_X) and (Y, ρ_Y), let $f : D \subset X \to Y$ be a mapping. f is said to be continuous at $a \in D$ if $\forall\ x \in D$, for each $\epsilon > 0$ there exists $\delta > 0$ such that $\rho_X(a, x) < \delta \Rightarrow \rho_Y(f(a), f(x)) < \epsilon$.

1.6.3   Definition: continuity on topological spaces

Let (X, F) and (y, V) be topological spaces and let f be a mapping of X into Y. The mapping f is said to be continuous (or F-V continuous) if $f^{-1}(G)$ is F-open whenever G is V-open. That is, the mapping f is continuous if and only if the inverse image under f of every V-open set is an F-open set.

1.6.4   Theorem (characterization of continuity)

Let (X, ρ_X) and (Y, ρ_y) be metric spaces and let $f : X \to Y$ be a mapping. Then the following statements are equivalent:

(i) f is continuous on X.

(ii) For each $x \in X, f(x_n) \to f(x)$ for every sequence $\{x_n\} \subset X$ with $x_n \to x$.

(iii) $f^{-1}(G)$ is open in X whenever G is open in Y.

(iv) $f^{-1}(F)$ is closed in X whenever F is closed in Y.

(v) $f(\overline{A}) \subset \overline{f(A)}\ \ \forall\ A \subset X$

(vi) $\overline{f^{-1}(B)} \subset f^{-1}(\overline{B}),\ \forall\ B \subset Y$.

1.6.5   Definition: homeomorphism

Let (X, ρ_X) and (Y, ρ_Y) be metric spaces. A mapping $f : X \to Y$ is said to be a *homeomorphism* if

(i) f *is bijective*

(ii) f *is continuous*

(iii) f^{-1} *is continuous*

If a *homeomorphism* from X to Y exists, we say that the spaces X and Y are homeomorphic.

1.6.6   Theorem

Let I_1 and I_2 be any two open intervals. Then (I_1, U_1) and (I_2, U_2) are homeomorphic, and a homeomorphism exists between the spaces (I_1, U_{I_1}) and (I_2, U_{I_2}).

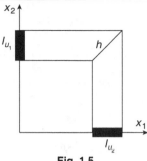

Fig. 1.5

1.6.7 Definition: covering, subcovering, τ-open covering

A collection $\mathcal{C} = \{S_\alpha : \alpha \in \lambda\}$ of subsets of a set X is said to be a covering of X if $\cup\{S_\alpha : \alpha \in \lambda\} = X$. If $\mathcal{C}_1$ is a covering of X, and $\mathcal{C}_2$ is a covering of X such that $\mathcal{C}_2 \subseteq \mathcal{C}_1$, then $\mathcal{C}_2$ is called a subcovering of $\mathcal{C}_1$. Let (X, ζ) be a topological space. A covering $\mathcal{C}$ of X is said to be a ζ-open covering of X if every member of $\mathcal{C}$ is a τ-open set. A covering $\mathcal{C}$ is said to be finite if $\mathcal{C}$ has only a finite number of members.

1.6.8 Definition: a compact topological space

A topological space (X, ζ) is said to be compact if every ζ-open covering of X has a finite subcovering.

Note 1.6.2. The outcome of the Heine-Borel theorem on the real line is taken as a definition of compactness in a topological space. The Heine-Borel theorem reads as follows: If X is a closed and bounded subset of the real line $\mathbb{R}$, then any class of open subsets of $\mathbb{R}$, the union of which contains X, has a finite subclass whose union also contains X.

1.6.9 Theorem

The space $(\mathbb{R}, U)$ is not compact. Therefore, no open interval is compact w.r.t. the U-topology.

1.6.10 Definition: finite intersection property

Let (X, ζ) be a topological space and $\{F_\lambda | \lambda \in \Lambda\}$ be a class of subsets such that the intersection of finite number of elements of $\{F_\lambda | \lambda \in \Lambda\}$ is non-void, i.e., $\bigcap\limits_{k=1}^{n} F_{\lambda_k} \neq \Phi$ irrespective of whatever manner any finite number of λ_ks $\{\lambda_k | k = 1, 2, \ldots, n\}$ is chosen from Λ. Then, $F = \{F_\lambda | \lambda \in \Lambda\}$ is said to have the Finite Intersection Property (FIP).

1.6.11 Theorem

The topological space (X, ζ) is compact if every class of closed subsets $\{F_\lambda | \lambda \in \Lambda\}$ possessing the finite intersection property has non-void *intersection. In other words, if all F_λ's are closed and $\bigcap\limits_{k=1}^{n} F_{\lambda_k} \neq \Phi$ for a finite subcollection $\{\lambda_1, \lambda_2, \ldots, \lambda_n\}$, then $\bigcap\limits_{\lambda \in \Lambda} F_\lambda \neq \Phi$. The converse result is also true, i.e., if every class of closed subsets of (X, ζ) having the FIP has non-void intersection, then (X, ζ) is compact.*

1.6.12   Theorem

    A continuous image of a compact space is compact.

1.6.13   Definition: compactness in metric spaces

    A metric space being a topological space under the metric topology, the definition of compactness as given in definition 1.6.5 is valid in metric spaces. However, the concept of compactness in a metric space can also be introduced in terms of sequences and can be related to completeness.

1.6.14   Definition: a compact metric space

    A metric space (X, ρ) is said to be compact if every infinite subset of X has at least one limit point.

Remark 1.6.1. A set $K \subseteq X$ is then compact if the space (K, ρ) is compact. K is compact if and only if every sequence with values in K has a subsequence which converges to a point in K.

1.6.15   Definition: relatively compact

    If X is a metric space and K is a subset of X such that its closure $\overline{K}$ is compact, then K is said to be relatively compact.

Lemma 1.6.1. A compact subset of a metric space is closed and bounded. The converse of the lemma is in general false.

1.6.16   Example

    Consider e_n in l_2 where $e_1 = (1, 0, 0, \ldots, 0)$, $e_2 = (0, 1, 0, \ldots, 0)$, $e_3 = (0, 0, 1, \ldots 0)$. This sequence is bounded, since $\rho(\theta, e_n) = 1$ where ρ stands for the metric in l_2. Its terms constitute a point set that is closed because it has no limit point. Hence the sequence is not compact.

Lemma 1.6.2. Every compact metric space is complete.

1.6.17   Definition: sequentially compact, totally bounded

    (i) A metric space X is said to be sequentially compact if every sequence in X has a convergent subsequence.

    (ii) A metric space X is said to be totally bounded if for every $\epsilon > 0$, X contains a finite net called an ϵ-net such that the finite set of open balls of radius $\epsilon > 0$ and centers in the $\epsilon - net$ covers X.

Lemma 1.6.3. X is totally bounded if and only if every sequence in X has a Cauchy sequence.

1.6.18 Theorem

In a metric space (X, ρ), the following statements are equivalent:

X is compact.

X is sequentially compact.

X is complete and totally bounded.

1.6.19 Corollary

For $1 \leq p < \infty$ and $x, y \in \mathbb{R}^n(\mathbb{C}^n)$, consider $\rho_p(x, y) =$

$$\left(\sum_{j=1}^{n} |x_j - y_j|^p \right)^{\frac{1}{p}}.$$

(a) (Heine-Borel) A subset of $\mathbb{R}^n(\mathbb{C}^n)$ is **compact** if and only if it is closed and bounded.

(b) (Bolzano-Weierstrass) Every bounded sequence in $\mathbb{R}^n(\mathbb{C}^n)$ has a convergent subsequence.

Proof: Since a bounded subset of $\mathbb{R}^n(\mathbb{C}^n)$ is totally bounded, part (a) follows from theorem 1.6.18. Since the closure of a bounded set of $\mathbb{R}^n(\mathbb{C}^n)$ is complete and totally bounded, part (b) follows from theorem 1.6.18.

1.6.20 Theorem

In a metric space (X, ρ) the following statements are true: X is totally bounded $\Rightarrow X$ is separable.

X is compact $\Rightarrow X$ is separable.

X is separable $\Rightarrow X$ is second countable.

1.6.21 Theorem

If f is a real-valued continuous function defined on a metric space (X, ρ), then for any compact set $A \subseteq X$ the values $Sup[f(x), x \in A]$, $Inf[f(x), x \in A]$ are finite and are attained by f at some points of A.

Remark 1.6.2. If a continuous function $f(x)$ is defined on some set M that is not compact, then $\underset{x \in M}{\text{Sup }} f(x)$ and $\underset{x \in M}{\text{Inf }} f(x)$ need not be attained. For example, consider the set of all functions $x(t)$ such that $x(0) = 0$, $x(1) = 1$ and $|x(t)| \leq 1$. The continuous functional $f(x) = \int_0^1 x^2(t)dt$, though continuous on M, does not attain the g.l.b. on M. For if $x(t) = t^n$, $f(x) = \frac{1}{2^{2n+1}} \rightarrow 0$, as $n \rightarrow \infty$. Hence $Inf[f(x)] = 0$. But the form of $f(x)$

indicates that $f(x) > 0$ for every $x = x(t)$ continuous curve that joins the points $(0,0)$ and $(1,1)$. The fallacy is that the set of curves considered is not compact even if the set is closed and bounded in C $([0,1])$.

1.6.22   Definition: uniformly bounded, equicontinuous

Let (X, ρ) be a complete metric space. The space of continuous real-valued functions on X with the metric, $\rho(x, y) = \max[|f(x) - g(x)| x \in X]$ is a complete metric space, which we denote by $C([X])$. A collection $\mathcal{F}$ of functions on a set X is said to be uniformly bounded if there is an $M > 0$ such that $|f(x)| \leq M \ \forall \ x \in X$ and all $f \in \mathcal{F}$. For subsets of $\mathcal{C}([X])$, uniform boundedness agrees with boundedness in a metric space, i.e., a set $\mathcal{F}$ is uniformly bounded if and only if it is contained in a ball.

A collection $\mathcal{F}$ of functions defined on a metric space X is called equicontinuous if for each $\epsilon > 0$, there is a $\delta > 0$ such that $\rho(x, x') < \delta \Rightarrow |f(x) - f(x')| < \epsilon$ for all $x, x' \in X$ and for $f \in \mathcal{F}$. It may be noted that the functions belonging to the equicontinuous collection are uniformly continuous.

1.6.23   Theorem (Arzela-Ascoli)

If (X, ρ) is a compact metric space, a subset $K \subseteq C(X)$ is relatively compact if and only if it is uniformly bounded and equicontinuous.

1.6.24   Theorem

If f is continuous on an open set D, then for every $(x_0, y_0) \in D$ the differential equation $\dfrac{dy}{dx} = f(x, y)$ has a local solution passing through (x_0, y_0).

Problems

1. Show that if (X, ζ) is a topological space such that X has only a finite number of points, then (X, ζ) is compact.

2. Which of the following subspaces of (R, U) are compact: (i) J, (ii) $[0, 1]$, (iii) $[0, 1] \cup [2, 3]$, (iv) the set of all rational numbers or (v) $[2, 3[$.

3. Show that a continuous real or complex function defined on a compact space is bounded. More generally, show that a continuous real or complex function mapping compact space into any metric space is bounded.

4. Show that if D is an open connected set and a differential equation $\dfrac{dy}{dx} = f(x, y)$ is such that its solutions form a simple covering of D, then f is the limit of a sequence of continuous functions.

5. Prove the Heine-Borel theorem: A subspace (Y, U_Y) of (R, U) is compact if and only if Y is bounded and closed.

6. Show that the subset I_2 of points $\{x_n\}$ such that $|x_n| \leq \dfrac{1}{n}, n = 1, 2, \ldots$ is compact.

7. Show that the unit ball in $C([0,1])$ of points $[x : \max |x(t)| \leq 1, t \in [0,1]]$ is not compact.

8. Show that X is compact if and only if for any collection $\mathcal{F}$ of closed sets with the property that $\bigcap_{i=1}^{n} F_i \neq \Phi$ for any finite collection in $\mathcal{F}$, it follows that $\cap\{F_\lambda, F_\lambda \in \mathcal{F}\} \neq \Phi$.

CHAPTER 2

NORMED LINEAR SPACES

If a linear space is simultaneously a metric space, it is called a metric linear space. The normed linear spaces form an important class of metric linear spaces. *Furthermore, in each space there is defined a notion of the distance from an arbitrary element to the null element or origin, that is, the notion of the size of an arbitrary element.* This gives rise to the concept of the norm of an element x or $||x||$ and finally to that of the normed linear space.

2.1 Definitions and Elementary Properties

2.1.1 Definition

Let E be a linear space over $\mathbb{R}$ (or $\mathbb{C}$).

To every element x of the linear space E, let there be assigned a unique real number called *the norm* of this element and denoted by $||x||$, satisfying the following properties (*axioms of a normed linear space*):

(a) $||x|| > 0$ and $||x|| = 0 \Leftrightarrow x = 0$;

(b) $||x + y|| \leq ||x|| + ||y||$ (*triangle inequality*), $\forall\, y \in X$

(c) $||\lambda x|| = |\lambda|\, ||x||$ (*homogeneity of the norm*), $\forall\, \lambda \in \mathbb{R}$ (or $\mathbb{C}$)

The normed linear space E is also written as $(E, ||\cdot||)$.

58

Remark 2.1:

1. Properties (b) and (c) $\Rightarrow$ property (a).

$$0 = ||x - x|| \leq ||x|| + || -x|| = 2||x||$$

which yields $||x|| \geq 0$.

2. If we regard x in a normed linear space E, as a vector, its length is $||x||$ and the length $||x-y||$ of the vector $x-y$ is the distance between the end points of the vectors x and y. Thus in view of (a) and (b) and the definition 2.1.1, we say all vectors in a normed linear space E have positive lengths except the zero vector. The property (c) states that the length of one side of a triangle can never exceed the sum of the lengths of the other two sides (Fig. 2.1(a)).

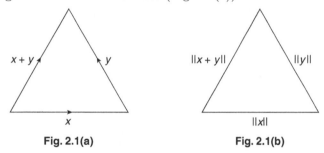

<table>
<tr><td>**Fig. 2.1(a)**</td><td>**Fig. 2.1(b)**</td></tr>
</table>

3. In a normed linear space a metric (*distance*) can be introduced by $\rho(x,y) = ||x - y||$. It is clear that this distance satisfies all the metric axioms.

2.1.2 *Examples*

1. The <u>n-dimensional Euclidean space $\mathbb{R}^n$</u> and   <u>the unitary space $\mathbb{C}^n$ are normed linear spaces.</u>

   Define the sum and product of the elements by a scalar as in Sec. 1.4.2, Ex. 2. The norm of $x \in \mathbb{R}^n$ is defined by

$$||x|| = \left(\sum_{i=1}^{n} |\xi_i|^2 \right)^{1/2}.$$

   This norm satisfies all the axioms of 2.1.1. Hence $\mathbb{R}^n$ and $\mathbb{C}^n$ are normed linear spaces.

2. **Space $C([0,1])$.**
   We define the addition of functions and multiplication by a scalar in the usual way.

   We set $||x|| = \max_{t \in [0,1]} |x(t)|$. It is clear that the axioms of a normed linear space are satisfied.

3. **Space l_p:** We define the addition of elements and multiplication of elements by a scalar as indicated earlier.

We set for $x = \{\xi_i\} \in l_p$, $||x||$ as $||x|| = \left(\sum_{i=1}^{p} |\xi_i|^p \right)^{1/p}$.

The axioms (a) and (c) of 2.1.1 are satisfied.

If $y = \{\eta_i\}$, it can be shown by making an appeal to Minkowskii's inequality ([1.4.5])

$$\left(\sum_{i=1}^{p} |\xi_i + \eta_i|^p \right)^{1/p} \le \left(\sum_{i=1}^{p} |\xi_i|^p \right)^{1/p} + \left(\sum_{i=1}^{p} |\eta_i|^p \right)^{1/p} < \infty$$

where $x(t)$ and $y(t)$ are p-th power summable.

Thus l_p is a normed linear space.

4. **Space $L_p([0, 1])$, $p > 1$**

We set $||x|| = \left(\int_0^1 |x(t)|^p dt \right)^{1/p}$ for $x(t) \in L_p([0, 1])$.

If $y(t) \in L_p([0, 1])$ then

$$\left(\int_0^1 |y(t)|^p \right)^{1/p} < \infty$$

where $x(t)$ and $y(t)$ are Riemann integrable. Then by *Minkowski's inequality* for integrals, we have the inequality (c) of (2.1.1).

5. **Space l_∞:** Let $x = \{\xi_i\}$, such that $|\xi_i| \le C_i$ where C_i for each i is a real number. Setting $||x|| = \sup_i |\xi_i|$, we see that all the norm axioms are fulfilled. Hence l_∞ is a normed linear space.

6. **Space $C^k([0, 1])$:** Consider a space of functions $x(t)$ defined and continuous on $[0, 1]$ and having their continuous derivatives upto the order k, i.e., $x(t) \in C^k[0, 1]$. The norm on this function space is defined by

$$||x|| = \max \left(\max_t |x(t)|, \max_t |x^1(t)|, \ldots, \max_t |x^k(t)| \right).$$

Then $||x + y|| = \max(\max_t |x(t)| + y(t)|, \max_t |x^1(t) + y^1(t)|,$

$$\ldots, \max_t |x^k(t) + y^k(t)|)$$

It is clear that all the axioms of 2.1.1 are fulfilled. Therefore $C^k([0, 1])$ is a normed linear space.

2.1.3 Induced metric, convergence in norm, Banach space

Induced metric: In a normed linear space a metric (distance) can be introduced by, $\rho(x, y) = ||x - y||$. It may be seen that the metric defined above satisfies all the axioms of a metric space. Thus the normed linear space $(E, || \cdot ||)$ can be treated as a metric space if we define the metric in the above manner.

Convergence in norm: After introducing the metric, we define the convergence of a sequence of elements $\{x_n\}$ to x namely, $x = \lim x_n$ or $x_n \to x$ if $||x_n - x|| \to 0$ as $n \to \infty$. Such a convergence in a normed linear space is called *convergence in norm*.

Banach Space: A real (complex) *Banach space* is a real (complex) normed linear space that is complete in the sense of convergence in norm.

2.1.4 Examples

1. The n-dimensional Euclidean space $\mathbb{R}^n$ is a Banach Space Defining the sum of elements and the product of elements by a scalar (real or complex) in the usual manner, the norm is defined as

$$||x|| = \left(\sum_{i=1}^{n} \xi_i^2 \right)^{1/2}$$

where $x = \{\xi_i\}$. By example 1, 2.1.2 we see that $\mathbb{R}^n$ is a normed linear space. If $x_m = \{\xi_i^{(m)}\}$ and if $\xi_i^{(m)} \to \xi_i$ as $m \to \infty$, $\forall\, i$, then,

$$\left(\sum_{i=1}^{n} |\xi_i^{(m)} - \xi_i|^2 \right)^{1/2} \to 0 \text{ as } m \to \infty$$

or in other words $||x_m - x|| \to 0$ as $m \to \infty$ where $x = \{\xi_i\}$. Since $x \in \mathbb{R}^n$ it follows that $\mathbb{R}^n$ is a complete normed linear space or $\mathbb{R}^n$ is a Banach space.

2. l_p is a Banach space($1 \leq p < \infty$)

Defining the sum of the elements and the product of elements by a scalar as in 1.3 and taking the norm as

$$||x|| = \left(\sum_{i=1}^{p} |\xi_i|^p \right)^{1/p}$$

where $x = \{\xi_i\}$, we proceed as follows.

Let $x_n = \{\xi_i^{(n)}\} \in l_p$ and $\xi_i^{(n)} \to \xi_i$ as $n \to \infty$. Then $||x_n - x|| \to 0$ as $n \to \infty$ where $x = \{\xi_i\}$. Now by Minkowski's inequality [1.4.5] we have

$$\left(\sum_{i=1}^{\infty} |\xi_i|^p\right)^{1/p} \leq \left(\sum_{i=1}^{\infty} |\xi_i^n|^p\right)^{1/p} + \left(\sum_{i=1}^{\infty} |\xi_i^{(x)} - \xi_i|^p\right)^{1/p}$$

or, $||x|| \leq ||x_n|| + ||x - x_n|| < \infty$ for n sufficiently large.

Hence $x \in l_p$, $1 \leq p < \infty$ and the space is a Banach space.

3. $C([0, 1])$ is a Banach space

$C([0, 1])$ is called the function space. Consider the linear space of all scalar valued (real or complex) continuous functions defined in $C([0, 1])$.

Let $\{x_n(t)\} \subset C([0, 1])$ and $\{x_n(t)\}$ be a Cauchy sequence in $C([0, 1])$. $C([0, 1])$ being a normed linear space [see 2.1],

$$||x_m - x_n|| = \max_t |x_m(t) - x_n(t)| < \epsilon \; \forall \; m, n \in N. \tag{2.1}$$

Therefore, for any fixed $t = t_0 \in [0, 1]$, we get

$$|x_m(t_0) - x_n(t_0)| < \epsilon \; \forall \; m, n \in N.$$

This shows that $\{x_m(t_0)\}$ is a Cauchy sequence in $\mathbb{R}(\mathbb{C})$. But $\mathbb{R}(\mathbb{C})$ being complete, the sequence converges to a limit in $\mathbb{R}(\mathbb{C})$. In this way, we can assign to each $t \in [0, 1]$ a unique $x(t) \in \mathbb{R}(\mathbb{C})$. This defines a (pointwise) function x on $[0, 1]$. Now, we show that $x \in C([0, 1])$ and $x_m \to x$.

Letting $n \to \infty$, we have from (2.1)

$$|x_m(t) - x(t)| < \epsilon \; \forall \; m \geq N \text{ and } t \in [0, 1] \tag{2.2}$$

This shows that the sequence $\{x_m\}$ of continuous functions converges uniformly to the function x on $[0, 1]$, and hence the limit function x is a continuous function on $[0, 1]$. As such, $x \in C([0, 1])$. Also from (2.2), we have

$$\max_t |x_m(t) - x(t)| < \epsilon, \; \forall \; m \geq N$$
$$\Rightarrow \quad ||x_m - x|| \leq \epsilon \; \forall \; m \geq N$$
$$\Rightarrow \quad x_m \to x \in C([0, 1]).$$

Hence $C[0, 1]$ is a Banach space.

4. l_∞ is a Banach space

Let $\{x_m\}$ be a Cauchy sequence in l_∞ and let $x_m = \{\xi_i^m\} \in l_\infty$. Then

$$\sup_i |\xi_i^{(m)} - \xi_i^{(n)}| \leq \epsilon, \ \forall \ m, n \geq N.$$

This gives $|\xi_i^m - \xi_i^n| < \epsilon \ \forall \ m, n \geq N \ (i = 1, 2, \ldots)$.

This shows that for each i, $\{\xi_i^m\}$ is a Cauchy sequence in $\mathbb{R}(\mathbb{C})$. Since $\mathbb{R}(\mathbb{C})$ is complete, $\{\xi_i^m\}$ converges in $\mathbb{R}(\mathbb{C})$. Let $\xi_i^m \to \xi_i$ as $m \to \infty$, and let $x = (\xi_1, \xi_2, \ldots, \xi_n, \ldots, \ldots)$.

Let $m \to \infty$, then

$$|\xi_i^{(m)} - \xi_i| < \epsilon \ \forall \ m \geq N \ (i = 1, 2, \ldots) \tag{2.3}$$

Since $x_m \in l_\infty$, there is a real number M_m such that $|\xi_i^m| \leq M_m, \ \forall \ i$.

Therefore, $|\xi_i| \leq |\xi_i^m| + |\xi_i - \xi_i^m| \leq M_m + \epsilon, \ \forall \ m \geq N, \ i = 1, 2, \ldots$

Since the RHS is true for each i and is independent for each i, it follows that $\{\xi_i\}$ is a bounded sequence of numbers and thus $x \in l_\infty$. Furthermore, it follows from (2.3),

$$||x_m - x||_\infty < \epsilon \ \forall \ m > N.$$

Hence $x_m \to x \in l_\infty$ and l_∞ is a *Banach Space*.

5. Incomplete normed linear space

Let X be a set of continuous functions defined on a closed interval $[a, b]$. For $x \in X$ let us take $||x||$ as

$$||x|| = \int_a^b |x(t)| dt \tag{2.4}$$

The metric induced by (2.4) for $x, y \in X$ is given by

$$\rho(x, y) = \int_a^b |x(t) - y(t)| dt \tag{2.5}$$

In note 1.4.10, we have shown that the metric space (X, ρ) is not complete, i.e., given a Cauchy sequence $\{x_m\}$ in (X, ρ), x_m does not converge to a point in X. Hence the normed linear space $(X, || \cdot ||)$ is not complete.

6.   An incomplete normed linear space and its completion $L_2([a, b])$

The linear space of all continuous real-valued functions on [a, b] forms a normed linear space X with norm defined by

$$||x|| = \left(\int_a^b (x(t))^2 dt \right)^{1/2} \tag{2.6}$$

Let, us consider $\{x_m\}$ as follows:

$$x_m(t) = \begin{cases} 0 & \text{if } t \in \left[0, \frac{1}{2}\right] \\ m\left(t - \frac{1}{2}\right) & \text{if } t \in \left[\frac{1}{2}, a_m\right] \\ 1 & \text{if } t \in [a_m, 1] \end{cases}$$

where $a_m = \dfrac{1}{2} + \dfrac{1}{m}$.

Let us take $a = 0$ and $b = 1$ and $n > m$.

Hence, $||x_n - x_m||^2 = \displaystyle\int_0^1 [x_n(t) - x_m(t)]^2 dt$

$$= \int_{\frac{1}{2}}^{\frac{1}{2}+\frac{1}{n}} |x_m(t) - x_n(t)|^2 dt + \int_{\frac{1}{2}+\frac{1}{n}}^{\frac{1}{2}+\frac{1}{m}} |x_n(t) - x_m(t)| dt$$

$$= \triangle ABC = \left(\frac{1}{m} - \frac{1}{n}\right) < \frac{1}{m} \text{ [see figure 2.1(c)]}.$$

The Cauchy sequence does not converge to a point in $(X, || \cdot ||)$. For every $x \in X$,

$$||x_n - x|| = \int_0^1 |x_n(t) - x(t)|^2 dt$$

$$= \int_0^{1/2} |x(t)|^2 dt + \int_{1/2}^{a_m} |x_n(t) - x(t)|^2 dt + \int_{a_m}^1 |1 - x(t)|^2 dt.$$

Since the integrands are non-negative, $x_n \to x$ in the space $(X, || \cdot ||)$ implies that $x(t) = 0$ *if* $t \in [0, \frac{1}{2}[$, $x(t) = 1$ if $t \in]\frac{1}{2}, 1]$. Since it is impossible for a continuous function to have this property $\{x_n\}$ does not have a limit in X.

The space X can be completed by Theorem 1.4.5. The completion is denoted by $L_2([0, 1])$. This is a Banach space. In fact the norm on X and the operations of a Linear $\mathbb{R}$ space can be extended to the completion of X. This process can be seen in Theorem 2.1.10 in the next section. In general for any $p \geq 1$, the Banach space $L_p[a, b]$ is the completion of the normed linear spaces which consists of all continuous real-valued functions on $[0, 1]$ and the norm defined by

$$||x||_p = \left(\int_0^1 |x(t)|^p dt \right)^{1/p}.$$

With the help of Lebesgue integrals the space $L_p([0, 1])$ can also be obtained in a direct way by the use of Lebesgue integral and Lebesgue measurable functions x on $[0, 1]$ such that the Lebesgue integral of $|x|^p$ on $[0, 1]$ exists and is finite. The elements of $L_p([0, 1])$ are equivalent classes of those functions, where x is equivalent to y if the Lebesgue integral of $|x-y|^p$ over $[0, 1]$ is zero. We discuss these (Lebesgue measures) in Chapter Ten. Until then the development will take place without the use of measure theory.

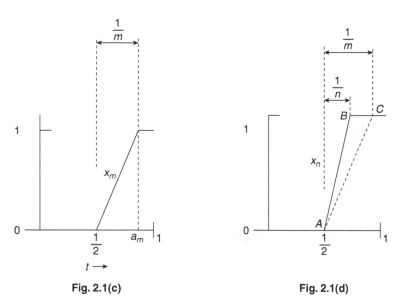

Fig. 2.1(c)   Fig. 2.1(d)

7. Space s

Every normed linear space can be reduced to a metric space. However, every metric cannot always be recovered from a norm. Consider the space with the metric ρ defined by

$$\rho(x, y) = \sum_{i=1}^{\infty} \frac{1}{2^i} \frac{|\xi_i - \eta_i|}{1 + |\xi_i - \eta_i|},$$

where $x = \{\xi_i\}$ and $y = \{\eta_i\}$ belong to s. This metric cannot be obtained from a norm. This is because this metric does not have the two properties that a metric derived from a norm possesses. The following lemma identifies the existence of these two properties.

2.1.5 Lemma (translation invariance)

A metric ρ induced by a norm on a normed linear space E satisfies:

(a) $\rho(x + a, y + a) = \rho(x, y)$ (b) $\rho(\alpha x, \alpha y) = |\alpha|\rho(x, y)$ for all $x, y, a \in E$ and every scalar α.

Proof: We have

$$\rho(x + a, y + a) = ||x + a - (y + a)|| = ||x - y|| = \rho(x, y)$$
$$\rho(\alpha x, \alpha y) = ||\alpha x - \alpha y|| = |\alpha| \, ||x - y|| = |\alpha|\rho(x, y).$$

Problems

1. Show that the norm $||x||$ of x is the distance from x to 0.

2. Verify that the usual length of a vector in the plane or in three dimensional space has properties (a), (b) and (c) of a norm.

3. Show that for any element x of a normed linear space $||x|| \geq 0$ follows from axioms (b) and (c) of a normed linear space.

4. Given $x = \{\xi_i\} \subset \mathbb{R}^n$, show that $||x|| = \left(\sum_{i=1}^{n} |\xi_i|^2 \right)^{1/2}$ defines a norm on $\mathbb{R}^n$.

5. Let E be the linear space of all ordered triplets $x = \{\xi_1, \xi_2, \xi_3\}, y = \{\eta_1, \eta_2, \eta_3\}$ of real numbers. Show that the norms on E are defined by,
$$||x||_1 = |\xi_1| + |\xi_2| + |\xi_3|, \quad ||x||_2 = \{\xi_1^2 + \xi_2^2 + \xi_3^2\}^{1/2},$$
$$||x||_\infty = \max\{|\xi_1|, |\xi_2|, |\xi_3|\}$$

6. Show that the norm is continuous on the metric space associated with a normed linear space.

7. In case $0 < p < 1$, show with the help of an example that $||\cdot||_p$ does not define a norm on $l_p^{(n)}$ unless $n = 1$ where $||x||_p^{(n)} = \left(\sum_{i=1}^{n} |\xi_i|^p \right)^{\frac{1}{p}}$, $x = \{\xi_i\}$.

8. Show that each of the following defines a norm on $\mathbb{R}^2$.

(i) $||x||_1 = \dfrac{|x_1|}{a} + \dfrac{|x_2|}{b}$, (ii) $||x||_2 = \left(\dfrac{x_1^2}{a^2} + \dfrac{x_2^2}{b^2}\right)$,

(iii) $||x||_\infty = \max\left\{\dfrac{|x_1|}{a} + \dfrac{|x_2|}{b}\right\}$

where a and b are two fixed positive real numbers and $x = (x_1, x_2) \in \mathbb{R}^2$. Draw a closed unit sphere $(||x|| = 1)$ corresponding to each of these norms.

9. Let $|| \cdot ||$ be a norm on a linear space E. If $x + y \in E$ and $||x + y|| = ||x|| + ||y||$, then show that $||sx + ty|| = s||x|| + t||y||$, for all $s \geq 0$, $t \geq 0$.

10. Show that a non-empty subset A of $\mathbb{R}^n$ is bounded $\Leftrightarrow$ there exists a real number K such that for each $x = (x_1, x_2, \ldots, x_n)$ in A we have $|x_i| \leq K$ for each subscript i.

11. Show that the real linear space $C([-1, 1])$ equipped with the norm given by

$$||x||_1 = \int_{-1}^{1} |x(t)|dt,$$

where the integral is taken in the sense of Riemann, is an incomplete normed linear space [Note: $||x||_1$ is precisely the area of the region enclosed within the integral $t = -1$ and $t = 1$].

12. Let E be a linear space and ρ the metric on E such that

$$\rho(x, y) = \rho(x - y, 0)$$

and $\rho(\alpha x, 0) = |\alpha|\rho(x, 0) \ \forall \ x, y \in E$ and $\alpha \in \mathbb{R}(\mathbb{C})$

Define $||x|| = \rho(x, 0)$, $x \in E$. Prove that $|| \cdot ||$ is a norm on E and that ρ is the metric induced by the norm $|| \cdot ||$ on E.

13. Let E be a linear space of all real valued functions defined on $[0, 1]$ possessing continuous first-order derivatives. Show that $||f|| = |f(0)| + ||f'||_\infty$ is a norm on E that is equivalent to the norm $||f||_\infty + ||f'||_\infty$.

2.1.6 Lemma

In a normed linear space E,

$$|\,||x|| - ||y||\,| \leq ||x - y||, \quad x, y \in E.$$

Proof: $||x|| = ||(x - y) + y|| \leq ||x - y|| + ||y||$.

Hence $||x|| - ||y|| \leq ||x - y||$.

Interchanging x with y,

$$||y|| - ||x|| \le ||y - x||.$$

Hence $\quad | \, ||x|| - ||y|| \, | \le ||x - y||.$

2.1.7   Lemma

The $|| \cdot ||$ is a continuous mapping of E into $\mathbb{R}$ where E is a Banach space.

Let $\{x_n\} \subset E$ and $x_n \to x$ as $n \to \infty$, it then follows from lemma 2.1.4 that

$$| \, ||x_n|| - ||x|| \, | \le ||x_n - x|| \to 0 \text{ as } n \to \infty.$$

Hence the result follows.

2.1.8   Corollary

Let E be a complete normed linear space over $\mathbb{R}(\mathbb{C})$. If $\{x_n\}, \{y_n\} \subset E$, $\alpha_n \in \mathbb{R}(\mathbb{C})$ and $x_n \to x$, $y_n \to y$ respectively as $n \to \infty$ and $\alpha_n \in \mathbb{R}(\mathbb{C})$ then (i) $x_n + y_n \to x + y$ (ii) $\alpha_n x_n \to x$ as $x \to \alpha$.

Proof: Now, $||(x_n + y_n) - (x + y)|| \le ||x_n - x|| + ||y_n - y|| \to 0$ as $n \to \infty$

Hence $x_n + y_n \to x + y$ as $n \to \infty$.

$$||\alpha_n x_n - \alpha x|| \le ||\alpha_n(x_n - x) + (\alpha_n - \alpha)x|| \le |\alpha| \, ||x_n - x|| + |\alpha_n - \alpha| \, ||x|| \to 0$$

because $\{\alpha_n\}$ being a convergent sequence is bounded and $||x||$ is finite.

2.1.9   Summable sequence

Definition: A sequence $\{x_n\}$ in a normed linear space E is said to be _summable_ to the limit sum s if the sequence $\{s_m\}$ of the partial sums of the series $\displaystyle\sum_{n=1}^{n} x_n$ converges to s in E, i.e.,

$$||s_m - s|| \to 0 \text{ as } m \to \infty \text{ or } \left\| \sum_{n=1}^{m} x_n - s \right\| \to 0 \text{ as } m \to \infty.$$

In this case we write $s = \displaystyle\sum_{n=1}^{\infty} x_n$. $\{x_n\}$ is said to be absolutely _summable_ if $\displaystyle\sum_{n=1}^{\infty} ||x_n|| < \infty.$

It is known that for a sequence of real (complex) numbers *absolute summability* implies *summability*. But this is not true in general for sequences in normed linear spaces. But in a Banach space every absolutely summable sequence in E implies that the sequence is summable. The converse is also true. This may be regarded as a characteristic of a Banach space.

2.1.10 Theorem

A normed linear space E is a Banach space if and only if every absolutely summable sequence in E is summable in E.

Proof: Assume that E is a Banach space and that $\{x_n\}$ is an absolutely summable sequence in E. Then,

$$\sum_{n=1}^{\infty} ||x_n|| = M < \infty,$$

i.e., for each $\epsilon > 0 \; \exists \; a \; K$ such that $\sum_{n=1}^{K} ||x_n|| < \epsilon$, i.e.,

$$||s_n - s_m|| = \left\| \sum_{k=m+1}^{n} x_k \right\| \le \sum_{k=m+1}^{n} ||x_k||$$

$$\le \sum_{n=K}^{\infty} ||x_k||, \; \le \epsilon, \; n, m > K.$$

In the above, $s_n = \sum_{k=1}^{n} ||x_k||$.

Thus s_n is a Cauchy sequence in E and must converge to some element s in E, since E is complete. Hence $\{x_n\}$ is summable in E.

Conversely, let us suppose that each absolutely summable sequence in E is summable in E. We need to show that E is a Banach space. Let $\{x_n\}$ be a Cauchy sequence in E. Then for each k, $\exists$ an integer n_k such that

$$||x_n - x_m|| < \frac{1}{2^k} \; \forall \; n, m \ge n_k.$$

We may choose n_k such that $n_{k+1} > n_k$. Then $\{x_{n_k}\}$ is a subsequence of $\{x_n\}$.

Let us set $y_0 = x_{n_1}$, $y_1 = x_{n_2} - x_{n_1}, \ldots, y_k = x_{n_{k+1}} - x_{n_k}, \ldots$. Then

(a) $\displaystyle\sum_{n=0}^{k} y_n = x_{n_{k+1}}$, (b) $||y_k|| < \dfrac{1}{2^k}$, $k \geq 1$.

Thus $\displaystyle\sum_{k=0}^{\infty} ||y_k|| \leq ||y_0|| + \sum_{k=1}^{\infty} \dfrac{1}{2^k} = ||y_0|| + 1 < \infty.$

Thus the sequence $\{y_k\}$ is absolutely summable and hence summable to some element x in E. Therefore by (a) $x_{n_k} \to x$ as $k \to \infty$. Thus the Cauchy sequence $\{x_n\}$ in E has a convergent subsequence $\{x_{n_k}\}$ converging to x. Now, if a subsequence in a Cauchy sequence converges to a limit then the whole sequence converges to that limit. Thus, the space is complete and is therefore a Banach space.

2.1.11   Ball, sphere, convex set, segment of a straight line

Since normed linear spaces can be treated as metric spaces, all concepts introduced in metric spaces (e.g., balls, spheres, bounded set, separability, compactness, linear dependence of elements, linear subspace, etc.) have similar meanings in normed linear spaces. Therefore, theorems proved in metric spaces using such concepts can have parallels in normed linear spaces.

Definition: ball, sphere

Let $(E, || \cdot ||)$ be a normed linear space.

(i) The set $\{x \in E : ||x - x_0|| < r\}$, denoted by $B(x_0, r)$, is called the *open ball* with centre x_0 and radius r.

(ii) The set $\{x \in E : ||x - x_0|| \leq r\}$, denoted by $\overline{B}(x_0, r)$, is called a *closed ball* with centre x_0 and radius r.

(iii) The set $\{x \in E : ||x - x_0|| = r\}$, denoted by $S(x_0, r)$, is called a *sphere* with centre x_0 and radius r.

Note 2.1.1.

1. An open ball is an open set.

2. A closed ball is a closed set.

3. Given $r > 0$

$$B(0,r) = \{x \in E : ||x|| < r\} = \left\{x \in E : \left|\left|\frac{x}{r}\right|\right| < 1\right\}$$

$$= \{ry \in E : ||y|| < 1\} \quad \text{where } y = \frac{x}{r}$$

$$= rB(0,1).$$

Therefore, in a normed linear space, without any loss of generality, we can consider $B(0,1)$, the ball centred at zero with a radius of 1. The ball $B(0,1)$ is called *the unit open ball in E.*

Definition: convex set, segment of a straight line

A set of elements of a linear space E having the form $y = tx$, $x \in E$, $x \neq 0$, $-\infty < t < \infty$ is called a *real line* defined by the given element x and a set of elements of the form

$$y = x_1 + (1 - \lambda)x_2, \quad x_1, x_2 \in X, \ 0 \leq \lambda \leq 1$$

is called a segment joining the points x_1 and x_2. A set X in E is called a *convex set* if

$$x_1, x_2 \in X \Rightarrow \lambda x_1 + (1 - \lambda)x_2, \quad x_1, x_2 \in E, \ 0 \leq \lambda \leq 1.$$

2.1.12 Lemma. In a normed linear space an open (closed) ball is a convex set

Let $x_1, x_2 \in B(x_0, r)$, i.e., $||x_1 - x_0|| < r$, $||x_2 - x_0|| < r$.

Let us select any element of the form,

$$y = \lambda x_1 + (1 - \lambda)x_2, \ 0 < \lambda < 1$$

Then $||y - x_0|| = ||\lambda x_1 + (1 - \lambda)x_2 - x_0||$

$$\leq ||\lambda(x_1 - x_0) + (1 - \lambda)(x_2 - x_0)||$$

$$\leq \lambda ||x_1 - x_0|| + (1 - \lambda)||x_2 - x_0||$$

$$< r.$$

Thus, $y \in B(x_0, r)$.

Note 2.1.2

(i) For any point $x \neq \theta$, a ball of radius $r > ||x||$ with its centre in the origin, contains the point x.

(ii) Any ball of radius $r' < ||x||$ with centre in the origin does not contain this point.

In order to have geometrical interpretations of different abstract spaces $l_p^{(n)}$, the n-dimensional pth summable spaces, we draw the shapes of unit balls for different values of p.

Examples: unit closed balls in $\mathbb{R}^2$ with different norms: Given $x = (x_1, x_2)$

(i) $||x||_{1/2} = (|x_1|^{1/2} + |x_2|^{1/2})^{1/2}$

(ii) $||x||_1 = (|x_1| + |x_2|)$

(iii) $||x_2||_2 = (|x_1|^2 + |x_2|^2)^{1/2}$

(iv) $||x||_4 = (|x_1|^4 + |x_2|^4)^{1/4}$

(v) $||x||_\infty = \max(|x_1|, |x_2|)$

Problems

1. Show that for the norms in examples (ii), (iii), (iv) and (v) the unit spheres reflect what is shown in figure below:

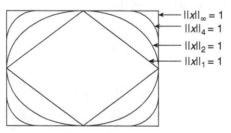

$$\begin{aligned} ||x||_\infty &= 1 \\ ||x||_4 &= 1 \\ ||x||_2 &= 1 \\ ||x||_1 &= 1 \end{aligned}$$

Fig. 2.2

2. Show that the closed unit ball is a convex set.

3. Show that $\varphi(x) = (\sqrt{|\xi_1|} + \sqrt{|\xi_2|})^{1/2}$ does not define a norm on the linear space of all ordered pairs $x = \{\xi_1, \xi_2\}$ of real numbers. Sketch the curve $\varphi(x) = 1$ and compare it with the following figure 2.3.

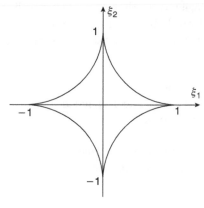

Fig. 2.3

4. Let ρ be the metric induced by a norm on a linear space $E \neq \Phi$. If ρ_1 is defined by

$$\rho_1(x, y) = \begin{cases} 0 & x = y \\ 1 + \rho(x, y) & x \neq y \end{cases}$$

then prove that ρ_1 can not be obtained from a norm on E.

5. Let E be a normed linear space. Let X be a convex subset of E. Show that (i) the interior X^0 and (ii) the closure $\overline{X}$ of X are convex sets. Show also that if $\overline{X} \neq \Phi$, then $\overline{X} = \overline{X}^0$.

2.2 Subspace, Closed Subspace

Since the normed linear space E is a special case of linear space, all the concepts introduced in a linear space (e.g., linear dependence and independence of elements, linear subspace, decomposition of E into direct sums, etc.) have a relevance for E.

Definition: subspace
A set X of a normed linear space E is called a *subspace* if

(i) X is a linear space with respect to vector addition and scalar multiplication as defined in E (1.3.2).

(ii) X is equipped with the norm $||\cdot||_X$ induced by the norm $||\cdot||$ on E i.e., $||x||_X = ||x||$, $\forall\, x \in X$.

We may write this subspace $(X, ||\cdot||_X)$ simply as X.

Note 2.2.1. It is easy to see that X is a normed linear space. Furthermore the metric defined on X by the norm coincides with the restriction to X of the metric defined on E by its norm. Therefore X is a subspace of the metric space E.

Definition: closed subspace
A subspace X of a normed linear space E is called a *closed subspace* of E if X is a closed metric space.

Definition: subspace of a Banach space
Given E a Banach space, a subspace X of E is said to be a *subspace* of Banach space E.

Examples

1. The space c of convergent numerical sequences in $\mathbb{R}$ (or $\mathbb{C}$) is a closed subspace of l_∞, the space of bounded numerical sequences in $\mathbb{R}$ ($\mathbb{C}$).

2. c_0, the space of all sequences converging to 0, is a closed subspace of c.

3. The space $\mathbf{P}[0,1]$ is a subspace of $C[0,1]$, but it is not closed.

   $\mathbf{P}[0,1]$ is spanned by the elements $x_0 = 1$, $x_1 = t, \ldots, x_n = t^n, \ldots$

   Then $\mathbf{P}[0,1]$ is a set of all polynomials, but $\overline{\mathbf{P}[0,1]} \neq C[0,1]$.

2.2.1   Theorem (subspace of a Banach space)

A subspace X of a Banach space E is complete if and only if the set X is closed in E.

Proof: Let X be complete. Let it contain a limit point x of X. Then, every open ball $B(x, 1/n)$ contains points of X (other than x). The open ball $B(x, 1/n)$ where n is a positive integer, contains a point x_n of X, other than x. Thus $\{x_n\}$ is a sequence in X such that

$$||x_n - x|| < \frac{1}{n}, \ \forall \ n,$$

$\Rightarrow \ \lim_{n \to \infty} x_n = x$ in X

$\Rightarrow \ \{x_n\}$ is a Cauchy sequence in E and therefore in X.

However, X being complete, it follows that $x \in X$. This proves that X is closed.

On the other hand, let X be closed, in which case it contains all of its limiting points. Hence every Cauchy sequence will converge to some point in X. Otherwise the subspace X will not be closed.

Examples:

4. Consider the space Φ of sequences

$$x = (\xi_1, \xi_2, \ldots, \xi_n, 0, \ldots) \text{ in } \mathbb{R}(\mathbb{C}),$$

where $\xi_n \neq 0$ for only finite values of n. Clearly, $\hat{\Phi} \subset c_0 \subset l_\infty$ and $\hat{\Phi} \neq c_0$. It may be noted that c_0 is the closure of $\hat{\Phi}$ in $(l_\infty, \|\cdot\|_\infty)$. Thus $\hat{\Phi}$ is not closed in l_∞ and hence $\hat{\Phi}$ is an *incomplete normed linear space* equipped with the norm induced by the norm $\|\cdot\|_\infty$ on l_∞.

5. For every real number $p \geq 1$, we have,

$$\hat{\Phi} \subset l_p \subset c_0$$

It may be noted that c_0 is the closure of l_p in c_0 and $l_p \neq c_0$. Thus, l_p is not closed in c_0 and hence l_p is an incomplete normed linear space when induced by the $\|\cdot\|$ in c_0.

Problems

1. Show that the closure $\overline{X}$ of a subspace X of a normed linear space E is again a subspace of E.

2. If $n \geq m \geq 0$, prove that $\mathbb{C}^n(\{a, b\}) \subset \mathbb{C}^m([a, b])$ and that the space $\mathbb{C}^n([a, b])$ with the norm induced by the norm on $\mathbb{C}^m([a, b])$ is not closed.

3. Prove that the intersection of an arbitrary collection of non-empty closed subspaces of the normed linear space. E is a closed subspace of E.

4. Show that $c \subset l_\infty$ is a vector subspace of l_∞ and so is c_0.

5. Let X be a subspace of a normed linear space E. Then show that X is **nowhere dense** in E (i.e., the interior of the closure of X is empty) if and only if X is nowhere dense in E.

6. Show that c is a nowhere dense subspace of m.

2.3　Finite　Dimensional　Normed　Linear Spaces and Subspaces

Although infinite dimensional normed linear spaces are more general than finite dimensional normed linear spaces, finite dimensional normed linear spaces are more useful. This is because in application areas we consider the finite dimensional spaces as subspaces of infinite dimensional spaces. Quite a number of interesting results can be derived in the case of finite dimensional spaces.

2.3.1　Theorem

All finite-dimensional normed linear spaces of a given dimension n are isomorphic to the n-dimensional Euclidean space E_n, and are consequently, isomorphic to each other.

Proof: Let E be an n-dimensional normed linear space and let $e_1, e_2, \ldots, e_n$ be the basis of the space. Then any element $x \in E$ can be uniquely expressed in the form $x = \xi_1 e_1 + \xi_2 e_2 + \cdots \xi_n e_n$. Corresponding to $x \in E$, let us consider the element $\tilde{x} = \{\xi_1, \xi_2, \ldots, \xi_n\}$ in the n-dimensional Euclidean space. The correspondence established in this manner between x and $\tilde{x}$ is one-to-one. Moreover, let $y \in E$ be of the form

$$y = \eta_1 e_1 + \eta_2 e_2 + \cdots + \eta_n e_n.$$

Then $y \in E$ is in one-to-one correspondence with $\bar{y} \in E_n$ where $\bar{y} = \{\eta_1, \eta_2, \ldots, \eta_n\} \in E_n$. It is apparent that

$$x \leftrightarrow \tilde{x}, \text{ and } y \leftrightarrow \tilde{y} \text{ implies } x + y \leftrightarrow +\tilde{x} + \tilde{y}$$

and　$\lambda x \leftrightarrow \lambda \tilde{x}, \ \lambda \in \mathbb{R}(\mathbb{C})$

To prove that E and E_n are isomorphic, we go on to show that the linear mapping from E onto E_n is mutually continuous.

For any $x \in E$, we have

$$\|x\| = \left\| \sum_{i=1}^{n} \xi_i e_i \right\| \leq \sum_{i=-1}^{n} |\xi_i| \, \|e_i\|$$

$$= \left(\sum_{i=1}^{n} \|e_i\|^2 \right)^{1/2} \left(\sum_{i=1}^{n} |\xi_i|^2 \right)^{1/2}$$

$$= \beta \|\tilde{x}\| \text{ where } \beta = \left(\sum_{i=1}^{n} |e_i|^2 \right)^{1/2}$$

In particular, for all $x, y \in E$,

$$||x - y||_E \leq \beta ||\tilde{x} - \tilde{y}||_{E_n} \tag{2.7}$$

Next we establish a reverse inequality. We note that the unit sphere $S(0,1) = \{||\tilde{x} - 0|| = 1\}$ in E_n is compact. We next prove that the function

$$f(\tilde{x}) = f(\xi_1, \ldots, \xi_n) = ||x|| = ||\xi_1 e_1 + \xi_2 e_2 + \cdots + \xi_n e_n||$$

defined on $S(0,1)$ is continuous. Now, a continuous function defined on a compact set attains its extremum.

Since all the ξ_i's cannot vanish simultaneously on S and since $e_1, e_2, \ldots, e_n$ are linearly independent, $f(\xi_1, \xi_2, \ldots, \xi_n) > 0$.

Now, $|f(\xi_1, \xi_2, \ldots, \xi_n) - f(\eta_1, \eta_2, \ldots, \eta_n)| = |\,||x|| - ||y||\,| \leq ||x - y||_E \leq \beta ||\tilde{x} - \tilde{y}||_{E_n}$.

The above shows that f is a continuous function.

Now, since the unit ball $S(0,1)$ in E_n is compact and the function $f(\xi_1, \xi_2, \ldots, \xi_n)$ defined on it is continuous, it follows that $f(\xi_1, \ldots, \xi_n)$ has a minimum on S. Hence,

$$f(\xi_1, \xi_2, \ldots, \xi_n) > r \text{ where } r > 0,$$

or, $f(\tilde{x}) = ||x|| \geq \gamma$.

Hence for any $\tilde{x} \in E_n$,

$$f(\tilde{x}) = ||x|| = ||\tilde{x}|| \left\| \sum_{i=1}^{n} \frac{\xi_i e_i}{\sqrt{\sum_{i=1}^{n} \xi_i^2}} \right\| \geq \gamma ||\tilde{x}||$$

or in other words

$$||x - y|| \geq \gamma ||\tilde{x} - \tilde{y}||. \tag{2.8}$$

From (2.7) and (2.8) it follows that the mapping of E onto E_n is one-to-one.

The mapping from E onto E_n is one-to-one and onto. Both the mapping and its inverse are continuous. Thus, the mapping is a homeomorphism. The homeomorphism between E and E_n implies that in a finite dimensional

Banach space the convergence in norm reduces to a coordinatewise convergence, and such a space is always complete.

The following lemma is useful in deriving various results. Very roughly speaking, it states that with regard to linear independence of vectors, we cannot find a linear combination that involves large number of scalars but represents a small vector.

2.3.2   Lemma (linear combination)

Let $\{e_1, e_2, \ldots, e_n\}$ be a linearly independent set of vectors in a normed linear space E (of any finite dimension). In this case there is a number $c > 0$ such that for every choice of scalars $\alpha_1, \alpha_2, \ldots, \alpha_n$ we have

$$||\alpha_1 e_1 + \alpha_2 e_2 + \cdots + \alpha_n e_n|| \geq c(|\alpha_1| + |\alpha_2| + \cdots + |\alpha_n|), \ c > 0 \quad (2.9)$$

Proof: Let $S = |\alpha_1| + |\alpha_2| + \cdots + |\alpha_n|$. If $S = 0$, all α_j are zero so that the above inequality holds for all c. *Let $S > 0$. Writing $\beta_i = \alpha_i/S$ (2.9) is equivalent to the following inequality,*

$$||\beta_1 e_1 + \beta_2 e_2 + \cdots + \beta_n e_n|| \geq c \quad\quad (2.10)$$

Note that $\displaystyle\sum_{i=1}^{n} |\beta_i| = 1$.

Hence it suffices to prove the existence of a $c > 0$ such that (2.10) holds for every n-tuples of scalars $\beta_1, \beta_2, \ldots, \beta_n$ with $\displaystyle\sum_{i=1}^{n} |\beta_i| = 1$.

Suppose that this is false. Then there exists a sequence $\{y_m\}$ of vector

$$y_m = \beta_1^{(m)} e_1 + \cdots + \beta_n^{(m)} e_n, \ \left(\sum_{i=1}^{n} |\beta_i^{(m)}| = 1\right)$$

such that $||y_m|| \to 0$, as $m \to \infty$.

Since $\left(\displaystyle\sum_{i=1}^{n} |\beta_i^{(m)}| = 1\right)$, we have $|\beta_i^{(m)}| \leq 1$. Hence for each fixed i the sequence

$$(\beta_i^{(m)}) = (\beta_i^{(1)}, \beta_i^{(2)}, \ldots)$$

is bounded. Consequently, by the Bolzano-Weierstrass theorem, $\{\beta_i^{(m)}\}$ has a convergent subsequence. Let β_i denote the limit of the subsequence and let $\{y_{1,m}\}$ denote the corresponding subsequence of $\{y_m\}$. By the same argument, $\{y_{1,m}\}$ has a subsequence $\{y_{2,m}\}$ for which the corresponding subsequence of scalars $\beta_2^{(m)}$ converges. Let β_2 denote the limit. Continuing in this way after n steps, we obtain a subsequence, $\{y_{n,m}\} = \{y_{n,1}, y_{n,2}, \ldots\}$ of $\{y_m\}$, the terms are of the form,

$$y_{n,m} = \sum_{i=1}^{n} \gamma_i^{(m)} e_i,$$

$\sum_{i=1}^{n} |\gamma_i^{(m)}| = 1$ with scalars $\gamma_i^{(m)} \to \beta_i$ as $m \to \infty$

$$y_{n,m} \to y = \sum_{i=1}^{n} \beta_i e_i$$

where $\sum_i |\beta_i| = 1$, so that not all β_i can be zero.

Since $\{e_1, e_2, \ldots, e_n\}$ is a linearly independent set, we thus have $y \neq 0$. On the other hand, $y_{n,m} \to y$ implies $||y_{n,m}|| \to ||y||$, by the continuity of the norm. Since $||y_m|| \to 0$ by assumption and $\{y_{n,m}\}$ is a subsequence of $\{y_m\}$, we must have $||y_{n,m}|| \to 0$. Hence $||y|| = 0$, so that $y = 0$ by (b) of 2.1.1. This contradicts the fact that $y \neq 0$ and the lemma is proved.

Using the above lemma we prove the following theorem.

2.3.3   Theorem (completeness)

Every finite dimensional *subspace* of a normed linear space E is complete. In particular, every finite dimensional normed linear space is complete.

Let us consider an arbitrary Cauchy sequence $\{y_m\}$ in X, a *subspace* of E and let the dimension of X be n. Let $\{e_1, e_2, \ldots, e_n\}$ be a basis for X. Then each y_m can be written in the form,

$$y_m = \alpha_1^{(m)} e_1 + \alpha_2^{(m)} e_2 + \cdots + \alpha_n^{(m)} e_n$$

Since $\{y_m\}$ is a Cauchy sequence, for every $\epsilon > 0$, there is an N such that $||y_m - y_p|| < \epsilon$ when $m, p > N$. From this and the lemma 2.3.2, *we have for some $c > 0$*

$$\epsilon > ||y_m - y_p|| = \left\| \sum_{i=1}^{n} (\alpha_i^{(m)} - \alpha_i^p) e_i \right\| \geq c \sum_{i=1}^{n} |\alpha_i^{(m)} - \alpha_i^p| \text{ for all } m, p \geq N$$

On division of both sides by c, we get

$$|\alpha_i^m - \alpha_i^p| \leq \sum_{i=1}^{n} |\alpha_i^m - \alpha_i^p| < \frac{\epsilon}{c} \text{ for all } m, p > N.$$

This shows that $\{\alpha_i^{(m)}\}$ is a Cauchy sequence in $\mathbb{R}(\mathbb{C})$ for $i = 1, 2, \ldots, n$. Hence the sequence converges. Let α_i denote the limit. Using these n limits $\alpha_1, \alpha_2, \ldots, \alpha_n$, let us construct y as

$$y = \alpha_1 e_1 + \alpha_2 e_2 + \cdots + \alpha_n e_n$$

Here $y \in X$ and

$$||y_m - y|| = \left\| \sum_{i=1}^{n} (\alpha_i^{(m)} - \alpha_i) e_i \right\| \leq \sum_{i=1}^{n} |\alpha_i^{(m)} - \alpha_i| \, ||e_i||.$$

Since $\alpha_i^{(m)} \to \alpha_i$ as $m \to \infty$ for each i, $y_m \to y$ as $m \to \infty$. This shows that $\{y_m\}$ is convergent in X. Since $\{y_m\}$ is an arbitrary Cauchy sequence in X it follows that X is complete.

2.3.4   Theorem (closedness)

Every finite dimensional subspace X of a normed linear space E is closed in E. If the subspace X of E is closed, then it is closed in E and the theorem is true. By theorem 2.3.3, X is complete. X being a complete normed linear space, it follows from theorem 2.2.1 that X is closed.

2.3.5   Equivalent norms

A norm on a linear space E induces a topology, called norm topology on E.

Definition 1: Two norms on a normed linear space are said to be equivalent if they induce the same norm topology or if any open set in one norm is also an open set in the other norm. Alternatively, we can express the concept in the following form:

Definition 2: Two norms $||\cdot||$ and $||\cdot||'$ on the same linear space E are said to be *equivalent norms* on E if the identity mapping $I_E : (E, ||\cdot||) \to (E, ||\cdot||')$ is a *topological homoeomorphism* of $(E, ||\cdot||)$ onto $(E, ||\cdot||')$.

Theorem: Two norms $||\cdot||$ and $||\cdot||'$ on the same normed linear space

E are equivalent if and only if $\exists$ positive constants α_1 and α_2 such that $\alpha_1||x|| \leq ||x||' \leq \alpha_2||x|| \ \forall \ x \in E$.

Proof: In view of definition 2 above, we know that $|| \cdot ||$ and $|| \cdot ||'$ are equivalent norms on E

$\Leftrightarrow$ the identity mapping I_E is a topological isomorphism of $(E, || \cdot ||)$ onto $(E, || \cdot ||')$.

$\Leftrightarrow \exists$ constants $\alpha_1 > 0$ and $\alpha_2 > 0$ such that $\alpha_1||x|| \leq ||I_E x||' \leq \alpha_2||x||, \ \forall \ x \in E,$

$\Leftrightarrow \alpha_1||x|| < ||x||' < a_2||x||, \ \forall \ x \in E.$

This completes the proof.

Note 2.3.1. The relation 'norm equivalence' is an equivalence relation among the norms on E. The special feature of a finite dimensional normed linear space is that all norms on the space are equivalent, or in other words, all norms on E lead to the same topology for E.

2.3.6   Theorem (equivalent norms)

On a finite dimensional normed linear space any norm $|| \cdot ||$ is equivalent to any norm $|| \cdot ||'$.

Proof: Let E be a n-dimensional normed linear space and let $\{e_1, e_2, \ldots, e_n\}$ be any basis in E. Then for every $x \in E$ we can find some scalars $\alpha_1, \alpha_2, \ldots, \alpha_n$, *not all zeros* such that

$$x = \alpha_1 e_1 + \alpha_2 e_2 + \cdots + \alpha_n e_n. \tag{2.11}$$

Then by lemma 2.3.2, we can find a constant $c > 0$ such that

$$||x|| = \left\| \sum_{i=1}^{n} \alpha_i e_i \right\| \geq c \left(\sum_{i=1}^{n} |\alpha_i| \right) \tag{2.12}$$

On the other hand,

$$||x||' = \left\| \sum_{i=1}^{n} \alpha_i e_i \right\|$$

$$\leq \sum_{i=1}^{n} |\alpha_i| \, ||e_i||' \tag{2.13}$$

$$\leq k_1 \sum_{i=1}^{n} |\alpha_i|$$

where $k_1 = \max \|e_i\|'$.

Hence $\|x\|' < \alpha_2 \|x\|$ (2.14)

where $\alpha_2 = \dfrac{k_1}{c}$

Interchanging $\|x\|$ and $\|x\|'$ we obtain as in the above

$$\|x\|' \geq \alpha_1 \|x\|$$

where $\alpha_1 = \left(\dfrac{k_2}{c}\right)^{-1}$, $k_2 = \max \|e_i\|$.

Problems

1. Let E_1 be a closed subspace and E_2 be a finite dimensional subspace of a normed linear space E. Then show that $E_1 + E_2$ is closed in E.

2. Show that equivalent norms on a vector space E induces the same topology on E.

3. If $\| \cdot \|$ and $\| \cdot \|'$ are equivalent norms on a normed linear space E, show that the Cauchy-sequences in $(E, \| \cdot \|)$ and $(E, \| \cdot \|')$ are the same.

4. Show that a finite dimensional normed linear space is separable. (A normed linear space is said to be separable if it is separable as a metric space.)

5. Show that a subset $L = \{u_1, u_2, \ldots, u_n\}$ of a normed linear space E is *linearly independent* if and only if for every $x \in$ span L, there exists a unique $(\alpha_1, \alpha_2, \ldots \alpha_n) \in \mathbb{R}^n(\mathbb{C}^n)$ such that $x = \alpha_1 u_1 + \alpha_2 u_2 + \alpha_3 u_3 + \cdots + \alpha_n u_n$.

6. Show that a Banach space is finite dimensional if and only if every subspace is closed.

7. Let $1 \leq p \leq \infty$. Prove that *a unit ball* in l_p is convex, closed and bounded but not compact.

8. Let $\mathbf{A}$ be the vector space generated by the functions $1, \sin x, \sin^2 x, \sin^3 x, \ldots$ defined on $[0,1]$. That is, $f \in \mathbf{A}$ if and only if there is a nonnegative integer k and real numbers $\alpha_1, \alpha_2, \alpha_3, \ldots, \alpha_n$ (all depending on f such that $f(x) = \displaystyle\sum_{n=0}^{\infty} \alpha_n \sin^n x$ for each $x \in [0,1]$).

   Show that $\mathbf{A}$ is an algebra and $\mathbf{A}$ is dense in $C([0,1])$ with respect to the uniform metric (A vector space $\mathbf{A}$ of real-valued functions is called an **algebra** of functions whenever the product of any two functions in $\mathbf{A}$ is in $\mathbf{A}$ (Aliprantis and Burkinshaw [1])).

9. Let E_1 be a compact subset and E_2 be a closed subset of a normed linear space such that $E_1 \cap E_2 = \Phi$. Then show that $(E_1 + B(0, r)) \cap E_2 = \Phi$ for some $r > 0$.

10. Let $1 \leq p \leq \infty$. Show that the closed unit ball in l_p is convex, closed and bounded, but not compact.

11. Let E denote the linear space of all polynomials in one variable with coefficients in $\mathbb{R}(\mathbb{C})$. For $p \in E$ with $p(t) = \alpha_0 + \alpha_1 t + \alpha_2 t^2 + \cdots + \alpha_n t^n$, let

$$||p|| = \sup\{|p(t)| : 0 \leq t \leq 1\},$$
$$||p||_1 = |\alpha_0| + |\alpha_1| + \cdots + |\alpha_n|,$$
$$||p||_\infty = \max\{|\alpha_0|, |\alpha_1|, \ldots, |\alpha_n|\}.$$

Then show that $||\ \ ||$, $||p||_1$, $||p||_\infty$ are norms on E, $||p|| \leq ||p||_1$ and $||p||_\infty \leq ||p||_1$ for all $p \in E$.

12. Show that equivalent norms on a linear space E induce the same topology for E.

13. If two norms $||\cdot||$ and $||\cdot||_0$ on a linear space are equivalent, show that (i) $||x_n - x|| \to 0$ implies (ii) $||x_n - x||_0 \to 0$ (and vice versa).

2.3.7   Riesz's lemma

Let Y and Z be subspaces of a normed linear space X and suppose that Y is closed and is a proper subset of Z. Then for every real number θ in the interval $(0, 1)$ there is a $z \in Z$ such that $||z|| = 1$, $||z - y|| \geq \theta$ for all $y \in Y$.

Proof: *Take any $v \in Z - Y$ and denote its distance from Y by d (fig. 2.4),*

$$d = \inf_{y \in Y} ||v - y||.$$

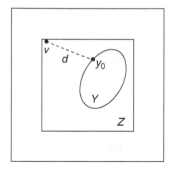

Fig. 2.4

Clearly, $d > 0$ since Y is closed. We now take any $\theta \in (0, 1)$. By the definition of an infinum there is a $y_0 \in Y$ such that

$$d \leq ||v - y_0|| \leq \frac{d}{\theta} \qquad (2.15)$$

(note that $\frac{d}{\theta} > d$ since $0 < \theta < 1$).

Let, $z = c(v - y_0)$ where $c = \dfrac{1}{||v - y_0||}$.

Then $||z|| = 1$ and we shall show that $||z - y|| \geq \theta$ for every $y \in Y$.

Now, $||z - y|| = ||c(v - y_0) - y|| = c||(v - y_0) - c^{-1}y|| = c||v - y_1||$ where $y_1 = y_0 + c^{-1}y$.

The form of y_1 shows that $y_1 \in Y$. Hence $||v - y_1|| \geq d$, by the definition of d. Writing c out and using (2.15), we have

$$||z - y|| = c||v - y_1|| \geq cd = \frac{d}{||v - y_0||} \geq \frac{d}{\frac{d}{\theta}} = \theta.$$

Since $y \in Y$ was arbitrary, this completes the proof.

2.3.8   Lemma

Let E_x and E_y be Banach spaces, A be compact and $R(A)$ be closed in E_y, then the range of A is finite dimensional.

Proof: Let, if possible, $\{z_1, z_2, \ldots\}$ be an infinite linearly independent subset of $R(A)$ and let $Z_n = \mathrm{span}\{z_1, z_2, \ldots\}$, $n = 1, 2, \ldots$.

Z_n is finite dimensional and is therefore a closed subspace of Z_{n+1}. Also, $Z_n \neq Z_{n+1}$, since $\{z_1, z_2, \ldots, z_{n+1}\}$ is linearly independent. By the Riesz lemma (2.3.7), there is a $\overline{y}_n \in Z_{n+1}$, such that

$$||\overline{y}_n|| = 1 \quad \text{and} \quad \mathrm{dist}(\overline{y}_n, Z_n) \geq \frac{1}{2}.$$

Now, $\{\overline{y}_n\}$ is a sequence in $\{\overline{y} \in R(A) : ||\overline{y}|| \leq 1\}$ having no convergent subsequence. This is because $||\overline{y}_n - \overline{y}_m|| \geq \frac{1}{2}$ for all $m \neq n$. Hence the set $\{y \in R(A) : ||y|| \leq \delta\}$ cannot be compact. Hence $R(A)$ is finite dimensional.

Problems

1. Prove that if E is a finite dimensional normed linear space and X is a proper subspace, there exists a point on the unit ball of E at a distance from X.

2. Let E be a normed linear space. Show that the Riesz lemma with $\theta = 1$ holds if and only if for every closed proper subspace X of E, there is a $x \in E$ and $y_0 \in X$ such that $||x - y_0|| = \text{dist}\,(x, X) > 0$.

3. Let $E - \{x \in C([0,1]) : x(0) = 0\}$ with the sup norm and $X = \left\{ x \in E : \int_0^1 x(t)dt = 0 \right\}$. Then show that X is a proper closed subspace of E. Also show that there is no $x \in E$ with $||x||_\infty = 1$ and $\text{dist}\,(x, X) = 1$.

2.4   Quotient Spaces

In this section we consider an useful method of constructing a new Banach space from a given Banach space. Earlier, in section 1.3.7 we constructed quotient space over a linear space. Because a Banach space is also a linear space, we can similarly construct a quotient space by introducing a norm consistent with the norm of the given Banach space.

2.4.1   Theorem

Let E be a normed linear space over $\mathbb{R}(\mathbb{C})$ and let L be a closed subspace of E.

Define $|| \cdot ||_q : E/L \to \mathbb{R}$ by

$$||x + L||_q = \inf\{||x + m|| : m \in L\}.$$

Then $(E/L, || \cdot ||_q)$ is a normed linear space. Furthermore, if E is a Banach space, then E/L is a Banach space.

Proof: We first show that $|| \cdot ||_q$ defines a norm on E/L. We note first that $||x + L||_q \geq 0, \forall\, x \in E$.

Next, if $x + L = L$, then $||x + L||_q = ||0 + L||_q = 0$

Conversely, let $||x + L||_q = 0$ for some $x \in E$. Then there exists a sequence $\{m_k\} \subset L$ such that

$$\lim_{k \to \infty} \|x + m_k\|_q = 0 \text{ i.e. } -m_k \to x \text{ in } E \text{ as } k \to \infty.$$

Now $x \in L$ as L is closed.

Hence $x + L = L$

Thus $\|x + L\|_q = 0 \Rightarrow x + L = L$.

Further, for $x, y \in E$, we have,

$$\begin{aligned}
\|(x + L) + (y + L)\|_q &= \|(x + y) + L\|_q \\
&= \inf\{\|(x + y) + m\| : m \in L\} \\
&= \inf\{\|(x + m_1) + (y + m_2)\|, \ m_1, m_2 \in L\} \\
&< \inf\{\|x + m_1\| : m_1 \in L\} \\
&\qquad\qquad\qquad + \inf\{\|y + m_2\| : m_2 \in L\} \\
&= \|x + L\|_q + \|y + L\|_q
\end{aligned}$$

This proves the triangle inequality.

Now, for $x \in L$ and $\alpha \in \mathbb{R}(\mathbb{C})$, with $\alpha \neq 0$, we have,

$$\begin{aligned}
\|\alpha(x + L)\|_q &= \|\alpha x + \alpha L\|_q = \|\alpha x + L\|_q \\
&= \inf\{\|\alpha x + m\| : m \in L\} \\
&= \inf\left\{\|\alpha(x + m')\| : m' = \frac{m}{\alpha} \in L\right\} \\
&= |\alpha| \inf\{\|x + m'\| : m' : L\} \\
&= |\alpha| \, \|x + L\|_q
\end{aligned}$$

Thus we conclude that $(E/L, \|\cdot\|_q)$ is a normed linear space.

We will next suppose that E is a Banach space. Let $(x_n + L)$ be a Cauchy sequence in E/L.

We next show that $\{x_n + L\}$ contains a convergent subsequence $\{x_{n_k} + L\}$. Let the subsequence $\{x_{n_k} + L\}$ be such that

$$\|(x_{n_2} + L) - (x_{n_1} + L)\|_q < \frac{1}{2}$$

$$\|(x_{n_3} + L) - (x_{n_2} + L)\|_q < \frac{1}{2^2}$$

$$\|(x_{n_{k+1}} + L) - (x_{n_k} + L)\|_q < \frac{1}{2^k}$$

Let us choose any vector $y_1 \in x_{n_1} + L$. Next choose $y_2 \in x_{n_2} + L$ such that $||y_2 - y_1|| < \frac{1}{2}$. We then find $y_3 \in x_{n_3} + L$ such that $||y_3 - y_2|| < \frac{1}{2^2}$. Proceeding in this way, we get a sequence $\{y_k\}$ in E such that

$$x_{n_k} + L = y_k + L \quad \text{and} \quad ||y_{k+1} - y_k|| < \frac{1}{2^k} \quad (k = 1, 2, \ldots)$$

Then for $p = 1, 2, \ldots$

$$||y_{k+p} - y_k|| \leq \sum_{i=1}^{p} ||y_{k+i} - y_{k+i-1}|| < \sum_{i=1}^{p} \frac{1}{2^{k+i-1}} = \frac{1}{2^{k+p-1}}$$

Therefore, it follows that $\{y_k\}$ is a Cauchy sequence in E. However because E is complete, $\exists \, y \in E$ such that $\lim_{k \to \infty} ||y_k - y|| = 0$. Since

$$||(x_{n_k} + L) - (y + L)||_q = ||(y_k + L) - (y + L)||_q$$
$$= ||(y_k - y)|L||_q$$
$$\leq ||y_k - y||(||L||_q = 0)$$

it follows that

$$\lim_{k \to \infty} (x_{n_k} + L) = y + L \in E/L$$

Hence $\{x_n + L\}$ has a subsequence $\{x_{n_k} + L\}$ that converges to some element in E/L.

Then $\quad ||(x_n + L) - (y + L)||_q \leq ||(x_n + L) - (x_{n_k} + L)||_q$
$$+ ||(x_{n_k} + L) - (y + L)||_q$$
$$\leq ||(x_n - x_{n_k}) + L||_q$$
$$+ ||(x_{n_k} - y) + L||_q \to 0 \text{ as } k \to \infty.$$

Hence the Cauchy sequence $\{x_n + L\}$ converges in E/L and thus E/L is complete.

Problems

1. Let M be a closed subspace of a normed linear space E. Prove that the quotient mapping $x \in x + M$ of E onto the quotient space E/M is continuous and that it maps open subsets of E onto open subsets of E/M.

2. Let $X_1 = (X_1, ||\cdot||_1)$ and $X_2 = (X_2, ||\cdot||_2)$ be Banach spaces over the same scalar field $\mathbb{R}(\mathbb{C})$. Let $X = X_1 \times X_2$ be the Cartesian product of X_1 and X_2. Then show that X is a linear space over $\mathbb{R}(\mathbb{C})$. Prove that the following is a norm on X:

$$||(x_1, x_2)||_\infty = \max\{||x_1||_1; ||x_2||_2\}$$

3. Let M and N be subspaces of a linear space E and let $E = M + N$. Show that the mapping $y \to y + M$, which sends each y in N to $y + M$ in E/M, is an isomorphism of N onto E/M.

4. Let M be a closed subspace of a normed space E. Prove that if E is separable, then E/M is separable (A space E is separable if it has a *denumerable everywhere dense* set).

5. If E is a normed vector space and $M \in E$ is a Banach space, then show that if E/M is a Banach space, E itself is a Banach space.

2.5 Completion of Normed Spaces

Definition: Let E_1 and E_2 be *normed linear spaces* over $\mathbb{R}(\mathbb{C})$.

(i) A mapping $T : E_1 \to E_2$ (not necessarily linear) is said to be an isometry if it preserves norms, i.e., if

$$||Tx||_{E_2} = ||x||_{E_1} \ \forall \ x \in E_1$$

such an *isometry* is said to *imbed* E_1 into E_2.

(ii) Two spaces E_1 and E_2 are said to be *isometric* if there exists an *one-one (bijective) isometry* of E_1 into E_2. The spaces E_1 and E_2 are then called *isometric spaces*.

Theorem 2.5.1. Let $E_1 = (E_1, || \cdot ||_{E_1})$ be a normed linear space. Then there is a Banach space E_2 and an *isometry* T from E_1 onto a subspace E_2' of E_2 which is dense in E_2. The space E_2 is unique, except for isometries.

Proof: Theorem 1.4.25 implies the existence of a complete metric space $X_2 = (X_2, \rho_2)$ and an isometry $T : X_1 \to X_2' = T(X_1)$, where X_2' is dense in X_2 and X_2 is unique, except for isometries. In order to prove the theorem, we need to make X_2 a linear space E_2 and then introduce a suitable norm on E_2.

To define on X_2 the two algebraic operations of a linear space, we consider any $\tilde{x}, \tilde{y} \in X_2$ and any representatives $\{x_n\} \in \tilde{x}$ and $\{y_n\} \in \tilde{y}$. We recall that $\tilde{x}, \tilde{y}$ are equivalence classes of Cauchy sequences in E_1.

We set $z_n = x_n + y_n$. Then $\{z_n\}$ is a Cauchy sequence in E_1 since

$$||z_n - z_m|| = ||x_n + y_n - (x_m + y_m)|| \leq ||x_n - x_m|| + ||y_n - y_m||.$$

We define the sum $\hat{z} = \hat{x} + \hat{y}$ of $\hat{x}$ and $\hat{y}$ to be an equivalence class of which $\{z_n\}$ is representative; thus $\{z_n\} \in \hat{z}$. This definition is independent of the particular choices of Cauchy sequences belonging to $\tilde{x}$ and $\tilde{y}$. We know that if $\{x_n\} \sim \{x'_n\}$ and $\{y_n\} \sim \{y'_n\}$, then $\{x_n + y_n\} \sim \{x'_n + y'_n\}$, because,

$$||x_n + y_n - (x'_n + y'_n)|| \leq ||x_n - x'_n|| + ||y_n - y'_n||.$$

Similarly, we define $\alpha \tilde{x} \in X_2$, the product of a scalar α and $\tilde{x}$ to be the equivalence class for which $\{\alpha x_n\}$ is a representative. Moreover, this definition is independent *of the particular choice of a representative $\tilde{x}$*. The zero element of X_2 is the equivalence class containing all Cauchy sequences that converge to zero. We thus see that these algebraic operations have all the *properties required by the definition of a linear space* and therefore X_2 is a linear space. Let us call it the normed linear space E_2. From the definition it follows that on X'_2 [see theorem 1.4.25] the operations of linear space induced from E_2 agree with those induced from E_1 by means of T. *We call X'_2, a subspace of E_2 as E'_2.*

Furthermore, T induces on E'_2 a norm $|| \cdot ||_1$, value of which at every $\tilde{y} = Tx \in E'_2$ is $||\tilde{y}||_1 = ||x||$. *The corresponding metric on E'_2 is the restrictions of ρ_2 to E'_2 since T is isometric.* We can extend the norm $|| \cdot ||_1$ to E_2 by setting $||\tilde{x}||_2 = \rho(0, \tilde{x})$ for every $\tilde{x} \in E_2$. It is clear that $|| \cdot ||_2$ satisfies axiom (a) of subsection 2.1.1 and that the other two axioms (b) and (c) of the above *follow from those for $|| \cdot ||_1$ by a limiting process.*

The *space E_2* constructed as above is sometimes called the completion of the normed linear space E_1.

Definition: completion A *completion* of a normed linear space E_1 is any normed linear space E_2 that contains a dense subspace that is *isometric* to E_1.

Theorem 2.5.2. All completions of a normed linear space are isometric.

Proof: Let, if possible, $\tilde{E}$ and $\tilde{\tilde{E}}$ be two completions of a normed linear space E. In particular, we assume that $\tilde{E}$ and $\tilde{\tilde{E}}$ are complete and both contain E as a dense subset. We now define an isometry T between $\tilde{E}$ and $\tilde{\tilde{E}}$. For each $\tilde{x} \in \tilde{E}$, since E is dense in $\tilde{E}$, $\exists$ a sequence $\{x_n\}$ of points of E converging to $\tilde{x}$. But we may also consider $\{x_n\}$ as a Cauchy sequence in $\tilde{\tilde{E}}$ and $\tilde{\tilde{E}}$ being complete, it must converge to $\tilde{\tilde{x}} \in \tilde{\tilde{E}}$. *Define $T\tilde{x} = \tilde{\tilde{x}}$ by the construction.* In what follows we will show that this construction is independent of the particular sequence $\{x_n\}$ converging to $\tilde{x}$ and gives a one-to-one mapping of $\tilde{X}$ onto $\tilde{\tilde{X}}$. Clearly $Tx = x \ \forall \ x \in E$. Now, if

$\{x_n\} \to \tilde{x}$ in $\tilde{E}$ and $x_n \to \tilde{\tilde{x}}$ in $\tilde{\tilde{E}}$, then

$$||\bar{x}||_{\tilde{E}} = \lim_{n\to\infty} ||x_n|| \quad \text{and} \quad ||\tilde{\tilde{x}}||_{\tilde{\tilde{E}}} = \lim_{n\to\infty} ||x_n||.$$

Thus $||T\tilde{x}||_{\tilde{\tilde{E}}} = ||\tilde{x}||_{\tilde{E}}$. Hence T is isometric.

Corollary 2.5.1. The space $\tilde{E}$ in theorem 2.5.2 is unique except for isometries.

Example: The completion of the normed linear space $(\mathbf{P}[a,b], ||\cdot||_\infty)$ where $\mathbf{P}[a,b]$ is the set of all polynomials with real coefficients defined on the closed interval $[a,b]$, is the space $(C([a,b]), ||\cdot||_\infty)$.

CHAPTER 3

HILBERT SPACE

A Hilbert space is a Banach space endowed with a dot product or scalar product. A normed linear space has a norm, or the concept of distance, but does not admit the concept of the angle between two elements or two vectors. But an inner product space admits both the concepts such as the concept of distance or norm and the concept of orthogonality–in other words, the angle between two vectors. Just as a complete normed linear space is called a Banach space, a complete inner product space is called a Hilbert space. An inner product space is a generalisation of the n-dimensional Euclidean space to infinite dimensions.

The whole theory was initiated by the work of D. Hilbert (1912) [24] on integral equations. The currently used geometrical notation and terminology is analogous to that of Euclidean geometry and was coined by E. Schmidt (1908) [50]. These spaces have up to now been the most useful spaces in practical applications of functional analysis.

3.1 Inner Product Space, Hilbert Space

3.1.1 Definition: inner product space, Hilbert space

An *inner product space (pre-Hilbert Space)* is a linear (vector) space H with an inner product defined on H. A *Hilbert space* is a complete *inner product space* (complete in the metric defined by the inner product) (cf. (3.2) below). Hence an *inner product* on H is a mapping of $H \times H$ into the scalar field $K(\mathbb{R}$ or $C)$ of H; that is, with every pair of elements x and y there is associated a scalar which is written $\langle x, y \rangle$ and is called **the inner product** (or *scalar product*) of x and y, such that for all elements x, y and z and scalar α we have s

(a) $\langle x + y, z \rangle = \langle x, z \rangle + \langle y, z \rangle$

(b) $\langle \alpha x, y \rangle = \alpha \langle x, y \rangle$

(c) $\langle x, y \rangle = \overline{\langle y, x \rangle}$

(d) $\langle x, x \rangle \geq 0$,

$\qquad \langle x, x \rangle = 0 \Leftrightarrow x = 0.$

An **inner product** on H defines a **norm** on H given by

$$\|x\| = \langle x, x \rangle^{\frac{1}{2}} \geq 0 \qquad (3.1)$$

and a metric on H given by

$$\rho(x, y) = \|x - y\| = \sqrt{\langle x - y, x - y \rangle}. \qquad (3.2)$$

Hence **inner product spaces** are normed linear spaces, and **Hilbert spaces** are Banach spaces.

In (c) the bar denotes complex conjugation. In case,

$$x = (\xi_1, \xi_2, \ldots \xi_n, \ldots) \quad \text{and} \quad y = (\eta_1, \eta_2, \ldots \eta_n, \ldots)$$

$$\langle x, y \rangle = \sum_{i=1}^{\infty} \xi_i \overline{\eta_i}. \qquad (3.3)$$

In case H is a real linear space

$$\langle x, y \rangle = \langle y, x \rangle.$$

The proof that (3.1) satisfies the axioms (a) to (d) of a norm [see 2.1] will be given in section 3.2.

From (a) to (d) we obtain the formula,

$$\left.\begin{array}{l}
(\text{a}') \ \langle \alpha x + \beta y, z \rangle = \alpha \langle x, z \rangle + \beta \langle y, z \rangle \text{ for all scalars } \alpha, \beta. \\
(\text{b}') \ \langle x, \alpha y \rangle = \overline{\alpha} \langle x, y \rangle \\
(\text{c}') \ \langle x, \alpha y + \beta z \rangle = \overline{\alpha} \langle x, y \rangle + \overline{\beta} \langle x, z \rangle.
\end{array}\right\} \qquad (3.4)$$

3.1.2   Observation

It follows from (a') that the inner product is linear in the first argument, while (b') shows that the inner product is conjugate linear in the second argument. Consequently, the inner product is **sesquilinear**, which means that $1\frac{1}{2}$ times linear.

3.1.3   Examples

1. **$\mathbb{R}^n, n$ dimensional Euclidean space**

The space $\mathbb{R}^n$ is a **Hilbert space** with inner product defined by

$$\langle x, y \rangle = \sum_{i=1}^{n} \xi_i \eta_i \qquad (3.5)$$

where $x = (\xi_i, \xi_2, \ldots \xi_n)$, and $y = (\eta_1, \eta_2, \ldots \eta_n)$

In fact, from (3.5) it follows

$$\|x\|^2 = \langle x, x \rangle^{\frac{1}{2}} = \left(\sum_{i=1}^{n} \xi_i^2 \right)^{\frac{1}{2}}.$$

The metric induced by the norm takes the form

$$\rho(x, y) = \|x - y\| = \langle x - y, x - y \rangle^{\frac{1}{2}} = \{(\xi_1 - \eta_1)^2 + \cdots + (\xi_n - \eta_n)^2\}^{\frac{1}{2}}$$

The completeness was established in 1.4.16.

2. Unitary space $\mathbb{C}^n$

The unitary space $\mathbb{C}^n$ is a Hilbert space with inner product defined by

$$\langle x, y \rangle = \sum_{i=1}^{n} \xi_i \overline{\eta_i} \tag{3.6}$$

where $x = (\xi_1, \xi_2, \ldots \xi_n)$ and $y = (\eta_1, \eta_2, \ldots \eta_n)$.

From (3.6) we obtain the norm defined by

$$\|x\| = \langle x, x \rangle^{\frac{1}{2}} = \left(\sum_{i=1}^{n} |\xi_i|^2 \right)^{\frac{1}{2}}.$$

The metric induced by the norm is given by

$$\rho(x, y) = \|x - y\| = \left(\sum_{i=1}^{n} |\xi_i - \eta_i|^2 \right)^{\frac{1}{2}}.$$

Completeness was shown in 1.4.16.

Note 3.1.1. In (3.6) we take the conjugate $\overline{\eta_i}$ so that we have $\langle y, x \rangle = \overline{\langle x, y \rangle}$ which is the requirement of the condition $\langle c \rangle$, so that $\langle x, x \rangle$ is real.

3. Space l_2

l_2 is a Hilbert space with inner product defined by

$$\langle x, y \rangle = \sum_{i=1}^{\infty} \xi_i \overline{\eta_i} \tag{3.7}$$

where $\quad x = (\xi_i, \xi_2, \ldots \xi_n, \ldots) \in l_2 \quad$ and $\quad y = (\eta_1, \eta_2, \ldots \eta_n, \ldots) \in l_2.$

Since $\quad x, y \in l_2, \quad \sum_{i=1}^{\infty} |\xi_i|^2 < \infty \quad$ and $\quad \sum_{i=1}^{\infty} |\eta_i|^2 < \infty.$

By *Cauchy-Bunyakovsky-Schwartz inequality* [see theorem 1.4.3]

We have $\quad \langle x, y \rangle = \sum_{i=1}^{\infty} \xi_i \overline{\eta}_i \leq \left(\sum_{i=1}^{\infty} |\xi_i|^2 \right)^{1/2} \left(\sum_{i=1}^{\infty} |\eta_i|^2 \right)^{1/2} < \infty$ $\qquad$ (3.8)

From (3.7) we obtain the norm defined by

$$\|x\| = \langle x, x \rangle^{\frac{1}{2}} = \left(\sum_{i=1}^{\infty} |\xi_i|^2 \right)^{\frac{1}{2}}.$$

Using the metric induced by the norm, **for l_2,** we see that all the axioms 3.1.1(a)–(d) are fulfilled.

4. Space $L_2([a, b])$

The inner product is defined

$$\langle x, y \rangle = \int_a^b x(t) y(t) dt \qquad (3.9)$$

for $x(t), y(t) \in L_2([a, b])$ i.e., $x(t), y(t)$ are Riemann square integrable.

The norm is then

$$\|x\| = \langle x, x \rangle^{\frac{1}{2}} = \left(\int_a^b x(t)^2 dt \right)^{\frac{1}{2}} \quad \text{where} \quad x(t) \in L_2([a, b]).$$

Using the metric induced by the norm we can show that $L_2([a, b])$ is a Hilbert space.

In the above $x(t)$ is a real-valued function.

In case $x(t)$ and $y(t)$ are complex-valued functions we can define the inner product $\langle x, y \rangle$ as $\langle x, y \rangle = \int_a^b x(t) \overline{y}(t) dt$ with the norm given by

$$\|x(t)\| = \left(\int_a^b |x(t)|^2 dt \right)^{\frac{1}{2}} \quad \text{because } x(t) \overline{x}(t) = |x(t)|^2.$$

Note 3.1.2. l_2 is the prototype of the Hilbert space. It was introduced and studied by D. Hilbert in his work on integral equations. The axiomatic definition of a Hilbert space was given by J. Von Neumann in a paper on the mathematical foundation of quantum mechanics [30].

3.2 Cauchy-Bunyakovsky-Schwartz (CBS) Inequality

3.2.1 Lemma (Cauchy-Bunyakovsky-Schwartz inequality) (CBS)

If x, y are two elements of an inner product space, then

$$|\langle x, y \rangle| \leq \|x\| \|y\| \qquad (3.10)$$

The equality occurs if and only if $\{x, y\}$ is a linearly dependent set.

Proof: If $y = \theta$ then $\langle x, y \rangle = \langle x, theta \rangle = \overline{\langle \theta, x \rangle} = 0\overline{\langle x, x \rangle} = 0$ and the conclusion is clear. Let $y \neq \theta$, then, for any scalar λ, we have

$$0 \leq \|\lambda x + y\|^2 = \langle \lambda x + y, \lambda x + y \rangle = \lambda\overline{\lambda}\langle x, x \rangle + \lambda\langle x, y \rangle + \overline{\lambda}\langle y, x \rangle$$
$$+ \langle y, y \rangle = |\lambda|^2\langle x, x \rangle + \lambda\langle x, y \rangle + \overline{\lambda\langle x, y \rangle} + \langle y, y \rangle \; \forall \; \lambda$$

Let $\lambda = -\dfrac{\langle y, x \rangle}{\langle x, x \rangle}$.

Then the above inequality reduces to,

$$\langle y, y \rangle - \frac{|\langle x, y \rangle|^2}{\langle x, x \rangle} \geq 0 \quad \text{or,} \quad |\langle x, y \rangle| \leq \|x\|\|y\|.$$

Note 3.2.1. By using Cauchy-Bunyakovsky-Schwartz inequality we can show that the norm defined by a scalar product of an inner product space [cf. (3.1)] satisfied all the axioms of a normed linear space.

For x, y belonging to an inner product space, we obtain,

$$\|x + y\|^2 = \langle x + y, x + y \rangle = \langle x, x \rangle + \langle x, y \rangle + \langle y, x \rangle + \langle y, y \rangle$$
$$= \|x\|^2 + 2 \operatorname{Re} \langle x, y \rangle + \|y\|^2$$

since $\langle y, x \rangle = \overline{\langle x, y \rangle} \leq \|x\| \|y\|$.

Cauchy-Bunyakovsky-Schwartz inequality reduces the above inequality to

$$\|x + y\|^2 \leq \|x\|^2 + 2\|x\|\|y\| + \|y\|^2 = (\|x\| + \|y\|)^2$$

Hence, $\|x + y\| \leq \|x\| + \|y\|$, which is the triangular inequality of a normed linear space.

Thus, the distance introduced by the norm satisfies all the axioms of a metric space.

Thus a Hilbert space $\Rightarrow$ a Banach space $\Rightarrow$ a complete metric space.

3.3 Parallelogram Law

3.3.1 Parallelogram law

The *parallelogram law* states that the norm induced by a scalar product satisfies

$$\|x + y\|^2 + \|x - y\|^2 = 2(\|x\|^2 + \|y\|^2) \tag{3.11}$$

where x, y are elements of an inner product space.

Proof: $\|x + y\|^2 = \langle x + y, x + y \rangle = \langle x, x \rangle + \langle x, y \rangle + \langle y, x \rangle + \langle y, y \rangle$.

$$\|x - y\|^2 = \langle x - y, x - y \rangle = \langle x, x \rangle - \langle x, y \rangle - \langle y, x \rangle + \langle y, y \rangle.$$

Therefore, $\|x + y\|^2 + \|x - y\|^2 = 2(\|x\|^2 + \|y\|^2)$.

The term *parallelogram equality* is suggested by elementary geometry, as we shall see from the figure below. Since norm stands for the length of a vector, the parallelogram law states an important property of a parallelogram, i.e., *the sum of the squares of the lengths of the diagonals is equal to twice the sum of the squares of the lengths of the sides.*

Thus the parallelogram law generalises a known property of elementary geometry to an inner product space.

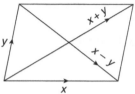

Fig. 3.1 Parallelogram with sides *x* and *y* in the plane

In what follows we give some examples of normed linear spaces which are not Hilbert spaces.

3.3.2 Space l_p

The space l_p with $p \neq 2$ is not an inner product space, hence not a Hilbert space. Hence we would like to show that the norm of l_p with $p \neq 2$ cannot be obtained from an inner product. We prove this by showing that the norm does not satisfy the parallelogram law. Let us take $x = (1, 1, 0 \ldots 0) \in l_p$ and $y = (1, -1, 0, 0 \ldots 0 \ldots) \in l^p$. Then $\|x\| = 2^{\frac{1}{p}}$. $\|y\| = 2^{\frac{1}{p}}$. Now $x + y = (2, 0, 0 \ldots)$, $\|x+y\| = 2$. Again $x - y = (0, 2, 0, 0 \ldots)$. Hence $\|x - y\| = 2$.

Thus the parallelogram law is not satisfied. Hence $l_p(p \neq 2)$ though a Banach space (cf. 2.1.4) is not a Hilbert space.

3.3.3 Space $C([0, 1])$

The space $C([0, 1])$ is *not an inner product space and hence not a Hilbert space.*

We show that the norm defined by

$$\|x\| = \max_{0 \leq t \leq 1} |x(t)|$$

cannot be obtained from an inner product. Let us take $x(t) = 1$ and $y(t) = t$. Hence $\|x\| = 1$, $\|y\| = 1$

$$\|x(t) + y(t)\| = \|1 + t\| = 2 \cdot \|x(t) - y(t)\| = \|1 - t\| = 1.$$

Thus $\|x + y\| + \|x - y\| = 3 \neq 4 = 2(\|x\| + \|y\|)$.

Hence the parallelogram law is not satisfied.

Thus $C([0, 1])$, although a complete normed linear space, i.e., a Banach space [cf. 2.1.4], is not a Hilbert space.

We know that the norm of a vector in an inner product space can be expressed in terms of the inner product: $\|x\| = \sqrt{\langle x, x \rangle}$. The inner product can also be recovered from the induced norm by the following formula known as **the polarization identity**:

$$\langle x, y \rangle = \begin{cases} \dfrac{1}{4}(\|x+y\|^2 - \|x-y\|^2) & \text{in } \mathbb{R} \\[2mm] \dfrac{1}{4}[(\|x+y\|^2 - \|x-y\|^2) + i(\|x+iy\|^2 - \|x-iy\|^2)] & \text{in } \mathbb{C} \end{cases}$$

$$(3.12)$$

Now, in $\mathbb{R}$,

$$\frac{1}{4}(\|x+y\|^2 - \|x-y\|^2) = \frac{1}{4}[\langle x+y, x+y \rangle - \langle x-y, x-y \rangle]$$

$$= \frac{1}{4}[\langle x, x \rangle + \langle y, y \rangle + \langle x, y \rangle + \langle y, x \rangle - \langle x, x \rangle - \langle y, y \rangle + \langle x, y \rangle + \langle y, x \rangle].$$

$$= \langle x, y \rangle \text{ since in } \mathbb{R}, \ \langle x, y \rangle = \langle y, x \rangle.$$

In $\mathbb{C}$

$$\frac{1}{4}[(\|x+y\|^2 - \|x-y\|^2) + i(\|x+iy\|^2 - \|x-iy\|^2)]$$

$$= \frac{1}{4}[\{\langle x, x \rangle + \langle y, y \rangle + \langle x, y \rangle + \langle y, x \rangle - \langle x, x \rangle - \langle y, y \rangle + \langle x, y \rangle + \langle y, x \rangle\}$$

$$+ i\{\langle x, x \rangle + \langle x, iy \rangle + \langle iy, x \rangle + i\bar{i}\langle y, y \rangle - \langle x, x \rangle - i\bar{i}\langle y, y \rangle + \langle x, iy \rangle + \langle iy, x \rangle\}]$$

$$= \frac{1}{4}[2\langle x, y \rangle + 2\langle \overline{x, y} \rangle + 2\langle x, y \rangle - 2\langle \overline{x, y} \rangle] = \langle x, y \rangle.$$

which is the **polarization identity**.

3.3.4   Theorem

A norm $\|\cdot\|$ on a linear space E is induced by an inner product $\langle , \rangle$ on it if and only the norm satisfies the *parallelogram law*. In that case the inner product $\langle , \rangle$ is given by the *polarization equality*.

Proof: Suppose that the norm is induced by the inner product. Then the parallelogram law holds true. Furthermore the inner product can be recovered from the norm by the polarization equality.

Conversely, let us suppose that $\|\cdot\|$ obeys the parallelogram law and $\langle , \rangle$ is defined by the polarization equality as given in (3.11). We have to show that $\langle , \rangle$ is an inner product and generalize $\|\cdot\|$ on E. Let us consider the formula (3.11) for the complex space.

(i) Then for all $x, y \in E$

$$\langle x, y \rangle = \frac{1}{4}[(\|x+y\|^2 - \|x-y\|^2 + i(\|x+iy\|^2 - \|x-iy\|^2)].$$

Putting $y = x \in E$ we get

$$\langle x, x \rangle = \frac{1}{4}[(4\|x\|^2 - 0 + i\|1+i\|^2\|x\|^2 - i\|1-i\|^2\|x\|^2]$$

$$= \frac{1}{4}[(4\|x\|^2 + 2i(\|x\|^2 - \|x\|^2)]$$

$$= \|x\|^2$$

Therefore inner product $\langle\ ,\ \rangle$ generates the norm $\|\cdot\|$.

(ii) For all $x, y \in E$ we have,

$$\langle\overline{y,x}\rangle = \frac{1}{4}[\|y + x\|^2 - \|y - x\|^2 - i(\|y + ix\|^2 - \|y - ix\|^2)]$$

$$= \frac{1}{4}[\|x + y\|^2 - \|x - y\|^2 + i(\|x + iy\|^2 - \|x - iy\|^2)] = \langle x, y \rangle.$$

(iii) Let $u, v, w \in X$. Then **parallelogram law** yields

$$\begin{cases} \|(u + v) + w\|^2 + \|(u + v) - w\|^2 = 2(\|u + v\|^2 + \|w\|^2) \\ \|(u - v) + w\|^2 + \|(u - v) - w\|^2 = 2(\|u - v\|^2 + \|w\|^2). \end{cases}$$

On substraction we get,

$$(\|(u + w) + v\|^2 - \|(u + w) - v\|^2) + (\|(u - w) + v\|^2 - \|(u - w) - v\|^2)$$
$$= 2(\|u + v\|^2 - \|u - v\|^2).$$

Using the **polarization identity**, we get

$$\begin{cases} \mathrm{Re}\ \langle u + w, v \rangle + \mathrm{Re}\ \langle u - w, v \rangle = 2\ \mathrm{Re}\ \langle u, v \rangle & (3.13) \\ \mathfrak{Im}\langle u + w, v \rangle + \mathfrak{Im}\langle u - w, v \rangle = 2\ \mathfrak{Im}\langle u, v \rangle. \end{cases}$$

Hence, $\langle u + w, v \rangle + \langle u - w, v \rangle = 2\langle u, v \rangle$ (3.14).

Putting $u + w = x$ $u - v = y$ and $v = z$, we obtain from (3.14)

$$\langle x, z \rangle + \langle y, z \rangle = 2 \left\langle \frac{x + y}{2}, z \right\rangle = \langle x + y, z \rangle$$

since on putting $w = u$, (3.13) reduces to $\mathrm{Re}\langle 2u, v \rangle = \mathrm{Re}2\langle u, v \rangle$ for all $u, v \in E$.

Thus condition (a) of 3.1 is proved.

(iv) Next we want to prove condition (b), i.e., $\langle \alpha x, y \rangle = \alpha \langle x, y \rangle$, for every complex scalar α and $\forall\ x, y \in E$. We shall prove it in stages.

Stage 1. Let $\lambda = m$, a positive integer, $m > 1$.

$$\langle mx, y \rangle = \langle (m - 1)x + x, y \rangle = \langle (m - 1)x, y \rangle + \langle x, y \rangle$$
$$= \langle (m - 2)x, y \rangle + 2\langle x, y \rangle$$

$$\vdots$$

$$= \langle x, y \rangle + (m - 1)\langle x, y \rangle = m\langle x, y \rangle.$$

Also for any positive integer n, we have

$$n \left\langle \frac{x}{n}, y \right\rangle = \left\langle n \left(\frac{1}{n} \right) x, y \right\rangle = \langle x, y \rangle$$

Hence $\left\langle \dfrac{x}{n}, y \right\rangle = \dfrac{1}{n} \langle x, y \rangle.$

If m is a negative integer, splitting m as $(m-1)+1$ we can show that (b) is true.

Stage 2. Let $\alpha = r = \dfrac{m}{n}$ be a rational number, m and n be prime to each other.

Then $r\langle x, y \rangle = \dfrac{m}{n} \langle x, y \rangle = m \left\langle \dfrac{x}{n}, y \right\rangle = \left\langle \dfrac{m}{n} x, y \right\rangle = \langle rx, y \rangle.$

Stage 3. Let α be a real number. Then there exists a sequence $\{r_n\}$ of rational numbers, such that $r_n \to \alpha$ as $n \to \infty$.

Hence $\qquad r_n \langle x, y \rangle \to \alpha \langle x, y \rangle$

But $\qquad r_n \langle x, y \rangle = \langle r_n x, y \rangle \quad$ and $\quad \|r_n x + y\| \to \|\alpha x + y\|$

Therefore, $\alpha \langle x, y \rangle = \langle \alpha x, y \rangle$ for any real α.

Stage 4. Let $\alpha = i$. Then, **the polarization identity** yields

$$\langle ix, y \rangle = \frac{1}{4}[\|ix + y\|^2 - \|ix - y\|^2 + i(\|i(x+y)\|)^2 - (\|i(x-y)\|)^2]$$
$$= \frac{i}{4}[\|x + y\|^2 - \|x - y\|^2 + i(\|x + iy\|^2 - \|x - iy\|^2)]$$
$$= i\langle x, y \rangle.$$

Stage 5. Finally, let $\alpha = p + iq$, be any complex number, then,

$$\alpha \langle x, y \rangle = p\langle x, y \rangle + iq\langle x, y \rangle = \langle px, y \rangle + \langle iqx, y \rangle$$
$$= \langle (p + iq)x, y \rangle = \langle \alpha x, y \rangle.$$

Thus we have shown that $\langle \, , \, \rangle$ is the inner product inducing the norm $\| \cdot \|$ on E.

3.3.5 *Lemma*

Let E be an inner product space with an inner product $\langle \, , \, \rangle$.

(i) The linear space E is uniformly convex in the norm $\| \cdot \|$, that is, for every $\epsilon > 0$, there is some $\delta > 0$, such that for all $x, y \in E$ with $\|x\| \le 1, \|y\| \le 1$ and $\|x - y\| \le \epsilon$, we have $\|x + y\| \le 2 - 2\delta$.

(ii) The scalar product is a continuous function with respect to norm convergence.

Proof: (i) Let $\epsilon > 0$. Given $x, y \in E$ with $\|x\| \le 1, \|y\| \le 1$ and $\|x-y\| \ge \epsilon$. Then $\epsilon \le \|x - y\| \le \|x\| + \|y\| \le 2$.

The parallelogram law gives

$$\|x + y\|^2 = \|x\|^2 + \|y\|^2 + [\langle x, y \rangle + \langle y, x \rangle]$$
$$= 2(\|x\|^2 + \|y\|^2) - [\|x\|^2 - \{\langle x, y \rangle + \langle y, x \rangle\} + \|y\|^2]$$

$$= 2(\|x\|^2 + \|y\|^2) - \|x - y\|^2 \le 4 - \epsilon^2.$$

Hence, $\|x + y\| \le \sqrt{4 - \epsilon^2} = 2 - 2\delta$ if $\delta = 1 - \left(1 - \frac{\epsilon^2}{4}\right)^{\frac{1}{2}}$.

(ii) Let $x, y \in E$ and $x_n \longrightarrow x, y_n \longrightarrow y$ where x_n, y_n are elements of E. Therefore $\|x_n\|$ and $\|y_n\|$ are bounded above and let M be an upper bound of both x_n and y_n.

Hence, $|\langle x_n, y_n \rangle - \langle x, y \rangle| = |\langle x_n, y_n \rangle - \langle x_n, y \rangle + \langle x_n, y \rangle - \langle x, y \rangle|$

$$= |\langle x_n, y_n - y \rangle + \langle x_n - x, y \rangle| \le M\|y_n - y\| + M\|x_n - x\|$$

using *Cauchy-Bunyakovsky-Schwartz inequality*, since $x_n \longrightarrow x$ and $y_n \longrightarrow y$ as $n \longrightarrow \infty$ we have from the above inequality,

$$\langle x_n, y_n \rangle \longrightarrow \langle x, y \rangle \quad \text{as } n \longrightarrow \infty.$$

This shows that $\langle x, y \rangle$, which is a function on $E \times E$ is continuous in both x and y.

3.4   Orthogonality

3.4.1   Definitions (orthogonal, acute, obtuse)

Let x, y be vectors in an inner product space.

(i) Orthogonal: x is said to be *orthogonal* to y or written as $x \perp y$ if $\langle x, y \rangle = 0$.

(ii) Acute: x is said to be *acute* to y if $\langle x, y \rangle \ge 0$.

(iii) Obtuse: x is said to be *obtuse* to y if $\langle x, y \rangle \le 0$.

In $\mathbb{R}^2$ or $\mathbb{R}^3$, x is orthogonal to y if the angle between the vectors is $90°$. Similarly when x is acute to y, the angle between x and y is less than or equal to $90°$. We can similarly explain when $\langle x, y \rangle \le 0$ the angle between x and y is greater than or equal to $90°$. This geometrical interpretation can be extended to infinite dimensions in an inner product space.

3.4.2   Definition: subspace

A non-empty subset X of the inner product space E is said to be a *subspace* of E if

(i) X is a (linear) subspace of E considered as a linear space.

(ii) X admits of a inner product $\langle \, , \, \rangle_X$ induced by the inner product $\langle \, , \, \rangle$ on E, i.e.,

$$\langle x, y \rangle_X = \langle x, y \rangle \quad \forall \, x, y \in E.$$

Note 3.4.1. A subspace X of an inner product E is itself an inner product space and the induced norm $\| \cdot \|_X$ on X coincides with the induced norm $\| \cdot \|$ on E.

3.4.3   Closed subspace

A subspace X of an inner product space E is said to be a *closed subspace* of E if X is closed with respect to $\| \cdot \|_X$ induced by the $\| \cdot \|$ on E.

Note 3.4.2. Given a Hilbert space H, when we call X a subspace (closed subspace) of H we treat H as an inner product space and X its subspace (closed subspace).

Note 3.4.3. Every subspace of a *finite dimensional* inner product space is *closed*.

This is not true in general, as the following example shows.

3.4.4   Example

Consider the Hilbert space l_2 and let X be the subset of all finite sequences in l_2 given by

$$X = \{x = \{\xi_i\} \in l_2 : \xi_i = 0 \text{ for } i > N, N \text{ is some positive integer}\}.$$

X is a proper subspace of l_2, but X is dense in l_2. Hence $\overline{X} = l_2 \neq X$. Hence X is not closed in l_2.

Problems [3.1–3.4]

1. Let $\alpha_1, \alpha_2, \ldots \alpha_n$ be n strictly positive real numbers. Show that the function of two variables $\langle \cdot, \cdot \rangle : \mathbb{R}^n \times \mathbb{R}^n \longrightarrow \mathbb{R}$, defined by $\langle x, y \rangle = \sum_{i=1}^{n} \alpha_i x_i y_i$, is an inner product on $\mathbb{R}^n$.

2. Show that equality holds in the Cauchy-Bunyakovsky-Schwartz inequality, (i.e., $|\langle x, y \rangle| = \|x\| \|y\|$) if and only if x and y are linearly dependent.

3. Let $\langle \, , \, \rangle$ be an inner product on a linear space E. For $x \neq \theta$, $y \neq \theta$ in E define the angle between x and y as follows:

$$\theta_{x,y} = \text{arc cos} \frac{\text{Re } (x,y)}{\sqrt{(x,x)(y,y)}}, \ 0 \leq \theta_{x,y} \leq \pi$$

Then show that $\theta_{x,y}$ is well-defined and satisfies the identity

$$\langle x, x \rangle + \langle y, y \rangle - \langle x - y, x - y \rangle = 2\langle x, x \rangle^{\frac{1}{2}} \langle y, y \rangle^{\frac{1}{2}} \cos \theta_{x,y}.$$

4. Let $\|\cdot\|$ be a norm on a linear space E which satisfies the parallelogram law

$$\|x + y\|^2 + \|x - y\|^2 = 2(\|x\|^2 + \|y\|^2), \ x, y \in E$$

For $x, y \in E$ define $\langle x, y \rangle = \frac{1}{4}[\|x + y\|^2 - \|x - y\|^2 + i\|x + iy\|^2 - i\|x - iy\|^2]$

Then show that $\langle \, , \, \rangle$ is the unique inner product on E satisfying $\sqrt{\langle x, y \rangle} = \|x\|$ for all $x \in E$.

5. [Limaye [33]] Let X be a normed space over $\mathbb{R}$. Show that the norm satisfies the parallelogram law, if and only if in every plane through the origin, the set of all elements having norm equal to 1 forms an ellipse with its centre at the origin.

6. Let $\{x_n\}$ be a sequence in a Hilbert space H and $x \in H$ such that $\lim\limits_{n\to\infty} \|x_n\| = \|x\|$, and $\lim\limits_{n\to\infty} \langle x_n, x\rangle = \langle x, x\rangle$. Show that $\lim\limits_{n\to\infty} x_n = x$.

7. Let C be a convex set in a Hilbert space H, and $d = \inf\{\|x\|, x \in C\}$. If $\{x_n\}$ is a sequence in C such that $\lim\limits_{n\to\infty} \|x_n\| = d$, show that $\{x_n\}$ is a Cauchy sequence.

8. **(Pythagorean theorem)** (Kreyszig [30]). If $x \perp y$ is an inner product on E, show that (fig. 3.2),

$$\|x + y\|^2 = \|x\|^2 + \|y\|^2.$$

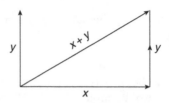

Fig. 3.2

9. (Appolonius' identity) (Kreyszig [30]). Verify by direct calculations that for any three elements x, y and z in an inner product space E,

$$\|z - x\|^2 + \|z - y\|^2 = \frac{1}{2}\|x - y\|^2 + 2\|z - \frac{1}{2}(x + y)\|^2.$$

Show that this identity can also be obtained from the parallelogram law.

3.5   Orthogonal Projection Theorem

In 1.4.7 we have defined the distance of a point x from a set A in a metric space E which runs as follows:

$$D(x, A) = \inf_{y \in A} \rho(x, y). \qquad (3.15)$$

In case E is a normed linear space, thus:

$$D(x, A) = \inf_{y \in A} \|x - y\| \qquad (3.16)$$

If $\hat{y}$ is the value of y for which the infimum is attained then, $D(x, A) = \|x - \hat{y}\|$, $\hat{y} \in A$. Hence $\hat{y}$ is the element in A closest to x. The existence of such an element $\hat{y}$ is not guaranteed and even if it exists it may not be unique. Such behaviour may be observed even if A happens to be a curve in $\mathbb{R}^2$. For example let A be an open line segment in $\mathbb{R}^2$.

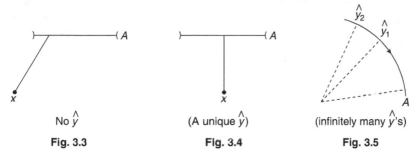

No $\hat{y}$	(A unique $\hat{y}$)	(infinitely many $\hat{y}$'s)
Fig. 3.3	**Fig. 3.4**	**Fig. 3.5**

Existence and uniqueness of points $\hat{y} \epsilon A$ satisfying (3.16) where the given set E, $A \subset \mathbb{R}^2$ is an open segment [in fig. 3.3 and in fig. 3.4] and is a circular arc [fig. 3.5].

The study of the problem of *existence and uniqueness of a point in a set closest to a given point* falls within the purview of the theory of optimization.

In what follows we discuss the orthogonal projection theorem in a Hilbert space which partially answers the above problem.

3.5.1 Theorem (orthogonal projection theorem)

If $x \in H$ (a Hilbert space) and L is some closed subspace of H, then x has a unique representation of the form

$$x = y + z, \tag{3.17}$$

with $y \in L$ and $z \perp L$ i.e. z is orthogonal to every element of L.

Proof: If $x \in L$, $y = x$ and $z = \theta$. Let us next say that $x \notin L$. Let $d = \inf\limits_{y \in L} \|x - y\|^2$, i.e., d is the square of the distance of x from L. Let $\{y_n\}$ be a sequence in L such that $d_n = \|x - y_n\|^2$ and let $d_n \to d$ as $n \to \infty$. Let h be any non-zero element of L. Then $y_n + \epsilon h \in L$ for any complex number ϵ.

Therefore, $\|x - (y_n + \epsilon h)\|^2 \geq d$ i.e. $\langle x - (y_n + \epsilon h), x - (y_n + \epsilon h)\rangle \geq d$

or, $\langle x - y_n, x - y_n \rangle - \epsilon \langle h, x - y_n \rangle - \langle x - y_n, \epsilon h \rangle + \epsilon \langle h, \epsilon h \rangle \geq d$

or, $\|x - y_n\|^2 - \epsilon \langle h, x - y_n \rangle - \bar{\epsilon} \langle x - y_n, h \rangle + |\epsilon|^2 \|h\|^2 \geq d,$

Let us put $\epsilon = \dfrac{\langle x - y_n, h \rangle}{\|h\|^2}$

The above inequality reduces to,

$$\|x - y_n\|^2 - \frac{|\langle x - y_n, h \rangle|^2}{\|h\|^2} \geq d,$$

or
$$\frac{|\langle x - y_n, h \rangle|^2}{\|h\|^2} \leq (d_n - d).$$

or
$$|\langle x - y_n, h \rangle| \leq \|h\| \sqrt{d_n - d}. \tag{3.18}$$

Inequality (3.18) is evidently satisfied for $h = 0$.

It then follows that
$$|\langle y_m - y_n, h \rangle| \leq |\langle x - y_n, h \rangle| + |\langle x - y_m, h \rangle|$$
$$\leq (\sqrt{d_n - d} + \sqrt{d_m - d})\|h\| \tag{3.19}$$

(3.19) yields
$$\frac{|\langle y_n - y_m, h \rangle|}{\|h\|} \leq (\sqrt{d_n - d} + \sqrt{d_m - d})$$

Taking supremum of LHS we obtain,
$$\|y_n - y_m\| \leq (\sqrt{d_n - d} + \sqrt{d_m - d}) \tag{3.20}$$

Since $d_n \to d$ as $n \to \infty$, the above inequality shows that $\{y_n\}$ is a Cauchy sequence. H being complete, $\{y_n\} \to$ some element $y \in H$. Since L is closed, $y \in L$. It then follows from (3.18) that
$$\langle x - y_n, h \rangle = 0 \tag{3.21}$$

where h is an arbitrary element of L.

Hence $x - y$ is perpendicular to any element $h \in L$, i.e. $x - y \perp L$. Setting $z = x - y$ we have
$$x = y + z \tag{3.22}$$

Next we want to show that the representation (3.22) is unique. If that be not so, let us suppose there exist y' and z' such that
$$x = y' + z' \tag{3.23}$$

It follows from (3.22) and (3.23) that
$$y - y' = z' - z. \tag{3.24}$$

Since L is a subspace and $y, y' \in L \Rightarrow y - y' \in L$.

Similarly $z - z' \perp L$. Hence (3.24) can be true if and only if $y - y' = z' - z = \theta$ showing that the representation (3.22) is unique. Otherwise we may note that $\|y - y'\|^2 = \langle y - y', z' - z \rangle = 0$ since $y - y'$ is $\perp$ to $z' - z$. Hence $y = y'$ and $z = z'$. In (3.22) y is called the **projection** of x on L.

3.5.2   Lemma

The collection of all elements z orthogonal to $L(\neq \Phi)$ forms a **closed subspace** M (say).

Let $z_1, z_2 \perp$ to L but $y \in L$ where L is non-empty.

Then, $\langle y, z_1 \rangle = 0$, $\langle y, z_2 \rangle = 0$.

Therefore, for scalars α, β, $\langle y, \alpha z_1 + \beta z_2 \rangle = \overline{\alpha}\langle y, z_1 \rangle + \overline{\beta}\langle y, z_2 \rangle = 0$.

Again, let $\{z_n\} \to z$ in H, where $\{z_n\}$ is orthogonal to L. Then, a scalar product being a continuous function

$$|\langle y, z_n \rangle - \langle y, z \rangle| = |\langle y, z_n - z \rangle| \leq \|y\|\|z_n - z\|, \ y \in L$$
$$\to 0 \quad \text{since} \ z_n \to z \text{ as } n \to \infty.$$

Therefore, $\langle y, z \rangle = \lim_{n \to \alpha} \langle y, z_n \rangle = 0$.

Hence, $z \perp L$.

Thus the elements orthogonal to L form a closed subspace M. We write $L \perp M$. M is called the **orthogonal complements** of L, and is written as $M = L^{\perp}$.

Note 3.5.1. In $\mathbb{R}^2$ if the line L is the subspace, the projection of any vector x with reference to a point 0 on L (the vector x not lying on L) is given by,

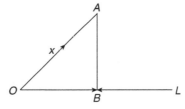

Fig. 3.6 Projection of *x* on *L*

Here, OB is the projection vector x on L. B is the point on L closest from A, the end point of x.

Next, we present a theorem enumerating the conditions under which the points in a set closest from a point (not lying on the set) can be found.

3.5.3 *Theorem*

Let H be a Hilbert space and L a closed, convex set in H and $x \in H - L$, then there is a unique $y_0 \in L$ such that

$$\|x - y_0\| = \inf_{y \in L} [\|x - y\|]$$

Proof: Let $d = \inf_{y \in L} \|x_0 - y\|^2$. Let us choose $\{y_n\}$ in L such that

$$\lim_{n \to \infty} \|x - y_n\|^2 = d \qquad (3.25)$$

Then by parallelogram law,

$$\|(y_m - x) - (y_n - x)\|^2 + \|(y_m - x) + (y_n - x)\|^2$$
$$= 2(\|y_m - x\|^2 + \|y_n - x\|^2)$$

or $\qquad \|y_m - y_n\|^2 = 2(\|y_m - x\|^2 + \|y_n - x\|^2) - 4\|\dfrac{y_m + y_n}{2} - x\|^2.$

Since L is convex, $y_m, y_n \in L \Rightarrow \dfrac{y_m + y_n}{2} \in L$

Hence $\quad \|\dfrac{y_m + y_n}{2} - x\|^2 \geq d.$

Hence, $\quad \|y_m - y_n\|^2 \leq 2\|y_m - x\|^2 + 2\|y_n - x\|^2 - 4d.$

Using (3.25) we conclude from above that $\{y_n\}$ is Cauchy.

Since L is closed, $\{y_n\} \to y_0$ (say) $\in L$ as $n \to \infty.$

Then $\qquad \|x - y_0\|^2 = d.$

For uniqueness, let us suppose that $\|x - z_0\|^2 = d$, where $z_0 \in L.$

Then, $\quad \|y_0 - z_0\|^2 = \|(y_0 - x) - (z_0 - x)\|^2$

$$= 2\|y_0 - x\|^2 + 2\|z_0 - x\|^2 - \|(y_0 - x) + (z_0 - x)\|^2$$

$$= 4d - 4\left\|\dfrac{y_0 + z_0}{2} - x\right\|^2 \leq 4d - 4d = 0$$

Hence $\quad y_0 = z_0.$

3.5.4   Theorem

Let H be a Hilbert space and L be a closed convex set in H and $x \in H - L$. Then there is a unique $y_0 \in L$ (i) such that $\|x - y_0\| = \inf\limits_{y \in L} \|x - y\|$ and (ii) $x - y_0 \in L^{\perp}.$

Theorem 3.5.3 guarantees the existence of a unique $y_0 \in L$ such that $\|x - y_0\| = \inf\limits_{y \in L} \|x - y\|.$

If $y \in L$ and ϵ is a scalar, then $y_0 + \epsilon y \in L$, so that

$$\|x - (y_0 + \epsilon y)\|^2 \geq \|y_0 - x\|^2$$

and $\quad -\bar{\epsilon}\langle x - y_0, y \rangle - \epsilon\langle y, x - y_0 \rangle + |\epsilon|^2\|y\|^2 \geq 0 \qquad (3.26)$

or $\quad |\epsilon|^2\|y\|^2 \geq 2 \operatorname{Re}\{\langle \bar{\epsilon}(x - y_0), y\rangle\}.$

If ϵ is real and $= \beta > 0$

$$2 \operatorname{Re} \langle x - y_0, y \rangle \leq \beta\|y\|^2 \qquad (3.27)$$

If $\epsilon = i\beta$, β real and > 0, and if $\langle x - y_0, y \rangle = \gamma + i\delta$ (3.26) yields,

$$i\beta(\gamma + i\delta) - i\beta(\gamma - i\delta) + \beta^2\|y\|^2 \geq 0 \quad \text{or,} \quad 2\beta\delta \leq \beta^2\|y\|^2$$

or, $\quad 2\operatorname{Im}\langle x - y_0, y\rangle \leq \beta\|y\|^2 \qquad (3.28)$

Since $\beta > 0$ is arbitrary, it follows from (3.27) and (3.28) that $\langle x - y_0, y \rangle = 0$ and $x - y_0$ is perpendicular to any $y \in L$. Thus $x - y_0 \perp L.$

Therefore $x - y_0 = z$ (say), where $z \in L^{\perp}$. Thus, $x = y_0 + z$, where $y_0 \in L$ and $z \in L^{\perp}$. y_0 is the vector in L at *minimum distance* from x. y_0 is called *the orthogonal projection* of x on L.

3.5.5 Lemma

In order that a linear subspace L is everywhere dense in a Hilbert space H, it is necessary and sufficient that an element exists which is different from zero and orthogonal to all elements of M.

Proof: Since M is everywhere dense in H, $x \perp M$ implies that $x \perp \overline{M}$. By hypothesis $\overline{M} = H$ and consequently, $x \perp H$, in particular $x \perp x$ implying $x = \theta$.

Conversely, let us suppose M is not everywhere dense in H. Then $\overline{M} \neq H$ and there is an $x \notin \overline{M}$ and $x \in H$. By theorem 3.5.1 $x = y + z$, $y \in \overline{M}$, $z \perp \overline{M}$, and since $x \notin \overline{M}$, it follows that $z \neq \theta$, which is a contradiction to our hypothesis. Hence $\overline{M} = H$.

3.6 Orthogonal Complements, Direct Sum

In 3.5.2, we have defined *the orthogonal complement* of a set L in a Hilbert space H as

$$L^1 = \{y | y \in H, \ y \perp x \text{ for every } x \in L\} \tag{3.29}$$

We write $(L^{\perp})^{\perp} = L^{\perp\perp}, (L^{\perp\perp})^{\perp} = L^{\perp\perp\perp}$ etc.

Note 3.6.1. (i) $\{\theta\}^{\perp} = X$ and $X^{\perp} = \{\theta\}$, i.e., θ is the only vector orthogonal to every vector.

Note that $\{\theta\}^{\perp} = \{x \in H : \langle x, \theta \rangle = 0\} = H$.

Since $\langle x, \theta \rangle = 0$ for all $x \in H$. Also if $x \neq \theta$, then $\langle x, x \rangle \neq 0$. Hence a non-zero vector x cannot be orthogonal to the entire space. Therefore, $H^{\perp} = \{\theta\}$.

(ii) If $L \neq \Phi$ is a subset of H, then the set $L^{\perp}$ is a closed subspace of H. Furthermore $L \cap L^{\perp}$ is either $\{\theta\}$ or empty (when $\theta \notin L$).

For the first part of the above see Lemma 3.5.2. For the second part, let us suppose $L \cap L^{\perp} \neq \Phi$ and let $x \in A \cap A^{\perp}$. Then we may have $x \perp x$. Hence $x = \theta$.

(iii) If A is a subset of H, then $A \subseteq A^{\perp\perp}$. Let $x \in A$. Then, $x \perp A^{\perp}$ which means that $x \in (A^{\perp})^{\perp}$.

(iv) If A and B are subsets of H such that $A \subseteq B$, then $A^{\perp} \supset B^{\perp}$.

Let $x \in B^{\perp}$, then $\langle x, y \rangle = 0 \ \forall \ y \in B$ and therefore $\forall \ y \in A$ since $A \subseteq B$. Thus $x \in B^{\perp} \Rightarrow x \in A^{\perp}$ that is, $B^{\perp} \subset A^{\perp}$.

(v) If $A \neq \Phi$ is a subset of H, then $A^{\perp} = A^{\perp\perp\perp}$. Changing A by $A^{\perp}$ in (iii) we get, $A^{\perp} \subseteq A^{\perp\perp\perp}$.

Since $A \subseteq A^{\perp\perp}$, it follows for (iv) that $A^{\perp} \supseteq A^{\perp\perp\perp}$ or $A^{\perp\perp\perp} \subseteq A^{\perp}$. Hence it follows from the above two inclusions that $A^{\perp} = A^{\perp\perp\perp}$.

3.6.1 Direct sum

In 1.3.9. we have seen that a linear space E can be expressed as the *direct sum* of its subspaces $X_1, X_2 \dots X_n$ if every $x \in E$ can be expressed *uniquely* in the form

$$x = x_1 + x_2 + \cdots + x_n, \ x_i \in X_i.$$

In that case we write $E = X_1 \oplus X_2 \oplus X_3 \cdots \oplus X_n$. In what follows we mention that using the *orthogonal projection theorem [cf. 3.5.1]* we can partition a Hilbert space H into two closed subspaces L and its *orthogonal complement* $L^\perp$.

i.e., $x = y + z$ where $x \in H, \ y \in L$ and $z \perp L$.

This representation is unique.

Hence we can write $H = L \oplus L^\perp$.   (3.30)

Thus **the orthogonal projection theorem** can also be stated as follows:

If L is a closed subspace of a Hilbert space H, then $H = L \oplus L^\perp$.
Proof: $L^\perp$ is a closed subspace of H (cf. Lemma 3.5.2). Therefore, L and $L^\perp$ are orthogonal closed subspace of H. We next want to show that $L + L^\perp$ is a closed subspace of H. Let $z \in \overline{L + L^\perp}$. Then there exists a sequence $\{z_n\}$ in $L + L^\perp$ such that $\lim_{n \to \infty} z_n = z$.

Since $z_n \in L + L^\perp$ can be uniquely represented as $z_n = x_n + y_n, \ x_n \in L, \ y_n \in L^\perp$.

Since $x_n \perp y_n, \ n \in \mathbb{N}$, by virtue of Pythagorean theorem we have

$$\|z_m - z_n\|^2 = \|(x_m - x_n) + (y_m - y_n)\|^2$$
$$= \|x_m - x_n\|^2 + \|y_m - y_n\|^2 \to 0 \text{ as } n \to \infty.$$

Hence $\{z_n\}$ is Cauchy and $\{x_n\}$ and $\{y_n\}$ are Cauchy sequences in L and $L^\perp$ respectively. Since L and $L^\perp$ are closed subspaces of H, they are complete. Hence $\{x_n\} \to x$ an element in L and $\{y_n\} \to y$, an element in $L^\perp$ as $n \to \infty$.

Thus, $z = \lim_{n \to \infty} z_n = \lim_{n \to \infty} (x_n + y_n) = x + y.$

Hence, $x + y \in L + L^\perp$ i.e. $L + L^\perp$ is a closed subspace of H.

Hence $L + L^\perp$ is complete. We next have to prove that $L + L^\perp = H$. If that be not so, let $L + L^\perp$ be a proper subspace of H. If $L = H$ then $L^\perp = \Phi$. On the other hand if $L^\perp = H, \ L = \Phi$. Hence $L + L^\perp = H$ is true in the above cases. Let us suppose that $L + L^\perp \neq \Phi$ is a complete proper subspace of H. We want to show that $(L + L^\perp)^\perp$ is $\neq \Phi$. Now $L + L^\perp$ is a convex set. Let $\theta \neq x \in H - [L^\perp]$. Then by theorem 3.5.4 there exists a $y_0 \in L + L^\perp$ s.t.

$$\|x - y_0\| = \inf_{y \in L + L^\perp} \|x - y\| = d > 0$$

and $x - y_0 \in L + L^\perp$. Let $z_0 = x - y_0$, $z_0 \neq \theta$, $z_0 \in H$ and $z_0 \in L + L^\perp$.
This gives,

$$\langle z_0, y + z \rangle = 0 \quad \forall \, y \in M \text{ and } z \in M^\perp.$$

$$\Rightarrow \quad \langle z_0, y \rangle + \langle z_0, z \rangle = 0 \quad \forall \, y \in M \text{ and } z \in M^\perp$$

In particular, taking $z = \theta$ and $y = \theta$ respectively,

$$\begin{cases} \langle z_0, y \rangle = 0 & \forall \, y \in M \\ \langle z_0, z \rangle = 0 & \forall \, z \in M^\perp \end{cases}$$

Consequently, $z_0 \in M \cap M^\perp$.

But $M \cap M^\perp = \{\theta\}$, because they are each closed subspaces.

Therefore its follows that $z_0 = \theta$, which contradicts the fact that $L + L^\perp \neq \Phi$ is a proper subspace of H.

Hence $H = L + L^\perp$. Since $L \cap L^\perp = \{\theta\}$, $H = L \oplus L^\perp$.

3.6.2 Projection operator

Let H be a Hilbert space. L is a closed subspace of H. Then by *orthogonal projection theorem* for every $x \in H$, $y \in L$ and $z \in L^\perp$, x can be uniquely represented by $x = y + z$. y is called the *projection* of x on L. We write $y = Px$, P is called the *projection mapping* of H onto L, i.e., $P : H \xrightarrow{\text{onto}} L$.

Since $\quad z = x - y = x - Px = (I - P)x.$

Thus $\quad PH = L \quad (I - P)H = L^\perp$

Now, $\quad Py = y,\ y \in L \quad$ and $\quad Pz = 0,\ z \in L^\perp.$

Thus the *range* of P and its *null space* are mutually orthogonal. Hence the projection mapping P is called an *orthogonal projection*.

3.6.3 Theorem

A subspace L of a Hilbert space H is closed if and only if $L = L^{\perp\perp}$.

If $L = L^{\perp\perp}$, then L is a closed subspace of H, because $L^{\perp\perp}$ is already a closed subspace of H [see note 3.6.1].

Conversely let us suppose that L is a closed subspace of H. For any subset L of H, we have $L \subseteq L^{\perp\perp}$ [see note 3.6.1]. So it remains to prove that $L^{\perp\perp} \subseteq L$.

Let $x \in L^{\perp\perp}$. By projection theorem, $x = y + z$, $y \in L$ and $z \in L^\perp$. Since $L \subseteq L^{\perp\perp}$, it follows that $y \in L^{\perp\perp}$.

$L^{\perp\perp}$ being a subspace of H, $z = x - y \in L^{\perp\perp}$.

Hence $z \in L^\perp \cap L^{\perp\perp}$. As such $z \perp z$, i.e., $z = \theta$. Hence, $z = x - y = \theta$. Thus $x \in L$. Hence $L^{\perp\perp} \subseteq L$. This proves the theorem.

3.6.4   Theorem

Let L be a non-empty subset of a Hilbert space H. Then, span L is dense in H if and only if $L^{\perp} = \{\theta\}$.

Proof: We assume that span L is dense in H. Let $M = $ span L so that $\overline{M} = H$. Now, $\{\theta\} \subset L^{\perp}$. Let $x \in L^{\perp}$ and since $x \in H = \overline{M}$, there exists a sequence $\{x_n\} \subseteq M$ such that

$$\lim_{n \to \infty} x_n = x.$$

Now, since $x \in L^{\perp}$ and $L^{\perp} \perp M$, we have, $\langle x, x_n \rangle = 0$, $\forall$ n.

Since the inner product is a continuous function, proceeding to the limits in the above, we have $\langle x, x \rangle = 0 \Rightarrow x = \theta$.

Thus, $\{L^{\perp}\} \subseteq \{\theta\}$. Hence $L^{\perp} = \theta$.

Conversely let us suppose that $L^{\perp} = \{\theta\}$. Let $x \in M^{\perp}$. Then $x \perp M$ and in particular $x \perp L$. This verifies that $x \in L^{\perp} = \{\theta\}$.

Hence $M^{\perp} = \{\theta\}$. But $\overline{M}$ is a closed subspace of H. Hence by the projection theorem $H = \overline{M}$.

Problems [3.5 and 3.6]

1. Find the projections of the position vector of a point in a two-dimensional plane, along the initial line and the line through the origin, perpendicular to the initial line.

2. Let O_x, O_y, O_z be mutually perpendicular axes through a fixed point O as origin. Let the spherical polar coordinates of P be (r, θ, ϕ). Find the projections of the position vector of P WRT O as the fixed point, on the axes O_x, O_y, O_z respectively.

3. Let $x_1, x_2, \ldots x_n$ satisfy $x_i \neq \theta$ and $x_i \perp x_j$ if $i \neq j$, $i, j = 1, 2, \ldots n$. Show that the x_i's are linearly independent and extend the Pythagorean theorem from 2 to n dimensions.

4. Show that if M and N are closed subspaces of a Hilbert space H, then $M + N$ is closed provided $x \perp y$ for all $x \in M$ and $y \in N$.

5. Let H be a Hilbert space, $M \subseteq H$ a convex subset, and $\{x_n\}$ a sequence in M such that $\|x_n\| \to d$ as $n \to \infty$ where $d = \inf_{x \in M} \|x\|$.
   Show that $\{x_n\}$ converges in H.

6. If $M \subseteq H$, a Hilbert space, show that $M^{\perp\perp}$ is the closure of the span of M.

7. If M_1 and M_2 are closed subspaces of a Hilbert space H, then show that $(M_1 \cap M_2)^{\perp}$ equals the closure of $M_1^{\perp} \oplus M_2^{\perp}$.

8. In the usual Hilbert space $\mathbb{R}^2$ find $L^{\perp}$ if

   (i) $L = \{x\}$ where x has two non-zero components x_1 and x_2.
   (ii) $L = \{x, y\} \subset \mathbb{R}^2$ is a linearly independent set.

Hints: (i) If $y \in L^{\perp}$ with components y_1, y_2, then $x_1 y_1 = x_2 y_2 = 0$.

(ii) If L is a linearly independent set then $x \neq ky$, $k \neq 0$ and a scalar.

9. Show that equality holds in the Cauchy-Bunyakovsky-Schwartz inequality (i.e. $|\langle x, y \rangle| = \|x\|\|y\|$) if and only if x and y are linearly dependent vectors.

3.7 Orthogonal System

Orthogonal system plays a great role in expressing a vector in a Hilbert space in terms of mutually orthogonal vectors. The consequences of the *orthogonal projection theorem* are germane to this approach. In the two-dimensional plane any vector can be expressed as the sum of the projections of the vector in two mutually perpendicular directions. If e_1 and e_2 are two unit vectors along the perpendicular axes Ox and Oy and P has co-ordinates (x_1, x_2) [fig. 3.7], then the position vector r of P is given by

$$r = x_1 e_1 + x_2 e_2.$$

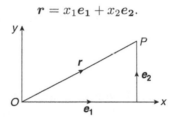

Fig. 3.7 Expressing a vector in terms of two perpendicular vectors

The above expression of the position vector can be extended to n dimensions.

3.7.1 Definition: orthogonal sets and sequences

A set L of a Hilbert space H is said to be an *orthogonal set* if its elements are *pairwise* orthogonal. An *orthonormal set* is an orthogonal subset $L \subset H$ whose elements e_i, e_j satisfy the following condition $(e_i, e_j) = \delta_{ij}$ where δ_{ij} is the Kronecker symbol in $\delta_{ij} = \begin{cases} 0, & i \neq j \\ 1, & i = j. \end{cases}$

If the orthogonal or orthonormal set is countable, then we can call the said set an *orthogonal or orthonormal sequence* respectively.

3.7.2 Example

$\{e^{2\pi i n x}\}$, $n = 0, 1, 2, \ldots$ is an example of an orthonormal sequence in the complex space $L_2([0, 1])$.

3.7.3 Theorem (Pythagorean)

If $a_1, a_2, \ldots a_m$ are mutually orthogonal elements of the Hilbert space H, then

$$\|a_1 + a_2 + \cdots + a_k\|^2 = \|a_1\|^2 + \cdots + \|a_k\|^2$$

Proof: Since $a_1, a_2, \ldots a_k$ are   mutually   orthogonal   elements,

$$\left.\langle a_i, a_j \rangle \right\} \begin{array}{ll} = \|a_i\|^2 & \text{if } \ i = j \\ = 0 & \text{if } \ i \neq j. \end{array}$$

$$\|a_1 + a_2 + \cdots + a_k\|^2 = \langle a_1 + a_2 + \cdots + a_k, \ a_1 + a_2 + \cdots + a_k \rangle$$

$$= \sum_{i=1}^{k} \sum_{j=1}^{k} \langle a_i, a_j \rangle = \sum_{i=1}^{k} \langle a_i, a_i \rangle = \sum_{i=1}^{k} \|a_i\|^2.$$

since $\langle a_i, a_j \rangle = 0$ for $i \neq j$.

3.7.4   Lemma (linear independence)

An orthonormal set is linearly independent.

Proof: Let $\{e_1, e_2, \ldots e_n\}$ be an orthonormal set. If the set is not linearly independent, we can find scalars $\alpha_1, \alpha_2, \ldots \alpha_n$ not all zeroes such that

$$\alpha_1 e_1 + \alpha_2 e_2 + \cdots \alpha_n e_n = \theta.$$

Taking scalar products of both sides with α_j, we get $\alpha_j (e_i, e_j) = 0$ since $(e_i, e_j) = 0, \ i \neq j$.

Therefore, $\alpha_j = 0$, for $j = 1, 2, \ldots n$. Hence $\{e_i\}$ is linearly independent.

3.7.5   Examples

1. $\mathbb{R}^n$ (n dimensional Euclidean space): In $\mathbb{R}^n$ the orthonormal set is given by $e_1 = (1, 0 \cdots 0)$, $e_2 = (0, 1, \ldots 0), \ldots e_n = (0, 0, \ldots 1)$ respectively.

2. l_2 space: In l_2 space the orthonormal set $\{e_n\}$ is given by, $e_1 = (1, 0, \ldots 0 \cdots)$, $e_2 = (0, 1, 0, \ldots 0 \cdots)$, $e_n = (0, 0, 0, \ldots 0, 1 \ldots)$ respectively.

3. $C([0, 2\pi])$: The inner product space of all continuous real-valued functions defined on $[0, 2\pi]$ with the inner product given by: $\langle u, v \rangle = \int_0^{2\pi} u(t) v(t) dt$.

We want to show that $\{u_n(t)\}$ where $u_n(t) = \sin nt$ is an orthogonal sequence.

$$\langle u_n(t), u_m(t) \rangle = \int_0^{2\pi} u_n(t) u_m(t) dt = \int_0^{2\pi} \sin nt \sin mt \, dt$$

$$= \begin{cases} 0 & n \neq m \\ \pi & n = m = 1, 2, \ldots \end{cases}$$

Hence $\|u_n(t)\| = \sqrt{\pi}$.

Hence, $\left\{ \dfrac{1}{\sqrt{\pi}} \sin nt \right\}$ is an orthonormal sequence.

On the other hand if we take $v_n = \cos nt$

$$\langle v_n, v_m \rangle = \int_0^{2\pi} \cos nt \cos mt \, dt$$

$$= \begin{cases} 0, & m \neq n \\ \pi, & m = n = 1, 2 \cdots \\ 2\pi, & m = n = 0. \end{cases}$$

Hence $\|v_n\| = \sqrt{\pi}$ for $n \neq 0$.

Therefore, $\dfrac{1}{\sqrt{2\pi}}, \dfrac{\cos t}{\sqrt{\pi}}, \dfrac{\cos nt}{\sqrt{\pi}}$ is an orthonormal sequence.

3.7.6 *Gram-Schmidt orthonormalization process*

Theoretically, every element in a Hilbert space H can be expressed in terms of any linearly independent set. But the orthonormal set in H has an edge over other linearly independent sets in this regard. If $\{e_i\}$ is a linearly independent set and $x \in H$, a Hilbert space, then x can be expressed in terms of e_i as follows:

$$x = \sum_{i=1}^{n} c_i e_i,$$

In case $e_i, e_j, i, j = 1, 2, \ldots n$ are mutually orthogonal, then $c_i = \langle x, e_i \rangle$, since $\langle e_i, e_j \rangle = 0$ for $i \neq j$.

Thus, in the case of an orthonormal system it is very easy to find the coefficients $c_i, i = 1, 2, \ldots n$.

Another advantage of the orthonormal set is as follows. Suppose we want to add $c_{n+1} e_{n+1}$ to x so that $\overline{x} = x + c_{n+1} e_{n+1} \in$ span $\{e_1, e_2, \ldots e_{n+1}\}$.

Now, $\langle \overline{x}, e_j \rangle = \langle x, e_j \rangle + c_{n+1} \langle e_{n+1}, e_j \rangle = \langle x, e_j \rangle = c_j, \; j = 1, 2, \ldots n.$

$\langle \overline{x}, e_{n+1} \rangle = \langle x, e_{n+1} \rangle + c_{n+1} \langle e_{n+1}, e_{n+1} \rangle = c_{n+1},$

since $x \in$ span $\{e_1, e_2, \ldots e_n\}$. Thus determination of c_{n+1} does not depend on the values $c_1, c_2, \ldots c_n$.

In what follows we explain the **Gram-Schmidt ortho-normalization process**.

Let $\{a_n\}$ be a (finite or countably infinite) set of vectors in a Hilbert space H. The problem is to convert the set into an orthonormal set $\{e_n\}$ such that span $\{e_i, e_2, \ldots e_n\} =$ span $\{a_1, a_2, \ldots a_n\}$, for each n. A few steps are written down.

Step 1. Normalize a_1 which is necessarily non-zero so that $e_1 = \dfrac{a_1}{\|a_1\|}$.

Step 2. Let $g_2 = a_2 - \langle a_2, e_1 \rangle$ so that $\langle g_2, e_1 \rangle = 0$.

Here take, $e_2 = \dfrac{g_2}{\|g_2\|}$, so that $e_2 \perp e_1$.

Here, $\|g_2\| \neq 0$, because otherwise $g_2 = 0$ and hence a_2 and a_1 are linearly dependent, which is a contradiction. Since g_2 is a linear combination of a_2 and a_1, we have span $\{a_2, a_1\}$ = span $\{e_1, e_2\}$.

Let us assume by way of induction that

$$g_m = a_m - \langle a_m, e_1 \rangle e_1 - \langle a_m, e_2 \rangle e_2 \cdots \langle a_m, e_{m-1} \rangle e_{m-1}$$

and $g_m \neq 0$

then $\langle g_m, e_1 \rangle = 0, \ldots \langle g_m, e_{m-1} \rangle = 0$ i.e. $g_m \perp e_1, e_2, \ldots e_{m-1}$.

We take $e_m = \dfrac{g_m}{\|g_m\|}$.

We assume span $\{a_1, a_2, \ldots a_{m-1}, a_m\}$ = span $\{e_1, e_2, \ldots e_{m-1}, e_m\}$

Next we take

$$g_{m+1} = a_{m+1} - \langle a_{m+1}, e_1 \rangle e_1 - \langle a_{m+1}, e_2 \rangle e_2 - \langle e_{m+1}, e_m \rangle e_m.$$

Now, $\langle g_{m+1}, e_1 \rangle = \langle g_{m+1}, e_2 \rangle \cdots = \langle g_{m+1}, e_m \rangle = 0.$

Thus $g_{m+1} \perp e_1, e_2, \ldots e_m.$

$g_{m+1} = \theta \Rightarrow a_{m+1}$ is a linear combination of $e_1, e_2, \ldots e_m$

$\Rightarrow a_{m+1}$ is a linear combination of $a_1, a_2, \ldots a_m$ which contradicts the hypothesis that $\{a_1, a_2, \ldots a_{m+1}\}$ are linearly independent.

Hence, $g_{m+1} \neq \theta$, i.e., $\|g_{m+1}\| \neq 0$.

We write, $e_{m+1} = \dfrac{g_{m+1}}{\|g_{m+1}\|}$.

Hence, $e_1, e_2, \ldots e_{m+1}$ form an orthonormal system.

3.7.7   Examples

1.   (Orthonormal polynomials) Let $L_{2,\rho}([a, b])$ be the space of square-summable functions with weight functions $\rho(t)$. Let us take a linearly independent set $1, t, t^2 \cdots t^n \cdots$ in $L_{2,\rho}([a, b])$. If we orthonormalize the above linearly independent set, we get Chebyshev system of polynomials, $p_0 = \text{const.}, p_1(t), p_2(t), \ldots, p_n(t), \ldots$ which are orthonormal with weight $\rho(t)$, i.e.,

$$\int_a^b \rho(t) p_i(t) p_j(t) dt = \delta_{ij}.$$

We mention below a few types of orthogonal polynomials.

	$\rho(t)$	Polynomial
(i) $a = -1$, $b = 1$	1	**Legendre polynomials**
(ii) $a = -\infty$, $b = \infty$	e^{-t^2}	**Hermite polynomials**
(iii) $a = 0$, $b = \infty$	e^{-t}	**Laguerre polynomials**

(i) Legendre polynomials

Let us take $g_1 = e_1 = 1$, $\langle g_1, g_1 \rangle = \int_{-1}^{1} dt = 2$, i.e., $\|g_1\| = \sqrt{2}$

$$e_1 = \frac{g_1}{\|g_1\|} = \frac{1}{\sqrt{2}} = \sqrt{\frac{2n+1}{2}} P_n(t)\Big|_{n=0}$$

where

$$P_n(t) = \frac{1}{2^n n!} \frac{d^n}{dt^n} (t^2 - 1)^n. \tag{3.31}$$

Take,

$$g_2 = a_2 - \frac{\langle a_2, g_1 \rangle}{\|g_1\|} g_1 = t - \frac{1}{2} \int_{-1}^{1} t \, dt = t.$$

$$\|g_2\|^2 = \int_{-1}^{1} t^2 dt = \frac{2}{3}, \quad e_2 = \frac{g_2}{\|g_2\|} = \sqrt{\frac{3}{2}} t.$$

Thus

$$e_2 = \sqrt{\frac{2n+1}{2}} P_2(t)\Big|_{n=1}.$$

It may be noted that

$$P_n(t) = \frac{1}{2^n n!} \frac{d^n}{dt^n} [t^{2n} - {}^n C_1 t^{2n-1} + \cdots (-1)^n]$$

applying binomial theorem

$$= \sum_{j=0}^{N} (-1)^j \frac{(2n-2j)!}{2^n j! (n-j)! (n-2j)!} t^{n-2j} \tag{3.32}$$

where $N = \dfrac{n}{2}$ if n is even and $N = \dfrac{(n-1)}{2}$ if n is odd.

Next, let us take $g_3 = a_3 - \langle a_3, e_1 \rangle e_1 - \langle a_3, e_2 \rangle e_2$

$$= t^2 - \frac{1}{2} \int_{-1}^{1} t^2 dt - \frac{3}{2} t \int_{-1}^{1} t^3 dt = t^2 - \frac{1}{3}.$$

$$\|g_3\|^2 = \int_{-1}^{1} \left(t^2 - \frac{1}{3} \right)^2 dt = \frac{8}{45}.$$

$$e_3 = \sqrt{\frac{5}{2}} \cdot \frac{1}{2} (3t^2 - 1) = \sqrt{\frac{2n+1}{2}} P_n(t)\Big|_{n=2}$$

We next want to show that

$$\|P_n\| = \left[\int_{-1}^{1} P_n^2(t) dt \right]^{\frac{1}{2}} = \sqrt{\frac{2}{2n+1}} \tag{3.33}$$

Let us write $v = t^2 - 1$. The function v^n and its derivatives $(v^n)^1, \ldots (v^n)^{(n-1)}$ are zero at $t = \pm 1$ and $(v^n)^{(2n)} = (2n)!$. Integrating n times by parts, we thus obtain from (3.31)

$$(2^n n!)^2 \|P_n\|^2 = \int_{-1}^{1} (v^n)^{(n)} (v^n)^{(n)} dt$$

$$= (v^n)^{(n-1)}(v^n)^{(n)}\Big|_{-1}^{1} - \int_{-1}^{1}(v^n)^{(n-1)}(v^n)^{(n+1)}dt.$$

$$= (-1)^n(2n)!\int_{-1}^{1} v^n dt = 2(2n)!\int_{-1}^{1}(1-t^2)^n dt$$

$$= 2(2n)!\int_{0}^{\frac{\pi}{2}} \cos^{2n+1}\alpha\, d\alpha \quad (t = \sin\alpha)$$

$$= \frac{2^{2n+1}(n!)^2}{2n+1}.$$

Thus $\left\|\sqrt{\dfrac{2n+1}{2}}P_n\right\| = 1.$

Next, $\langle P_m, P_n\rangle = \dfrac{1}{2^{m+n}m!n!}\int_{-1}^{1}((t^2-1)^m)^{(m)}((t^2-1)^n)^{(n)}dt$

$$= \frac{1}{2^{m+n}m!n!}\int_{-1}^{1}(v^m)^{(m)}(v^n)^{(n)}dt$$

where $m > n$ (suppose) and $v = t^2 - 1$.

$$= \frac{1}{2^{m+n}m!n!}\left[(v^m)^{(m)}(v^n)^{(n-1)}\Big|_{-1}^{+1}\right.$$

$$\left. -2m\int_{-1}^{1}(v^m)^{(m-1)}(v^n)^{(n+1)}dt\right]$$

$$= \frac{(-1)^n 2m\cdots 2(m-n)}{2^{m+n}m!n!}\int_{-1}^{1}(v^m)^{(m-n)}dt = 0 \quad m > n.$$

A similar conclusion is drawn if $n > m$.

Thus, $\{P_n(t)\}$ as given in (3.31) is an orthonormal sequence of polynomials.

The *Legendre polynomials* are solutions of the important *Legendre differential equation*

$$(1-t^2)P_n'' - 2tP_n' + n(n+1)P_n = 0 \tag{3.34}$$

and (3.32) can also be obtained by applying the *power series method* to (3.34).

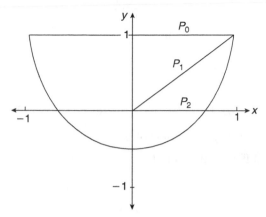

Fig. 3.8 Legendre polynomials

Remark 1. Legendre polynomials find frequent use in applied mathematics, especially in *quantum mechanics*, numerical analysis, theory of approximations etc.

(ii) Hermite polynomials

Since $a = -\infty$ and $b = \infty$, we consider $L_2([-\infty, \infty])$ which is also a Hilbert space. Since the interval of integration is infinite, we need to introduce a weight function which will make the integral convergent. We take the weight function $w(t) = e^{-\frac{t^2}{2}}$ so that the Gram-Schmidt orthonormalization process is to be applied to $w, wt, wt^2 \ldots$ etc.

$$\|a_0\|^2 = \int_{-\infty}^{\infty} w^2(t)dt = \int_{-\infty}^{\infty} e^{-t^2}dt = 2\int_0^{\infty} e^{-t^2}dt = \sqrt{\pi},$$

$$e_0 = \frac{w}{(\sqrt{\pi})^{\frac{1}{2}}} = \frac{e^{-\frac{t^2}{2}}}{(\sqrt{\pi})^{\frac{1}{2}}}$$

Take $\quad g_1 = a_1 - \dfrac{\langle a_1, g_0 \rangle}{\|g_0\|^2} g_0 = wt - \dfrac{[\int_{-\infty}^{\infty} w^2 t dt]}{\sqrt{\pi}}$

$$= wt - \frac{\int_{-\infty}^{\infty} te^{-\frac{3t^2}{2}}dt}{\sqrt{\pi}} = wt.$$

$$\|g_1\|^2 = \int_{-\infty}^{\infty} w^2 t^2 dt = \int_0^{\infty} t^2 e^{-t^2}dt$$

$$= \frac{1}{2} 2 \int_0^{\infty} z^{\frac{3}{2}-1} e^{-z}dz \quad \text{putting } z = t^2.$$

$$= \Gamma\left(\frac{3}{2}\right) = \frac{1}{2}\sqrt{\pi} \quad \text{since } \Gamma(n+1) = n\Gamma(n).$$

so that $\quad e_1(t) = \dfrac{g_1}{\|g_1\|} = \dfrac{\sqrt{2}te^{-\frac{t^2}{2}}}{(\sqrt{\pi})^{\frac{1}{2}}} = \sqrt{\dfrac{1}{2 \cdot 1\sqrt{\pi}}}e^{-\frac{t^2}{2}} \cdot 2t$

We want to show that $e_n(t) = \sqrt{\dfrac{1}{2^n n! \sqrt{\pi}}} e^{-\frac{t^2}{2}} H_n(t), \ n \geq 1$ $\qquad$ (3.35)

and $\qquad e_0(t) = \dfrac{e^{-\frac{t^2}{2}}}{(\sqrt{\pi})^{\frac{1}{2}}} H_0(t)$

where $\qquad H_0(t) = 1, \ H_n(t) = (-1)^n e^{t^2} \dfrac{d^n}{dt^n}(e^{-t^2}), \ n = 1, 2, 3, \dots$ $\qquad$ (3.36)

H_n are called *Hermite polynomials* of order n. $H_0(t) = 1$.

Performing the differentiations indicated in (3.36) we obtain

$$H_n(t) = n! \sum_{j=0}^{N} (-1)^j \frac{2^{n-2j}}{j!(n-2j)!} t^{n-2j} \qquad (3.37)$$

where $N = \dfrac{n}{2}$ if n is even and $N = \dfrac{(n-1)}{2}$ if n is odd. The above form can also be written as

$$H_n(t) = \sum_{j=0}^{N} \frac{(-1)^j}{j!} n(n-1) \cdots (n-2j+1)(2t)^{n-2j} \qquad (3.38)$$

Thus $\qquad e_0(t) = \dfrac{e^{-\frac{t^2}{2}}}{(\sqrt{\pi})^{\frac{1}{2}}} H_0(t)$

$$e_1(t) = \frac{\sqrt{2} t e^{-\frac{t^2}{2}}}{(\sqrt{\pi})^{\frac{1}{2}}} = \frac{e^{-\frac{t^2}{2}}(2t)}{(2\sqrt{\pi})^{\frac{1}{2}}} = \sqrt{\frac{1}{2^n n! \sqrt{\pi}}} e^{-\frac{t^2}{2}} H_n(t)\big|_{n=1}.$$

(3.36) yields explicit expressions for $H_n(t)$ as given below for a few values of n.

$\qquad H_0(t) = 1 \qquad\qquad\qquad\qquad H_1(t) = 2t$

$\qquad H_2(t) = 4t^2 - 2 \qquad\qquad\qquad H_3(t) = 8t^3 - 12t$

$\qquad H_4(t) = 16t^4 - 48t^2 + 12 \quad H_5(t) = 32t^5 - 160t^3 + 120t.$

We next want to show that $\{e_n(t)\}$ *is orthonormal* where $e_n(t)$ *is given by (3.35) and* $H_n(t)$ *by (3.37)*,

$$\langle e_n(t), e_m(t) \rangle = \sqrt{\frac{1}{2^{n+m} n! m! (\sqrt{\pi})^2}} \int_{-\infty}^{\infty} e^{-t^2} H_m H_n \, dt.$$

Differentiating (3.38) we obtain for $n \geq 1$,

$$H_n'(t) = 2n \sum_{j=0}^{M} \frac{(-1)^j}{j!} (n-1)(n-2) \cdots (n-2j)(2t)^{n-1-2j}$$

$$= 2n H_{n-1}(t)$$

where $N = \dfrac{(n-2)}{2}$ if n is even and $N = \dfrac{(n-1)}{2}$ if N is odd.

Let us assume $m \leq n$ and $u = e^{-t^2}$. Integrating m times by parts we obtain from the following integral,

$$(-1)^n \int_{-\infty}^{\infty} e^{-t^2} H_m(t) H_n(t) dt$$

$$= \int_{-\infty}^{\infty} H_m(t)(u)^{(n)} dt$$

$$= H_m(t)(u)^{(n-1)} \Big|_{-\infty}^{\infty} - 2m \int_{-\infty}^{\infty} H_{m-1}(u)^{n-1} dt$$

$$= -2m \int_{-\infty}^{\infty} H_{m-1}(u)^{(n-1)} dt$$

$$= \cdots \cdots$$

$$= (-1)^m 2^m m! \int_{-\infty}^{\infty} H_0(t)(u)^{(n-m)} dt$$

$$= (-1)^m 2^m m! (u)^{(n-m-1)} \Big|_{-\infty}^{\infty}$$

$$= 0 \quad \text{if} \quad n > m \text{ because as } t \to \pm\infty, t^2 \to \infty \text{ and } u \to 0.$$

This proves orthogonality of $\{e_m(t)\}$, when $m = n$

$$(-1)^n \int_{-\infty}^{\infty} e^{-t^2} H_n^2(t) dt = (-1)^n 2^n n! \int_{-\infty}^{\infty} H_0(t) e^{-t^2} dt$$

$$= (-1)^n 2^n n! \sqrt{\pi}.$$

Hence $\int_{-\infty}^{\infty} e^{-t^2} H_n^2(t) dt = 2^n n! \sqrt{\pi}.$

This proves *orthonormality* of $\{e_n\}$.

Hermite polynomials H_n satisfy the *Hermite differential equations* $H_n'' - 2t H_n' + 2n H_n = 0$.

Like Legendre polynomials, Hermite polynomials also find applications in *numerical analysis, approximation theory, quantum mechanics* etc.

(iii) Laguerre polynomials

We consider $L_2([0, \infty))$ and apply Gram-Schmidt process to the sequence defined by

$$e^{-\frac{t}{2}}, \quad te^{-\frac{t}{2}}, \quad t^2 e^{-\frac{t}{2}}.$$

Take $g_0 = e^{-\frac{t}{2}}, \|g_0\|^2 = \int_0^{\infty} e^{-t} dt = 1.$

$$e_0 = \frac{g_0}{\|g_0\|} = e^{-\frac{t}{2}}$$

$$g_1 = te^{-\frac{t}{2}} - \langle te^{-\frac{t}{2}}, e^{-\frac{t}{2}} \rangle e^{-\frac{t}{2}}$$

$$\langle te^{-\frac{t}{2}}, e^{-\frac{t}{2}}\rangle = \int_0^\infty te^{-t}dt = \frac{te^{-t}}{-1}\Big|_0^\infty + \int_0^\infty e^{-t}dt = 1.$$

$$g_1 = (t-1)e^{-\frac{t}{2}}$$

$$\|g_1\|^2 = \langle (t-1)e^{-\frac{t}{2}}, (t-1)e^{-\frac{t}{2}}\rangle = \int_0^\infty (t-1)^2 e^{-t}dt = 1.$$

$$e_1(t) = (t-1)e^{-\frac{t}{2}}.$$

Let us take $e_n(t) = e^{-\frac{t}{2}} L_n(t), n = 0, 1, 2$ \hfill (3.39)

where the Laguerre polynomials of order n is defined by

$$L_0(t) = 1, \quad L_n(t) = \frac{e^t}{n!}\frac{d^n}{dt^n}(t^n e^{-t}), \quad n = 1, 2 \tag{3.40}$$

i.e. $\quad L_n(t) = -\sum_{j=0}^n \frac{(-1)^j}{j!}\binom{n}{j} t^j$ \hfill (3.41)

Explicit expressions for the first few Laguerre polynomials are

$$L_0(t) = 1, \qquad\qquad L_1(t) = 1 - t$$

$$L_2(t) = 1 - 2t + \frac{1}{2}t^2 \qquad L_3(t) = 1 - 3t + \frac{3}{2}t^2 - \frac{1}{6}t^3.$$

$$L_4(t) = 1 - 4t + 3t^2 - \frac{2}{3}t^3 + \frac{1}{24}t^4$$

The *Laguerre polynomials* L_n are solutions of the *Laguerre differential equations*

$$tL_n'' + (1-t)L_n' + nL_n = 0. \tag{3.42}$$

In what follows we find out $e_2(t)$ by Gram-Schmidt process.

We know $g_2(t) = a_2 - \langle a_2, e_1\rangle e_1 - \langle a_2, e_0\rangle e_0.$

$$a_2(t) = t^2 e^{-\frac{t}{2}}; \quad \langle a_2, e_1\rangle = \int_0^\infty t^2 \cdot (t-1) \cdot e^{-t}dt$$

$$= \int_0^\infty (t^3 - t^2)e^{-t}dt$$

$$= 2\int_0^\infty t^2 e^{-t}dt = 4.$$

$$\langle a_2, e_0\rangle = \int_0^\infty t^2 e^{-t}dt = 2.$$

$$g_2(t) = t^2 e^{-\frac{t}{2}} - 4(t-1)e^{-\frac{t}{2}} - 2e^{-\frac{t}{2}} = (t^2 - 4t + 2)e^{-\frac{t}{2}}.$$

$$\|g_2(t)\|^2 = \int_0^\infty (t^2 - 4t + 2)^2 e^{-t}dt$$

$$= \int_0^\infty (t^4 - 8t^3 + 16t^2 + 4 + 4t^2 - 16t)e^{-t}dt$$

$$= \int_0^\infty (t^4 - 8t^3 + 20t^2 - 16t + 4)e^{-t}dt$$

$$= 4.$$

$$e_2(t) = \left(1 - 2t + \tfrac{1}{2}t^2\right)e^{-t}.$$

The orthogonality $\{L_m(t)\}$ can be proved as before.

3.7.8 *Fourier series*

Let L be a linear subspace **of a Hilbert space** spanned by $e_1, e_2, \ldots e_n \ldots$ and let $x \in L$. Therefore, there is a linear combination $\sum\limits_{i=1}^{n} \alpha_i e_i$ for every $\epsilon > 0$ such that $\left\| x - \sum\limits_{i=1}^{n} \alpha_i e_i \right\| < \epsilon$.

Hence $\left\| x - \sum\limits_{i=1}^{n} \alpha_i e_i \right\|^2 = \left\langle x - \sum\limits_{i=1}^{n} \alpha_i e_i, x - \sum\limits_{i=1}^{n} \alpha_i e_i \right\rangle$

$$= \langle x, x \rangle - \left\langle x, \sum_{i=1}^{n} \alpha_i e_i \right\rangle - \left\langle \sum_{i=1}^{n} \alpha_i e_i, x \right\rangle + \left\langle \sum_{i=1}^{n} \alpha_i e_i, \sum_{i=1}^{n} \alpha_i e_i \right\rangle$$

$$= \|x\|^2 - \sum_{i=1}^{n} \overline{\alpha_i} \langle x, e_i \rangle - \sum_{i=1}^{n} \alpha_i \langle e_i, x \rangle + \sum_{i=1}^{n} \sum_{j=1}^{n} \alpha_i \overline{\alpha_j} \langle e_i, e_j \rangle$$

$$= \|x\|^2 - \sum_{i=1}^{n} \overline{\alpha_i} d_i - \sum_{i=1}^{n} \alpha_i \overline{d_i} + \sum_{i=1}^{n} |\alpha_i|^2 \quad \text{where } d_i = \langle x, e_i \rangle.$$

Therefore, $\left\| x - \sum \alpha_i e_i \right\|^2 = \|x\|^2 - \sum\limits_{i=1}^{n} |d_i|^2 + \sum\limits_{i=1}^{n} |\alpha_i - d_i|^2.$

The numbers d_i are called *Fourier coefficients of the element x* with respect to the orthonormal system $\{e_i\}$.

The expression on the RHS for different values of α_i takes its least value when $\alpha_i = d_i$, the ith *Fourier coefficient of x.*

Hence, $\quad 0 \le \left\| x - \sum\limits_{i=1}^{n} d_i e_i \right\|^2 = \|x\|^2 - \sum\limits_{i=1}^{n} |d_i|^2 < \epsilon. \qquad (3.43)$

ϵ being arbitrary small, it follows that $x = \lim\limits_{n \to \infty} \sum\limits_{i=1}^{n} d_i e_i = \sum\limits_{i=1}^{\infty} d_i e_i.$

The convergence of the series $\sum\limits_{i=1}^{\infty} |d_i|^2$ also follows from (3.43) and moreover

$$\sum_{i=1}^{\infty} |d_i|^2 = \|x\|^2. \qquad (3.44)$$

Next let x be an arbitrary element in the Hilbert space H. Let y be the projection of x on L.

Then $y = \sum_{i=1}^{\infty} d_i e_i$ where $d_i = \langle y, e_i \rangle = \langle x, e_i \rangle$ and $\sum_{i=1}^{\infty} |d_i|^2 = \|y\|^2$.

Since $x = y + z$, $y \in L$, $z \perp L$, it follows from the **Pythagorean theorem**,

$$\|x\|^2 = \|y\|^2 + \|z\|^2 \geq \|y\|^2. \tag{3.45}$$

Consequently for any element x in H, the inequality,

$$\sum_{i=1}^{\infty} |d_i|^2 \leq \|x\|^2 \tag{3.46}$$

holds for any $x \in H$ where $d_i = \langle x, e_i \rangle$, $i = 1, 2, 3 \ldots$. The above inequality is called **Bessel inequality**.

3.8   Complete Orthonormal System

It is known that in two-dimensional Euclidean space, $\mathbb{R}^2$, any vector can be uniquely expressed in terms of two mutually orthogonal vectors of unit norm or, in other words, any non-zero vector x can be expressed uniquely in terms of bases e_1, e_2. Similarly, in $\mathbb{R}^3$ we have three terms, e_1, e_2, e_3, each having unit norm, and the bases are pairwise orthonormal. This concept can be easily extended to n-dimensional Euclidean space where we have $n$pair-wise orthonormal bases $e_1, e_2, \ldots e_n$. But the question that invariably comes up is whether this concept can be extended to infinite dimensions. Or, in other words, the question arises as to whether an infinite dimensional space, like an inner product space, can contain a 'sufficiently large' set of orthonormal vectors such that any element $x \in H$ (a Hilbert space) can be uniquely represented by said set of orthonormal vectors.

Let $H \neq \Phi$ be a Hilbert space. Then the collection $\mathcal{C}$ of all orthonormal subsets of H is clearly non-empty. It can be easily seen that the class $\mathcal{C}$ can be partially ordered under 'set inclusion' relation.

3.8.1   Definition: complete orthonormal system

V.A. Steklov first introduced the concept of a **complete orthonormal system**. An orthonormal system $\{e_i\}$ in H is said to be a *complete orthonormal system* if there is no non-zero $x \in H$ such that x is orthogonal to every element in $\{e_i\}$. In other words, a complete orthonormal system cannot be extended to a larger orthonormal system by adding new elements to $\{e_i\}$. Hence, a *complete orthonormal* system is *maximal* with respect to inclusion.

3.8.2   Examples

1. The unit vectors $e_i = (1, 0, 0)$, $e_2 = (0, 1, 0)$, $e_3 = (0, 0, 1)$ in the directions of three axes of rectangular coordinate system form a *complete orthonormal system*.

2. In the Hilbert space l_2, the sequence $\{e_n\}$ where $e_n = \{\delta_{nj}\}$, is a *complete orthonormal system*.

3. The orthonormal system in example 3.7.5(3), i.e., the sequence $u_n = \left\{ \dfrac{1}{\sqrt{2\pi}}, \dfrac{\cos t}{\sqrt{\pi}}, \dfrac{\cos 2t}{\sqrt{\pi}} \cdots \right\}$, although an orthonormal set is not complete for $\langle \sin t, u_n \rangle = 0$. But the system $\dfrac{1}{\sqrt{2\pi}}, \dfrac{1}{\sqrt{\pi}} \cos t, \dfrac{1}{\sqrt{\pi}} \sin t, \dfrac{1}{\sqrt{\pi}} \cos 2t, \ldots$ is a complete orthonormal system.

3.8.3   Theorem

Every Hilbert space $H \neq \{\theta\}$ contains a complete orthonormal set.

Consider the family $\mathcal{C}$ of orthonormal sets in H partially ordered by set inclusion. For any non-zero $x \in H$, the set $\left\{ \dfrac{x}{\|x\|} \right\}$ is an orthonormal set. Therefore, $\mathcal{C} \neq \Phi$. Now let us consider any totally ordered subfamily in $\mathcal{C}$. The union of sets in this subfamily is clearly an orthonormal set and is an upper bound for the totally ordered subfamily. Therefore, by Zorn's lemma (1.1.4), we conclude that $\mathcal{C}$ has a maximal element which is a *complete orthonormal set* in H.

The next theorem provides another characterization of a complete orthonormal system.

3.8.4   Theorem

Let $\{e_i\}$ be an orthonormal set in a Hilbert space H. Then $\{e_i\}$ is a complete orthonormal set if and only if it is impossible to adjoin an additional element $e \in H$, $e \neq \theta$ to $\{e_i\}$ such that $\{e_i, e\}$ is an orthonormal set in H.

Proof: Suppose $\{e_i\}$ is a complete orthonormal set. Let it be possible to adjoin an additional vector $e \in H$ of unit norm $e \neq \theta$, such that $\{e, e_i\}$ is an orthonormal set in H i.e. $e \perp \{e_i\}$, e is a non-zero vector of unit norm. But this contradicts the fact that $\{e_i\}$ is a *complete orthonormal set*. On the other hand, let us suppose that it is impossible to adjoin an additional element $e \in H$, $e \neq \theta$, to $\{e_i\}$ such that $\{e, e_i\}$ is an orthonormal set. Or, in other words, there exists no non-zero $e \in H$ of unit norm such that $e \perp \{e_i\}$. Hence the system $\{e_i\}$ is a *complete orthonormal system*.

In what follows we define a *closed orthonormal system* in a Hilbert space and show that it is the same as a *complete orthonormal system*.

3.8.5   Definition: closed orthonormal system

An orthonormal system $\{e_i\}$ in H is said to be **closed** if the subspace L spanned by the system coincides with H.

3.8.6   Theorem

A Fourier series with respect to a closed orthonormal system, constructed for any $x \in H$, converges to this element and for

every $x \in H$**, the Parseval-Steklov equality**

$$\sum_{i=1}^{\infty} d_i^2 = \|x\|^2 \tag{3.47}$$

holds.

Proof: Let $\{e_i\}$ be a closed orthonormal system. Then the subspace spanned by $\{e_i\}$ coincides with H.

Let $x \in H$ be any element. Then the *Fourier series* (c.f. 3.7.8) WRT the closed system is given by

$$x = \sum_{i=1}^{\infty} d_i e_i \quad \text{where} \quad d_i = \langle x, e_i \rangle,$$

Then by the relation (3.43), we have,

$$\|x\|^2 = \sum_{i=1}^{\infty} d_i^2 \quad \text{where} \quad d_i = \langle x, e_i \rangle \quad \text{and} \quad \|e_i\|^2 = 1.$$

3.8.7   Corollary

An orthonormal system is complete if and only if the system is closed.

Let $\{e_i\}$ be a complete orthonomal system in H. If $\{e_i\}$ is not closed, let the subspace spanned by $\{e_i\}$ be L where $L \neq H$. If any non-zero x is not orthogonal to L, then x is also not orthogonal to $\overline{L}$. Thus $\Phi \equiv H - \overline{L}$. This contradicts that $\{e_i\}$ is not closed.

Conversely, let us suppose that the orthonormal system $\{e_i\}$ is closed. Then, by theorem 3.8.6, we have for any $x \in H$,

$$\|x\|^2 = \sum_{i=1}^{\infty} d_i^2 \quad \text{where} \quad d_i = \langle x, e_i \rangle.$$

If $x \perp e_i$, $i = 1, 2, \ldots$, that is, $e_i = 0$, $i = 1, 2, \ldots$, then $\|x\| = 0$. This implies that the system $\{e_i\}$ is complete.

3.8.8   Definition: orthonormal basis

A closed orthonormal system in a Hilbert space H is also called an **orthonormal basis** in H.

Note 3.8.1 We note that completeness of $\{e_n\}$ is tantamount to the statement that each $x \in L_2$ has Fourier expansion,

$$x(t) = \frac{1}{\sqrt{2\pi}} \sum_{n=-\infty}^{\infty} x_n e^{\text{int}} \tag{3.48}$$

It must be emphasized that the expansion is not to be interpreted as saying that the series converges pointwise to the function $x(t)$.

One can conclude that the partial sums of 3.48 is the vector $u_n(t) = \frac{1}{\sqrt{2\pi}} \sum_{k=-\infty}^{n} x_k e^{ikt}$ converges to the vector x in the sense of L_2, i.e.,

$$\|u_n(t) - x(t)\| \longrightarrow 0 \quad \text{as } n \longrightarrow \infty.$$

This situation is often expressed by saying that x is the limit in the mean of u_n's.

3.8.9 Theorem

In a separable Hilbert space H a complete orthonormal set is enumerable.

Proof: Let H be a separable Hilbert space, then $\exists$ an enumerable set M, everywhere dense in H, i.e., $\overline{M} = H$. Let $\{e_\lambda |_{\lambda \in \Lambda}\}$ be a complete orthonormal sets in H. If possible let the set $\{e_\lambda |_{\lambda \in \Lambda}\}$ be not enumerable then $\exists$ an element $e_\beta \in \{e_\lambda |_{\lambda \in \Lambda}\}$ such that $e_\beta \notin M$. Since $\overline{M} = H$, $e_\beta \in \overline{M}$, since M is dense in H, $\exists$ a separable set $\{x_n\} \subseteq M$ such that $\lim_{n \to \infty} x_n = e_\beta$.

Since $\{e_\lambda |_{\lambda \in \Lambda}\}$ is a complete set in H and $x_n \in M \subset H$, then x_n can be expressed as

$$x_n = \sum_{\substack{\lambda \in \Lambda \\ \lambda \neq \beta}} C_\lambda^{(n)} e_\lambda$$

$$e_\beta = \lim_{n \to \infty} x_n = \lim_{n \to \infty} \sum_{\substack{\lambda \in \Lambda \\ \lambda \neq \beta}} C_\lambda^{(n)} e_\lambda.$$

The above relation shows that e_β is a linear combination of $\left\{e_\lambda |_{\substack{\lambda \in \Lambda \\ \lambda \neq \beta}}\right\}$, which is a contradiction, since $\{e_\lambda |_{\lambda \in \Lambda}\}$ is an orthonormal set and hence linearly independent.

Hence $\{e_\lambda |_{\lambda \in \Lambda}\}$ cannot be non-enumerable.

Problems [3.7 and 3.8]

1. Let $\{a_1, a_2, \ldots a_n\}$ be an orthogonal set in a Hilbert space H, and $\alpha_1, \alpha_2, \ldots \alpha_n$ be scalars such that their absolute values are respectively 1. Show that

$$\|\alpha_1 a_1 + \cdots + \alpha_n a_n\| = \|a_1 + a_2 + \cdots + a_n\|.$$

2. Let $\{e_n\}$ be an orthonormal sequence in a Hilbert space H. If $\{a_n\}$ be a sequence of scalars such that $\sum_{i=1}^{\infty} |\alpha_i|^2$ converges then show that $\sum_{i=1}^{\infty} \alpha_i e_i$ converges to an $x \epsilon H$ and $\alpha_n = \langle x, e_n \rangle$, $\forall n \in \mathbb{N}$.

3. Let $\{e_1, e_2, \ldots e_n\}$ be a finite orthonormal set in a Hilbert space H and let x be a vector in H. If $p_1, p_2, \ldots p_n$ are arbitrary scalers show that $\|x - \sum_{i=1}^{n} p_i e_i\|$ attains its minimum value $\Rightarrow p_i = \langle x, e_i \rangle$ for each i.

4. Let $\{e_n\}$ be a orthonormal sequence in a Hilbert space H. Prove that

$$\sum_{n=1}^{\infty} |\langle x, e_n \rangle \langle y, e_n \rangle| \leq \|x\| \|y\|, \quad x, y \in H.$$

5. Show that on the unit disk $B(|z| < 1)$ of the complex plane $z = x + iy$, the functions

$$g_k(x) = \left(\frac{k}{\pi}\right)^{\frac{1}{2}} z^{k-1} \ (k = 1, 2, 3)$$

form an orthonormal system under the usual definition of a scalar product in a complex plane.

6. Let $\{e_\alpha | \lambda \in \Lambda\}$ be an orthonormal set in a Hilbert space H.

   (a) If x belongs to the closure of span $\{e_n\}$, then show that

   $$x = \sum_{n=1}^{\infty} \langle x, e_n \rangle e_n \quad \text{and} \quad \|x\|^2 = \sum_{n=1}^{\infty} |\langle x, e_n \rangle|^2$$

   where $\{e_1, e_2, \ldots\} = \{e_n : \langle x, e_n \rangle \neq 0\}$.

   (b) Prove that the span $\{e_n\}$ is dense in H if and only if every x in H has a Fourier expression as above and for every x, y in H, the identity

   $$\langle x, y \rangle = \sum_{n=1}^{\infty} \langle x, e_n \rangle \langle e_n, y \rangle$$

   holds, where $\{e_i, e_2, \ldots\} = \{e_\alpha : \langle x, e_\alpha \rangle \neq 0 \text{ and } \langle y, e_\alpha \rangle \neq 0\}$.

7. Let L be a complete orthonormal set in a Hilbert space H. If $\langle u, x \rangle = \langle v, x \rangle$ for all $x \in L$, show that $u = v$.

8. The **Haar system** in $L_2([0, 1])$ is defined as follows:

$$\phi_{(0)}^{(a)}(x) = 1 \ x \in [0, 1]$$

$$\phi_1^{(0)}(x) = \begin{cases} 1, & x \in \left[0, \frac{1}{2}\right) \\ -1, & x \in \left(\frac{1}{2}, 1\right] \\ 0, & x = \frac{1}{2} \end{cases}$$

and for $m = 1, 2, \ldots;$ $\quad K = 1, \ldots 2^m$

$$\phi_m^{(K)}(x) = \begin{bmatrix} \sqrt{2^m}, & x \in \left(\dfrac{K-1}{2^m}, \dfrac{K-\frac{1}{2}}{2^m} \right) \\[3mm] -\sqrt{2^m}, & x \in \left(\dfrac{K-\frac{1}{2}}{2^m}, \dfrac{K}{2^m} \right) \\[3mm] 0, & x \in [0,1] \sim \left[\dfrac{K-1}{2^m}, \dfrac{K}{2^m} \right] \end{bmatrix}$$

and at that finite set of points at which $\phi_m^{(K)}(x)$ has not yet been defined, let $\phi_m^{(K)}(x)$ be the average of the left and right limits of $\phi_m^{(K)}(x)$ as x approaches the point in question. At 0 and 1, let $\phi_m^{(K)}(x)$ assume the value of the one-sided limit. Show that the **Haar system** given by

$$\{\phi_0^{(0)}, \phi_1^{(0)} \cdots \phi_m^{(K)}, \ m = 1, 2, \ldots; K = 1, \ldots 2^m\}$$

is orthonormal in $L_2([0,1])$.

9. If H has a denumerable orthonormal basis, show that every orthonormal basis for H is denumerable.

10. Show that the Legendre differential equation can be written as
$[(1 - t^2)P_n']' = -n(n+1)P_n$.

Multiply the above equation by P_n and the corresponding equation in P_m by P_n. Then subtracting the two and integrating resulting equation from -1 to 1, show that $\{P_n\}$ in an orthogonal sequence in $L_2([-1,1])$.

11. **(Generating function)** Show that

$$\frac{1}{\sqrt{1 - 2tw + w^2}} = \sum_{n=0}^{\infty} P_n(t)w^n.$$

12. **(Generating function)** Show that

$$\exp(2wt - w^2) = \sum_{n=0}^{\infty} \frac{1}{n!} H_n(t)w^n.$$

The function on the left is called a generating function of the Hermite polynomials.

13. Given $H_0(t) = 1,$ $H_n(t) = (-1)^n e^{t^2} \dfrac{d^n}{dt^n}(e^{-t^2})$ $n = 1, 2, \ldots$

Show that $H_{n+1}(t) = 2tH_n(t) - H_n'(t)$.

14. Differentiating the generating function in problem 12 W.R.T. t, show that $H_n'(t) = 2nH_{n-1}(t),$ $(n \geq 1)$ and using problem 13, show that H_n satisfies the Hermite differential equation.

3.9   Isomorphism between Separable Hilbert Spaces

Consider a separable Hilbert space H and let $\{e_i\}$ be a complete orthonormal system in the space. If x is some element in H then we can assign to this element a sequence of numbers $\{c_1, c_2, \ldots c_n\}$, the Fourier coefficients. As shown earlier the series $\sum\limits_{i=1}^{\infty} |c_i|^2$ is convergent and consequently the sequence $\{c_1, c_2, \ldots c_n, \ldots\}$ can be treated as some element $\tilde{x}$ of the complex space l_2. Thus to every element $x \in H$ there can be assigned some element $\tilde{x} \in l_2$. Moreover, the assumption on the completeness of the system implies

$$\|x\|_H = \left(\sum_{i=1}^{\infty} |c_i|^2\right)^{\frac{1}{2}} = |\tilde{x}\|_{l_2} \tag{3.49}$$

where the subscripts H and l_2 denote the respective spaces whose norms are taken.

Moreover, it is clear that if $x \in H$ corresponds to $\tilde{x} \in l_2$, and $y \in H$ corresponds to $\tilde{y} \in l_2$, then $x \pm y$ corresponds to $\tilde{x} \pm \tilde{y}$. It then follows from (3.49) that

$$\|x - y\|_M = \|\tilde{x} - \tilde{y}\|_{l_2}. \tag{3.50}$$

Let us suppose that $\overline{z} = \{\zeta_i\}$ is an arbitrary element in l_2. We next consider in H the elements $z_n = \sum\limits_{i=1}^{n} \zeta_i e_i$, $n = 1, 2, \ldots$

We have then $\|z_n - z_m\|^2 = \|\sum\limits_{i=m+1}^{n} \zeta_i e_i\|^2 = \sum\limits_{i=m+1}^{n} |\zeta_i|^2.$

Now, $\|z_m - z_n\| \to 0$ as $n, m \to \infty.$

Then $\{z_n\}$ is a Cauchy sequence in the sense of the metric in H and by virtue of completeness of H, converges to some element z of the space, since

$$\langle z, e_i \rangle = \lim_{n \to \infty} \langle z_n, e_i \rangle = \zeta_i, \ i = 1, 2, \ldots.$$

It therefore follows that the ζ_i are Fourier coefficients of z w.r.t. the chosen orthonormal system $\{e_i\}$. Thus each element $\tilde{z} \in l_2$ is assigned to some element $z \in H$. In the same manner, corresponding to every $\tilde{z} \in l_2$ we can find a $z \in H$. Thus a one-to-one correspondence between the elements of H and l_2 is established. The formula (3.50) shows that the correspondence between H and L_2 is an isometric correspondence. Now, if $x \in H$ corresponds to $\overline{x} \in l_2$ and $y \in H$ corresponds to $\tilde{y} \in l_2$, $x \pm y \in H_2$ corresponds to $\tilde{x} \pm \tilde{y} \in l_2$. Again $\lambda x \in H$ corresponds to $\lambda \tilde{x} \in l_2$ for any scalar λ. Since $\|x \pm y\|_H = \|\tilde{x} \pm \tilde{y}\|_{l_2}$ and $\|\lambda x\|_H = \|\lambda \tilde{x}\|_{l_2}$, it follows that the correspondence between H and l_2 is both isometric and isomorphic. We thus obtain the following theorem.

3.9.1 Theorem

Every complex (real) separable Hilbert space is isomorphic and isometric to a complex (real) space l_2. Hence all complex (real) separable Hilbert spaces are isomorphic and isometric to each other.

CHAPTER 4

LINEAR OPERATORS

There are many operators, such as matrix operator, differential operator, integral operator, etc. which we come across in applied mathematics, physics, engineering, etc. The purpose of this chapter is to bring the above operators under one umbrella and call them 'linear operators'. The continuity, boundedness and allied properties are studied. If the range of the operators is $\mathbb{R}$, then they are called functionals. Bounded functionals and space of bounded linear functionals are studied. The inverse of an operator is defined and the condition of the existence of an inverse operator is investigated. The study will facilitate an investigation into whether a given equation has a solution or not. The setting is always a Banach space.

4.1 Definition: Linear Operator

We know of a mapping from one space onto another space. In the case of vector spaces, and in particular, of normed spaces, a mapping is called an **operator**.

4.1.1 Definition: linear operator

Given two topological linear spaces (E_x, τ_x) and (E_y, τ_y) over the same scalar field (real or complex) an operator A is defined on E_x with range in E_y.

We write $y = Ax$; $x \in E_x$ and $y \in E_y$.

The operator A is said to be **linear** if:

(i) it is *additive*, that is,

$$A(x_1 + x_2) = Ax_1 + Ax_2, \text{forall } x_1, x_2 \in E_x \qquad (4.1)$$

(ii) it is *homogeneous*, that is,

$$A(\lambda x) = \lambda Ax,$$

forall $x \in E_x$ and every real (complex) λ whenever E_x is real (complex).

Observe the **notation**. Ax is written instead of $A(x)$; this simplification is standard in functional analysis.

4.1.1a Example

Consider a real square matrix (a_{ij}) of order n, $(i, j = 1, 2, \ldots n)$. The equations,

$$\eta_i = \sum_{j=1}^{n} a_{ij}\xi_j \quad (i = 1, 2, \ldots, n)$$

can be written in the compact form,

$$y = Ax$$

where $y = (\eta_1, \eta_2, \ldots, \eta_n) \in E_y$, $A = (a_{ij})_{\substack{i=1,\ldots,n \\ j=1,\ldots,n}}$,

$$x = (\xi_1, \xi_2, \ldots, \xi_n) \in E_x$$

If $x^1 = (\xi_1^1, \xi_2^1, \ldots, \xi_n^1)$ and $x^2 = (\xi_1^2, \xi_2^2, \ldots, \xi_n^2)$ and $y^1 = (\eta_1^1, \eta_2^1, \ldots, \eta_n^1)$ and $y^2 = (\eta_1^2, \eta_2^2, \ldots, \eta_n^2)$ such that

$$Ax^1 = y^1 \Rightarrow \sum_{j=1}^{n} a_{ij}\xi_j^1 = \eta_i^1, \quad i = 1, 2, \ldots, n$$

$$Ax^2 = y^2 \Rightarrow \sum_{j=1}^{n} a_{ij}\xi_j^2 = \eta_i^2, \quad i = 1, 2, \ldots, n$$

Then $A(x^1 + x^2) = \left(\sum_j a_{ij}(\xi_j^1 + \xi_j^2)\right) = \left(\sum_j a_{ij}\xi_j^1\right) + \left(\sum_j a_{ij}\xi_j^2\right)$

$$= Ax^1 + Ax^2 = y^1 + y^2.$$

The above shows that A is additive.

The fact that A is homogeneous can be proven in a similar manner.

4.1.2a Example

Let $k(t, s)$ be a continuous function of t and s, $a \leq t, s \leq b$. Consider the integral equation

$$y(t) = \int_a^b k(t, s)x(s)ds.$$

If $x(s) \in C([a, b])$, then $y(t) \in C([a, b])$.

The above equation can be written as

$$y = Ax \quad \text{where } A = \int_a^b k(t, s)x(s)ds.$$

The operator A maps the space $C([a,b])$ into itself.

Let $x_1(s), x_2(s) \in C([a,b])$, then,

$$A(x^1 + x^2) = \int_a^b k(t,s)(x_1(s) + x_2(s))ds$$

$$= \int_a^b k(t,s)x_1(s)ds + \int_a^b k(t,s)x_2(s)ds = Ax_1 + Ax_2$$

Moreover, $A(\lambda x) = \int_a^b k(t,s)(\lambda x(s))ds = \lambda \int_a^b k(t,s)x(s)ds = \lambda Ax$.

Thus A is additive and homogeneous.

4.1.3a Example

Let $E_x = C^1([a,b]) = \{x(t) : x(t) \text{ is continuously differentiable in } a < t < b, \frac{dx}{dt} \text{ is continuous in } (a,b)\}$.

Define the norm as

$$\|x\| = \sup_{a \leq t \leq b} \left[|x(t)| + \left| \frac{dx}{dt} \right| \right] \tag{4.2}$$

Let $x \in C^1([a,b])$, then $y = Ax = \frac{dx}{dt} \in C([a,b])$

$$\|y\| = \sup_{a \leq t \leq b} |y(t)| = \sup_{a \leq t \leq b} \left| \frac{dx}{dt} \right|$$

Since the sup of (4.2) exists, $\sup |y(t)|$ also exists. Moreover the operator $A = \frac{dx}{dt}$ is linear for

$$\begin{array}{ll} x_1, x_2 \in C^1([a,b]) & A(x_1 + x_2) = Ax_1 + Ax_2 \\ \lambda \in \mathbb{R}(\mathbb{C}) & A(\lambda x) = \lambda Ax \end{array} \Rightarrow$$

4.1.4　Continuity

We know that the continuity of A in the case of a metric space means that there is a $\delta > 0$ such that the collection of images of elements in the ball $B(x, \delta)$ lie in $B(Ax, \epsilon)$.

4.1.1b Example

Let us suppose in example 4.1.1a $\{\xi_i^{(m)}\}$ is convergent.

Then $\eta_i^{(m)} - \eta_i^{(p)} = \sum_{j=1}^n a_{ij}(\xi_j^{(m)} - \xi_j^{(p)})$

Hence by Cauchy-Bunyakovsky-Schwartz inequality (1.4.3)

$$\sqrt{\sum_{i=1}^n (\eta_i^{(m)} - \eta_i^{(p)})^2} \leq \sqrt{\sum_{i=1}^n \sum_{j \neq 1}^n a_{ij}^2} \sqrt{\sum_{j=1}^n (\xi_i^{(m)} - \xi_i^{(p)})^2}$$

where $x_m = \{\xi_i^{(m)}\}$, $y_m = Ax_m = \{\eta_i^{(m)}\}$.

Since $\displaystyle\sum_{i=1}^{n}\sum_{j=1}^{n} a_{ij}^2$ is finite, convergence of $\{\xi_i^{(m)}\}$ implies convergence of $\{\eta_i^{(m)}\}$. Hence continuity of A is established.

4.1.2b Example

Consider Example 4.1.2a.

Let $\{x_n(t)\}$ converge to $x(t)$ in the sense of convergence in $C([0,1])$, i.e., converges uniformly in $C[(0,1]]$. Now in the case of uniform convergence we can take the limit under the integral sign.

It follows that $\displaystyle\lim_{m}\int_a^b K(t,s)x_m(s)ds = \int_a^b K(t,s)x(s)ds$ i.e. $\displaystyle\lim_{m\to\infty} Ax_m = Ax$ and the continuity of A is proved.

4.1.3b Example

We refer to the example 4.1.3a. The operator A, in this case, although additive and homogeneous, is not **continuous**. This is because the derivative of a limit element of a uniformly convergent sequence of functions need not be equal to the limit of the derivative of these functions, even though all these derivatives exist and are continuous.

4.1.4′ Example

Let A be a continuous linear operator. Then

(i) $A(\theta) = \theta$ (ii) $A(-z) = -Az$ for any $z \in E_x$

Proof: (i) for any $x, y, z \in E_x$, put $x = y + z$ and consequently $y = x - z$. Now,

$$Ax = A(y + z) = Ay + Az = Az + A(x - z)$$

Hence, $A(x - z) = Ax - Az$ (4.3)

Putting $x = z$, we get $A(\theta) = \theta$.

(ii) Taking $x = \theta$ in (4.3) we get

$$A(-z) = -Az$$

4.1.5 Theorem

If an additive operator $\mathbf{A}$, mapping a real linear space E_x into a real linear space E_y s.t. $y = Ax$, $x \in E_x$, $y \in E_y$, be continuous at a point $x^* \in E_x$, then it is continuous on the entire space E_x.

Proof: Let x be any point of E_x and let $x_n \to x$ as $n \to \infty$. Then $x_n - x + x^* \to x^*$ as $n \to \infty$.

Since A is continuous at x^*,

$$\lim_{n \to \infty} A(x_n - x + x^*) = Ax^*$$

However, $A(x_n - x + x^*) = Ax_n - Ax + Ax^*$,

since A is additive in nature.

Therefore, $\lim_{n \to \infty} (Ax_n - Ax + Ax^*) = Ax^*$ or $\lim_{n \to \infty} Ax_n = Ax$

where x is any element of E_x.

4.1.6   Theorem

An additive and continuous operator A defined on a real linear space is homogeneous.

Proof: (i) Let $\alpha = m$, a positive integer. Then

$$A(\alpha x) = \underbrace{Ax + Ax + \cdots + Ax}_{m \text{ terms}} = mAx$$

(ii) Let $\alpha = -m$, m is a positive integer.

By sec. 4.1.4

$$A(\alpha x) = A(-mx), \ m \text{ positive integer}$$
$$= -A(mx) = -mAx = \alpha Ax$$

(iii) Let $\alpha = \dfrac{m}{n}$ be a rational number m and n prime to each other, then

$$A\left(\frac{m}{n}x\right) = mA\left(\frac{x}{n}\right)$$

Let $\dfrac{x}{n} = \xi$, n is an integer. Then $x = n\xi$.

$$Ax = A(n\xi) = nA\xi = nA\left(\frac{x}{n}\right)$$

Hence $A\left(\dfrac{x}{n}\right) = \dfrac{1}{n}Ax$ i.e. $A(\alpha x) = A\left(\dfrac{m}{n}x\right)$

$$= mA\left(\frac{x}{n}\right) = \frac{m}{n}Ax = \alpha Ax.$$

If $\alpha = -\dfrac{m}{n}$ where $\dfrac{m}{n} > 0$, then also $A(\alpha x) = \alpha Ax$.

Let us next consider α to be an irrational number. Then we can find a sequence of rational number $\{s_i\}$ such that $\lim_{i \to \infty} s_i = \alpha$.

s_i being a rational number, then

$$A(s_i x) = s_i Ax \qquad\qquad (4.4)$$

since A is continuous at αx, α is a real number, and since $\lim_{i \to \infty} s_i x = \alpha x$ we have,

$$\lim_{i \to \infty} A(s_i x) = A(\alpha x)$$

Again, $\lim_{i \to \infty} s_i = \alpha$.

Hence taking limits in (4.4), we get,

$$A(\alpha x) = \alpha A x.$$

4.1.7   The space of operators

The algebraic operations can be introduced on the set of linear continuous operators, mapping a linear space E_x into a linear space E_y. Let A and B map E_x into E_y.

For any $x \in E_x$, we define the addition by

$$(A + B)x = Ax + Bx$$

and the scalar multiplication by

$$(\lambda A)x = \lambda A x.$$

Thus, we see the set of linear operators defined on E_x is closed w.r.t. addition and scalar multiplication. Hence, the set of linear operators mapping E_x into E_y is a linear space.

In particular if we take $B = -A$, thus

$$(A + B)x = (A - A)x = \mathbf{0} \cdot x = Ax - Ax = \theta$$

Thus $\mathbf{0}$, the null operator, is an element of the said space. The limit of a sequence is defined in a space of linear operators by assuming for example that $A_n \to A$ if $A_n x \to A x$ for every $x \in E_x$.

This space of continuous linear operators will be discussed later.

4.1.8   The ring of continuous linear operators

Let E be a linear space over a scalar field $\mathbb{R}(\mathbb{C})$. We next consider the space of continuous linear operators mapping E into itself. Such a **space** we denote by $(\boldsymbol{E \to E})$. The product of two linear operators A and B in $(E \to E)$ is denoted by $AB = A(B)$, i.e., $(AB)x = A(Bx)$ for all $x \in E$.

Let $x_n \to x$ in E. A and B being continuous linear operators, $Ax_n \to Ax$ and $Bx_n \to Bx$.

Since AB is the product of A and B in $(E \to E)$,

$$ABx_n - ABx = A(Bx_n - Bx).$$

Since $Bx_n \to Bx$ as $n \to \infty$ or $Bx_n - Bx \to \theta$ as $n \to \infty$ and A is a continuous linear operator, $A(Bx_n - Bx) \to \theta$ as $n \to \infty$. Thus, AB is a continuous linear operator.

Since A maps $E \to E$ and is continuous linear $A^2 = A \cdot A \in (E \to E)$. Let us suppose $A^n \in (E \to E)$ for any finite n. Then

$$A^{n+1} = A \cdot A^n \in (E \to E)$$

It can be easily seen that, if A, B and $C \in (E \to E)$ respectively, then

$$(AB)C = A(BC), \quad (A+B)C = AC + BC \quad \text{and also} \quad C(A+B) = CA + CB.$$

Moreover, there exists an **identity operator** I, defined by $Ix = x$ for all x and such that $AI = IA = A$ for every operator A. In general $AB \neq BA$, the set $(E \to E)$ is a **non-commutative ring with identity**.

4.1.9 Example

Consider the linear space $\mathbb{P}$ of all polynomials $p(s)$, with real coefficients. The operator A is defined by

$$y(t) = \int_0^1 tsp(s)ds = Ap \quad \text{and} \quad y(t) = t\frac{dp}{dt} = Bp$$

$$ABp = \int_0^1 ts^2 \frac{dp(s)}{ds}ds = t\left[s^2 p(s)\big|_0^1 - 2\int_0^1 sp(s)ds\right]$$

$$= t\left[p(1) - 2\int_0^1 sp(s)ds\right]$$

$$BAp = t\frac{d}{dt}\left[\int_0^1 tsp(s)ds\right] = t\int_0^1 sp(s)ds$$

Thus $AB \neq BA$.

4.1.10 Function of operator

The operator $A^n = \underbrace{A \cdot A \cdots \cdot A}_{n \text{ terms}}$ represents a simple example of an **operator function**. A more general function, namely of the polynomial function of operator,

$$p_n(A) = a_0 I + a_1 A + a_2 A^2 + \cdots + a_n A^n.$$

4.2 Linear Operators in Normed Linear Spaces

Let E_x and E_y be two normed linear spaces. Since a normed linear space is a particular case of a topological linear space, the definition of a linear operator mapping a topological linear space E_x into a topological linear space E_y holds good in case the spaces Ex and E_y reduce to normed linear spaces. Theorems 4.1.5 and 4.1.6 also remain valid in normed linear spaces.

4.2.1 Definition: continuity of a linear operator mapping E_x into E_y

Since convergence in a normed linear space is introduced through the convergence in the induced metric space, we define convergence of an operator A in a normed linear space as follows:

Given A, a linear operator mapping a normed linear space E_x into a normed linear space E_y, A is said to be **continuous** at $x \in E_x$ if $||x_n - x||_{E_x} \to 0$ as $n \to \infty \Rightarrow ||Ax_n - Ax||_{E_y} \to 0$ as $n \to \infty$.

4.1.1c Example

Now refer back to example 4.1.1a. Let, $x_m = \{\xi_i^{(m)}\}$. $y_m = \{\eta_i^{(m)}\}$. Then $y_m = Ax_m$. Let $x_m \to x$ as $m \to \infty$. We assume that $x_m \in E_x$, the n-dimensional Euclidean space. Then

$$\sum_{i=1}^{n}(\xi_i^{(m)})^2 < \infty.$$

Now, by Cauchy-Bunyakovsky-Schwartz's inequality, (Sec. 1.4.3)

$$y - y_m = \{\eta_i - \eta_i^{(m)}\} = \left\{\sum_{i=1}^{n}\sum_{j=1}^{n} a_{ij}(\xi_j - \xi_j^{(m)})\right\}$$

or

$$|\eta_i - \eta_i^{(m)}| = \left|\sum_{j=1}^{n} a_{ij}(\xi_j - \xi_j^{(m)})\right| \le \left(\sum_{j=1}^{n} a_{ij}^2\right)^{1/2}.$$

$$\left(\sum_{j=1}^{n}|\xi_j - \xi_j^{(m)}|^2\right)^{1/2}$$

Thus

$$\left(\sum_{i=1}^{n}|\eta_i - \eta_i^{(m)}|^2\right)^{1/2} \le \left(\sum_{i=1}^{n}\sum_{j=1}^{n} a_{ij}^2\right)^{1/2}\left(\sum_{j=1}^{n}|\xi_j - \xi_j^{(m)}|^2\right)^{1/2}$$

Since $\sum_{i=1}^{n}\sum_{j=1}^{n} a_{ij}^2 < \infty$

$$||x_m - x||^2 \longrightarrow 0 \Longrightarrow \sum_{i=1}^{\infty}|\xi_i - \xi_i^{(m)}|^2 \longrightarrow 0$$

$$\Longrightarrow \sum_{i=1}^{n}|\eta_i - \eta_i^{(m)}|^2 \longrightarrow 0$$

$$\Longrightarrow ||Ax_m - Ax||^2 \longrightarrow 0 \text{ as } m \to \infty.$$

This shows that A is continuous.

4.1.2c Example

We consider example 4.1.2a in the normed linear space $C([a,b])$. Since $x(s) \in C([a,b])$ and $K(t,s)$ is continuous in $a \leq t,\, s \leq b$, it follows that $y(t) \in C([a,b])$.

Let $x_n, x \in C([a,b])$ and $||x_n - x|| \longrightarrow 0$ as $n \to \infty$, where $||x|| = \max\limits_{a \leq t \leq b} |x(t)|$.

Now, $\quad ||y_n(t) - y(t)|| = \max\limits_{a \leq t \leq b} |y_n(t) - y(t)|$

$$= \max\limits_{a \leq t \leq b} \left| \int_a^b |K(t,s)(x_n(s) - x(s))ds \right|$$

$$\leq [b-a] \max\limits_{a \leq t,s \leq b} |K(t,s)| \cdot \max\limits_{a \leq s \leq b} |x_n(s) - x(s)|$$

$$= [b-a] \max\limits_{a \leq t,s \leq b} |K(t,s)| \, ||x_n(s) - x(s)|| \longrightarrow 0 \text{ as } n \to \infty.$$

or, $\quad ||Ax_n - Ax|| \longrightarrow 0$ as $||x_n - x|| \to 0$,

showing that A is continuous.

4.2.2 Example

Let $A = (a_{ij})$, $i,j = 1,2,\ldots$, respectively.

Let $x = \{\xi_i\}$, $i = 1,2,\ldots$ Then $y = Ax$ yields

$$\eta_i = \sum_{j=1}^{\infty} a_{ij}\xi_j \quad \text{where } y = \{\eta_i\}$$

Let us suppose

$$K = \sum_{i=1}^{\infty} \sum_{j=1}^{\infty} |a_{ij}|^q < \infty, \, q > 1 \tag{4.5}$$

and $\quad x \in l_p$ i.e. $\sum\limits_{i=1}^{\infty} |\xi_i|^p \leq \infty.$

For $x_1 = \{\xi_i^{(1)}\} \in l_p$, $x_2 = \{\xi_i^{(2)}\} \in l_p$ it is easy to show that A is linear. Then, using Hölder's inequality (1.4.3)

$$\sum_{i=1}^{n} |\eta_i|^q \leq \sum_{i=1}^{n} \left\{ \left(\sum_{j=1}^{\infty} |a_{ij}|^q \right)^{1/q} \left(\sum_{j=1}^{\infty} |\xi_j^p| \right)^{1/p} \right\}^q$$

$$= ||x||^q \sum_{i=1}^{n} \sum_{j=1}^{\infty} |a_{ij}|^q$$

$$\leq ||x||^q \sum_{i=1}^{\infty} \sum_{j=1}^{\infty} |a_{ij}|^q$$

Hence $||y|| = \left(\sum_{i=1}^{\infty} |\eta_i|^q \right)^{1/q} \leq \left(\sum_{i=1}^{\infty} \sum_{j=1}^{\infty} |a_{ij}|^q \right)^{1/q} \cdot ||x||$ where $\dfrac{1}{p} + \dfrac{1}{q} = 1.$

Hence, using (4.5), we can say that $x \in l_p \Longrightarrow y \in l_q$.

Let now $||x_m - x||_p \longrightarrow 0$, i.e., $x_m \longrightarrow x$ in l_p as $m \longrightarrow \infty$, where $x_n = \{\xi_i^{(n)}\}$ and $x = \{\xi_i\}$.

Now, using Hölder's inequality (sec. 1.4.3)

$$||Ax_m - Ax||_q^q = \sum_{i=1}^{\infty} \left| \sum_{j=1}^{\infty} a_{ij}(\xi_j^{(m)} - \xi_j) \right|^q$$

$$\leq \left(\sum_{i=1}^{\infty} \sum_{j=1}^{\infty} |a_{ij}|^q \right) \cdot \left(\sum_{j=1}^{\infty} |\xi_j^{(m)} - \xi_j|^p \right)^{q/p}$$

$$= K \left(\sum_{j=1}^{\infty} |\xi_j^{(m)} - \xi_j|^p \right)^{q/p}$$

$$= K||x_m - x||_p^q$$

Hence $||x_m - x||_p \longrightarrow 0 \Longrightarrow ||Ax_m - Ax||_q \longrightarrow 0.$

Hence A is linear and continuous.

4.2.3 Definition: bounded linear operator

Let A be a linear operator mapping E_x into E_y where E_x and E_y are linear operators over the scalar field $\mathbb{R}(\mathbb{C})$. A is said to be bounded if there exists a constant $K > 0$ s.t.

$$||Ax||_{E_x} \leq K||x||_{E_x} \quad \text{for all} \ \ x \in E_x$$

Note 4.2.1. The definition 4.2.3 of a bounded linear operator is not the same as that of an ordinary real or complex function, where a **bounded** function is one whose range is a bounded set.

We would next show that a bounded linear operator and a continuous linear operator are one and the same.

4.2.4 Theorem

In order that an additive and homogeneous operator A be continuous, it is necessary and sufficient that it is bounded.

Proof: (*Necessity*) Let A be a continuous operator. Assume that it is not bounded. Then, there is a sequence $\{x_n\}$ of elements, such that

$$||Ax_n|| > n||x_n|| \tag{4.6}$$

Let us construct the elements

$$\xi_n = \frac{x_n}{n||x_n||}, \text{ i.e., } ||\xi_n|| = \frac{||x_n||}{n||x_n||} \longrightarrow 0 \text{ as } n \to \infty$$

Therefore, $\xi_n \longrightarrow \xi = \theta$ as $n \to 0$.

On the other hand,

$$||A\xi_n|| = \frac{1}{n||x_n||}||Ax_n|| > 1 \tag{4.7}$$

Now, A being continuous, and since $\xi_n \to \theta$ as $n \to \infty$

$$\xi_n \to \theta \text{ as } n \to \infty \Longrightarrow A\xi_n \longrightarrow A \cdot \theta = \theta$$

This contradicts (4.7). Hence, A is bounded.

(Sufficiency) Let the additive operator A be bounded i.e. $||Ax|| \leq K||x|| \; \forall \; x \in E_x$.

Let $x_n \longrightarrow x$ as $n \to \infty$ i.e. $||x_n - x|| \longrightarrow 0$ as $n \to \infty$.

Now, $||Ax_n - Ax|| = ||A(x_n - x)|| \leq K||x_n - x|| \to 0$ as $n \to \infty$.

Hence, A is continuous at x_n.

4.2.5 Lemma

Let a given linear (not necessarily bounded) operator A map a Banach space E_x into a Banach space E_y. Let us denote by E_n the set of those $x \in E_x$ for which $||Ax|| < n||x||$. Then E_x is equal to $\bigcup\limits_{n=1}^{\infty} E_n$ and at least one E_n is everywhere dense.

Proof: Since $||A \cdot \theta|| < n||\theta||$, the null element belongs to every E_n for every n. Again, for every x, we can find a n say n' such that $||Ax|| < n'||x||$, $n' > \dfrac{||Ax||}{||x||}$.

Therefore, every x belongs to same E_n.

Hence $E_x = \bigcup\limits_{n=1}^{\infty} E_n$. E_x, being a Banach space, can be reduced to a complete metric space. By theorem 1.4.19, a complete metric space is a set of the second category and hence cannot be expressed as a countable union of nowhere dense sets. Hence, at least one of the sets of E_x is everywhere dense.

To actually construct such a set in E_n we proceed as follows. Let $E_{\hat{n}}$ be a set which is everywhere dense in E_x. Consequently there is a ball $B(x_0, r)$ containing $B(x_0, r) \cap E_{\hat{n}}$, everywhere dense.

Let us consider a ball $B(x_1, r_1)$ lying completely inside $B(x_0, r)$ and such that $x_1 \in E_{\hat{n}}$. Take any element x with norm $||x|| = r_1$. Now $x_1 + x \in B(x_1, r_1)$. Since $\overline{B(x_1, r_1)} \subseteq \overline{E_{\hat{n}}}$, there is a sequence $\{y_k\}$ of elements in $B(x_1, r_1) \cap E_{\hat{n}}$ such that $y_k \longrightarrow x_1 + x$ as $k \to \infty$. Therefore,

$x_k = y_k - x_1 \longrightarrow x$. Since $||x|| = r_1$, there is no loss of generality if we assume $r_1/2 < ||x_k||$.

Since y_k and $x_1 \in E_{\hat{n}}$,

$$||Ax_k|| = ||Ay_k - Ax_1|| \leq ||Ay_k|| + ||Ax_1||$$
$$\leq \hat{n}(||y_k|| + ||x_1||)$$

Besides $||y_k|| = ||x_k + x_1|| \leq ||x_k|| + ||x_1||$
$$\leq r_1 + ||x_1||$$

Hence, $||Ax_k|| \leq \hat{n}(r_1 + 2||x_1||) \leq \dfrac{2\hat{n}(r_1 + ||x||)}{r_1}||x_k||$ since $\dfrac{r_1}{2} \leq ||x_k||$.

Let $\bar{n}$ be the least integer greater than $\dfrac{2\hat{n}(r_1 + 2||x_1||)}{r_1}$, then $||Ax_k|| \leq \hat{n}||x_k||$, implying that all $x_k \in E_{\bar{n}}$.

Thus, any element x with norm equal to r_1 can approximate elements in $E_{\bar{n}}$. Let x be any element in E_x. Then $\hat{x} = r_1\left(\dfrac{x}{||x_1||}\right)$ satisfies $||\hat{x}|| = r_1$. Hence there is a sequence $\{\hat{x}_k\} \in E_{\bar{n}}$, which converges to $\hat{x}$. Then $\hat{x} = r_1\left(\dfrac{x}{||x_1||}\right)$ satisfies $||\hat{x}|| = r_1$. Then $x_k = \hat{x}_k\dfrac{||x_1||}{r_1} \to x$, as $k \to \infty$.

$$||Ax_k|| = \dfrac{||x_1||}{r_1}||A\hat{x}_k|| \leq \dfrac{||x_1||}{r_1} \cdot \bar{n}||\hat{x}_k|| = \bar{n}||x_k||$$

Thus $x_k \in E_{\bar{n}}$. Consequently $E_{\bar{n}}$ is everywhere dense in E_x.

4.2.6 Definition: the norm of an operator

Let A be a bounded linear operator mapping E_x into E_y. Then we can find $K > 0$ such that

$$||Ax||_{E_y} \leq K||x||_{E_x} \tag{4.8}$$

The smallest value of K, say M, for which the above inequality holds is called **the norm of A** and is denoted by $||A||$.

4.2.7 Lemma

The operator A has the following two properties:

(i) $||Ax|| \leq ||A|| \, ||x||$ for all $x \in E_x$, (ii) for every $\epsilon > 0$, there is an element x_ϵ such that

$$||Ax_\epsilon|| > (||A|| - \epsilon)||x_\epsilon||$$

Proof: (i) By definition of the norm of A, we can write,

$$||A|| = \inf\{K : K > 0 \text{ and } ||Ax||_{E_y} \leq K||x||_{E_x}, \forall \, x \in E_x\} \tag{4.9}$$

Hence $||Ax||_{E_y} \leq ||A|| \, ||x||_{E_x}$

(ii) Since $||A||$ is the infimum of K, in (4.9)

$$\exists \, x_\epsilon \in E_x, \text{ s.t. } ||Ax_\epsilon|| > (||A|| - \epsilon)||x_\epsilon||$$

4.2.8   Lemma

(i) $\|A\| = \sup\limits_{x \neq \theta} \dfrac{\|Ax\|_{E_y}}{\|x\|_{E_x}}$ $\qquad\qquad\qquad\qquad\qquad$ (4.10)

(ii) $\|A\| = \sup\limits_{\|x\| \leq 1} \|Ax\|_{E_y}$ $\qquad\qquad\qquad\qquad\qquad$ (4.11)

Proof: It follows from (4.9) that

$$\sup_{x \neq \theta} \frac{\|Ax\|_{E_y}}{\|x\|_{E_x}} \leq \|A\| \qquad\qquad\qquad (4.12)$$

Again (ii) of lemma 4.2.7 yields

$$\frac{\|Ax_\epsilon\|}{\|x_\epsilon\|} > (\|A\| - \epsilon)$$

Hence, $\sup\limits_{x \neq \theta} \dfrac{\|Ax\|_{E_y}}{\|x\|_{E_x}} \geq \|A\|$ $\qquad\qquad\qquad\qquad$ (4.13)

It follows from (4.12) and (4.13) that

$$\sup_{x \neq \theta} \frac{\|Ax\|_{E_y}}{\|x\|_{E_x}} = \|A\|$$

Next, if $\|x\|_{E_x} \leq 1$, it follows from (i) of lemma 4.2.7,

$$\sup_{\|x\| \leq 1} \|Ax\|_{E_y} \leq \|A\| \qquad\qquad\qquad (4.14)$$

Again, we obtain from (ii) of lemma 4.2.7

$$\|Ax_\epsilon\| > (\|A\| - \epsilon)\|x_\epsilon\|$$

Put $\zeta_\epsilon = \frac{x_\epsilon}{\|x_\epsilon\|}$, then

$$\|A\zeta_\epsilon\| = \frac{1}{\|x_\epsilon\|}\|Ax_\epsilon\| > \frac{1}{\|x_\epsilon\|}(\|A\| - \epsilon)\|x_\epsilon\|$$

Since $\|\zeta_\epsilon\| = 1$, it follows that

$$\sup_{\|x\| \leq 1} \|Ax\| \geq \|A\zeta_\epsilon\| > \|A\| - \epsilon$$

Hence, $\sup\limits_{x \leq 1} \|Ax\| \geq \|A\|$. $\qquad\qquad\qquad\qquad$ (4.15)

Using (4.14) and (4.15) we prove the result.

4.2.9   Examples of operators

Examples of operators include the identity operator, the zero operator, the differential operator, and the integral operator. We discuss these operators in more detail below.

4.2.10   Identity operator

The *identity operator* $I_{E_x} : E_x \to E_x$ is defined by $I_{E_x} x = x$ for all $x \in E_x$. In case E_x is a non-empty normed linear space, the operator I is bounded and the norm $||I|| = 1$.

4.2.11   Zero operator

The zero operator $\mathbf{0} : E_x \to E_y$ where E_x and E_y are normed linear spaces, is bounded and the norm $||\mathbf{0}||_{E_y} = 0$.

4.2.12   Differential operator

Let E_x be the normed linear space of all polynomials on $J = [0,1]$ with norm given by $||x|| = \max_{t \in J} |x(t)|$. A differential operator $\mathbf{A}$ is defined on E_x by $Ax(t) = x'(t)$ where the prime denotes differentiation WRT t.

The operator is linear for differentiable $x(t)$ and $y(t) \in E_x$ for we have,

$$A(x(t) + y(t)) = (x(t) + y(t))' = x'(t) + y'(t) = Ax(t) + Ay(t)$$

Again $A(\lambda x(t)) = (\lambda x(t))' = \lambda x'(t) = \lambda A(x(t))$ where λ is a scalar.

If $x_n(t) = t^n \quad Ax(t) = nt^{n-1}, t \in J$

then $||x_n(t)|| = 1$ and $Ax_n(t) = x'_n(t) = nt^{n-1}$ so that

$$||Ax_n|| = n \quad \text{and} \quad ||Ax_n|| / ||x_n|| = n.$$

Since n is any positive integer we cannot find any fixed number M s.t.

$$||Ax_n|| / ||x_n|| \leq M$$

Hence A is not bounded.

4.2.13   Integral operator

Refer to example 4.1.2.

$A : C([a,b]) \to C([a,b])$, for $x(s) \in C([a,b])$,

$$Ax = \int_a^b K(t,s)x(s)ds$$

where $K(t,s)$ is continuous in $a \leq t, s \leq b$. Therefore

$$y(t) = Ax(s) \in C([a,b])$$

$$||y(t)|| = \max_{a \leq t \leq b} |y(t)| = \max_{a \leq t \leq b} \left| \int_a^b K(t,s)x(s)ds \right|$$

$$\leq \max_{a \leq t \leq b} \int_a^b |K(t,s)|ds \cdot \max_{a \leq s \leq b} |x(s)| = \max_{a \leq t \leq b} \int_0^1 |K(t,s)|ds \cdot ||x||$$

Thus, $||A|| \leq \max_{a \leq t \leq b} \int_a^b |K(t,s)|ds$ \hfill (4.16)

Since $\displaystyle\int_a^b |K(t,s)| ds$ is a continuous function, it attains the maximum at some point t_0 of the interval $[a,b]$. Let us take,

$$z_0(s) = \operatorname{sgn} K(t_0, s)$$

where $\operatorname{sgn} z = \bar{z}/|z|$, $z \in \mathbb{C}$.

Let $x_n(s)$ be a continuous function, such that $|x_n(s)| \leq 1$ and $x_n(s) = z_0(s)$ everywhere except on a set E_n of measure less than $1/2Mn$, where $M = \max_{t,s} |K(t,s)|$. Then, $|x_n(s) - z_0(s)| \leq 2$ everywhere on E_n.

We have $\displaystyle\left| \int_a^b K(t,s) z_0(s) ds - \int_a^b K(t,s) x_n(s) ds \right|$

$$\leq \int_a^b |K(t,s)|\, |x_n(s) - z_0(s)| ds$$

$$= \int_{E_n} |K(t,s)|\, |x_n(s) - z_0(s)| ds$$

$$\leq 2 \max_{t,s} |K(t,s)| \cdot \frac{1}{2Mn} = \frac{1}{n}$$

Thus $\displaystyle\int_a^b K(t,s) z_0(s) ds \leq \int_a^b K(t,s) x_n(s) ds + \frac{1}{n}$

$$\leq ||A||\, ||x_n|| + \frac{1}{n}$$

for $t \in [a,b]$, putting $t = t_0$

$$\int_a^b |K(t_0, s)| ds \leq ||A||\, ||x_n|| + \frac{1}{n}$$

Since $||x_n|| \leq 1$, the preceeding inequality in the limit as $n \to \infty$ gives rise to

$$\int_a^b |K(t,s)| ds \leq ||A||,$$

i.e., $\displaystyle\max_t \int_a^b |K(t,s)| ds \leq ||A|| \qquad (4.17)$

From (4.16) and (4.17) it follows that

$$||A|| = \max_t \int_0^1 |K(t,s)| ds$$

4.2.14   Theorem

A bounded linear operator A_0 defined on a linear subset X, which is everywhere dense in a normed linear space E_x with values in a complete normed linear space E_y can be extended to the entire space with preservation of norm.

Proof: A can be defined on E_x such that $Ax = A_0 x$, $x \in X$ and $||A||_{E_x} = ||A_0||_X$.

Let x be any element in E_x not belonging to X.

Since X is everywhere dense in E_x, there is a sequence $\{x_n\} \subset X$ s.t. $||x_n - x|| \to 0$ as $n \to \infty$ and hence $||x_n - x_m|| \to 0$ as $n, m \to \infty$. However then

$$||A_0 x_n - A_0 x_m|| = ||A_0(x_n - x_m)|| \le ||A_0||_X ||x_n - x_m||$$
$$\to 0 \text{ as } n, m \to \infty, \ \{A_0 x_n\}$$

is a Cauchy sequence and converges by the completeness of E_y to some limit denoted by Ax. Let $\{\xi_n\} \subset X$ be another sequence convergent to x. Evidently $||x_n - \xi_n|| \to \theta$ as $n \to \infty$. Hence, $||A_0 x_n - A_0 \xi_n|| \to 0$ as $n \to \infty$. Consequently $A_0 \xi_n \longrightarrow Ax$, implying that A is defined uniquely by the elements of E_x. If $x \in X$ select $x_n = x$ for all n. Then $Ax = \lim\limits_{n \to \infty} A_0 x_n = A_0 x$.

The operator A is additive, since

$$A(x_1 + x_2) = \lim_{n \to \infty} A_0(x_n^{(1)} + x_n^{(2)}) = \lim_{n \to \infty} A_0 x_n^{(1)} + \lim_{n \to \infty} A_0 x_n^{(2)} = Ax_1 + Ax_2$$

We will next show that the operator is bounded

$$||A_0 x_n|| \le ||A_0||_X ||x_n||$$

Making $n \to \infty$ we have

$$||Ax|| \le ||A_0||_X ||x||$$

Dividing both sides by $||x||$ and taking the supremum

$$||A||_{E_x} \le ||A_0||_X$$

But the norm of A over E_x cannot be smaller than the norm of A_0 over X, therefore we have

$$||A||_{E_x} = ||A||_X$$

The above process exhibits the completion by continuity of a bounded linear operator from a dense subspace to the entire space.

Problems [4.1 & 4.2]

1. Let A be an nth order square matrix, i.e., $A = (a_{ij})_{\substack{i=1,\ldots n \\ j=1,\ldots n}}$. Prove that A is linear, continuous and bounded.

2. Let B be a bounded, closed domain in $\mathbb{R}'$ and let $x = (x_1, x_2, \ldots, x_n)^T$. We denote by $\mathbb{R}'$ the space $C([B])$ of functions $f(x)$ which are continuous on B. The function $\phi(x)$ and the n-dimensional

vector $p(x)$ are fixed members of $C([B])$. The values of $p(x)$ lie in B for all $x \in B$. Show that T_1 and T_2 given by $T_1 f(x) = \phi(x) f(x)$ and $T_2(f(x)) = f(p(x))$ are linear operators.

3. Show that the matrix

$$A = \begin{pmatrix} a_{11} & a_{11} & \cdots & a_{1n} \\ a_{21} & a_{22} & \cdots & a_{2n} \\ \vdots & & & \\ a_{n1} & a_{n2} & \cdots & a_{nn} \end{pmatrix}$$

is a bounded linear operator in $\mathbb{R}_p^n$ for $p = 1, 2, \infty$ and that for

$$p = 1, \quad ||A||_1 \leq \max_j \sum_{i=1}^n |a_{ij}|$$

$$p = 2, \quad ||A||_2 \leq \max_j \left(\sum_{\substack{i=1 \\ j=1}}^n |a_{ij}|^2 \right)^{1/2}$$

$$p = \infty, \quad ||A||_\infty = \max_j \sum_{j=1}^n |a_{ij}|$$

4. Let $E = C([a,b])$ with $|| \cdot ||_\infty$. Let $t_1, \ldots, t_n$ be different points in $[a,b]$ and $\phi_1, \ldots, \phi_n$ be such that $\phi_j(t_i) = \delta_{ij}$ for $i, j = 1, 2, \ldots, n$.

Let $P : E = \longrightarrow E$ be denoted by

$$Px = \sum_{j=1}^n x(t_j)\phi_j, \quad x \in E$$

Then show that P is a *projection operator* onto span $\{\phi_1, \phi_2, \ldots, \phi_k\}$, $P \in (E \to E)$ and

$$||P|| = \sup_{a \leq t \leq b} \sum_{j=1}^n |u_j(t)|$$

The projection operator P above is called the interpolatory projection onto the span of $\{\phi_1, \phi_2, \ldots, \phi_n\}$ corresponding to the 'nodes' $\{t_1, \ldots, t_n\}$.

[For 'Projection operator' see Lemma 4.5.3]

4.3 Linear Functionals

If the range of an operator consists of real numbers then the operator is called a **functional**. In particular if the functional is **additive** and **homogeneous** it is called a **linear functional**.

Thus, a functional $f(x)$ defined on a linear topological space E is said to be linear, if

(i) $f(x_1 + x_2) = f(x_1) + f(x_2) \ \forall \ x_1, x_2 \in E$ and

(ii) $f(x_n) \longrightarrow f(x)$ as $x_n \longrightarrow x$ and $x_n, x \in E$ in the sense of convergence in a linear space E.

4.3.1 Similarity between linear operators and linear functionals

Since the range of a linear functional $f(x)$ is the real line $\mathbb{R}$, which is a Banach space, the following theorems which hold for a linear operator mapping a Banach space into another Banach space are also true for linear functionals defined on a Banach space.

4.3.2 Theorem

If an additive functional $f(x)$, defined on a normed linear space E on $\mathbb{R}(\mathbb{C})$, is continuous at a single point of this space, then it is also continuous linear on the whole space E.

4.3.3 Theorem

Every linear functional is homogeneous.

4.3.4 Definition

A linear functional defined on a normed linear space E over $\mathbb{R}(\mathbb{C})$ is said to be **bounded** if there exists a constant $M > 0$ such that

$$|f(x)| \leq M||x|| \ \forall \ x \in E.$$

4.3.5 Theorem

An additive functional defined on a linear space E over $\mathbb{R}(\mathbb{C})$ is linear if and only if it is bounded.

The smallest of the constants M in the above inequality is called the **norm** of the functional $f(x)$ and is denoted by $||f||$.

Thus $|f(x)| \leq ||f|| \, ||x||$

Thus $||f|| = \sup_{||x|| \leq 1} |f(x)|$

or in other words $||f|| = \sup_{||x||=1} |f(x)| = \sup_{x \neq \theta} \dfrac{|f(x)|}{||x||}$

4.3.6   Examples

1. **Norm:** The norm $||\cdot|| : E_x \to \mathbb{R}$ on a normed linear space $(E_x, ||\cdot||)$ is a functional on E_x which is not linear.

2. **Dot Product:** *Dot Product,* with one factor kept fixed, defines a functional $f : \mathbb{R}^n \longrightarrow \mathbb{R}$ by means of

$$f(x) = a \cdot x = \sum_{i=1}^{n} a_i x_i$$

where $a = \{a_i\}$ and $x = \{x_i\}$, f is linear and bounded.

$$|f(x)| = |a \cdot x| \le ||a|| \, ||x||$$

Therefore, $\sup\limits_{x \neq \theta} \dfrac{|f(x)|}{||x||} \le ||a||$ for $x = a$, $f(a) = a \cdot a = ||a||^2$

Hence, $\dfrac{f(a)}{||a||} = ||a||$, i.e., $||f|| = ||a||$.

3. **Definite Integral:** The definite integral is a number when we take the integral of a single function. But when we consider the integral over a class of functions in a function space, then the integral becomes a functional.

Let us consider the functional

$$f(x) = \int_a^b x(t)dt, \quad x \in C([a,b])$$

f is additive and homogeneous.

Now, $|f(x)| = |\int_a^b x(t)dt| \le \sup\limits_{t} |x(t)| \int_a^b dt = (b-a)||x||$

Therefore, $|f(x)| \le (b-a)||x||$ or, $\sup\limits_{x \in C[(a,b)]} |f(x)| = ||f(x)|| \le (b-a)||x||$

showing that f is bounded.

Now, $\sup\limits_{x \in C[(a,b)]} \dfrac{|f(x)|}{||x||} = ||f(x)|| \le (b-a)$

Next we choose $x = x_0 = 1$, so that $||x_0|| = 1$.

Again, $f(x_0) = \int_a^b x_0(t)dt = \int_a^b 1 \cdot dt = (b-a)$ or, $\dfrac{f(x_0)}{||x_0||} = (b-a)$.

Hence, $\sup\limits_{x \neq \theta} \dfrac{||f(x)||}{||x||} = (b-a)$.

4. **Definite integral on $C([a,b])$;**

If $K(\cdot, \cdot)$ is a continuous function on $[a,b] \times [a,b]$

and $\quad F(x)(s) = \displaystyle\int_a^b K(s,t)x(t)dt, \quad x \in C([a,b]), \ s \in [a,b]$

then $\quad |F(x)(s)| \leq \displaystyle\int_a^b |K(s,t)|\,|x(t)|dt$

$$\leq \int_a^b |K(s,t)| \sup_t |x(t)|dt$$

$$= ||x|| \int_a^b |K(s,t)|dt \leq \sup_s \int_a^b |K(s,t)|dt||x||$$

$$||F|| = \sup_{x \in \theta} \frac{|Fx(s)|}{||x||} \leq \sup_{s \in [a,b]} \left\{ \int_a^b |K(s,t)|dt; \ s \in [a,b] \right\}$$

Since $|K(s,t)|$ is a continuous function of s defined over a compact interval $[a,b]$, $\exists$ a $s_0 \in [a,b]$ s.t.

$$|K(s_0,t)| = \sup_{s \in [a,b]} |K(s,t)|$$

Thus, $||F|| = \displaystyle\sup_{x \in [a,b]} \int_a^b |K(s_0,t)|dt$

4.3.7 *Geometrical interpretation of norm of a linear functional*

Consider in a normed linear space E over $\mathbb{R}(\mathbb{C})$ a linear functional $f(x)$. The equation $f(x) = c$, where $f(x) = \sum_{i=1}^{\infty} c_i x_i$, is called a **hyperplane**. This is because in n-dimensional Euclidean space E_n, such an equation of the form $f(x) = c$ represents a n-dimensional plane.

Now, $|f(x)| \leq ||f||\,||x||$. If $||x|| \leq 1$, i.e., x lies in a unit ball, then $f(x) \leq ||f||$. Thus the hyperplane $f(x) = ||f||$ has the property that all the unit balls $||x|| \leq 1$ lie completely to the left of this hyperplane (because $f(x) < ||f||$ holds for the points of the ball $||x|| < 1$). The plane $f(x) = ||f||$ is called the **support of the ball** $||x|| \leq 1$. The points on the surface of the ball $||x|| = 1$ lie on the hyperplane. All other points within the ball $||x|| = 1$ lie on one side of the hyperplane. Such a hyperplane is also called a **supporting hyperplane**.

Problems

1. Find the norm of the linear functional f defined on $C([-2,2])$ by

$$f(x) = \int_{-2}^0 x(t)dt - \int_0^2 x(t)dt$$

[Ans. $||f|| = 4$]

2. Let f be a bounded linear functional on a complex normed linear space. Show that $\bar{f}$, although bounded, is not linear (the bar denotes the complex conjugate).

3. The space $C'([a, b])$ is the normed linear space of all continuously differentiable functions on $J = [a, b]$ with norm defined by

$$||x|| = \max_{t \in J} |x(t)| + \max_{t \in J} |x'(t)|$$

   Show that all the axioms of a norm are satisfied.

   Show that $f(x) = x'(\alpha)$, $\alpha = (a + b)/2$ defines a bounded linear functional on $C'([a, b])$.

4. Given $F(x)(s) = \int_0^1 \dfrac{st}{2 - st} x(t) dt$, $x \in C([0, 1])$, $s \in [0, 1]$, show that $||F|| = 1$.

5. If X is a subspace of a vector space E over $\mathbb{R}$ and f is a linear functional on E such that $f(X)$ is not the whole of $\mathbb{R}$, show that $f(y) = 0$ for all $y \in X$.

4.4   The Space of Bounded Linear Operators

In what follows we want to show that the set of bounded linear operators mapping a normed linear space $(E_x, || \cdot ||_{E_x})$ into another normed linear space $(E_y, || \cdot ||_{E_y})$ forms again a normed linear space.

4.4.1   *Definition: space of bounded linear operators*

Let two bounded linear operators, A_1 and A_2, map a normed linear space $(E_x, || \cdot ||_{E_x})$ into the normed linear space $(E_y, || \cdot ||_{E_y})$.

We can define addition and scalar multiplication by

$$(A_1 + A_2)x = A_1 x + A_2 x,$$
$$A_1(\lambda x) = \lambda A x \quad \text{for all scalars } \lambda \in \mathbb{R}(\mathbb{C})$$

This set of linear operators forms a linear space denoted by $(E_x \to E_y)$.

We would next show that $(E_x \to E_y)$ is a normed linear space. Let us define the norm of $\boldsymbol{A}$ as $||A|| = \sup\limits_{||x|| < 1} ||Ax||_{E_y}$. Then $||A|| \geq 0$. If $||A|| = 0$, i.e., if $\sup\limits_{||x||_{E_x} \leq 1} ||Ax||_{E_y} = 0$ then $||Ax||_{E_y} = 0$ for all x s.t. $||x||_{E_x} \leq 1$.

However because of the homogeneity of A, $Ax = 0$ for all x and therefore $A = 0$.

Now, $||\lambda A|| = \sup\limits_{||x|| \leq 1} ||\lambda A x|| = |\lambda| \sup\limits_{||x|| \leq 1} ||Ax|| = |\lambda|\, ||A||$

$||A + B|| = \sup\limits_{||x|| \leq 1} ||(A + B)x|| \leq \sup\limits_{||x|| \leq 1} ||Ax|| + \sup\limits_{||x|| \leq 1} ||Bx||$

$\qquad\qquad = ||A|| + ||B||$

Thus, the space of **bounded linear operators** is a normed linear space.

4.4.2 Theorem

If E_y is complete, the space of bounded linear operators is also complete and is consequently a Banach space.

Proof: Let us be given a Cauchy sequence $\{A_n\}$ of linear operators. Then with respect to the norm is a sequence of linear operators such that $||A_n - A_m|| \to 0$ as $n, m \to \infty$. Hence, $||A_n x - A_m x|| \leq ||A_n - A_m|| \, ||x|| \to 0$, as $n, m \to \infty$, for any x.

Therefore the sequence $\{A_n x\}$ of elements of E_y is a Cauchy sequence for any fixed x. Now, since E_y is complete $\{A_n x\}$ has some limit, y. Thus, every $y \in E_y$ is associated with some $x \in E_x$ and we obtain some operator A defined by the equation $Ax = y$. Such an operator A is additive and homogeneous. Because

(i) $\quad A(x_1 + x_2) = \lim_{n \to \infty} A_n(x_1 + x_2) = \lim_{n \to \infty} A_n x_1 + \lim_{n \to \infty} A_n x_2$
$$= Ax_1 + Ax_2, \quad \text{for } x_1, x_2 \in E_x$$

(ii) $\quad A(\lambda x_1) = \lim_{n \to \infty} A_n(\lambda x_1) = \lambda \lim_{n \to \infty} A_n x_1$
$$= \lambda A x_1, \ \forall \ \lambda \in \mathbb{R}(\mathbb{C})$$

To show that A is bounded, we note that

$$||A_n - A_m|| \to 0 \quad \text{as } n \to \infty$$

Hence $||A_n|| - ||A_m|| \to 0$, i.e., $\{||A_n||\}$ is a Cauchy sequence and is therefore bounded. Then there is a constant k s.t. $||A_n|| \leq k$ for all n. Consequently, $||A_n x||_{E_y} \leq k||x||_{E_x}$ for all n and hence

$$||Ax||_{E_y} = \lim_{n \to \infty} ||A_n x||_{E_y} \leq k||x||_{E_x}$$

Hence, it is proved that A is bounded. Since in addition, A is additive and homogeneous, A is a bounded linear operator.

Next, we shall prove that A is the limit of the sequence $\{A_n\}$ in the sense of **norm convergence**, in a space of linear operators.

Because of the convergence of $\{A_n x\}$, there is an index n_0 for every $\epsilon > 0$ s.t.

$$||A_{n+p} x - A_n x||_{E_y} < \epsilon \tag{4.18}$$

for $n \geq n_0$, $p \geq 1$ and all x with $||x|| \leq 1$.

Taking the limit in (4.18) as $p \to \infty$, we get

$$||Ax - A_n x||_{E_y} < \epsilon \text{ for } n \geq n_0,$$

and all x with $||x|| \leq 1$. Hence for $n \geq n_0$,

$$||A_n - A|| = \sup_{||x|| \leq 1} ||(A_n - A)x||_{E_y} < \epsilon$$

Consequently, $A = \lim\limits_{n \to \infty} A_n$ in the sense of norm convergence in a space of bounded linear operators and completeness of the space is proved.

We next discuss the composition of two bounded linear operators, each mapping $E_x \to E_x$, E_x being a normed linear space.

4.4.3 Theorem

Let E_x be a normed linear space over $\mathbb{R}(\mathbb{C})$. If $A, B \in (E_x \longrightarrow E_x)$, then

$$AB \in (E_x \longrightarrow E_x) \quad \text{and} \quad ||AB|| \le ||A|| \, ||B||$$

Proof: Since $A, B : (E_x \longrightarrow E_x)$, we have

$$
\begin{aligned}
(AB)(\alpha x + \beta y) &= A(B(\alpha x + \beta y)) = A(\alpha Bx + \beta By) \\
&= \alpha(AB)x + \beta(AB)y \; \forall \, x, y \in E_x, \; \forall \, \alpha, \beta \in \mathbb{R}(\mathbb{C})
\end{aligned}
$$

Furthermore, A and B being bounded operators

$$x_n \longrightarrow x \Rightarrow AB(x_n) = A(Bx_n) \longrightarrow ABx$$

showing that AB is bounded and hence continuous.

$$||(AB)(x)||_{E_x} = ||A(Bx)||_{E_x} \le ||A|| \, ||Bx||_{E_x} \le ||A|| \, ||B|| \, ||x||_{E_x}$$

Hence, $\quad ||AB|| = \sup\limits_{x \ne \theta} \dfrac{||(AB)x||_{E_x}}{||x||_{E_x}} \le ||A|| \, ||B||$

4.4.4 Example

1. Let $A \in (\mathbb{R}^m \longrightarrow \mathbb{R}^n)$ where both $\mathbb{R}^m$ and $\mathbb{R}^n$ are normed by l_1 norm.

Then $||A||_{l_1} = \max\limits_{1 \le j \le n} \sum\limits_{i=1}^{n} |a_{ij}|$ 　　　　　　　　　　　　(4.19)

Proof: For any $x \in \mathbb{R}^n$,

$$
\begin{aligned}
||Ax||_{l_1} &= \sum_{i=1}^{m} \Big| \sum_{j=1}^{n} a_{ij} x_j \Big| \le \sum_{i=1}^{m} \sum_{j=1}^{n} |a_{ij}| \, |x_j| \\
&\le \sum_{j=1}^{n} |x_j| \sum_{i=1}^{m} |a_{ij}| \\
&\le \Big(\max_{1 \le j \le n} \sum_{i=1}^{m} |a_{ij}| \Big) ||x||_1
\end{aligned}
$$

or $\quad ||A|| = \sup\limits_{x \ne \theta} \dfrac{||Ax||_1}{||x||_1} \le \max\limits_{1 \le j \le n} \sum\limits_{i=1}^{n} |a_{ij}|$.

We have to next show that there exists some $x \in \mathbb{R}^n$ s.t. the RHS in the above inequality is attained.

Let k be an index for which the maximum in (4.18) is attained; then

$$\|Ae^k\|_1 = \sum_{i=1}^{m} |a_{ik}| = \max_{1 \le j \le n} \sum_{i=1}^{m} |a_{ij}|,$$

where, $e^k = \begin{Bmatrix} 0 \\ 0 \\ \vdots \\ 1 \\ \vdots \\ 0 \end{Bmatrix}$, i.e. the maximum in (4.18) is attained for the kth

coordinate vector.

Problem

1. Let $A \in (\mathbb{R}^n \longrightarrow \mathbb{R}^m)$ where $\mathbb{R}^n$ are normed by l_∞ norm.

   Show that (i) $\|A\|_{l_\infty} = \max_{1 \le i \le m} \sum_{j=1}^{n} |a_{ij}|$

   (ii) $\|A\|_{l_2} = \sqrt{\lambda}$ where λ is the maximum eigenvalue of $A^T A$.

   [Hint. $\|Ax\|^2 = (Ax)^T Ax = (x^T A^T Ax)$]

2. (Limaye [33]) Let $A = (a_{ij})$ be an infinite matrix with scalar entries and

$$\alpha_{p,r} = \begin{cases} \sup\limits_{j=1,2,\ldots} \left(\sum\limits_{i=1}^{\infty} |a_{ij}|^r \right)^{1/r} & \text{if } p = 1,\ 1 \le r < \infty \\[2em] \sup\limits_{i=1,2,\ldots} \left(\sum\limits_{j=1}^{\infty} |a_{ij}|^q \right)^{1/q} & \text{if } 1 < p \le \infty,\ \dfrac{1}{p} + \dfrac{1}{q} = 1,\ r = \infty \\[2em] \sup\limits_{i,j=1,2,\ldots} |a_{ij}| & \text{if } p = 1,\ r = \infty \end{cases}$$

If $\alpha_{pr} < \infty$ then show that A defines a continuous linear map from l_p to l_r and its operator norm equals $\alpha_{p,r}$.

[Note that, $\alpha_{1,1} = \alpha_1 = \sup\limits_{j=1,2,\ldots} \left(\sum\limits_{i=1}^{\infty} |a_{ij}| \right)$ and $\alpha_{\infty,\infty} = \alpha_\infty = \sup\limits_{i=1,2,\ldots} \left\{ \sum\limits_{j=1}^{\infty} |a_{ij}| \right\}$].

4.5   Uniform Boundedness Principle

4.5.1   Definition: uniform operator convergence

Let $\{A_n\} \subset (E_x \longrightarrow E_y)$ be a sequence of bounded linear operators, where E_x and E_y are complete normed linear spaces. $\{A_n\}$ is said to **converge uniformly** if $\{A_n\}$ converges in the **sense of norm**, i.e., $\|A_n - A_m\| \to 0$ as $n, m \longrightarrow \infty$.

4.5.2   Lemma

$\{A_n\}$ converges uniformly if and only if $\{A_n\}$ converges uniformly for every $x \in E_x$.

Proof: Let us suppose that $\{A_n\}$ converges uniformly, i.e., $\|A_n - A_m\| \longrightarrow 0$ as $n, m \to \infty$.

Hence, $\sup\limits_{x \neq \theta} \dfrac{\|(A_n - A_m)x\|}{\|x\|} \to 0$ as $n, m \to \infty$.

or in other words, given $\epsilon > 0$, there exists an $r > 0$, s.t.

$$\|A_n x - A_m x\| < \frac{\epsilon}{r} \cdot r$$

where $x \in B(0, r)$ and $n, m \geq n_0(\epsilon/r)$.

Hence the uniform convergence of $\{A_n x\}$ for any $x \in B(0, r)$ is established. Conversely, let us suppose that $\{A_n x\}$ is uniformly convergent for any $x \in B(0, 1)$. Hence, $\sup\limits_{\|x\| \leq 1} \|A_n x - A_m x\| \to 0$ as $n, m \to \infty$. Or, in other words, $\|A_n - A_m\| \to 0$ as $n, m \to \infty$. Using theorem 4.4.2, we can say

$$A_n \longrightarrow A \in (E_x \to E_y) \text{ as } n \to \infty$$

4.5.3   Definition: pointwise convergence

A sequence of bounded linear operators $\{A_n\}$ is said to **converge pointwise** to a linear operator A if, for every fixed x, the sequence $\{A_n x\}$ converge to Ax.

4.5.4   Lemma

If $\{A_n\}$ converge *uniformly* to A then $\{A_n\}$ converge *pointwise*.

Proof: $\|A_n - A\| \to 0$ as $n \to \infty \Rightarrow \sup\limits_{x \neq \theta} \dfrac{\|(A_n - A)x\|}{\|x\|} \to 0$ as $n \to \infty$.

Hence, if $\|x\| \leq r$, for $\epsilon/r > 0$, $\exists$ as n_0 s.t. for all $n \geq n_0$

$$\|A_n x - Ax\| < \frac{\epsilon}{r}\|x\| \leq \frac{\epsilon}{r} \cdot r = \epsilon,$$

i.e., $\{A_n\}$ is pointwise convergent.

On the other hand if $\{A_n\}$ is pointwise convergent, $\{A_n\}$ is not necessarily uniformly convergent as is evident from the example below.

Let H be a Hilbert space with basis $\{e_1, e_2, \ldots, e_n\}$ and for $x \in H$ let A_n denote the **projection of** x on H_n, the n-dimensional subspace spanned by $e_1, e_2, \ldots, e_n$. Then

$$A_n x = \sum_{i=1}^{n} \langle x, e_i \rangle e_i \longrightarrow \sum_{i=1}^{\infty} \langle x, e_i \rangle e_i = x$$

for every $x \in H$. Hence, $A_n \to I$ is the sense of pointwise convergence.

On the other hand, for $\epsilon < 1$, any n and $p > 0$,

$$||A_{n+p} e_{n+1} - A_n e_{n+1}|| = ||e_{n+1} - 0|| = ||e_{n+1}|| = 1 > \epsilon$$

Hence, the uniform convergence of the sequence $\{A_n\}$ in the unit ball $||x|| < 1$ of the space H does not hold.

4.5.5 Theorem

If the spaces E_x and E_y are complete, then the space of bounded linear operators is also complete in the sense of pointwise convergence.

Proof: Let $\{A_n\}$ of bounded linear operators converge pointwise. Since $\{A_n\}$ is a Cauchy sequence for every x, there exists a limit $y = \lim\limits_{n \to \infty} A_n x$ for every x.

Since E_y is complete $y \in E_y$. This asserts the existence of an operator A such that $Ax = y$. That A is linear can be shown as follows

for, $x_1, x_2 \in E_x$,

$$A(x_1 + x_2) = \lim_{n \to \infty} A_n(x_1 + x_2) = \lim_{n \to \infty} (A_n x_1 + A_n x_2)$$
$$= Ax_1 + Ax_2, \text{ which shows } A \text{ is additive.}$$

Again, for $\lambda \in \mathbb{R}(\mathbb{C})$,

$$A(\lambda x) = \lim_{n \to \infty} A_n(\lambda x) = \lambda \lim_{n \to \infty} A_n x$$
$$= \lambda Ax, \text{ showing } A \text{ is homogeneous.}$$

That A is bounded can be proved by making an appeal to Banach-Steinhans theorem. (4.5.6)

4.5.6 Uniform boundedness principle

Uniform boundedness principle was discovered by Banach and Steinhaus. It is one of the basic props of functional analysis. Earlier Lebesgue first discovered the principle in his investigations of Fourier series. Banach and Steinhans isolated and developed it as a general principle.

4.5.7 Theorem (Banach and Steinhaus)

If a sequence of bounded linear operators is a Cauchy sequence at every point x of a Banach space E_x, the sequence $\{||A_n||\}$ of these operators is uniformly bounded.

Proof: Let us suppose the contrary. We show that the assumption implies that the set $\{||A_n x||\}$ is not bounded on any closed ball $B(x_0, \epsilon)$.

Any $x \in B(x_0, \epsilon)$ can be written as

$$x_0 + \frac{\epsilon}{||\xi||}\xi \quad \text{for any } \xi \in E_x$$

In fact, if $||A_n x|| \leq C$ for all n and if all x is in some ball $B(x_0, \epsilon)$, then

$$||A_n \left(x_0 + \frac{\epsilon}{||\xi||}\xi \right) || \leq C$$

or, $\quad \dfrac{\epsilon}{||\xi||}||A_n \xi|| - ||A_n x_0|| \leq C$

or, $\quad ||A_n \xi|| \leq \dfrac{C + ||A_n x_0||}{\epsilon}||\xi||$

Since the norm sequence $\{||A_n x_0||\}$ is bounded due to $\{A_n x_0\}$ being a convergent sequence, it follows that

$$||A_n \xi|| \leq C_1 ||\xi|| \quad \text{where } C_1 = \frac{C + ||A_n x_0||}{\epsilon}.$$

The above inequality yields

$$||A_n|| = \sup_{\xi \neq 0} \frac{||A_n \xi||}{||\xi||} \leq C_1$$

But this contradicts the hypothesis, because

$$||A_n x|| \leq C \ \forall \ x \in B(x_0, \epsilon) \Rightarrow ||A_n|| \leq C_1$$

Next, let us suppose $\overline{B}(x_0, \epsilon_0)$ is any closed ball in E_x. The sequence $\{||A_n x||\}$ is not bounded on it.

Hence, there is an index n_0 and an element

$$x_1 \in B_0(x_0, \epsilon_0) \text{ s.t. } ||A_{n_1} x_1|| > 1$$

By continuity of the operator A_{n_1}, then the above inequality holds in some closed ball $\overline{B}_1(x_1, \epsilon_1) \subset B_0(x_0, \epsilon_0)$. (see fig. 4.1) The sequence $\{||A_n x||\}$ is again not bounded on $\overline{B}(x_1, \epsilon_1)$, and therefore there is an index n_2 s.t. $n_2 > 0$, and an element $x_2 \in \overline{B}_1(x_1, \epsilon_1)$ s.t. $||A_{n_2} x_2|| > 2$.

Since A_{n_2} is continuous the above inequality must hold in some ball $\overline{B}_2(x_2, \epsilon_2) \subseteq \overline{B}_1(x_1, \epsilon_1)$ and so on. If we continue this process and let $\epsilon_n \to 0$ as $n \to \infty$, there is a point $\overline{x}$ belonging to all balls $\overline{B}_r(x_n, \epsilon_n)$. At this point

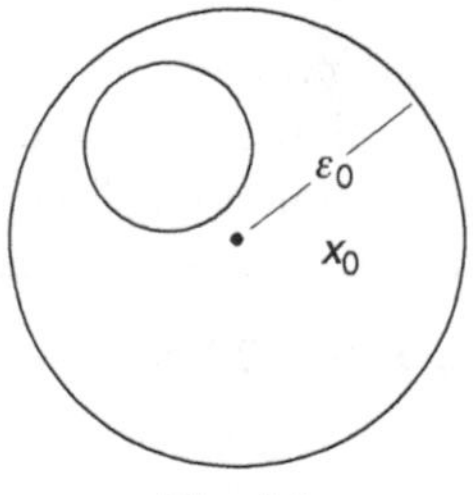

$$||A_{n_k} \overline{x}|| \geq k$$

Fig. 4.1

which contradicts the hypothesis that $\{A_n x\}$ converges for all $x \in E_x$. Hence the theorem.

Now we revert to the operator

$$Ax = \lim_{n \to \infty} A_n x$$

The inequality $\|A_n x\| \leq M\|x\|$, $n = 1, 2, \ldots$ holds good.

Now, given $\epsilon > 0$, there exists $n_0(> 0)$ s.t. for $n \geq n_0$

$$\|Ax\| \leq \|A_n x\| + \|(A - A_n)x\| \leq (M + \epsilon)\|x\| \quad \text{for} \quad n \geq n_0$$

Making $\epsilon \to 0$, we get

$$\|Ax\| \leq M\|x\|$$

4.5.8 Theorem (uniform boundedness principle)

Let E_x and E_y be two Banach spaces. Let $\{A_n\}$ be a sequence of bounded linear operators mapping $E_x \longrightarrow E_y$ s.t.

(i) $\{A_n x\}$ is a bounded subset of E_y for each $x \in E_x$,

Then the sequence $\{\|A_n\|\}$ of norms of these operators is bounded.

Proof: Let $S_k = \{x : \|A_n x\| \leq k, \ x \in E_x\}$.

Clearly, S_k is a subset of E_x.

Since $\{\|A_n x\|\}$ is bounded for any x, E_x can be expressed as

$$E_x = \bigcup_{k=1}^{\infty} S_k$$

Since E_x is a Banach space, it follows by the Baire's Category theorem 1.3.7 that there exists at least one S_{k_0} with non-empty interior and thus contains a closed ball $B_0(x_0, r_0)$ with centre x_0 and radius $r_0 > 0$.

Thus $\|A_n x_0\| \leq k_0$

Now, $x = x_0 + r_0 \dfrac{\xi}{\|\xi\|}$ would belong to the $B_0(x_0, r_0)$ for every $\xi \in E_x$.

Thus, $\left\| r_0 \dfrac{A_n \xi}{\|\xi\|} + A_n x_0 \right\| \leq k_0$

Hence, $\dfrac{r_0}{\|\xi\|}\|A_n \xi\| - \|A_n x_0\| \leq k_0$

or $\quad \|A_n \xi\| \leq \dfrac{k_0 + \|A_n x_0\|}{r_0}\|\xi\| \leq \dfrac{2k_0}{r_0}\|\xi\|$

Thus, $\|A_n\| = \sup_{\xi \neq 0} \dfrac{\|A_n \xi\|}{\|\xi\|} \leq \dfrac{2k_0}{r_0}$

Hence, $\{\|A_n\|\}$ is bounded.

4.5.9   Remark

The theorem does not hold unless E_x is a Banach space, as it follows from the following example.

4.5.10   Examples

1. Let $E_x = \{x = \{x_j\} : \text{only a finite number of } x_j\text{'s are non-zero}\}$.

$$\|x\| = \sup_j |x_j|$$

Consider the mapping, which is a linear functional defined by $f_n(x) = \sum_{j=1}^{n} x_j$ then

$$|f_n(x)| = \left| \sum_{j=1}^{n} x_j \right| \leq \sum_{j=1}^{n} |x_j| \leq \sum_{j=1}^{m} |x_j|, \text{ since } x_i = 0 \text{ for } i > m.$$

$$|f_n(x)| \leq \sum_{j=1}^{m} |x_j| \leq m\|x\|$$

Hence, $\{|f_n(x)|\}$ is bounded.

Now, $|f_n(x)| = \left| \sum_{j=1}^{n} x_j \right| \leq \sum_{j=1}^{n} |x_j| \leq n\|x\|.$

Hence $\|f_n\| \leq n$

Next, consider the element $\xi = \{\xi_1, \xi_2, \dots, \xi_i, \dots\}$

where $\xi_i = 1 \quad 1 \leq i \leq n$

$\qquad = 0 \quad i > n.$

$\|\xi\| = 1$

$f(\xi_i) = \sum_{n} \xi_i = n = n\|\xi\|$

$\dfrac{|f_n(\xi)|}{\|\xi\|} = n$

Thus $\{f_n(x)\}$ is bounded but $\{\|f_n\|\}$ is *not bounded*. This is because $E_x = \{x = \{x_j\} : \text{only for a finite number of } x_j\text{'s is non-zero}\}$ is **not** a Banach space.

2. Let E_x be the set of polynomials $x = x(t) = \sum_{n=0}^{\infty} p_n t^n$ where $p_n = 0$ for $n > N_x$.

Let $\|x\| = \max[|p_n|, \ n = 1, 2, \dots]$

Let $f_n(x) = \sum_{k=0}^{n-1} p_k$. The functionals f_n are continuous linear functionals on E_x. Moreover, for every $x = p_0 + p_1 t + \cdots + p_m t^m$, it is clear that for every n, $|f_n(x)| \leq (m+1)||x||$, so that $\{|f_n(x)|\}$ is bounded. For $f_n(x)$ we choose $x(t) = 1 + t + \cdots + t^n$.

Now, $||f_n|| = \sup_{x \neq \theta} \dfrac{|f_n(x)|}{||x||} \geq n$ since $||x|| = 1$ and $|f_n(x)| = n$.

Hence $\{||f_n||\}$ is unbounded.

4.6   Some Applications

4.6.1   *Lagrange's interpolation polynomial*

Lagrange's interpolation formula is to find the form of a given function in a given interval, when the values of the function is known at not necessarily equidistant interpolating points within the said interval. In what follows, we want to show that although the Lagrangian operator converges pointwise to the identity operator, it is not uniformly convergent.

For any function f defined on the interval $[0,1]$ and any partition $0 \leq t_1 < t_2 < \cdots t_n \leq 1$ of $[0,1]$, there is a polynomial of degree $(n-1)$ which interpolates to f at the given points, i.e., takes the values $f(t_i)$ at $t = t_i$, $i = 1, 2, \ldots, n$. This is called the Lagrangian interpolation polynomial and is given by

$$L_n f = \sum_{k=1}^{n} w_k^{(n)}(t) f(t_k^n) \tag{4.20}$$

where $w_k^{(n)}(t) = \dfrac{p_n(t)}{p_n'(t)(t - t_k^n)}$ \tag{4.21}

and $p_n(t) = \displaystyle\prod_{k=1}^{n}(t - t_k^n).$ \tag{4.22}

4.6.2   *Theorem*

We are given some points on the segment $[0,1]$ forming the infinite triangular matrix,

$$T = \begin{bmatrix} t_1^1 & 0 & 0 & \cdots & 0 \\ t_1^2 & t_2^2 & 0 & \cdots & 0 \\ t_1^3 & t_2^3 & t_3^3 & \cdots & 0 \\ \cdots & \cdots & \cdots & \cdots & \cdots \\ \cdots & \cdots & \cdots & \cdots & \cdots \end{bmatrix}. \tag{4.23}$$

For a given function, $f(t)$ defined on $[0,1]$, we construct the Lagrangian interpolation polynomial $L_n f$ whose partition points are the points of nth row of (4.23),

$$L_n f = \sum_{k=1}^{n} w_k^{(n)}(t) f(t_k^n)$$

where $\quad w_k^{(n)} = \dfrac{p_n(t)}{p_n'(t)(t - t_k^n)}, \quad p_n(t) = \prod_{k=1}^{n} (t - t_k^n).$

For every choice of the matrix (4.23), there is a continuous function $f(t)$ s.t. $L_n f$ does not uniformly converge to $f(t)$ as $n \to \infty$.

Proof: Let us consider L_n as an operator mapping the function $f(t) \in C([0,1])$ into the elements of the same space and put

$$\lambda_n = \max_t \lambda_n(t) \quad \text{where} \quad \lambda_n(t) = \sum_{k=1}^{n} |w_k^{(n)}(t)|$$

Now, $\quad \|L_n f\| = \max_t \left| \sum_{k=1}^{n} w_k^{(n)}(t) f(t_k^n) \right|$

$$\leq \max_t \sum_{k=1}^{n} |w_k^{(n)}(t)| \max_t |f(t_k^n)|$$

$$= \lambda_n \|f\|, \quad \text{where} \quad \lambda_n = \max_t \sum_{k=1}^{n} |w_k^{(n)}(t)|$$

On $C([0,1]) \quad \|f\| = \max_t |f(t)|$

$$\|L_n\| = \sup_{f \neq \theta} \frac{\|L_n f\|}{\|f\|} \leq \lambda_n$$

Since $\lambda_n(t)$ is a continuous function defined on a closed and bounded set $[0,1]$, the supremum is attained.

Hence, $\|L_n\| = \lambda_n$.

On the other hand, the Bernstein inequality (See Natanson [40]) $\lambda_n > \dfrac{\ln n}{8\sqrt{\pi}}$ holds.

Consequently $\|L_n\| \longrightarrow \infty$ as $n \to \infty$.

This proves the said theorem, because if $L_n f \longrightarrow f$ uniformly for all $f(t) \in C([0,1])$, then the norm $\|L_n\|$ must be bounded.

4.6.3 *Divergence of Fourier series of continuous functions*

In section 7 of Chapter 3 we have introduced, Fourier series and Fourier coefficients in a Hilbert space. In $L_2([-\pi, \pi])$ the orthonormal set can be taken as

$$e_n(t) = \left\{ \frac{e^{int}}{\sqrt{2\pi}}, \ n = 0, 1, 2, \ldots \right\}$$

If $x(t) \in L_2([-\pi, \pi])$ then $x(t)$ can be written as

$$\sum_{n=-\infty}^{\infty} x_n e_n$$

where $x_n = \langle x, e_n \rangle = \frac{1}{\sqrt{2\pi}} \int_{-\pi}^{\pi} x(t) e^{-int} dt$

$$= \frac{1}{\sqrt{2\pi}} \int_{-\pi}^{\pi} x(t) \cos nt \, dt - \frac{1}{\sqrt{2\pi}} \int_{-\pi}^{\pi} x(t) \sin nt \, dt$$

$$= c_n - i d_n \ (\text{say})$$

Then, $\bar{x}_n = c_n + i d_n$

Thus $\displaystyle\sum_{n=-\infty}^{\infty} x_n e_n = x_0 e_0 + \sum_{n=1}^{\infty} \frac{2 c_n \cos nt}{\sqrt{2\pi}} + \sum_{n=1}^{\infty} \frac{2 d_n \sin nt}{\sqrt{2\pi}}$ (4.24)

Here, c_n and d_n are called the Fourier coefficient of $x(t)$.

Note 4.6.1. It may be noted that the completeness of $\{e_n\}$ is equivalent to the statements that each $x \in L_2$ has the Fourier expansion

$$x(t) = \frac{1}{\sqrt{2\pi}} \sum_{n=-\infty}^{\infty} x_n e^{int}$$ (4.25)

It must be emphasized that this expansion is not to be interpreted as saying that the series **converges pointwise to** the function.

One can conclude that the partial sums in (4.25), i.e., the vector

$$u_n(t) = \frac{1}{\sqrt{2\pi}} \sum_{k=-\infty}^{n} x_k e^{ikt}$$ (4.26)

converges to the vector x **in the sense of L_2**, i.e.,

$$\|u_n(t) - x(t)\| \longrightarrow 0 \text{ as } n \to \infty$$

This situation is often expressed by saying that x is the **limit in the mean** of u_n's.

4.6.4 Theorem

Let $E = \{x \in C([-\pi, \pi]) : x(\pi) = x(-\pi)\}$ with the sup norm. Then the Fourier series of every x in a dense subset of E diverges at 0. We recall (see equation (4.25)) that

$$x(t) = x_0 + \sum_{n=1}^{\infty} \frac{2c_n \cos nt}{\sqrt{2\pi}} + \sum_{n=1}^{\infty} \frac{2d_n \sin nt}{\sqrt{2\pi}} \tag{4.27}$$

where $x_0 = \dfrac{1}{\sqrt{2\pi}} \displaystyle\int_{-\pi}^{\pi} x(t)dt$ \hfill (4.28)

$$c_n = \int_{-\pi}^{\pi} x(t) \cos nt\, dt \tag{4.29}$$

$$d_n = \int_{-\pi}^{\pi} x(t) \sin nt\, dt \tag{4.30}$$

For $x(t) \in E$

$$\|x\| = \max_{-\pi \le t \le \pi} |x(t)| \tag{4.31}$$

Let us take operator $A_n = u_n$ where $u_n(x)$ is the value at $t = 0$ of the nth partial sum of the Fourier series of x, since for $t = 0$ the sine terms are zero and the cosine is one we see from (4.28) and (4.29) that

$$u_n(x(0)) = \frac{1}{\sqrt{2\pi}} \left[x_0 + 2 \sum_{m=1}^{n} c_m \right]$$

$$= \frac{1}{\sqrt{2\pi}} \int_{-\pi}^{\pi} x(t) \left[2 \left\{ \frac{1}{2} + \sum_{m=1}^{n} \cos mt \right\} \right] dt \tag{4.32}$$

Now, $2 \sin \dfrac{1}{2}t \displaystyle\sum_{m=1}^{n} \cos mt = \sum_{m=1}^{n} 2 \sin \dfrac{1}{2}t \cos mt$

$$= \sum_{m=1}^{n} \left[\sin\left(m + \frac{1}{2}\right)t - \sin\left(m - \frac{1}{2}\right)t \right]$$

$$= -\sin \frac{1}{2}t + \sin\left(n + \frac{1}{2}\right)t$$

It may be noted that, except for the end terms, all other intermediate terms in the summation vanish in pairs.

Dividing both sides by $\sin \frac{1}{2}t$ and adding 1 to both sides, we have

$$1 + 2 \sum_{m=1}^{n} \cos mt = \frac{\sin\left(n + \frac{1}{2}\right)t}{\sin \frac{1}{2}t}$$

Consequently, the expression for $u_n(x)$ can be written in the simple form:

$$u_n(x) = \frac{1}{\sqrt{2\pi}} \int_{-\pi}^{\pi} x(t) D_n(t)\, dt, \text{ where } D_n(t) = \frac{\sin\left(n + \frac{1}{2}\right)t}{\sin \frac{1}{2}t}$$

It should be noted that $u_n(x)$ is a linear functional in $x(t)$.

We would next show that u_n is bounded. It follows from (4.29) and the above integral,

$$|u_n(x)| \leq \frac{1}{\sqrt{2\pi}} \int_{-\pi}^{\pi} |x(t)| |D_n(t)| dt$$

$$\leq \frac{1}{\sqrt{2\pi}} \int_{-\pi}^{\pi} |D_n(t)| dt ||x||$$

Therefore, $\quad ||u_n|| = \sup_{x \neq \theta} \frac{|u_n(x)|}{||x||} \leq \frac{1}{\sqrt{2\pi}} \int_{-\pi}^{\pi} |D_n(t)| dt$

$$= \frac{1}{\sqrt{2\pi}} ||D_n(t)||_1$$

where $|| \cdot ||_1$ denotes L_1-norm.

Actually the equality sign holds, as we shall prove.

Let us write $|D_n(t)| = y(t)D_n(t)$ where $y(t) = +1$ at every t at which $D_n(t) \geq 0$ and $y(t) = -1$ otherwise $y(t)$ is not continuous, but for any given $\epsilon > 0$ it may be modified to a continuous x of norm 1, such that for this x we have

$$\left| u_n(x) - \frac{1}{\sqrt{2\pi}} \int_{-\pi}^{\pi} |D_n(t)| dt \right| = \frac{1}{\sqrt{2\pi}} \left| \int_{-\pi}^{\pi} (x(t) - y(t)) D_n(t) dt \right| < \epsilon$$

If follows that

$$||u_n|| = \frac{1}{\sqrt{2\pi}} \int_{-\pi}^{\pi} |D_n(t)| dt$$

We next show that the sequence $\{||u_n||\}$ is unbounded. Since $\sin u \leq u$, $0 \leq u \leq \pi$, we note that

$$\left\{ \int_{-\pi}^{\pi} \left| \frac{\sin \left(n + \frac{1}{2}\right) t}{\sin \frac{1}{2} t} \right| dt \right\} \geq 4 \left\{ \int_{0}^{\pi/2} \left| \frac{\sin(2n + 1)u}{u} \right| du \right\}$$

$$= 4 \int_{0}^{\frac{(2n+1)\pi}{2}} \frac{|\sin v|}{v} dv$$

$$= 4 \sum_{k=0}^{2n} \int_{k\frac{\pi}{2}}^{(k+1)\frac{\pi}{2}} \frac{|\sin v|}{v} dv$$

$$\geq 4 \sum_{k=0}^{2n} \frac{1}{(k+1)\frac{\pi}{2}} \int_{\frac{k\pi}{2}}^{\frac{(k+1)\pi}{2}} |\sin v| dv$$

$$= \frac{8}{\pi} \sum_{k=0}^{2n} \frac{1}{(k+1)} \longrightarrow \infty \text{ as } n \to \infty$$

Because the harmonic series $\sum_{k=1}^{\infty} \frac{1}{k}$ diverges.

Hence $\{\|u_n\|\}$ is unbounded. Since E is complete this implies that there exists no $c > 0$ and finite such that $\|u_n(x)\| \leq c$ holds for all x. Hence there must be an $x' \in E$ such that $\{\|u_n(x')\|\}$ is unbounded. This implies that the Fourier series of that x' diverges at $t = 0$.

Problems [4.5 & 4.6]

1. Let E_x be a Banach space and E_y a normed linear space. If $\{T_n\}$ is a sequence in $(E_x \to E_y)$ such that $Tx = \lim_{n\to\infty} T_n x$ exists for each x in E_x, prove that T is a continuous linear operator.

2. Let E_x and E_y be normed linear spaces and $A : E_x \to E_y$ be a linear operator with the property that the set $\{\|Ax_n\| : n \in \mathbb{N}\}$ is bounded whenever $x_n \to \theta$ in E_x. Prove that $A \in (E_x \to E_y)$.

3. Given that E_x and E_y are Banach spaces and $A : E_x \to E_y$ is a bounded linear operator, show that either $A(E_x) = E_y$ or is a set of the first category in E_y.

   [Hint: A set $X \subseteq E_x$ is said to be of the **first category** in E_x if it is the union of countably many nowhere dense sets in E_x].

4. If E_x is a Banach space, and $\{f_n(x)\}$ a sequence of continuous linear functionals on E_x such that $\{|f_n(x)|\}$ is bounded for every $x \in E_x$, then show that the sequence $\{\|f_n\|\}$ is bounded.

   [Hint: Consult theorem 4.5.7]

5. Let E_x and E_y be normed linear spaces. E be a bounded, complete convex subset of E_x. A mapping A from E_x to E_y is called **affine** if

   $$A(\lambda a + (1 - \lambda)b) = \lambda A(a) + (1 - \lambda)A(b) \text{ for all } 0 < \lambda < 1,$$

   and $a, b \in E$.

   Let $\mathcal{F}$ be a set of continuous affine mappings from E_x to E_y. Then show that either the set $\{\|A(x)\| : A \in \mathcal{F}\}$ is unbounded for each x in some dense subset of E or else $\mathcal{F}$ is uniformly bounded in E_y.

6. Let E_x and E_y be Banach spaces and $A_n \in (E_x \longrightarrow E_y)$, $n = 1, 2, \ldots$. Then show that there is some $A \in (E_x \longrightarrow E_y)$ such that $A_n x \to Ax$ for every $x \in E_x$ if and only if $\{A_n\}$ converges for every x in some set whose span is dense in E_x and the set $\{\|A_n\| : n = 1, 2, \ldots\}$ is bounded.

7. Show that there exists a dense set Ω of $E_x = \{x \in C([-\pi, \pi]) : x(\pi) = x(-\pi)\}$ such that the Fourier series of every $x \in \Omega$ diverges at every rational number in $[-\pi, \pi]$.

4.7   Inverse Operators

In what follows, we introduce the notion of the inverse of a linear operator and investigate the conditions of its existence and uniqueness. This is, in other words, searching the conditions under which a given system of equations will have a solution or if the solution at all exists it is unique.

4.7.1 Definition: domain of an operator

Let a linear operator A map a subspace E of a Banach space E_x into a Banach space E_y. The subspace E on which the operator A is defined is called the **domain** of A and is denoted by $\mathcal{D}(A)$.

4.7.2 Example

Let $A : C^2([0,1]) \longrightarrow C[(0,1)]$ where $Au = f$, $A = -\frac{d^2}{dx^2}$, $u(0) = u(1) = 0$ and $f \in C([0,1])$.

Here, $\mathcal{D}(A) = \{u(x)|u(x) \in C^2([0,1]),\ u(0) = u(1) = 0\}$

4.7.3 Example

Let a linear operator A map $\mathcal{D}(A) \subset E_x \longrightarrow E_y$, where E_x and E_y are Banach spaces. The range of A is the subspace of E_y into which $\mathcal{D}(A)$ is mapped to, and the range of A is denoted by $\mathcal{R}(A)$.

4.7.4 Example

Let $y(s) = \int_0^s k(s,t)x(t)dt$ where $x(t) \in C([0,1])$, $K(s,t) \in C([0,1]) \times C([0,1])$.

$$\mathcal{R}(A) = \{y : y \in C([0,1]),\ y = Ax\}$$

4.7.5 Definition: null space of a linear operator

The null space of a linear operator A is defined by the set of elements of E, which is mapped into the null element and is denoted by $N(A)$.

Thus $N(A) = \{x \in E : Ax = \theta\}$.

4.7.6 Example

Let $A : R^2 \longrightarrow R^2$

$$A = \begin{pmatrix} 2 & 1 \\ 4 & 2 \end{pmatrix}$$

$$N(A) = \left\{ x \in R^2 : x_1 = -\frac{1}{2}x_2,\ \text{where } x = \begin{pmatrix} x_1 \\ x_2 \end{pmatrix} \right\}$$

4.7.7 Definition: left inverse and right inverse of a linear operator

A linear continuous operator B is said to be a **left inverse** of A if $BA = I$.

A linear continuous operator C is said to be a **right inverse** of A if $AC = 1$.

4.7.8 Lemma

If A has a left inverse B and a right inverse C, then $B = C$.

For $B = B(AC) = (BA)C = C$.

If A has a left inverse as well as a right inverse, then A is said to have an **inverse** and the inverse operator is denoted by A^{-1}.

Thus if A^{-1} exists, then by definition $A^{-1}A = AA^{-1} = I$.

4.7.9 Inverse operators and algebraic equations

Let E_x and E_y be two Banach spaces and A be an operator s.t. $A \in (E_x \longrightarrow E_y)$.

We want to know when one can solve

$$Ax = y \qquad (4.33)$$

Here y is a known element of the linear space E_y and $x \in E_x$ is unknown.

If $\mathcal{R}(A) = E_y$, we can solve (4.31) for each $y \in E_y$.

If $N(A)$ consists only one element then the solution is unique. Thus if $\mathcal{R}(A) = E_y$ and $N(A) = \{\theta\}$, we can assign to each $y \in E_y$ the unique solution of (4.33). This assignment gives the inverse operator A^{-1} of A. We next show that A^{-1} if it exists is linear. Let $x = A^{-1}(y_1 + y_2) - A^{-1}y_1 - A^{-1}y_2$. Then A being linear

$$Ax = AA^{-1}(y_1 + y_2) - A \cdot A^{-1}y_1 - A \cdot A^{-1}y_2$$
$$= y_1 + y_2 - y_1 - y_2 = \theta$$

Thus, $x = A^{-1}Ax = A^{-1}\theta = \theta$, i.e., $A^{-1}(y_1 + y_2) = A^{-1}y_1 + A^{-1}y_2$, proving A^{-1} to be additive. Analogously the homogeneity of A^{-1} is established.

Note 4.7.1. It may be noted that the **continuity** of the operator A in some topology does not necessarily imply the **continuity** of its inverse, i.e., an operator inverse to a bounded linear operator is not necessarily a bounded linear operator. In what follows we investigate sufficient conditions for the existence of the inverse to a linear operator.

4.7.10 Theorem (Banach)

Let a linear operator A map a normed linear (Banach) space E_x onto a normed linear (Banach) space E_y, satisfying for every $x \in E_x$ the condition

$$||Ax|| \geq m||x||, \; m > 0 \qquad (4.34)$$

m, some constant. Then the inverse bounded linear operator A^{-1} exists.

Proof: The condition (4.34) implies that A maps E_x onto E_y in a one-to-one fashion. If $Ax_1 = y$ and $Ax_2 = y$, then $A(x_1 - x_2) = \theta$ yields

$m||x_1 - x_2|| \leq ||A(x_1 - x_2)|| = 0$, whenever $x_1 = x_2$. Hence, there is a linear operator A^{-1}. The operator is bounded, as is evident from (4.34),

$$||A^{-1}y|| \leq \frac{1}{m}||A \cdot A^{-1}y|| = \frac{1}{m}||y||,$$

for every $y \in E_y$.

4.7.11 Theorem (Banach)

Let A and B be two bounded linear operators mapping a normed linear space E into itself, so that A and B are conformable for multiplication. Then
$$||AB|| \leq ||A||\,||B||.$$

If further $\{A_n\} \to A$ and $\{B_n\} \to B$ as $n \to \infty$, then $A_n B_n \longrightarrow AB$ as $n \to \infty$.

Proof: For any $x \in E$,

$$||ABx|| \leq ||A||\,||Bx|| \leq ||A||\,||B||\,||x||$$

or, $$||AB|| = \sup_{x \neq \theta} \frac{||ABx||}{||x||} < ||A||\,||B||.$$

Hence $\quad ||AB|| \leq ||A||\,||B||$

Now, since $\{A_n\}$ and $\{B_n\}$ are bounded linear operators,

$$||A_n B_n - AB|| = ||A_n B_n - A_n B|| + ||A_n B - AB||$$
$$\leq ||A_n||\,||B_n - B|| + ||A_n - A||\,||B||$$

$\to 0$ as $n \to \infty$, since $A_n \to A$ and $B_n \to B$ as $n \to \infty$

4.7.12 Theorem

Let a bounded linear operator A map E into E and let $||A|| \leq q < 1$. Then the operator $I + A$ has an inverse, which is a bounded linear operator.

Proof: In the space of operators with domain E and range as well in E, we consider the series

$$I - A + A^2 - A^3 + \cdots + (-1)^n A^n + \cdots \qquad (4.35)$$

Since $||A^2|| \leq ||A|| \cdot ||A|| = ||A||^2$ and analogously $||A^n|| \leq ||A||^n$, it follows for the partial sums S_n of the series (4.35), that

$$||S_{n+p} - S_n|| = ||(-1)^{n+1}A^{n+1} + (-1)^{n+2}A^{n+2} + \cdots + (-1)^{n+p}A^{n+p}||$$
$$\leq ||A^{n+1}|| + ||A^{n+2}|| + \cdots + ||A^{n+p}||$$
$$\leq q^{n+1} + q^{n+2} + \cdots + q^{n+p}$$
$$= q^{n+1}\frac{(1 - q^p)}{(1 - q)} \to 0 \text{ as } n \to \infty, \text{ since } p > 0,\ 0 < q < 1.$$

Hence, $\{S_n\}$ is a Cauchy sequence and the space of operators being complete, the sequence $\{S_n\}$ converges to a limit.

Let S be the sum of the series.

Then
$$S(I + A) = \lim_{n \to \infty} S_n(I + A)$$
$$= \lim_{n \to \infty} (I + A)(I - A + A^2 + \cdots + (-1)^n A^n)$$
$$= \lim_{n \to \infty} (I - A^{n+1}) = I$$

Hence $S = (I + A)^{-1}$

Let $x_1 = (I + A)^{-1} y_1$, $x_2 = (I + A)^{-1} y_2$, $x_1, x_2, y_1, y_2 \in E$

Then $y_1 + y_2 = (I + A)(x_1 + x_2)$

or, $x_1 + x_2 = (I + A)^{-1} y_1 + (I + A)^{-1} y_2 = (I + A)^{-1}(y_1 + y_2)$.

Hence, S is a linear operator. Moreover,

$$\|S\| \leq \sum_{n=0}^{\infty} \|A^n\| \leq \sum_{n=0}^{\infty} q^n = \frac{1}{1-q}, \quad 0 < q < 1 \tag{4.36}$$

Then $(I + A)^{-1}$ is a bounded linear operator.

4.7.13   Theorem

Let $A \cdot \in (E_x \longrightarrow E_y)$, A have a bounded inverse with $\|A^{-1}\| \leq \alpha$, $\|A - C\| \leq \beta$ and $\beta\alpha < 1$, then C has a bounded inverse and

$$\|C^{-1}\| \leq \frac{\alpha}{(1 - \alpha\beta)}$$

Proof: Since $\|I - A^{-1}C\| = \|A^{-1}(A - C)\| \leq \alpha\beta < 1$ and $A^{-1}C = I - (I - A^{-1}C)$, it follows from theorem 4.7.12 that $A^{-1}C$ has a bounded inverse and hence C has a bounded inverse.

Hence, $\|C^{-1}\| = \|(A^{-1}C)^{-1}A^{-1}\| = \|(I - (I - A^{-1}C))^{-1}A^{-1}\|$
$$\leq \|(I - (I - A^{-1}C))^{-1}\| \, \|A^{-1}\|$$
$$\leq \alpha \cdot \sum_{n=0}^{\infty} (\alpha\beta)^n \quad \text{using (4.35) and noting that}$$
$$\qquad\qquad\qquad\qquad \|I - A^{-1}C\| \leq \alpha\beta$$
$$= \frac{\alpha}{1 - \alpha\beta}$$

4.7.14   Example

Consider the integral operator

$$Cx = x(s) - \int_0^1 K(s, t)x(t)dt \tag{4.36'}$$

with continuous kernel $K(s,t)$ which maps the space $C([0,1])$ into $C([0,1])$. Let $K_0(s,t)$ be a **degenerate** kernel close to $K(s,t)$. That is, $K_0(s,t)$ is of the form $\sum_{i=1}^{n} a_i(s)b_i(t)$. In such a case, the equation

$$Ax = x(s) - \int_0^1 K_0(s,t)x(t)dt = y \tag{4.37}$$

can be reduced to a system of algebraic equations and finding the solution of equation (4.37) can be reduced to finding the solution of the concerned algebraic equations. Let us assume that equation (4.37) has a solution.

In order to know whether the integral equation

$$Cx = y \tag{4.38}$$

has a solution, we frame the equation (4.37) in such a manner that

$$w = \max_{t,s} |K(t,s) - K_0(t,s)| < \frac{1}{r} \tag{4.39}$$

where $r = ||R||$, $R = (r_{ij})$ being the matrix associated with the solution $x_0(s) = Ry$ of the linear algebraic system generated from equation (4.37). It follows from (4.38), (4.37) and (4.39)

$$||C - A|| \leq w < \frac{1}{r} \tag{4.40}$$

It follows from theorem 4.7.13 that equation (4.37) with a continuous kernel has a solution; if $x(t)$ be the solution, then

$$||x(t) - x_0(t)|| \leq \frac{w}{1 - wr}r^2$$

The above inequality gives an estimate as to how much the solution of equation (4.38) differs from the solution of (4.37), the explicit form of the solution of (4.38) being not known.

Finally, we obtain the following theorem.

4.7.15 Theorem (Banach)

If a bounded linear operator A maps the whole of the Banach space E_x onto the whole of the Banach space E_y in a one-to-one manner, then there exists a bounded linear operator A, which maps E_x onto E_y.

Proof: The operator A being one-to-one and onto has an inverse A^{-1}. We need to show that A^{-1} is bounded.

Let $S_k = \{y \in E_y : ||A^{-1}y|| \leq k||y||\}$.

E_y can be represented as

$$E_y = \bigcup_{k=0}^{\infty} S_k$$

Since E_y is a complete metric space, by Baire's category theorem (th. 1.4.19) at least one of the sets S_k is everywhere dense. Let this set be S_n. Let us take an element $y \in E_y$. Let $||y|| = l$. Then there exists $y_1 \in S_n$ such that

$$||y - y_1|| \leq \frac{l}{2}, \quad ||y_1|| \leq l.$$

This is possible, since $\overline{B}(0, l) \cap S_n$ is everywhere dense in $\overline{B}(0, l)$ and $y \in \overline{B}(0, 1)$.

Moreover, we can find an element $y_2 \in S_n$ such that

$$||(y - y_1) - y_2|| \leq \frac{l}{2^2}, \quad ||y_2|| \leq \frac{l}{2},$$

Continuing this process, we can find element $y_k \in S_n$, such that

$$||y - (y_1 + y_2 + \cdots + y_k)|| \leq \frac{l}{2^k}, \quad ||y_k|| \leq \frac{l}{2^{k-1}}$$

Making $k \longrightarrow \infty$, we have, $y = \lim\limits_{k \to \infty} \sum\limits_{i=1}^{k} y_i$.

Let $x_k = A^{-1} y_k$, then

$$||x_k||_{E_x} = ||A^{-1} y_k||_{E_x} \leq n ||y_k||_{E_y} \leq \frac{nl}{2^{k-1}}$$

Expressing $s_k = \sum\limits_{i=1}^{k} x_i$

$$||s_{k+p} - s_k|| = || \sum_{i=k+1}^{k+p} x_i || \leq \frac{nl}{2^{k-1}} \left(1 - \frac{1}{2^p}\right) < \frac{nl}{2^{k-1}}$$

E_x being complete, $\{s_k\}$ converges to some element $x \in E_x$ as $k \to \infty$.

Hence, $x = \lim\limits_{k \to \infty} \sum\limits_{i=1}^{k} x_i = \sum\limits_{i=1}^{\infty} x_i$.

Moreover, A being continuous

$$Ax = A \left(\lim_{k \to \infty} \sum_{i=1}^{k} x_i \right) = \lim_{k \to \infty} \left(\sum_{i=1}^{k} A x_i \right)$$

$$= \lim_{k \to \infty} \sum_{i=1}^{k} y_i = y$$

Hence, $||A^{-1} y|| = ||x|| = \lim\limits_{k} || \sum\limits_{i=1}^{k} x_i || \leq \lim\limits_{k} \sum\limits_{i=1}^{k} ||x_i||$

$$\leq \sum_{i=1}^{\infty} \frac{nl}{2^{i-1}} = 2nl = 2n\|y\|$$

Since y is an arbitrary element of E_y, that A^{-1} is a bounded linear operator is proved.

Note 4.7.2. This is further to the note 4.7.1.

(i) A bounded linear operator A mapping a Banach space E_x into a Banach space E_y may have an inverse which is not bounded.

(ii) An unbounded linear operator mapping $E_x \longrightarrow E_y$ may have a bounded inverse.

4.7.16 Examples

1. Let $E = C([0, 1])$ and let $Au = \int_0^t u(\zeta)d\zeta$.

Thus, A is a bounded linear operator mapping $C([0, 1])$ into $C([0, 1])$.

$$A^{-1} \text{ given by } A^{-1}u = \frac{d}{dt}u(t)$$

is an unbounded operator defined on the linear subspace of continuously differentiable functions such that $u(0) = 0$.

2. Let $E = C([0, 1])$. The operator B is given by

$$Bu = f(x) \tag{4.41}$$

where $B = -\dfrac{d^2}{dx^2}$

$$\mathcal{D}_B = \{u : u \in C^2([0, 1]),\ u(0) = u(1) = 0\}$$

Integrating equation (4.41) twice, we obtain,

$$u(x) = -\int_0^x ds \int_0^s f(t)dt + C_1 x + C_2$$

The condition $u(0) = 0$ immediately gives $C_2 = 0$ and consequently

$$u(x) = -\int_0^x ds \int_0^s f(t)dt + C_1 x$$

We change the order of integration and use Fubini's theorem [see 10.5]. This leads to

$$u(x) = -\int_0^x f(t)dt \int_t^x ds + C_1 x$$

$$= -\int_0^x (x-t)f(t)dt + C_1 x \tag{4.42}$$

$$u(1) = 0 \Rightarrow C_1 = \int_0^1 (1-t)f(t)dt \tag{4.43}$$

Using (4.43), (4.42) reduces to

$$u(x) = \int_0^x t(1-x)f(t)dt + \int_x^1 x(1-t)f(t)dt$$

$$= \int_0^1 K(x,t)f(t)dt = B^{-1}f \tag{4.44}$$

$K(x,t)$, the kernel of the integral equation (4.44) is given by

$$K(x,t) = \begin{cases} t(1-x) & 0 \le t \le x \\ x(1-t) & x \le t \le 1 \end{cases}$$

We note that B is not a bounded operator. For example, take $u(x) = \sin n\pi x$, $u(x) \in \mathcal{D}(B)$

Now, $\dfrac{d^2}{dx^2} \sin n\pi x = -n^2\pi^2 \sin n\pi x.$

Then, $\left\| \dfrac{d^2}{dx^2}(\sin n\pi x) \right\| = n^2\pi^2 \longrightarrow \infty$ as $n \to \infty.$

On the other hand

$$\|B^{-1}f\| \le \int_0^1 |K(x,t)|dt\|f\| \le \frac{1}{8}\|f\|$$

Hence B^{-1} is bounded.

4.7.17   *Operators depending on a parameter*

Let us consider the equation of the form

$$Ax - \lambda x = y \quad \text{or,} \quad (A - \lambda I)x = y \tag{4.45}$$

Here, λ is a parameter. Such equations occur frequently in Applied Mathematics and Theoretical Physics.

4.7.18   *Definition: homogeneous equation, trivial solution*

In case $y = \theta$ in equation (4.45) then it is called a **homogeneous** equation. Thus

$$(A - \lambda I)x = \theta \tag{4.46}$$

is a homogeneous equation. This equation always has a solution $x = \theta$, called the **trivial solution**.

4.7.19   *Definition: resolvent operator, regular values*

In case $(A - \lambda I)$ has an inverse $(A - \lambda I)^{-1}$, the operator $R_\lambda = (A - \lambda I)^{-1}$ is called the resolvent of (4.45). Equation (4.46) will then have a unique solution $x = \theta$. Those λ for which equation (4.45) has a unique solution for every y and the operator R_λ is bounded, are called **regular values**.

4.7.20   Definition: eigenvector, eigenvalue or characteristic value, spectrum

If the homogeneous equation (4.46) or the equation $Ax = \lambda x$ has a non-trivial solution x, then that x is called **the eigenvector**. The values of λ corresponding to the non-trivial solution, the eigenvector, are called the **eigenvalues** or **characteristic values**.

The collection of all non-regular values of λ is called the **spectrum** of the operator A.

4.7.21   Theorem

If in the equation $(A - \lambda I)x = y$, $(1/|\lambda|)||A||q < 1$ holds for λ, thus $A - \lambda I$ has an inverse operator; moreover,

$$R_\lambda = -\frac{1}{\lambda}\left(1 + \frac{A}{\lambda} + \frac{A^2}{\lambda^2} + \cdots\right)$$

If λ is a regular value, then $\lambda + \Delta\lambda$ for $|\Delta\lambda| < ||(A - \lambda I)^{-1}||^{-1}$ is also a regular value. This implies that the collection of regular values is an open set and hence the spectrum of an operator is a closed set.

Proof: The equation $(A - \lambda I)x = y$ can be written as

$$-\left(I - \frac{A}{\lambda}\right)x = \frac{1}{\lambda}y$$

Thus, by theorem 4.7.12 we can say, if $\frac{1}{|\lambda|}||A|| = q < 1$, then $(I - A/\lambda)$ will have an inverse and the concerned values of λ will be its regular values.

If in theorem 4.7.13 we take $C = A - (\lambda + \Delta\lambda)I$ then if $||C - (A - \lambda I)|| = |\Delta\lambda| < ||(A - \lambda I)^{-1}||^{-1}$ then we conclude that $C = A - (\lambda + \Delta\lambda)I$ has an inverse and $(\lambda + \Delta\lambda)$ is also a regular value.

4.7.22   Example

Let us consider the Fredholm integral equation of the second kind:

$$\phi(x) - \lambda \int_a^b K(x, s)\phi(s)ds = f(x) \qquad (4.47)$$

If we denote $A\phi$ by $\int_a^b K(x, s)\phi(s)ds$, equation (4.47) can be written as

$$(I - \lambda A)\phi = f \qquad (4.48)$$

The solution of the equation (4.47) can be expressed in the form of an infinite series

$$\phi(x) = f(x) + \sum_{m=1}^{\infty} \lambda^m \int_a^b K_m(x, s)f(x)ds \quad \text{(see Mikhlin [37])} \qquad (4.49)$$

where the mth iterated kernel $K_m(x, s)$ is given by

$$K_m'(x, s) = \int_a^b K_{m-1}(x, t)K(t, s)dt \qquad (4.50)$$

By making an appeal to theorem 4.7.21, we can obtain the resolvent operator R_λ as

$$R_\lambda f = (I - \lambda A)^{-1}f = [I + \lambda A + \lambda^2 A^2 + \cdots + \lambda^p A^p + \cdots]f$$

where $A^p f = \int_a^b K_p(t, s)f(s)ds$ is the pth iterate of the kernel $K(t, s)$.

Thus if $\|A\| < \dfrac{1}{|\lambda|}$, we get

$$\phi(x) = f(x) + \lambda \int_a^b R(x, s, \lambda)f(s)ds$$

Here, $R(x, s, \lambda)$, the resolvent of the kernel $K(t, s)$ is defined by,

$$R(x, s, \lambda) = K(x, s) + \lambda K_2(x, s) + \cdots + \lambda^p K_{p+1}(x, s) + \cdots$$

Problems

1. Let E_x and E_y be normed linear spaces over $\mathbb{R}(\mathbb{C})$. $A : E_x \longrightarrow E_y$ is a given linear operator. Show that

   (i) A is bijective $\Longrightarrow$ A has an inverse $\Longrightarrow$ $N(A) = \{\theta\}$

   (ii) A^{-1}, if it exists, is a linear operator.

2. $L(\mathbb{R}^n)$ *denotes the space of linear operators mapping* $\mathbb{R}^n \to \mathbb{R}^n$. Suppose that the mapping $A : \mathcal{D} \subset \mathbb{R}^m \to L(\mathbb{R}^n)$ is continuous at a point $x_0 \in D$ for which $A(x_0)$ has an inverse. Then, show that there is a $\delta > 0$ and a $\nu > 0$ so that $A(x)$ has an inverse and that

   $$\|A(x)^{-1}\| \le \nu \quad \forall\, x \in D \cap \overline{B}(x_0, \delta)$$

   Hint: Use theorem 4.7.13

3. **(Sherman-Morrison-Woodbury Formula)** Let $A \in L(\mathbb{R}^n)$ have an inverse and let U, V map $\mathbb{R}^n \to \mathbb{R}^m$, $m \le n$. Show that $A + UV^T$ has an inverse if and only if $(I + (V^T)^{-1})U$ has an inverse and that

   $$(A + UV^T)^{-1} = A^{-1} - A^{-1}U(I + V^T A^{-1}U)^{-1}V^T A^{-1}$$

   Hint: Use theorem 4.7.13

4. If L is a bounded linear operator mapping a Banach space E into E, show that L^{-1} exists if and only if there is a bounded linear operator K in E such that K^{-1} exists and

$$||I - KL|| < 1$$

   If L^{-1} exists then show that

$$L^{-1} = \sum_{n=0}^{\infty} (I - KL)^n K \quad \text{and} \quad ||L^{-1}|| \leq \frac{||K^{-1}||}{1 - ||I - KL||}$$

5. Use the result of problem 4 to find the solution of the linear differential equation

$$\frac{dU}{dt} - \lambda U = f, \quad u(t) \in C^1[0,1], \quad f \in C([0,1]) \quad \text{and} \quad |\lambda| < 1$$

6. Let m be the space of bounded number sequences, i.e., for $x \in m \implies$ $x = \{\xi_i\}$, $|\xi_i| \leq K_x$, $||x|| = \sup_i |\xi_i|$.

   In m the *shift operator* E is defined by

$$Ex = (0, \xi_1, \xi_2, \xi_3, \ldots) \text{ for } x = (\xi_1, \xi_2, \xi_3, \ldots)$$

   Find $||E||$ and discuss the inversion of the *difference operator* $\Delta = E - I$.

   [Hint: Show that $\Delta x = \theta \implies x = 0$]

4.8 Banach Space with a Basis

If a space E has a denumerable basis, then it is a separable space. A denumerably everywhere dense set in a space with basis is a linear combination of the form $\sum_{i=1}^{\infty} r_i e_i$ with rational coefficients r_i. Though many separable Banach spaces have bases, it is **not** proved for certain that every separable Banach space has a **basis**.

Note 4.8.1.

1. It can be shown that a Banach space E is either finite dimensional or else it has a Hamel basis which is not denumerable and hence nonenumerable or uncountable.

2. An infinite dimensional separable Banach space has, in fact, a basis which is in one-to-one correspondence with the set of real numbers.

   Note 4.8.1. has exposed a severe limitation of the Hamel basis, that every element of the Banach space E must be a **finite** linear combination of the basic elements and has given rise to the concept of a new basis known as **Schauder** basis.

4.8.1 Definition: Schauder basis

Let E be a normed linear space. A **denumerable** subset $\{e_1, e_2, \ldots\}$ of E is called a **Schauder basis** for E if $\|e_n\| = 1$ for each n and if for every $x \in E$, there are unique scalars $\alpha_1, \alpha_2, \ldots$ in $\mathbb{R}(\mathbb{C})$ such that

$$x = \sum_{i=1}^{\infty} \alpha_i e_i.$$

In case E is finite dimensional and $\{a_1, a_2, \ldots, a_n\}$ is a Hamel basis, then $\{a_1/\|a_1\|, a_2/\|a_2\|, \ldots, a_n/\|a_n\|\}$ is a Schauder basis for E.

If $\{e_1, e_2, \ldots\}$ is a Schauder basis for E then for $n = 1, 2, \ldots$, let us define functionals $f_n : E \to \mathbb{R}(\mathbb{C})$ by $f_n(x) = \alpha_n$ for

$$x = \sum_{n=1}^{\infty} \alpha_n e_n \in E. \tag{4.51}$$

The uniqueness condition in the definition of a Schauder basis yields that each f_n is well-defined and linear on E. It is called the nth **coefficient functional** on E.

4.8.2 Definition: biorthogonal sequence

Putting $x = e_j$ in (4.51) we have

$$e_j = \sum_{i=1}^{\infty} f_i(e_j) e_i,$$

since e_i are linearly independent,

$$f_i(e_j) = \begin{cases} 1 & \text{if } i = j \\ 0 & \text{if } i \neq j \end{cases} \tag{4.52}$$

The two sequences $\{e_i\}$ and sequence of functionals $\{f_i\}$ such that, (4.52) is true, are called **biorthogonal sequences**.

4.8.3 Lemma

For every functional f defined on the Banach space E, we can find coefficients $c_i = f(e_i)$, where $\{e_i\}$ is a Schauder basis in E, such that

$$f = \sum_{i=1}^{\infty} f(e_i) f_i = \sum_{i=1}^{\infty} c_i f_i,$$

$\{f_i\}$ being a sequence of functionals defined on E and satisfying (4.51). **Proof:** For any functional f defined on E, it follows from (4.51)

$$f(x) = \sum_{i=1}^{\infty} f_i(x) f(e_i)$$

Writing, $f(e_i) = c_i$,

$$f(x) = \sum_{i=1}^{\infty} c_i f_i(x) \quad \text{or,} \quad f = \sum_{i=1}^{\infty} c_i f_i \qquad (4.53)$$

The representation (4.53) is unique. The series (4.53) converges for every $x \in E$.

Problems

1. Let $E = C([0,1])$. Consider in $C([0,1])$ the sequence of elements

$$t, (1-t), u_{00}(t), u_{10}(t), u_{11}(t), u_{20}(t), u_{21}(t), \ldots, u_{22}(t) \qquad (4.54)$$

where $u_{kl}(t)$, $k = 1, 2, \ldots$, $0 \le l \le 2^k$, are defined in the following way $u_{kl} = 0$, if t is located outside the interval $(l/2^k, l + 1/2^k)$ but inside of this interval $u_{kl}(t)$ has a graph in the form of a isosceles triangle with height equal to 1. [See figure 4.2]. Take a function $x(t) \in C([0,1])$ representable in the form of the series

$$x(t) = a_0(1-t) + a_1 t + \sum_{k=0}^{\infty} \sum_{l=0}^{2^k - 1} a_{kl} u_{kl}(t)$$

where $a_0 = x(0)$, $a_1 = x(1)$ and the coefficients a_{kl} admit a unique geometric construction as given in the following figure [see figure (4.2)]

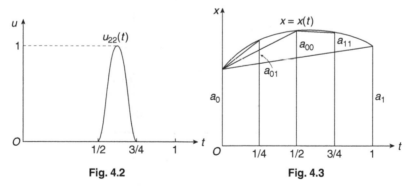

Fig. 4.2   Fig. 4.3

The graph of the partial sums of the above series

$$a_0(1-t) + a_1 t + \sum_{k=0}^{s-1} \sum_{l=0}^{2^k - 1} a_{kl} u_{kl}(t)$$

is an open polygon with $2^s + 1$ vertices lying on the curve $x = x(t)$ at the points with equidistant abscissae.

Show that the collection of functions in (4.54) forms a basis in $C([0,1])$.

2. Let $1 \le p < \infty$. For $t \in [0,1]$, let $x_1(t) = 1$.

$$x_2(t) = \begin{cases} 1 & \text{if } 0 \le t \le \dfrac{1}{2} \\ -1 & \text{if } \dfrac{1}{2} < t \le 1 \end{cases}$$

and for $n = 1, 2, \ldots, \; j = 1, \ldots, 2^n$

$$x_{2^n+j}(t) = \begin{cases} 2^{n/2}, & \text{if } (2j-2)/2^{n+1} \le t \le (2j-1)/2^{n+1} \\ -2^{n/2}, & \text{if } (2j-1)/2^{n+1} \le t \le 2j/2^{n+1} \\ 0, & \text{otherwise} \end{cases}$$

Show that the Haar system $\{x_1, x_2, x_3, \ldots\}$ is a Schauder basis for $L_2([0,1])$.

Note each x_n is a step function.

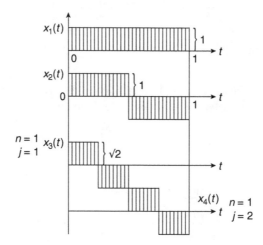

Fig. 4.4

CHAPTER 5

LINEAR
FUNCTIONALS

In this chapter we explore some simple properties of functionals defined on a normed linear space. We indicate how linear functionals can be extended from a subspace to the entire normed linear space and this makes the normed linear space richer by new sets of linear functionals. The stage is thus set for an adequate theory of conjugate spaces, which is an essential part of the general theory of normed linear spaces. The Hahn-Banach extension theorem plays a pivotal role in extending linear functionals from a subspace to an entire normed linear space. The theorem was discovered by H. Hahn (1927) [23], rediscovered in its present, more general form (5.2.2) by S. Banach (1929) [5]. The theorem was further generalized to complex spaces (5.1.8) by H.F. Bohnenblust and A. Sobezyk (1938) [8].

Besides the **Hahn-Banach extension theorem**, there is another important theorem discovered by Hahn-Banach which is known as **Hahn-Banach separation theorem**. While the Hahn-Banach extension theorem is analytic in nature, the Hahn-Banach separation theorem is geometric in nature.

5.1 Hahn-Banach Theorem

4.3 introduced 'linear functionals' and 4.4 'the space of bounded linear operators'. Next comes the notion of 'the space of bounded linear functionals'.

5.1.1 Definition: conjugate or dual space

The space of bounded linear functionals mapping a Banach space E_x into $\mathbb{R}$ is called **the conjugate or (dual)** of E_x and is denoted by E_x^*.

In theorem 4.2.13 we have seen how a bounded linear operator A_0

defined on a linear subspace X, which is *everywhere dense* in a complete normed linear space E_x with values in a complete normed linear space, can be extended to the entire space with preservation of norm. Hahn-Banach theorem considers such an extension even if X is *not necessarily dense* in E_x.

What follows is a few results as a prelude to proving the main theorem.

5.1.2   Lemma

Let L be a linear subspace of a normed linear space E_x and f be a functional defined on L. If x_0 is a vector **not** in L, and if $L_1 = (L, x_0)$, a set of elements of the form $x + tx_0$, $x \in L$ and t any real number, then f can be extended to a functional f_0 defined on L_1 such that $||f_0|| = ||f||$.

Proof: We assume that E_x is a real normed linear space. It is seen that L_1 is a linear subspace because $x_1 + t_1 x_0$ and $x_2 + t_2 x_0 \in L_1 \Rightarrow (x_1 + x_2) + (t_1 + t_2)x_0 \in L_1$ and $\lambda(x_1 + t_1 x_0) \in L_1$ etc. Any $u \in L_1$ has two representations of the form $x_1 + t_1 x_0$, and $x_2 + t_2 x_0$ respectively and that $t_1 \neq t_2$. Otherwise

$$x_1 + t_1 x_0 = x_2 + t_1 x_0 \Longrightarrow x_1 = x_2$$

Now, $x_1 + t_1 x_0 = x_2 + t_2 x_0$ or, $x_0 = \dfrac{x_1 - x_2}{t_2 - t_1}$

showing that $x_0 \in L$ since $x_1, x_2 \in L$. Hence $x_1 = x_2$ and $t_1 = t_2$, i.e., the representation of u is unique. Let us take two elements, x' and $x'' \in L$.

We have, $f(x') - f(x'') = f(x' - x'')$

$$\leq ||f|| \, ||x' - x''||$$

$$\leq ||f||(||x' + x_0|| + ||x'' + x_0||)$$

Thus, $f(x') - ||f|| \, ||x' + x_0|| \leq f(x'') + ||f|| \, ||x'' + x_0||$

Since x' and x'' are arbitrary in M, independent of each other,

$$\sup_{x \in L}\{f(x) - ||f|| \, ||x + x_0||\} \leq \inf_{x \in L}\{f(x) + ||f|| \, ||x + x_0||\}$$

Consequently, there is a real number c, satisfying the inequality,

$$\sup_{x \in L}\{f(x) + ||f|| \, ||x + x_0||\} \leq c \leq \inf_{x \in L}\{f(x) + ||f|| \, ||x + x_0||\} \qquad (5.1)$$

Now any element $u \in L_1$ has the form $u = x + tx_0$, $x \in M$ and $t \in \mathbb{R}$. Let us define a new functional $f_0(u)$ on L, such that

$$f_0(u) = f(x) - tc \qquad (5.2)$$

c is some fixed real number satisfying (5.1).

Note that $u = x + tx_0$, $x \in M$, $x_0 \notin M$.

If in particular $u \in L$ then $t = 0$ and $u = x$.

Hence f and f_0 coincide on L.

Now let $u_1 = x_1 + t_1 x_0$, $u_2 = x_2 + t_2 x_0$.

Then, $f_0(u_1 + u_2) = f(x_1 + x_2) - (t_1 + t_2)c$

$$= f(x_1) - t_1 c + f(x_2) - t_2 c$$

$$= f_0(u_1) + f_0(u_2)$$

Thus, $f_0(u)$ is additive. To show that $f_0(u)$ is bounded and has the same norm as that of $f(x)$ we consider two cases:

(i) For $t > 0$, it follows from (5.1) and (5.2) that

$$|f_0(u)| = t \left| f \left(\frac{x}{t} \right) - c \right| \leq t \left\{ ||f|| \, \left\| \frac{x}{t} + x_0 \right\| \right\}$$

$$= ||f|| \, ||x + t x_0|| = ||f|| \, ||u||$$

Hence, $|f_0(u)| \leq ||f|| \, ||u||$ \hfill (5.3)

(ii) For $t < 0$, then (5.1) yields

$$f(x/t) - c \geq -||f|| \, \left\| \frac{x}{t} + x_0 \right\| = -\frac{1}{|t|} ||f|| \, ||x + t x_0||$$

$$= \frac{1}{t} ||f|| \, ||u||$$

Hence, $f_0(u) = t \left\{ f \left(\frac{x}{t} \right) - c \right\} \leq ||f|| \, ||u||$

That is, we get back (5.3).

Hence, inequality (5.3) remains valid for all $u \in (L, x_0) = L_1$. Thus it follows from (5.3) that $||f_0|| \leq ||f||$. However, since the functional f_0 is an extension of f from L to L_1, where $||f_0|| \geq ||f||$. Hence $||f_0|| = ||f||$.

Note that we have determined the norm of the functional f_0 with respect to that linear subspace on which f_0 is defined. Thus, the functional $f(x)$ is extended to $L_1 = (L, x_0)$ with presentation of norm.

5.1.3   Theorem (the Hahn-Banach theorem)

Every linear functional $f(x)$ defined on a linear subspace L of a normed linear space E can be extended to the entire space with preservation of norm. That is, we can construct a linear function $F(x)$ defined on E such that

(i) $F(x) = f(x)$ for $x \in L$, (ii) $||F||_E = ||f||_L$

Proof: Let us first suppose the space E is separable. Let N be a countable everywhere dense set in E. Let us select those elements of this set which do not fall in L and arrange them in the sequence $x_0, x_1, x_2, \ldots, x_n, \ldots$

By virtue of lemma 5.1.2 we can extend the functional $f(x)$ successively to the subspaces $(L, x_0) = L_1$, $(L_1, x_1) = L_2$ and so on and ultimately construct a certain functional f_w defined on the linear subspace L_w, which is everywhere dense in E and is equal to the union of all L_n.

Moreover, $||f_w|| = ||f||$. Since L_w is everywhere dense in E, we can apply theorem 4.2.13 to extend the functional f_w by continuity to the entire space E and obtain the functional F defined on E such that

$$F(x)|_L = f(x) \quad \text{and} \quad ||F||_E = ||f||_L$$

Alternatively, in case the space is not separable, we can proceed as follows. Consider all possible extensions of the functional with preservation of the norm. Such extensions always exist. We next consider the set Φ of these extensions and introduce a partial ordering as detailed below. We will say $f' < f''$ if a linear subspace L' on which f' is defined is contained in the linear subspace L'' on which f'' is defined and if $f'(x) = f''(x)$ for $x \in L'$. Evidently $f' < f''$ has all the properties of ordering.

Now, let $\{f_\alpha\}$ be an arbitrary, totally ordered subset of the set Φ. This subset has an upper bound, which is the functional f^*, defined on a linear subspace $L^* = \cup_\alpha L_\alpha$. L_α is the domain of f_α and $f^*(x) = f_{\alpha_0}(x)$ if $x \in L^*$ is an element of L_{α_0}. Hence f^* is a linear functional and $||f^*|| = ||f||$, that is, $f^* \in \Phi$. Thus, it is seen that all the hypotheses of Zorn's lemma (1.1.4) are satisfied and Φ has a maximal element F. This functional is defined on the entire span E. If that is not so, the functional can be further extended, contradicting the fact that F is the maximal element in Φ.

Hence, the proof is completed.

Note 5.1.1 Since the constant c satisfying (5.1) may be arbitrarily preassigned, and hence there may not be a single maximal element in Φ, the extension of a linear functional by the Hahn-Banach theorem **is generally not unique.**

Note that Hahn-Banach theorem is a potential source of generating different linear functionals in a Banach space or a normed linear space. The two theorems which are offshoots of the Hahn-Banach theorem give rise to various applications.

5.1.4   *Theorem*

Let E be a normed linear space and $x_0 \neq \theta$ be a fixed element in E. There exists a linear functional $f(x)$, defined on the entire space E, such that

(i) $||f|| = 1$ and (ii) $f(x_0) = ||x_0||$

Consider the set of elements $\{tx_0\} = L$, where t runs through all positive real numbers.

The set L is a subspace of E, spanned by x_0.

A functional $f_0(x)$, defined on L has the following form: if $x = tx_0$, then

$$f_0(x) = t||x_0|| \tag{5.4}$$

Then, $f_0(x_0) = ||x_0||$ and $|f_0(x)| = |t|||x_0|| = ||x||$

Thus, $\dfrac{|f_0(x)|}{||x||} = 1$, $\forall\ x$, i.e., $\sup\limits_{x \neq \theta} \dfrac{|f_0(x)|}{||x||} = 1$, i.e., $||f_0|| = 1$

Now, if the functional $f_0(x)$ is extended to the entire space with preservation of norm, we get a functional $f(x)$ having the required properties.

5.1.5 Theorem

Given a subspace L and an element $x_0 \notin L$ in a normed linear space E. Let $d > 0$ be the distance from x_0 to L, i.e.,

$$d = \inf_{x \in L} ||x - x_0||$$

Then there is a functional $f(x)$ defined everywhere on E and such that

(i) $f(x) = 0$, $x \in L$, (ii) $f(x_0) = 1$ and (iii) $||f|| = \dfrac{1}{d}$

Proof: Consider a set $(L; x_0)$. Each of its elements is uniquely representable in the form $u = x + tx_0$, where $x \in L$ and t is real. Let us construct the functional $f_0(u)$ by the following rule. If $u = x + tx_0$, define $f_0(u) = t$. Evidently $f_0(x) = 0$, if $x \in L$ and $f_0(x_0) = 1$.

To determine $||f_0||$ we have, for $t \neq 0$,

$$|f_0(u)| = |t| = \frac{|t|\,||u||}{||u||} = \frac{|t|\,||u||}{||x + tx_0||}$$

$$= \frac{||u||}{\left\|\frac{x}{t} + x_0\right\|} = \frac{||u||}{\left\|x_0 - \left(-\frac{x}{t}\right)\right\|} \leq \frac{||u||}{d} \quad \text{since } \left\|x_0 + \frac{x}{t}\right\| \geq d$$

Thus, $||f_0|| \leq \dfrac{1}{d}$ $\tag{5.5}$

Furthermore, there is a sequence $\{x_n\} \subset L$ s.t.

$$\lim_{n \to \infty} ||x_n - x_0|| = d$$

Then we have $|f_0(x_n - x_0)| \leq ||f_0||\,||x_n - x_0||$

Since $|f_0(x_n - x_0)| = |f_0(x_n) - f_0(x_0)| = 1$, $\quad x_n \in L$

$$1 \leq ||f_0||\,||x_n - x_0||$$

Hence, by taking the limit we get,

$$1 \leq ||f_0||d \quad \text{or} \quad ||f_0|| \geq \frac{1}{d} \tag{5.6}$$

Then, by extending $f_0(x)$ to the entire space with preservation of norm we obtain the functional $f(x)$ with the required property.

5.1.6   Geometric interpretation

The conclusion of the above theorem admits geometric interpretation as follows.

Through every point x_0 on the surface of the ball $||x|| \leq r$ a **supporting hyperplane** can be drawn.

Note that $||f|| = \sup\limits_{x \neq \theta} \dfrac{|f(x)|}{||x||}$

For a functional $f(x)$ on the ball $||x|| \leq r$,

$$||f|| = \sup\limits_{x \neq \theta} \frac{|f(x)|}{||x||} \geq \frac{\sup |f(x)|}{r}$$

Consider the hyperplane $f(x) = r||f||$.

For points x_0 on the surface of the ball, $||x_0|| = r$, $f(x_0) = r||f||$ and for other points of the ball

$$f(x) \leq r||f||$$

Thus, $f(x) = r||f||$ is a supporting hyperplane.

5.1.7   Definition: sublinear functional

A real valued function p on a normed linear space E is said to be **sublinear**, if it is

(i) **subadditive**, i.e.,

$$p(x + y) \leq p(x) + p(y), \ \forall \ x, y \in E \tag{5.7}$$

and (ii) **positive homogeneous**, i.e.,

$$p(\alpha x) = \alpha p(x), \ \forall \ \alpha \geq 0 \text{ in } \mathbb{R} \text{ and } x \in E \tag{5.8}$$

5.1.8   Example

$|| \cdot ||$ is a sublinear functional.

For, $||x + y|| \leq ||x|| + ||y||, \ \forall \ x, y \in E$

$||\alpha x|| = \alpha ||x||, \ \forall \ \alpha \geq 0 \text{ in } \mathbb{R} \text{ and } x \in E$

The Hahn-Banach theorem (5.1.3) can be further generalized in the following theorem. Let a functional $f(x)$ defined on a subspace L of a normed linear space E and be majorized on L by a sublinear functional $p(x)$ defined on E. Then f can be extended from L to E without losing the linearity and the majorization, so that the extended functional F on E is still linear and majorized by p. Here we have taken E to be real.

5.1.9 Theorem Hahn-Banach theorem (using sublinear functional)

Let E be a normed linear space and p a sublinear functional defined on E. Furthermore, let f be a linear functional defined on a subspace L of E and let f satisfy,

$$f(x) \le p(x), \quad \forall\, x \in L \tag{5.9}$$

Then f can be extended to a linear functional F satisfying,

$$F(x) \le p(x), \quad \forall\, x \in E \tag{5.10}$$

$F(x)$ is a linear functional on E and

$$F(x) = f(x), \quad \forall\, x \in E$$

Proof: The proof comprises the following steps:

(i) The set $\mathfrak{F}$ of all linear extensions g of f satisfying $g(x) \le p(x)$ on $\mathcal{D}(g)$ can be partially ordered and Zorn's lemma yields a maximal element F on $\mathfrak{F}$.

(ii) F is defined on the entire space E.

To show that $\mathcal{D}(F)$ is all of E, the arguments will be as follows. If $\mathcal{D}(F)$ is not all of E, choose a $y_1 \in E - \mathcal{D}(F)$ and consider the subspace Z_1 of E spanned by $\mathcal{D}(F)$ and y_1, and show that any $x \in Z_1$ can be uniquely represented by $x = y + \alpha y_1$. A functional g_1 on Z_1, defined by

$$g_1(y + \alpha y_1) = F(y) + \alpha c$$

is linear and a proper extension of F, i.e., $\mathcal{D}(g_1)$ is a proper subset of $\mathcal{D}(F)$. If, in addition, we show that $g_1 \in F$, i.e.,

$$g_1(x) \le p(x) \quad \forall\, x \in \mathcal{D}(g_1),$$

then the fact that F is the *maximal element* of $\mathfrak{F}$ is *contradicted*.

For details of the proof see Kreyszig [30].

5.1.10 Theorem

Let E be a normed linear space over $\mathbb{R}$ and f is a linear functional defined on a subspace L of E. Then theorem 5.1.9 $\Longrightarrow$ theorem 5.1.3.

Proof: Let $p(x) = ||f||\,||x||$, $x \in E$.

Then p is a sublinear functional on E and $f(x) \le p(x)$, $x \in E$.

Hence, by theorem 5.1.9, it follows that there exists a real linear functional F on E such that

$$F(x) = f(x)\ \forall\, x \in L \quad \text{and} \quad F(x) \le ||f||\,||x||, \quad \forall\, x \in E$$
$$F(x) = f(x) \le ||f||\,||x||\ \forall\, x \in E \Longrightarrow ||F||_E \le ||f||_L$$

On the other hand, take $x_1 \in L$, $x_1 \neq \theta$, Then

$$||F|| \geq \frac{|F(x_1)|}{||x_1||} = \frac{|f(x_1)|}{||x_1||} \implies ||F|| \geq ||f||$$

This completes the proof when E is a real normed linear space.

5.2 Hahn-Banach Theorem for Complex Vector and Normed Linear Space

5.2.1 *Lemma*

Let E be a normed linear space over $\mathbb{C}$. Regarding E as a linear space over $\mathbb{R}$, consider a real-linear functional $u : E \to \mathbb{R}$. Define

$$f(x) = u(x) - iu(ix), \quad x \in E$$

Then f is a complex-linear functional on E.

Proof: As u is real and linear, f is also real, linear.

Now, since u is linear

$$f(ix) = u(ix) - iu(i \cdot ix) = iu(x) + iu(ix)$$
$$= i[u(x) - iu(ix)] = if(x) \; \forall \; x \in E$$

Hence, f is complex-linear.

5.2.2 *Hahn-Banach theorem (generalized)*

Let E be a real or complex vector space and p be a real-valued functional on E which is subadditive, i.e., for all $x, y \in E$

$$p(x + y) \leq p(x) + p(y) \tag{5.11}$$

and for every scalar α satisfies,

$$p(\alpha x) = |\alpha| p(x) \tag{5.12}$$

Moreover, let F be a linear functional defined on a subspace L of E and satisfy,

$$|f(x)| \leq p(x) \quad \forall \; x \in L \tag{5.13}$$

Then f has a linear extension F from L to E satisfying

$$|F(x)| \leq p(x) \quad \forall \; x \in E \tag{5.14}$$

Proof: (a) *Real Vector Space:* Let E be real. Then (5.13) yields $f(x) \leq p(x) \; \forall \; x \in L$. It follows from theorem 5.1.9 that f can be extended to a linear functional F from L to E such that

$$F(x) \leq p(x) \; \forall \; x \in E \tag{5.15}$$

(5.12) and (5.15) together yield

$$-F(x) = F(-x) \leq p(-x) = |-1|p(x) = p(x) \quad \forall \, x \in E \qquad (5.16)$$

From (5.15) and (5.16) we get (5.14).

(b) *Complex Vector Space:* Let E be complex. Then L is a complex vector space, too. Hence, f is complex-valued and we can write,

$$f(x) = f_1(x) + if_2(x) \quad \forall \, x \in L \qquad (5.17)$$

where $f_1(x)$ and $f_2(x)$ are real valued. Let us assume, for the time being, E and L as real vector spaces and denote them by E_r and L_r respectively. We thus restrict scalars to real numbers. Since f is linear on L and f_1 and f_2 are real-valued, f_1 and f_2 are linear functionals on L. Also, $f_1(x) \leq |f(x)|$ because the real part of a complex quantity cannot exceed the absolute value of the whole complex quantity.

Hence, by (5.13),

$$f_1(x) \leq p(x) \quad \forall \, x \in L_r$$

Hence, by the Hahn-Banach theorem 5.1.8 f_1 can be extended to a functional F_1 from L_r to E_r, such that

$$F_1(x) \leq p(x) \quad \forall \, x \in L_r \qquad (5.18)$$

We next consider f_2. For every $x \in L$,

$$i[f_1(x) + if_2(x)] = if(x) = f(ix) = f_1(ix) + if_2(ix)$$

The real parts on both sides must be equal.

Hence, $f_2(x) = -f_1(ix) \quad \forall \, x \in L$ \qquad (5.19)

Just as in (5.18) $f_1(x)$ has been extended to $F_1(x)$, $\forall \, x \in E$, $f_2(x) = -[f_1(ix)]$ can be extended exploiting Hahn-Banach theorem 5.18, to $F_2(x) = [-F_1(ix)] \, \forall \, x \in E$.

Thus we can write

$$F(x) = F_1(x) - iF_1(ix) \quad \forall \, x \in E \qquad (5.20)$$

(i) We would next prove that F is a *linear* functional on the *complex* vector space E.

For real c, d, $c + id$ is a complex scalar.

$$
\begin{aligned}
F((a+ib)x) &= F_1(ax+ibx) + iF_2(ax+ibx) \\
&= F_1(ax+ibx) - iF_1(iax-bx) \\
&= aF_1(x) + bF_1(ix) - i[aF_1(ix) - bF_1(x)] \\
&= (a+ib)F_1(x) - i[(a+ib)F_1(ix)] \\
&= (a+ib)[F_1(x) - iF_1(ix)] \\
&= (a+ib)[F_1(x) + iF_2(x)] \\
&= (a+ib)F(x)
\end{aligned}
$$

(ii) Next to be shown is that $|F(x)| \le p(x), \forall x \in E$.

It follows from (5.12) that $p(\theta) = 0$. Taking $y = -x$ in (5.11), we get

$$0 \le p(x) + p(-x) = 2p(x), \text{i.e., } p(x) \ge 0 \quad \forall x \in E$$

Thus, $F(x) = 0 \Longrightarrow F(x) \le p(x)$.

Let $F(x) \ne 0$. Then we can write using polar coordinates,

$$F(x) = |F(x)|e^{i\theta},$$

thus $|F(x)| = F(x)e^{-i\theta} = F(e^{-i\theta}x)$

Since $|F(x)|$ is real, the last expression is real and is equal to its *real* part.

Hence, by (5.12)

$$|F(x)| = F(e^{-i\theta}x) = F_1(e^{-i\theta}x) \le p(e^{-i\theta}x) = |e^{i\theta}|p(x) = p(x)$$

This completes the proof.

We next consider Hahn-Banach theorem (generalized) in the setting of a normed linear space E over $\mathbb{R}(\mathbb{C})$.

5.2.3 Hahn-Banach theorem (generalized form in a normed linear space)

Every bounded linear functional f defined on a subspace L of a normed linear space E over $\mathbb{R}(\mathbb{C})$ can be extended with preservation of norm to the entire space, i.e., there exists a linear functional $F(x)$ defined on E such that

(i) $F(x) = f(x)$ for $x \in L$; (ii) $||F|_E = ||f||_L$

Proof: If $L = \{\theta\}$ then $f = \theta$ and the extension is $F = \theta$. Let $L \ne \{\theta\}$. We want to use theorem 5.2.2. For that purpose we have to find out a suitable p.

We know, for all $x \in L$ $\quad |f(x)| \le ||f_L|| \, ||x||$.

Let us take $p(x) = ||f||_L \, ||x|| \, \forall x \in E$.

Thus, p is defined on all E.

Furthermore, $p(x + y) = ||f||_L||x + y|| \le ||f||_L(||x|| + ||y||)$

$$= p(x) + p(y), \quad \forall x, y \in E$$

$$p(\alpha x) = ||f||_L||\alpha x|| = |\alpha| \, ||f||_L||x||)$$

$$= |\alpha|p(x) \, \forall x \in E$$

Thus conditions (5.11) and (5.12) of theorem 5.2.2 are satisfied. Hence, the above theorem can be applied, and we get a linear functional F on E which is an extension of f and satisfies.

$$|F(x)| \leq p(x) = ||f||_L ||x|| \quad \forall \; x \in E$$

Hence, $||F||_E = \sup\limits_{x \neq \theta} \dfrac{|F(x)|}{||x||} \leq p(x) = ||f||_L \qquad (5.21)$

On the other hand, F being an extension of f,

$$||F||_E \geq ||f||_L \qquad (5.22)$$

Combining (5.21) and (5.22) we get,

$$||F||_E = ||f||_L$$

5.2.4 Hyperspace and related results

5.2.5 Definition: Hyperspace

A proper subspace E_0 of a normed linear space E is called a hyperspace in E if it is a **maximal** proper subspace of E. It may be noted that a proper subspace E_0 of E is maximal if and only if the span of $E_0 \cup \{a\}$ equals E for each $a \notin E_0$.

5.2.6 Remark 1

A hyperplane H is a translation of a hyperspace by a vector, i.e., H is of the form

$$H = x + E_0$$

where E_0 is a hyperspace and $x \in E$.

5.2.7 Theorem

A subspace of a normed linear space E is a hyperspace if and only if it is the null space of a non-zero functional.

Proof: We first show that null spaces of non-zero linear functionals are hyperspaces.

Let $f : E \longrightarrow \mathbb{R}(\mathbb{C})$ be a non-zero linear functional and $x_0 \in E$ be such that $f(x_0) \neq 0$. Then, for every $x \in E$, there exists a $u \in N(f)$ such that

$$x = u + \frac{f(x)}{f(x_0)} x_0, \quad u \in N(f)$$

If in particular $x_0 \in E/N(f)$, then

$$E = \mathrm{span} \; \{x_0; N(f)\}$$

Thus we see that null spaces of f are hyperspaces.

Conversely, let us suppose that H is a hyperspace of E and $x_0 \in E/H$ such that $E = \mathrm{span} \; \{x_0; H\}$. Then for every $x \in E$, there exists unique pair (λ, u) in $\mathbb{R}(\mathbb{C}) \times H$ such that $x = \lambda x_0 + u$.

Let us define $f(\lambda x_0 + u) = \lambda \in \mathbb{R}(\mathbb{C})$, $u \in H$.

Now, $f((\lambda + \mu)x_0 + u) = \lambda + \mu = f(\lambda x_0 + u) + f(\mu x_0 + u)$, $\lambda, \; \mu \in \mathbb{R}(\mathbb{C})$.

$$f(p(\lambda x_0 + u)) = p\lambda = pf(\lambda x_0 + u), \; p \in \mathbb{R}(\mathbb{C})$$

Thus, f is a linear functional.

Again, taking $\lambda = 1$, $f(x_0) = 1$, and when $\lambda = 0$, $f(u) = 0$ i.e. $u \in N(f)$, i.e., $N(f) = H$.

5.2.8   Remark

A subset $H \subseteq E$ is a hyperplane in E if and only if there exists a non-zero linear functional f and a scalar 'λ' such that

$$H = \{x \in E : f(x) = \lambda\}$$

since $\{x \in E : f(x) = \lambda\} = x_\lambda + N(f)$ for some $x_\lambda \in E$ with $f(x_\lambda) = \lambda$. Thus, hyperplanes are of the form H for some non-zero linear functional f and for some $\lambda \in \mathbb{R}(\mathbb{C})$.

5.2.9   Definition

If the scalar field is $\mathbb{R}$, a set X is said to be on *the left side* of a hyperplane H if

$$X \subseteq \{x \in E : f(x) \leq \lambda\}$$

and it is *strictly on the left side* of H if

$$X \subseteq \{x \in E : f(x) < \lambda\}$$

Similarly, X is said to on *the right side* of a hyperplane H if

$$X \subseteq \{x \in E : f(x) \geq \lambda\}$$

and *strictly on the right side* of a hyperplane H if

$$X \subseteq \{x \in E : f(x) > \lambda\}.$$

5.2.10   Theorem (Hahn-Banach separation theorem)

Let E be a normed linear space and X_1 and X_2 be nonempty disjoint convex sets with X_1 being an open set. Then there exists a functional $f \in E^*$ and a real number β, such that

$$X_1 \subseteq \{x \in E : Ref(x) < \beta\}, \quad X_2 \subseteq \{x \in E : Ref(x) \geq \beta\}$$

Before we prove the theorem we introduce few terminologies and a lemma.

5.2.11   Definition: absorbing set

Let E be a linear space. A set $X \subseteq E$ is said to be an **absorbing set** if for every $x \in E$, there exists $t > 0$ such that $t^{-1}x \in X$.

5.2.12   Definition: Minkowski functional

Let $X \subseteq E$ be a convex, absorbing set. Then, $\mu : E \longrightarrow \mathbb{R}$ is said to be a **Minkowski functional of X** if

$$\mu_X(x) = \inf\{t > 0 : t^{-1}x \in X\}$$

5.2.13 Remark

   (i) If X is an absorbing set, $\mu_X(x) < \infty$ for every $x \in E$.

   (ii) If X is an absorbing set then, $\theta \in X$ and

   (iii) If X is a normed linear space, then every open set containing θ is an absorbing set.

Proof: We prove (iii). Since an open set containing 'θ' contains an open ball $B(\theta, \epsilon)$, if $x \in B(\theta, \epsilon)$ then $||x|| < \epsilon$. Therefore, for any $\alpha > 1$

$$||\alpha^{-1}x|| \leq \frac{\epsilon}{\alpha} < \epsilon$$

Hence, every open set containing θ is an absorbing set.

5.2.14 Lemma

Let X be a convex, absorbing subset of a linear space E and let μ_X be the corresponding Minkowski functional. Then μ_X is a sublinear functional, [see 5.1.7] and

$$\{x \in E : \mu_X(x) < 1\} \subseteq X \subseteq \{x \in E : \mu_X(x) \leq 1\}. \tag{5.23}$$

Proof: For μ_X to be a sublinear functional, it has to satisfy the properties,

$$\mu_X(x + y) \leq \mu_X(x) + \mu_X(y), \quad \mu_X(\zeta x) = \zeta \mu(x)$$

for all $x, y \in E$ and for all $\zeta \geq 0$.

Let $x, y \in E$. Let $p > 0$, $q > 0$ be such that $p^{-1}x \in E$, $q^{-1}x \in E$. Then, using the convexity of X, we have,

$$(p + q)^{-1}(x + y) = \left(\frac{p}{p+q}\right) \cdot p^{-1}x + \left(\frac{q}{p+q}\right) \cdot q^{-1}y \in X$$

Hence, $\mu_X(x + y) \leq p + q$.

Taking infimum over all such p and q, it follows that

$$\mu_X(x + y) \leq \mu_X(x) + \mu_Y(y)$$

Next, to show that $\mu_X(\zeta x) = \zeta \mu_X(x)$ for all $x \in E$, and for all $\zeta \geq 0$. Let $x \in E$ and $\zeta > 0$. Let $p > 0$ be such that $\mu_X(\zeta x) \leq \zeta p$. Taking infimum over all $p > 0$ with $p^{-1}x \in X$, we have

$$\mu_X(\zeta x) \leq \zeta \mu_X(x) \tag{5.24}$$

Let us take ζx in place of x and let $p > 0$ be such that $p^{-1}(\zeta x) \in X$. Since $p^{-1}(\zeta x) = (\zeta^{-1}p)^{-1}x$, we have,

$$\mu_X(x) \leq \zeta^{-1}p$$

Taking infimum over all x such that $p > 0$, we obtain

$$\mu_X(x) \leq \zeta^{-1}\mu_X(\zeta x)$$

Thus, $\zeta\mu_X(x) \leq \mu_X(\zeta x)$ (5.25)

It follows from (5.24) and (5.25) that

$$\mu_X(\zeta x) = \zeta\mu_X(x), \quad \zeta \geq 0$$

To prove the last part of the lemma we proceed as follows. Let $x \in X$. Then $1 \in \{t > 0 : t^{-1}x \in X\}$

Then $\mu_X(x) \leq 1$.

Next, let us suppose that $x \in E$ be such that $\mu_X(x) < 1$

Then there exists $p_0 > 0$, such that $p_0 < 1$ with $p_0^{-1}x \in X$. Since X is convex and $\theta \in X$, we have

$$x = p_0(p_0^{-1}x) + (1 - p_0) \cdot \theta \in X$$

Thus, $\mu_X(x) < 1$ implies $x \in X$.

Hence (5.23) is proved.

5.2.15   Proof of Theorem 5.2.10

We prove the theorem for $\mathbb{R}$.

Let $x_1 \in X_1$ and $x_2 \in X_2$. Then $X_1 - X_2 = \{x_1 - x_2 : x_1 \in X_1, x_2 \in X_2\}$. Since X_1 and X_2 are each convex, $X_1 - X_2$ is convex. Thus $X_1 - X_2$ is non-empty and convex. We next show that $X_1 - X_2$ is open, given that X_1 is open. Since X_1 is open, it contains an open ball $B(x_1, \epsilon)$ where $x_1 \in X_1$. For, $x_2 \in X_2$,

$$B(x_1 - x_2, \epsilon) = B(x_1, \epsilon) - x_2 \subset X_1 - x_2$$

Thus, $X_1 - X_2$ is open.

Hence, $X_1 - X_2 = \bigcup_{\substack{x_1 \in X_1 \\ x_2 \in X_2}} (x_1 - x_2)$ is open in E.

Also, $\theta \notin X_1 - X_2$, since $X_1 \cap X_2 = \Phi$.

Let $X = X_1 - X_2 + u_0$ where $u_0 = x_2 - x_1$. Then X is an open convex set with $\theta \in X$. Hence, X is an absorbing set as well. Let μ_X be the Minkowski functional of X.

In order to obtain the required functional, we apply the theorem 5.1.9. Let $E_0 = $ span $\{u_0\}$, $p = \mu_X$ and the linear functional $f_0 : E \longrightarrow \mathbb{R}$ defined by

$$f_0(\lambda u_0) = \lambda, \quad \lambda \in \mathbb{R}$$ (5.26)

Since $X_1 \cap X_2 = \Phi$ and $u_0 = x_2 - x_1 \notin X$, by lemma 5.2.14, we have $\mu_X(u_0) \geq 1$ and hence,

$$f_0(\lambda u_0) = \lambda \leq \lambda\mu_X(u_0) = \mu_X(\lambda u_0), \quad \forall \lambda \in \mathbb{R}$$

Therefore, by theorem 5.1.9 f_0 has a linear extension $f : E \to R$ such that

$$f(x) \leq \mu_X(x), \quad \forall\, x \in E \tag{5.27}$$

Lemma 5.2.14 yields that $\mu_X(x) \leq 1$ for some $x \in X$.

Hence, it follows from (5.27) that $f(x) \leq 1$ for every $x \in X$.

Thus, $f(x) \geq -1$ for every $x \in (-X)$. Thus we have,

$$|f(x)| \leq 1 \,\forall\, x \in X \cap (-X)$$

Since $X \cap (-X)$ is an open set containing θ, and is in the preimage of any open set in the range of f, f is continuous.

Next to show that there exists $\beta \in \mathbb{R}$ such that

$$f(x_1) < \beta \leq f(x_2), \; \forall\, x_1 \in X_1, \; x_2 \in X_2$$

Since $f : E \longrightarrow R$ and $X_1, X_2 \subset E$, $f(X_1)$, $f(X_2)$ are intervals in $\mathbb{R}$. Given X_1 is open we next show that $f(X_1)$ is open.

Since f is non-zero there is some $a \in E$ such that $f(a) = 1$, $a \neq \theta$. Let $x \in X$. Since X_1 is open, it contains an open ball $B(x, \epsilon)$, $\epsilon > 0$.

If $|k| < \epsilon/||a||$, then $x - ka \in X_1$ so that $f(x - ka) \in f(X_1)$. Then

$$\left\{ k^1 \in \mathbb{R}, \quad |f(x) - k^1| < \frac{\epsilon}{||a||} \right\} \subset f(X_1)$$

Hence, $f(X_1)$ is open in $\mathbb{R}$.

Hence it is enough to show that $f(x_1) \leq f(x_2)$ for every $x_1 \in X_1$ and every $x_2 \in X_2$. Since $x_1 - x_2 + u_0 \in X$, and taking note (5.26) by lemma 5.2.14, we have

$$f(x_1) - f(x_2) + 1 = f(x_1 - x_2 + u_0) = \mu_X(x_1 - x_2 + u_0) \leq 1$$

Thus, we have $f(x_1) \leq f(x_2)$ for all $x_1 \in X_1$, $x_2 \in X_2$.

In case the scalar field is $\mathbb{C}$, we can prove the theorem by using lemma 5.2.1.

5.2.16   *Remark*

Geometrically, the above separation theorem says that the set X_1 lies on one side of the real hyperplane $\{x \in E : Ref(x) = \beta\}$ and the set X_2 lies on the other, since

$$X_1 \subset \{x \in E : Ref(x) < \beta\} \quad \text{and} \quad X_2 \subset \{x \in E : Ref(x) \geq \beta\}$$

Problems [5.1 and 5.2]

1. Show that a norm of a vector space E is a sublinear functional on E.

2. Show that a sublinear functional p satisfies (i) $p(0) = 0$ (ii) $p(-x) \geq -p(x)$.

3. Let L be a closed linear subspace of a normed linear space E, and x_0 be a vector not in L_0. Given d is the distance from x_0 to L, show that there exists a functional $f_0 \in E^*$ such that $f_0(L) = 0$, $f_0(x_0) = 1$ and $||f_0|| = \dfrac{1}{d}$.

4. (i) Let L be a linear subspace of the normed linear space E over $\mathbb{R}(\mathbb{C})$.

   (ii) Let $f : L \longrightarrow \mathbb{R}(\mathbb{C})$ be a linear functional such that $|f(x)| \leq \alpha||x||$ for all $x \in L$ and fixed $\alpha > 0$.

   Then show that f can be extended to a linear continuous functional $F : E \longrightarrow \mathbb{R}(\mathbb{C})$ such that

   $$|F(x)| \leq \alpha||x|| \ \forall \ x \in \mathbb{R}(\mathbb{C})$$

5. Every linear functional $F(x)$ defined on a linear subspace L of a normed linear space E can be extended to the entire space with preservation of the norm, that is, we can construct a linear functional $F(x)$, defined on E such that

   (i) $F(x) = f(x)$ for $x \in L$ (ii) $||F||_E = ||f||_L$.

   Prove the above theorem in case E is separable without using Zorn's lemma (1.1.4).

6. Let L be a closed subspace of a normed linear space E such that

   $$f(L) = 0 \Longrightarrow f(E) = 0, \quad \forall \ f \in E^*$$

   Prove that $L = E$.

7. Let E be a normed linear space. For every subspace L of E and every functional f defined on E, prove that there is a unique Hahn-Banach extension of f to E if and only if E^* is **strictly convex** that is, for $f_1 \neq f_2$ in E, with $||f_1|| = 1 = ||f_2||$ we have $||f_1 + f_2|| < 2$.

   [Hint: If F_1 and F_2 are extensions of f, show that $F_1 + F_2/2$ is also a continuous linear extension of f and the strict convexity condition is violated].

5.3 Application to Bounded Linear Functionals on $C([a, b])$

In this section we shall use theorem 5.1.3 for obtaining a general representation formula for bounded linear functionals on $C([a, b])$, where

$[a, b]$ is a fixed compact interval. In what follows, we use representations in terms of Riemann-Stieljes integral. As a sort of recapitulation, we mention a few definitions and properties of Riemann-Stieljes integration which is a generalization of Riemann integration.

5.3.1 Definitions: partition, total variation, bounded variation

A collection of points $P = [t_0, t_1, \ldots, t_n]$ is called a **partition** of an interval $[a, b]$ if

$$a = t_0 < t_1 < \cdots < t_n = b \text{ holds} \qquad (5.28)$$

Let $w : [a, b] \to \mathbb{R}$ be a function. Then the **(total) variation** Var (w) of w over $[a, b]$ is defined to be

$$\text{Var }(w) = \sup \left[\sum_{j=1}^{n} |w(t_j) - w(t_{j-1})| : P = [t_0, t_1, \ldots, t_n] \right.$$

$$\left. \text{is a partition of } [a, b] \right] \qquad (5.29)$$

The supremum being taken over all partitions 5.28 of the interval $[a, b]$.

If Var $(w) < \infty$ holds, then w is said to be a function of **bounded variation**.

All functions of bounded variation on $[a, b]$ form a normed linear space. A norm on this space is given by

$$||w|| = |w(a)| + \text{Var }(w) \qquad (5.30)$$

The normed linear space thus defined is denoted by $\boldsymbol{BV}([\boldsymbol{a}, \boldsymbol{b}])$, where BV suggests 'bounded variation'.

We now obtain the concept of a Riemann-Stieljes integral as follows. Let $x \in C([a, b])$ and $w \in BV([a, b])$. Let P_n be any partition of $[a, b]$ given by (5.28) and denote by $\eta(P_n)$ the length of a largest interval $[t_{j-1}, t_j]$ that is,

$$\eta(P_n) = \max(t_1 - t_0, t_2 - t_1, \ldots, t_n - t_{n-1}).$$

For every partition P_n of $[a, b]$, we consider the sum,

$$S(P_n) = \sum_{j=1}^{n} x(t_j)[w(t_j) - w(t_{j-1})] \qquad (5.31)$$

There exists a number $\mathfrak{I}$ with the property that for every $\epsilon > 0$ there is a $\delta > 0$ such that

$$\eta(P_n) < \delta \implies |\mathfrak{I} - S(P_n)| < \epsilon$$

$\mathfrak{I}$ is called the **Riemann-Stieljes** integral of x over $[a, b]$ with respect to w and is denoted by

$$\int_a^b x(t)dw(t) \qquad (5.32)$$

Thus, we obtain (5.32) as the limit of the sum (5.31) for a sequence $\{P_n\}$ of partitions of $[a, b]$ satisfying $\eta(P_n) \to 0$ as $n \to \infty$.

In case $w(t) = t$, (5.31) reduces to the familiar Riemann integral of x over $[a, b]$.

Also, if x is continuous on $[a, b]$ and w has a derivative which is integrable on $[a, b]$ then

$$\int_a^b x(t)dw(t) = \int_a^b x(t)w'(t)dt \tag{5.33}$$

We show that the integral (5.32) depends linearly on x, i.e., given $x_1, x_2 \in C([a, b])$,

$$\int_a^b [px_1(t) + qx_2(t)]dw(t) = p\int_a^b x_1(t)dw(t) + q\int_a^b x_2(t)dw(t)$$

where $p, q \in \mathbb{R}$

The integral also depends linearly on $w \in BV([a, b])$ because for all $w_1, w_2 \in BV([a, b])$ and scalars r, s

$$\int_a^b x(t)d(rw_1 + sw_2)(t) = r\int_a^b x(t)dw_1(t) + s\int_a^b x(t)dw_2(t)$$

5.3.2 *Lemma*

For $x(t) \in C([a, b])$ and $w(t) \in BV([a, b])$,

$$\left| \int_a^b x(t)dw(t) \right| \leq \max_{t \in [a,b]} |x(t)| \ \text{Var} \ (w) \tag{5.34}$$

If P_n is any partition of $[a, b]$

$$S(P_n) = \sum_{j=1}^n x(t_j)(w(t_j) - w(t_{j-1}))$$

$$\leqq \left| \sum_{j=1}^n x(t_j)(w(t_j) - w(t_{j-1})) \right|$$

$$\leqq \max_{t_j \in [a,b]} x(t_j) \sup \sum_{j=1}^n |w(t_j) - w(t_{j-1})|$$

$$= \max_{t \in [a,b]} |x(t)| \cdot \text{Var} \ (w)$$

Hence making $n \to \infty$, we get,

$$\left| \int_a^b x(t)dw(t) \right| \leq \max_{t \in [a,b]} |x(t)| \ \text{Var} \ (w) \tag{5.35}$$

The representation theorem for bounded linear functionals on $C([a, b])$ by F. Riesz (1909) [30] is discussed next.

5.3.3 Theorem (Riesz's representation theorem on functionals on $C([a, b])$)

Every bounded linear functional f on $C([a, b])$ can be represented by a Riemann-Stieljes integral

$$f(x) = \int_a^b x(t)\,dw(t) \qquad (5.36)$$

where w is of bounded variation on $[a, b]$ and has the total variation

$$\text{Var } (w) = ||f|| \qquad (5.37)$$

Proof: Let $M([a, b])$ be the space of functions bounded in the closed interval $[a, b]$. By making an appeal to Hahn-Banach theorem 5.1.3. we can extend the functional f from $C([a, b])$ to the normed linear space $M([a, b])$ that is defined by

$$||x|| = \sup_{t \in [a,b]} |x(t)|$$

Furthermore, F is a bounded linear functional and

$$||f||_{C([a,b])} = ||F||_{M([a,b])}$$

We define the function w needed in (5.36). For this purpose, we consider the function $u_t(\xi)$ as follows,

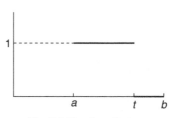

Fig. 5.1 The function u_t

$$u_t(\xi) = \begin{cases} 1 & \text{for } a \le \xi \le t \\ 0 & \text{otherwise} \end{cases} \qquad \text{[see figure 5.1]} \qquad (5.38)$$

Clearly, $u_t(\xi) \in M([a, b])$. We mention that $u_t(\xi)$ is called the characteristic function of the interval $[a, t]$. Using $u_t(\xi)$ and the functional F, we define w on $[a, b]$ by

$$w(a) = 0 \qquad w(t) = F(u_t(\xi)) \qquad t \in [a, b]$$

We show that this function w is of bounded variation and $\text{Var } (w) \le ||f||$.

For a complex quantity we can use the polar form. If fact setting, $\theta = \arg\xi$, we may write,

$$\xi = |\xi|e(\xi)$$

where $e(\xi) = \begin{cases} 1 & \text{if } \xi = 0 \\ e^{i\theta} & \text{if } \xi \ne 0 \end{cases}$

We see that if $\xi \ne 0$, then $|\xi| = \xi e^{-i\theta}$. Hence, for any ξ, zero or not, we have,

$$|\xi| = \xi \overline{e(\xi)} \qquad (5.39)$$

where the bar indicates complex conjugation.

In what follows we write,

$$\epsilon_j = \overline{e(w(t_j) - w(t_{j-1}))} \quad \text{and} \quad u_{t_j}(\xi) = u_j(\xi)$$

Then by (5.34), for any partition (5.26) we obtain,

$$\sum_{j=1}^{n} |w(t_j) - w(t_{j-1})| = |F(u_1)| + \sum_{j=2}^{n} |F(u_j) - F(u_{j-1})|$$

$$= \epsilon_1 F(u_1) + \sum_{j=2}^{n} \epsilon_j [F(u_j) - F(u_{j-1})]$$

$$= F\left(\epsilon_1 u_1 + \sum_{j=2}^{n} \epsilon_j [u_j - u_{j-1}] \right)$$

$$\leq ||F|| \left\| \epsilon_1 u_1 + \sum_{j=2}^{n} \epsilon_j [u_j - u_{j-1}] \right\|$$

Now, $u_j(\xi) = u_{t_j}(\xi) = 1$, $t_{j-1} < \xi \leq t_j$. Hence, on the right-hand side of the above inequality $||F|| = ||f||$ and the other factor $|| \cdots ||$ equals 1 because $|\epsilon_j| = 1$ and from the definition of $u_j(\xi)$'s we see that for each $t \in [a, b]$ only one of the terms $u_1, u_2 \cdots u_i, \ldots$ is not zero (and its norm is 1). On the left we can now take the supremum over all partitions of $[a, b]$.

Then we have,

$$\text{Var }(w) \leq ||f|| \tag{5.40}$$

Hence w is of bounded variation on $[a, b]$.

We prove (5.36) when $x \in C([a, b])$. For every partition P_n of the form (5.28) we define a function, which we denote simply by $z_n(t)$, keeping in mind that z_n depends on P_n, and not merely on n. $z_n(t)$ will be as follows:

$$z_n(t) = x(t_0)x(u_{t_1}(t)) + \sum_{j=2}^{n} x(t_{j-1})[u_{t_j}^{(t)} - u_{t_{j-1}}^{(t)}] \tag{5.41}$$

$z_n(t)$ is a step function.

Then $z_n \in M([a, b])$. By the definition of w,

$$F(z_n) = x(t_0)F(x_{t_1}) + \sum_{j=2}^{n} x(t_{j-1})[F(x_{t_j}) - F(x_{t_{j-1}})]$$

$$= x(t_0)w(t_1) + \sum_{j=2}^{n} x(t_{j-1})[w(t_j) - w(t_{j-1})]$$

$$= \sum_{j=1}^{n} x(t_{j-1})[w(t_j) - w(t_{j-1})] \tag{5.42}$$

where the last equality follows from $w(t_0) = w(a) = 0$. We now choose any sequence $\{P_n\}$ of partitions of $[a, b]$ such that $\eta(P_n) \to 0$ (It is to be kept in mind that t_j depends on the particular partition P_n). As $n \to \infty$ the sum on the right-hand side of (5.42) tends to the integral in (5.26) and (5.26) follows provided $F(z_n) \longrightarrow F(x)$, which equals $f(x)$ since $x \in C([a, b])$.

We need to prove that $F(z_n) \longrightarrow F(x)$. Keeping in mind the definition of $u_t(\xi)$ (fig. (5.1)), we note that (5.41) yields $z_n(a) = x(a) \cdot 1$ since the sum in (5.41) is zero at $t = a$. Hence $z_n(a) - x(a) = 0$. Moreover by (5.41) if $t_{j-1} \leq \xi < t_j$, then we obtain $z_n(t) = x(t_{j-1}) \cdot 1$. It follows that for those t,

$$|z_n(t) - x(t)| = |x(t_{j-1}) - x(t)|$$

Consequently, if $\eta(P_n) \to 0$, then $||z_n - x|| \to 0$ because x is continuous on $[a, b]$, hence uniformly continuous on $[a, b]$, since $[a, b]$ is compact in $\mathbb{R}$. The continuity of F now implies that $F(z_n) \to F(x)$ and $F(x) = f(x)$ so that

$$f(x) = \int_a^b x(t) dw(t)$$

It follows from (5.34) and (5.36) that

$$|f(x)| = \left| \int_a^b x(t) dw(t) \right| \leq \max_{t \in [a,b]} |x(t)| \mathrm{Var}\ (w) = ||x|| \mathrm{Var}\ (w)$$

Therefore, $\forall\ x \in C([a, b])$

$$||f|| = \sup_{x \neq \theta} \frac{|f(x)|}{||x||} \leq \mathrm{Var}\ (w) \tag{5.43}$$

It follows from (5.35) and (5.37) that

$$||f|| = \mathrm{Var}\ (w)$$

Note 5.3.1. We note that w in the theorem is not unique. Let us impose on w the following conditions

(i) w is zero at a and continuous from the right

$$w(a) = 0, \quad w(t+0) = w(t)\ (a < t < b)$$

Then, w will be unique [see A.E. Taylor [55]].

5.4 The General Form of Linear Functionals in Certain Functional Spaces

5.4.1 *Linear functionals on the n-dimensional Euclidean space* $\mathbb{R}^n$

Let f be a linear functional defined on E_n.

Now, $x \in E_n$ can be written as

$$x = \sum_{i=1}^{n} \xi_i e_i \quad \text{where } x = \{\xi_1, \xi_2, \dots, \xi_n\}$$

f being a linear functional

$$f(x) = f\left(\sum_{i=1}^{n} \xi_i e_i\right) = \sum_{i=1}^{n} \xi f(e_i) \tag{5.44}$$

$$= \sum_{i=1}^{n} \xi_i f_i \quad \text{where } f_i = f(e_i)$$

For $x = \{\xi_i\}$, let us suppose,

$$\phi(x) = \sum_{i=1}^{n} \xi_i \phi_i$$

where ϕ_i are arbitrary.

For $y = \{\eta_i\}$, we note that

$$\phi(x + y) = \sum_{i=1}^{n} (\xi_i + \eta_i)\phi_i = \sum_{i=1}^{n} \xi_i \phi_i + \sum_{i=1}^{n} \eta_i \phi_i = \phi(x) + \phi(y)$$

$$\phi(\lambda x) = \sum_{i=1}^{n} \lambda \xi_i \phi_i = \lambda \sum_{i=1}^{n} \xi_i \phi_i = \lambda \phi(x)$$

for all sectors λ.

Hence ϕ is a linear functional defined on an n-dimensional space. Since ϕ_i can be regarded as the components of an n-dimensional vector ϕ, the space $(\mathbb{R}^n)^*$, the dual of $\mathbb{R}^n$, is also an n-dimensional space with a metric, generally speaking, different from the metric of $\mathbb{R}^n$.

Let $||x|| = \max_i |\xi_i|$; then

$$|\phi(x)| = \left|\sum_{i=1}^{n} \xi_i \phi_i\right| \leq \sum_{i=1}^{n} |\xi_i||\phi_i| \leq \left(\sum_{i=1}^{n} |\phi_i|\right) ||x||,$$

Hence, $\|\phi\| = \sup\limits_{x \neq \theta} \dfrac{|\phi(x)|}{\|x\|} \leq \sum\limits_{i=1}^{n} |\phi_i|$ (5.45)

On the other hand, if we select an element $x_0 = \sum\limits_{i=1}^{n} \operatorname{sgn} \phi_i e_i \in \mathbb{R}^n$, then $\|x_0\| = 1$ and

$$\phi(x_0) = \sum_{i=1}^{n} \operatorname{sgn} \phi_i \cdot \phi_i(e_i) = \sum_{i=1}^{n} \operatorname{sgn} \phi_i \cdot \phi_i$$

$$= \sum_{i=1}^{n} |\phi_i| \|x_0\|$$

Hence, $\|\phi\| \geq \sum\limits_{i=1}^{n} |\phi_i|$ (5.46)

From (5.45) and (5.46), it follows that

$$\|\phi\| = \sum_{i=1}^{n} |\phi_i|$$

If an Euclidean metric is introduced, we can verify that the metric in $(\mathbb{R}^n)^*$ is also Euclidean.

5.4.2 The general form of linear functional on s

s is the space of all sequences of numbers.

Let $f(x)$ be a linear functional defined on s. Put

$$e_n = \{\xi_i^n\} \quad \text{where } \xi_i^n = 0, \; i \neq n \quad \text{and} \quad \xi_n^n = 1$$

Further, let $f(e_n) = u_n$. The convergence in s is coordinatewise. Therefore,

$$x = \lim_{n \to \infty} \sum_{k=1}^{n} \xi_k e_k = \sum_{n=1}^{\infty} \xi_k e_k$$

holds where $x = \{\xi_i\}$.

Because f is continuous,

$$f(x) = \lim_{m \to \infty} \sum_{k=1}^{m} \xi_k f(e_k) = \sum_{k=1}^{\infty} \xi_k u_k$$

Since this series must converge for every number sequence $\{\xi_k\}$, the u_k must be equal to zero from a certain index onwards and consequently

$$f(x) = \sum_{k=1}^{m} \xi_k u_k$$

Conversely, for any $x = \{\xi_i\}$, let $\phi(x)$ be given by

$$\phi(x) = \sum_{k=1}^{m} \xi_k u_k \tag{5.47}$$

where u_k's are real and arbitrary. For if $y = \{\eta_i\}$,

$$\phi(x+y) = \sum_{k=1}^{m} (\xi_k + \eta_k) u_k = \sum_{k=1}^{m} \xi_k u_k + \sum_{k=1}^{m} \eta_k u_k = \phi(x) + \phi(y)$$

Moreover, $\phi(\lambda x) = \sum_{k=1}^{m} \lambda \xi_k u_k = \lambda \sum_{k=1}^{m} \xi_k u_k = \lambda \phi(x)$.

Hence, $\phi(x)$ is a linear functional on s.

It therefore follows that every linear functional defined on s has the general form given by (5.47) where m and u_k, $k = 1, 2, \ldots, m$ are uniquely defined by (5.47).

5.4.3   The general form of linear functionals on l_p

Let $f(x)$ be a bounded linear functional defined on l_p. Since the elements $e_i = \{\xi_i^j\}$ where $\xi_i^j = 1$ for $i = j$ and $\xi_i^j = 0$ for $i \neq j$, form basis of l_p, every element $x \in l_p$ can be written in the form

$$x = \sum_{i=1}^{\infty} \xi_i e_i$$

Since $f(x)$ is bounded linear,

$$f(x) = \sum_{i=1}^{\infty} \xi_i f(e_i)$$

Writing $u_i = f(e_i)$, $f(x)$ takes the form,

$$f(x) = \sum_{i=1}^{\infty} u_i \xi_i \tag{5.48}$$

Let us put $x_n = \{\xi_i^{(n)}\}$, where

$$\xi_i^{(n)} = \begin{cases} |u_i|^{q-1} \operatorname{sgn} u_i, & \text{if } i \leq n \\ 0 & \text{if } i > n \end{cases}$$

q is chosen such that the equality $[(1/p) + (1/q)] = 1$ holds

$$f(x_n) = \sum_{i=1}^{n} u_i \xi_i^{(n)} = \sum_{i=1}^{n} |u_i|^{q-1} u_i \operatorname{sgn} u_i = \sum_{i=1}^{n} |u_i|^q \tag{5.49}$$

On the other hand,

$$f(x_n) \leq ||f|| \, ||x_n|| = ||f|| \left(\sum_{i=1}^{\infty} |\xi_i^{(n)}|^p \right)^{1/p}$$

$$= ||f|| \left(\sum_{i=1}^{n} |u_i|^{p(q-1)} \right)^{1/p} = ||f|| \left(\sum_{i=1}^{n} |u_i|^q \right)^{1/p}$$

Thus $\displaystyle\sum_{i=1}^{n} |u_i|^q \leq ||f|| \left(\sum_{i=1}^{n} |u_i|^q \right)^{1/p}$

Since $\dfrac{1}{p} + \dfrac{1}{q} = 1$

Thus, $\left(\displaystyle\sum_{i=1}^{n} |u_i|^q \right)^{1/q} \leq ||f||.$

Since the above is true for every n, it follows that

$$\left(\sum_{i=1}^{\infty} |u_i|^q \right)^{1/q} \leq ||f|| \tag{5.50}$$

Thus $\{u_i\} \in l_q$

Conversely, let us take an arbitrary sequence $\{v_i\} \in l_q$. Then, for $x = \{\xi_i\}$ let us write,

$$\phi(x) = \sum_{i=1}^{\infty} v_i \xi_i$$

To show that ϕ is a linear functional, we proceed as follows. For $y = \{\eta_i\}$,

$$\phi(x + y) = \sum_{i=1}^{\infty} v_i(\xi_i + \eta_i) \tag{5.51}$$

Since $x, y \in l_p$, $x + y \in l_p$, i.e., $\left(\displaystyle\sum_{i=1}^{\infty} |\xi + \eta_i|^p \right)^{1/p} < \infty.$

Since $\{v_i\} \in l_q$, $|\phi(x+y)| \leq \left(\displaystyle\sum_{i=1}^{\infty} |v_i|^q \right)^{1/q} \left(\sum_{i=1}^{\infty} |\xi_i + \eta_v|^p \right)^{1/p} < \infty.$

Hence, ϕ is additive and homogeneous.

$$|\phi(x)| \leq \left(\sum_{i=1}^{\infty} |v_i|^q \right)^{1/q} \left(\sum_{i=1}^{\infty} |\xi_i|^p \right)^{1/p} = M||x||$$

where $M = \left(\displaystyle\sum_{i=1}^{\infty} |v_i|^q \right)^{1/q}$

Thus, ϕ is a bounded linear functional. For calculating the norm of the functional we proceed as follows

$$|f(x)| \leq \left(\sum_{i=1}^{\infty} |u_i|^q \right)^{1/q} ||x||$$

Consequently, $||f|| \leq \left(\sum_{i=1}^{\infty} |u_i|^q \right)^{1/q}$ \hfill (5.52)

It follows from (5.50) and (5.52)

$$||f|| = \left(\sum_{i=1}^{\infty} |u_i|^q \right)^{1/q}$$

5.4.4 Corollary

Every linear functional defined on l_2 can be written in the general form

$$f(x) = \sum_{i=1}^{\infty} u_i \xi_i$$

where $\sum_{i=1}^{\infty} |u_i|^2 < \infty$ and $||f|| = \left(\sum_{i=1}^{\infty} |u_i|^2 \right)^{1/2}$

5.5   The General Form of Linear Functionals in Hilbert Spaces

Let H be a Hilbert space over $\mathbb{R}(\mathbb{C})$ and $f(x)$ a linear bounded functional defined on H. Let $N(f)$ denote the null space of $f(x)$, i.e., the space of zeroes of $f(x)$. Let $x_1, x_2 \in N(f)$. Then

$$f(\alpha x_1 + \beta x_2) = \alpha f(x_1) + \beta f(x_2) = 0 \ \ \forall \text{ scalars } \alpha, \beta$$

Hence, $N(f)$ is a subspace of H. Since f is a bounded linear functional for a convergent sequence $\{x_n\}$, i.e.,

$$x_n \to x \Rightarrow |f(x_n) - f(x)| \leq ||f|| \, ||x_n - x|| \to 0 \text{ as } n \to \infty$$

Hence, $N(f)$ is a closed subspace. Let $x \in H$ and $x \notin N(f)$. Let x_0 be the projection of x on the subspace $H - N(f)$, the orthogonal complement of H. Let $f(x_0) = \alpha$, obviously $\alpha \neq 0$. Put $x_1 = x_0/\alpha$. Then

$$f(x_1) = f\left(\frac{x_0}{\alpha} \right) = \frac{1}{\alpha} f(x_0) = 1$$

If now $x \in H$ is arbitrary and $f(x) = \beta$, then $f(x - \beta x_1) = f(x) - \beta f(x_1) = 0$. Let us put $x - \beta x_1 = z$ then $z \in N(f)$ and we have $x = \beta x_1 + z$. This equality shows that H is the orthogonal sum of $N(f)$ and the one-dimensional subspace spanned by x_1.

Since $z \in N(f)$ and $x_1 \in H - N(f)$,

$z \perp x_1$ or $\langle x - \beta x_1, x_1 \rangle = 0$ where $\langle \, , \, \rangle$ stands for the scalar product. Hence,

$$\langle x, x_1 \rangle = \beta \langle x_1, x_1 \rangle = \beta ||x_1||^2$$

Since $\beta = f(x)$, we have

$$f(x) = \left\langle x, \frac{x_1}{||x_1||^2} \right\rangle$$

If $\dfrac{x_1}{||x_1||^2} = u$, then

$$f(x) = \langle x, u \rangle, \tag{5.53}$$

i.e., we get the representation of an arbitrary functional as an inner product of the element x and a fixed element u. The element u is defined uniquely by f because if $f(x) = \langle x, v \rangle$, then $\langle x, u - v \rangle = 0$ for every $x \in H$, implying $u = v$.

Further (5.53) yields,

$$|f(x)| = |\langle x, u \rangle| \le ||x|| \, ||u||$$

which implies that

$$\sup_{x \ne \theta} \frac{|f(x)|}{||x||} \le ||u|| \quad \text{or} \quad ||f|| \le ||u|| \tag{5.54}$$

Since, on the other hand, $f(u) = \langle u, u \rangle = ||u||^2$, it follows that $||f||$ cannot be smaller than $||u||$, hence $||f|| = ||u||$.

Thus, every linear functional $f(x)$ in a Hilbert space H can be represented uniquely in the form $f(x) = \langle x, u \rangle$, where the element u is uniquely defined by the functional f. Moreover, $||f|| = ||u||$.

Problem

1. If l_1 is the space of real elements $x = \{\xi_i\}$ where $\displaystyle\sum_{i=1}^{\infty} |\xi_i| < \infty$, show that a linear functional f on l_1 can be represented in the form

$$f(x) = \sum_{k=1}^{\infty} c_k \xi_k$$

where $\{c_k\}$ is a bounded sequence of real numbers.

5.6   Conjugate Spaces and Adjoint Operators

In 5.1.1 we have defined the **conjugate (or dual) space** E^* of a Banach space E. We may recall that the *conjugate (or dual) space* E^* is the space of bounded linear functionals mapping the Banach space $E \to \mathbb{R}$. The idea that comes next is to find the characterisation, if possible, of a *conjugate or dual space*. In this case isomorphism plays a great role. We recall that two spaces E and E' are said to be **isomorphic** if, between their elements, there can be established a one-to-one correspondence, preserving the algebraic structure, that is such that

$$\left. \begin{array}{c} x \longleftrightarrow x' \\ y \longleftrightarrow y' \end{array} \right\} \Rightarrow \left\{ \begin{array}{c} x + y \longleftarrow x' + y' \\ \lambda x \longleftarrow \lambda x' \text{ for scalar } \lambda \end{array} \right. \tag{5.55}$$

5.6.1   Space $\mathbb{R}^n$: the dual space of $\mathbb{R}^n$ is $\mathbb{R}^n$

Let $\{e_1, e_2, \ldots, e_n\}$ be a basis in $\mathbb{R}^n$. Then any $x \in \mathbb{R}^n$ can be written as

$$x = \sum_{i=1}^{n} \xi_i e_i, \text{ where } x = \{\xi_i\}$$

Let f be a linear functional defined on $\mathbb{R}^n$.

Then $f(x) = \sum_{i=1}^{n} \xi_i f(e_i) = \sum_{i=1}^{n} \xi_i a_i$ where $a_i = f(e_i)$.

Now, by Cauchy-Bunyakovsky-Schwartz inequality (sec. 1.4.3)

$$|f(x)| \leq \left(\sum_{i=1}^{n} |\xi_i|^2 \right)^{1/2} \left(\sum_{i=1}^{n} |a_i|^2 \right)^{1/2} = \left(\sum_{i=1}^{n} |a_i|^2 \right)^{1/2} ||x||$$

where $||x|| = \left(\sum_{i=1}^{n} |\xi_i|^2 \right)^{1/2}$

Hence, $||f|| = \sup_{x \neq \theta} \dfrac{|f(x)|}{||x||} \leq \left(\sum_{i=1}^{n} |a_i|^2 \right)^{1/2}$ (5.56)

Taking $x = \{a_i\}$ we see that

$$f(x) = \sum_{i=1}^{n} |a_i|^2 = \left(\sum_{i=1}^{n} |a_i|^2 \right)^{1/2} ||x||$$

Hence, the upper bound in (5.56) is attained. That is

$$||f|| = \left(\sum_{i=1}^{n} |a_i|^2 \right)^{1/2}$$

This shows that the norm of f is the Euclidean norm and $||f|| = ||a||$ where $a = \{a_i\} \in \mathbb{R}$.

Hence, the mapping of $\mathbb{R}^n$ onto $\mathbb{R}^n$ defined by $f \longmapsto a = \{a_i\}$ where $a_i = f(e_i)$, is norm preserving and, since it is linear and bijective, it is an isomorphism.

5.6.2 Space l_1: the dual space of l_1 is l_∞

Let us take a Schauder basis $\{e_i\}$ for l_1, where $e_i = (\delta_{ij})$, δ_{ij} stands for the Kronecker δ-symbol.

Thus every $x \in l_1$ has a unique representation of the form

$$x = \sum_{i=1}^{\infty} \xi_i e_i \tag{5.57}$$

For any bounded linear functional f defined on l_1 i.e. for every $f \in l_1^*$ we have

$$f(x) = \sum_{i=1}^{\infty} \xi_i f(e_i) = \sum_{i=1}^{\infty} \xi_i a_i \tag{5.58}$$

where $a_i = f(e_i)$ are uniquely defined by f. Also, $||e_i|| = 1$, $i = 1, 2, \ldots$ and

$$|a_i| = ||f(e_i)|| \le ||f|| \, ||e_i|| = ||f|| \tag{5.59}$$

Hence $\sup_i |a_i| \le ||f||$. Therefore $\{a_i\} \in l_\infty$.

Conversely, for every $b = \{b_i\} \in l_\infty$ we can obtain a corresponding bounded linear functional ϕ on l_1. We can define ϕ on l_1 by

$$\phi(x) = \sum_{i=1}^{\infty} \xi_i b_i, \quad \text{where } x = \{\xi_i\} \in l_1$$

If $y = \{\eta_i\} \in l_1$, then

$$\phi(x + y) = \sum_{i=1}^{\infty} (\xi_i + \eta_i) \phi(e_i) = \sum_{i=1}^{\infty} \xi_i b_i + \sum_{i=1}^{\infty} \eta_i b_i$$
$$= \phi(x) + \phi(y) \text{ showing } \phi \text{ is additive}$$

For all scalars λ,

$$\phi(\lambda x) = \sum_{i=1}^{\infty} (\lambda \xi_i) \phi(e_i) = \lambda \sum_{i=1}^{\infty} \xi_i \phi(e_i)$$
$$= \lambda \phi(x), \text{ i.e., } \phi \text{ is homogeneous.}$$

Thus ϕ is homogeneous. Hence, ϕ is linear.

Moreover, $|\phi(x)| \le \sum_{i=1}^{\infty} |\xi_i \cdot b_i| \le \sup_i |b_i| \sum_{i=1}^{\infty} |\xi_i| = ||x|| \sup_i |b_i|$

Therefore, $||\phi|| = \sup\limits_{x \neq \theta} \dfrac{|\phi(x)|}{||x||} \leq \sup\limits_{i} |b_i| < \infty$ since $b = \{b_i\} \in l_\infty$. Thus ϕ is bounded linear and $\phi \in l_1^*$.

We finally show that the norm of f is the norm on the set l_∞. From (5.58), we have,

$$|f(x)| = \left| \sum_{i=1}^{\infty} \xi_i a_i \right| \leq \sup\limits_{i} |a_i| \sum_{i=1}^{\infty} |\xi_i| = ||x|| \sup\limits_{i} |a_i|$$

Hence, $||f|| = \sup\limits_{x \neq \theta} \dfrac{f(x)}{||x||} \leq \sup\limits_{i} |a_i|$.

It follows from (5.59) and this above inequality,

$$||f|| = \sup\limits_{i} |a_i|,$$

which is the norm on l_∞. Hence, we can write $||f|| = ||a||$ where $a = \{a_i\} \in l_\infty$. It shows that the bijective linear mapping of l_1^* onto l_∞ defined by $f \to a = \{a_i\}$ is an isomorphism.

5.6.3 Space l_p theorem

The dual space of l_p is l_q, here, $1 < p < \infty$ and q is the conjugate of p, that is, $\dfrac{1}{p} + \dfrac{1}{q} = 1$

Proof: A Schauder basis for l_p is $\{e_i\}$ where $e_i = \{\delta_{ij}\}$, δ_{ij} is the Kronecker δ symbol. Thus for every $x = \{\xi_i\} \in l_p$ we can find a unique representation of the form

$$x = \sum_{i=1}^{\infty} \xi_i e_i \tag{5.60}$$

We consider any $f \in l_p^*$ where l_p^* is the conjugate (or dual) space of l_p. Since f is linear and bounded,

$$f(x) = \sum_{i=1}^{\infty} \xi_i f(e_i) = \sum_{i=1}^{\infty} \xi_i a_i \tag{5.61}$$

where $a_i = f(e_i)$.

Let q be the conjugate of p i.e. $\dfrac{1}{p} + \dfrac{1}{q} = 2.$

Let $\bar{x}_n = \{\bar{\xi}_i^{(n)}\}$ with

$$\bar{\xi}_i^{(n)} = \begin{cases} |a_i|^q / a_i & \text{if } i \leq n \text{ and } a_i \neq 0 \\ 0 & \text{if } i > n \text{ or } a_i = 0 \end{cases} \tag{5.62}$$

Then $f(\bar{x}_n) = \sum\limits_{i=1}^{\infty} \bar{\xi}_i^{(n)} a_i = \sum\limits_{i=1}^{n} |a_i|^q$

Using (5.62) and that $(q-1)p = q$, it follows from the above,

$$f(\overline{x}_n) \leq \|f\| \, \|\overline{x}_n\| = \|f\| \left(\sum_{i=1}^{\infty} |\overline{\xi}^{(n)}|^p \right)^{1/p}$$

$$= \|f\| \left(\sum_{i=1}^{n} |a_i|^{p(q-1)} \right)^{1/p}$$

$$= \|f\| \left(\sum_{i=1}^{n} |a_i|^q \right)^{1/p}$$

Hence, $\quad f(\overline{x}_n) = \sum_{i=1}^{n} |a_i|^q \leq \|f\| \left(\sum_{i=1}^{n} |a_i|^q \right)^{1/p}$

Dividing both sides by the last factor, we get,

$$\left(\sum_{i=1}^{n} |a_i|^q \right)^{1-p^{-1}} = \left(\sum_{i=1}^{n} |a_i|^q \right)^{1/q} = \|f\| \tag{5.63}$$

Hence, on letting $n \to \infty$, we prove that,

$$\{a_i\} \in l_q$$

Conversely, for $b = \{b_i\} \in l_q$, we can get a corresponding bounded linear functional Φ on l_p. For $x = \{\xi_i\} \in l_p$, let us define Φ as,

$$\Phi(x) = \sum_{i=1}^{\infty} \xi_1 b_i \tag{5.64}$$

For $y = \{\eta_i\} \in l_p$, we have,

$$\Phi(x+y) = \sum_{i=1}^{\infty} (\xi_i + \eta_i) b_i = \sum_{i=1}^{\infty} \xi_i b_i + \sum_{i=1}^{\infty} \eta_i b_i$$

$$= \Phi(x) + \Phi(y) \leq \left[\left(\sum_{i=1}^{\infty} |\xi_i|^p \right)^{1/p} + \left(\sum_{i=1}^{\infty} |\eta_i|^p \right)^{1/p} \right] \times$$

$$\left(\sum_{i=1}^{\infty} |b_i|^q \right)^{1/q} < \infty$$

Also, $\quad \Phi(\alpha x) = \sum_{i=1}^{\infty} (\alpha \xi_i) b_i = \alpha \sum_{i=1}^{\alpha} \xi_i b_i = \alpha \Phi(x)$, for all scalars α.

Hence ϕ is linear.

To prove that Φ is bounded, we note that,

$$|\Phi(x)| = \left| \sum_{i=1}^{\infty} \xi_i b_i \right| \leq \left(\sum_{i=1}^{\infty} |\xi_i|^p \right)^{1/p} \left(\sum_{i=1}^{\infty} |b_i|^q \right)^{1/q}$$

$$= \left(\sum_{i=1}^{\infty} |b_i|^q \right)^{1/q} ||x||$$

Hence, Φ is bounded since $\{b_i\} \in l_q$.

Thus, Φ is a bounded linear functional. Finally, to show that the norm of f is the norm on the space l_q we proceed as follows: (5.61) yields,

$$|f(x)| = \left| \sum_{i=1}^{\infty} \xi_i a_i \right| \le \left(\sum_{i=1}^{\infty} |\xi_i|^p \right)^{1/p} \left(\sum_{i=1}^{\infty} |a_i|^q \right)^{1/q}$$

$$= ||x|| \left(\sum_{i=1}^{\infty} |a_i|^q \right)^{1/q}$$

Hence, $\quad ||f|| = \sup_{x \ne \theta} \dfrac{|f(x)|}{||x||} \le \left(\sum_{i=1}^{\infty} |a_i|^q \right)^{1/p}$ \hfill (5.65)

It follows from (5.63) and (5.65) that

$$||f|| = \left(\sum_{i=1}^{\infty} |a_i|^q \right)^{1/q} = ||a||_q \hfill (5.66)$$

Thus, $||f|| = ||a||_q$ where $a = \{a_i\} \in l_q$ and $a_i = f(e_i)$. The mapping of l_p^* onto l_q as defined $f \mapsto a$ is linear and bijective, and from (5.66) we see that it is norm preserving. Therefore, it is an isomorphism.

Note 5.6.1.

(i) It can be shown that l_q^* is **isomorphic** to l_p. Hence

(ii) $l_2^* = l_2$, i.e., l_2 is called a **self-conjugate space.**

(iii) A linear functional in a Hilbert space is spanned by elements of the same space. A Hilbert space is, therefore, **self-conjugate.**

5.6.4   Reflexive space

Let E be a normed linear space. In 5.1.2, we have defined E^*, the conjugate space of E as the space of bounded linear functionals defined on E. In the same manner we can introduce the concept of a conjugate space $(E^*)^*$ of a Banach space E^* and call the space E^{**} as the second conjugate of the normed linear space E. To be more specific, consider a bounded linear f defined on E, so that in $f(x)$, f remains fixed and x varies over E. We can also think of a situation where x is kept fixed and f is varying in E. For example, let

$$f(x) = \int_0^1 x(t) dg(t)$$

Then we have two cases: namely (i) $g(t)$ is fixed and $x(t)$ varying or (ii) $x(t)$ is fixed and $g(t)$ varies. Now, since $f(x) \in \mathbb{R}$, $f(x)$ can be treated as a functional F_x, defined on E^*, for fixed x and variable f. Hence, it is possible to write $f(x) = F_x(f)$. In what follows, we shall show that the mapping F_x is an isometric isomorphism of E onto a subspace of E^{**}.

5.6.5 *Theorem*

Let $E \neq \{\Phi\}$ be a normed linear space over $\mathbb{R}(\mathbb{C})$. Given $x \in E$, let

$$F_x(f) = f(x) \quad \forall f \in E^* \tag{5.67}$$

Then F_x is a bounded linear functional on E^*, i.e., $F_x \in E^{**}$.

Further, the mapping F_x is an isometric isomorphism of E onto the subspace $\hat{E} = \{F_x : x \in E\}$ of E^{**}.

Proof: The mapping F_x satisfies,

$$F_x(\alpha f_1 + \beta f_2) = (\alpha f_1 + \beta f_2)(x) = \alpha f_1(x) + \beta f_2(x)$$
$$= \alpha F_x(f_1) + \beta F_x(f_2) \tag{5.68}$$

$\forall f_1, f_2 \in E^*$ and $\alpha, \beta \in \mathbb{R}(\mathbb{C})$

Hence F_x is linear.

Also F_x is bounded, since

$$|F_x(f)| = |f(x)| \leq ||f|| \, ||x||, \quad \forall f \in E^*. \tag{5.69}$$

Consequently, $F_x \in E^{**}$. If $F_x \in E^{**}$ is not unique, let us suppose $F_{1x}(f) = F_{2x}(f)$ or $(F_{1x} - F_{2x})f = 0$. We keep x fixed and vary f. Since f is arbitrary $F_{1x} = F_{2x}$, showing that F_x is unique.

Thus to every $x \in E$, $\exists$ a unique $F_x \in E^{**}$ given by (5.67). This defines a function $\phi : E \to E^{**}$ given by

$$\phi(x) = F_x.$$

(i) We show that ϕ is linear.

For, $x, y \in E$ and $\alpha, \beta \in \mathbb{R}(\mathbb{C})$,

$$(\phi(\alpha x + \beta y))(f) = F_{\alpha x + \beta y}(f) = f(\alpha x + \beta y)$$
$$= (\alpha F_x + \beta F_y)(f)$$
$$= (\alpha \phi(x) + \beta \phi(y))(f), \quad \forall f \in E^*$$

Hence, $\phi(\alpha x + \beta y) = \alpha \phi(x) + \beta \phi(y)$

(ii) We next show that ϕ *preserves norm*.

For each $x \in E$, we have

$$||\phi(x)|| = ||F_x|| = \sup_{f \neq \theta} \left\{ \frac{|F_x(f)|}{||f||} : f \in E^* \right\}$$

$$= \sup_{f \neq \theta} \left\{ \frac{|f(x)|}{||f||} : f \in E^* \right\}$$

Now, by theorem 5.1.4, for every $x \in E$, $\exists$ a functional $g \in E^*$ such that $||g|| = 1$ and $g(x) = ||x||$.

Therefore, $||\phi(x)|| = ||F_x|| = \sup_{f \neq \theta} \dfrac{|f(x)|}{||f||} \geq \dfrac{|g(x)|}{||g||} = ||x||$

Using (5.69) we prove $||\phi(x)|| = ||x||$.

(iii) We next show that ϕ is *injective*. Let $x, y \in E$. Then

$$x - y \neq \theta \Rightarrow ||x - y|| \neq 0 \Rightarrow ||\phi(x - y)|| \neq 0$$
$$\Rightarrow ||\phi(x) - \phi(y)|| \neq 0 \Rightarrow \phi(x) \neq \phi(y)$$

We thus conclude that ϕ is an isometric isomorphism of E onto the subspace $\hat{E}(\phi(E))$ of E.

5.6.6   Definition

Let E be a normed linear space over $\mathbb{R}(\mathbb{C})$. The isometric isomorphism $\phi : E \to E^{**}$ defined by $\phi(x) = F_x$ is called the *natural embedding (or the canonical embedding)* of E into the second conjugate space E^{**}. The functional $F_x \in E^{**}$ is called the functional induced by the vector x. We refer to the functional of this type as *induced functional*.

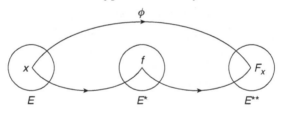

Fig. 5.2

5.6.7   Definition: reflexive normed linear space

A normed linear space E is said to be *reflexive* if the natural embedding ϕ maps the space E onto its second conjugate space E^{**}, i.e., $\phi(E) = E^{**}$.

Note 5.6.2.

(i) If E is a reflexive normed linear space, then E is isometrically isomorphic to E^{**} under the *natural embedding*.

(ii) If E is a reflexive normed linear space, since the second conjugate space E^{**} is always *complete*, the space E must be *complete*. Hence, completeness of the second conjugate space is a necessary condition for a normed linear space to be complete. However, this condition need not be sufficient (Example 4 (sec. 5.6.4)). Thus, it is clear that if E is not a Banach space, then we must have $\phi(E) \neq E^{**}$ and hence E is not reflexive.

5.6.8   Definitions: algebraic dual, topological dual

Algebraic dual (conjugate)

Given E_x, a topological linear space, the space of linear functionals mapping $E_x \to \mathbb{R}$ is called *the algebraic dual (conjugate)* of E_x.

Topological dual (conjugate)

On the other hand the space of **continuous** linear functionals mapping $E_x \to \mathbb{R}$ is called the *topological dual (conjugate)* of E_x.

5.6.9   Examples

1. $\mathbb{R}^n$, n-dimensional Euclidean space is **reflexive**.

In 5.4.1 we have seen that $\mathbb{R}^{n*} = \mathbb{R}^n$. Then,

$$(\mathbb{R}^n)^{**} = (\mathbb{R}^{n*})^* = (\mathbb{R}^n)^* = \mathbb{R}^n$$

Hence, $\mathbb{R}^n$ is reflexive.

Note 5.6.3 Every finite dimensional normed linear space is reflexive. We know that in a finite dimensional normed linear space E, every linear functional on E is bounded, so that the reflexivity of E follows.

2. **The space l_p $(p > 1)$**

In 5.6.3, we have seen that $l_p^* = l_q$, $\dfrac{1}{p} + \dfrac{1}{q} = 1$

Therefore, $l_P^{**} = (l_p^*)^* = (l_q)^* = l_p$

Hence, l_p is reflexive.

3. The space $C([0,1])$ is non-reflexive. For that see 6.2.

5.6.10   Theorem

A normed linear space is isometrically isomorphic to a dense subspace of a Banach space.

Proof: Let E be a normed linear space. If $\phi : E \to E^{**}$ be the natural embedding, then E and $\phi(E)$ are isometrically isomorphic spaces. But $\phi(E)$ is a dense subspace of $\overline{\phi(E)}$ and $\overline{\phi(E)}$ is a closed subspace of the Banach space E^{**}, it follows that $\overline{\phi(E)}$ itself is a Banach space. Hence E is isometrically isometric to the dense subspace $\phi(E)$ of the Banach space $\overline{\phi(E)}$.

We next discuss the relationship between separability and reflexivity of a normed linear space.

5.6.11   Theorem

Let E be a normed linear space and E^* be its dual. Then E^* is separable $\Rightarrow E$ is separable.

Proof: Since E^* is separable, $\exists$ a countable set $S = \{f_n : f_n \in E^*,\ n \in \mathbb{N}\}$ such that S is dense in E^*, i.e., $\overline{S} = E^*$.

For each $n \in \mathbb{N}$, choose $x_n \in E$ such that

$$||x_n|| = 1 \quad \text{and} \quad |f_n(x)| \geq \frac{1}{2}||f_n||$$

Let X be a closed subspace of E generated by the sequence $\{x_n\}$, i.e., $X = \overline{\text{span}}\{x_n \in E, n \in \mathbb{N}\}$.

Suppose $X \neq E$ then $\exists$ a point $x_0 \in E - X$. Theorem 5.1.5 yields that we can find a functional $\theta \neq g \in E^*$ such that $g(x_0) \neq 0$ and $g(X) = 0$.

Thus $\begin{cases} g(x_n) = 0, \quad n \in \mathbb{N} \\ \frac{1}{2}||f_n|| \leq |f_n(x_n)| = |(f_n - g)(x_n)| \leq ||f_n - g|| \end{cases}$

Therefore, $||g|| \leq ||f_n - g|| + ||f_n|| \leq 3||f_n - g|| \ \forall \ n \in \mathbb{N}$

But since $\overline{S} = E^*$, it follows that $g = 0$, which contradicts the assumption that $X \neq E$. Hence, $X = E$ and thus E is separable.

5.6.12   Theorem

Let E be a separable normed linear space. If the dual E^* is non-separable then E is non-reflexive.

Proof: Let E be reflexive if possible. Then, E^{**} is isometrically isomorphic to E under the natural embedding. Given E is separable, E^{**} will be separable. But, by theorem 5.6.11, E^* is separable, which contradicts our assumption. Hence, E is non reflexive.

5.6.13   Example

The space $(l_1, ||\cdot||_1)$ is not reflexive.

The space l_1 is separable.

Now, $(l_1)^* = l_\infty$. But l_∞ is not separable. By theorem 5.6.12 we can say that l_1 is non-reflexive.

5.6.14   Adjoint operator

We have, so far, talked about bounded linear operators and studied their properties. We also have discussed bounded linear functionals. Associated with linear operators are adjoint linear operators. Adjoint linear operators find much use in the solution of equations involving operators. Such equations arise in Physics, Applied Mathematics and in other areas.

Let A be a bounded linear operator mapping a Banach space E_x into a Banach space E_y, and let us consider the equation $Ax = y$, $x \in E_x$, $y \in E_y$.

If $g : E_y \to \mathbb{R}$ be a linear functional, then

$$g(y) = g(Ax) = \text{a functional of } x = f(x) \text{ (say)} \tag{5.70}$$

$f(x)$ is a functional on E_x.

We can see that f is linear. Let $x_1, x_2 \in E_x$ and $y_1, y_2 \in E_y$, such that

$$y_1 = Ax_1, \quad y_2 = Ax_2$$

Then $g(y_1 + y_2) = g(Ax_1 + Ax_2) = g(A(x_1 + x_2)) = f(x_1 + x_2)$.
Since g is linear,

$$f(x_1 + x_2) = g(y_1 + y_2) = g(y_1) + g(y_2) = f(x_1) + f(x_2) \qquad (5.71)$$

Thus, f is linear. Hence the functional $f \in E_x^*$ corresponds to some $g \in E_y^*$. This sets the definition of an **adjoint operator**. The correspondence so obtained forms a certain operator with domain E_y^* and range contained in E_x^*.

5.6.15 *Definition: adjoint operator*

Let A be a bounded linear operator mapping a normed linear space E_x into a normed linear space E_y, let $f \in E_x^*$ and $g \in E_y^*$ be given linear functionals, then the operator adjoint to A is denoted by A^* and is given by $f = A^* g$ [see figure 5.3]

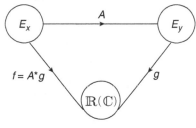

Fig. 5.3

5.6.16 *Examples*

1. Let A be an operator in $(\mathbb{R}^n \to \mathbb{R}^n)$, where $\mathbb{R}^n$ is an n-dimensional space. Then A is defined by a matrix (a_{ij}) of order n and equality $y = Ax$ where $x = \{\xi_1, \xi_2, \ldots, \xi_n\}$ and $y = \{\eta_1, \eta_2, \ldots, \eta_n\}$ such that

$$\eta_i = \sum_{j=1}^{n} a_{ij} \xi_j$$

Consider a functional $f \in \mathbb{R}^{n*} (= \mathbb{R}^n)$ since $\mathbb{R}^n$ is self-conjugate; $f = (f_1, f_2, \ldots, f_n)$, $f(x) = \sum_{i=1}^{n} f_i \xi_i$.

Hence, $f(Ax) = \sum_{i=1}^{n} f_i \eta_i = \sum_{i=1}^{n} f_i \sum_{j=1}^{n} a_{ij} \cdot \xi_j$

$$= \sum_{i=1}^{n} \sum_{j=1}^{n} a_{ij} f_i \xi_j = \sum_{j=1}^{n} \left(\sum_{i=1}^{n} a_{ij} f_i \right) \xi_j$$

$$= \sum_{j=1}^{n} \phi_j \xi_j \qquad (5.72)$$

where $\phi_j = \sum_{i=1}^{n} a_{ij} f_i$ (5.73)

The vector $\phi = (\phi_1, \phi_2, \ldots, \phi_n)$ is an element of $\mathbb{R}^n$ and is obtained from the vector $f = (f_1, f_2, \ldots, f_n)$ of the same space by the linear transformation

$$\phi = A^* f$$

where A^* is the transpose of the matrix A. Therefore, the transpose of A corresponds to the adjoint of the matrix A in the n-dimensional space.

2. Let us consider in $L_2([0,1])$ the integral

$$Tf = g(s) = \int_0^1 K(s,t) f(t) dt$$

$K(s,t)$ is a continuous kernel.

An arbitrary linear functional $\phi(g) \in L_2([0,1])$ will be of the form $\langle g, v \rangle$ where $v \in L_2([0,1])$ and $\langle \ , \ \rangle$ denotes scalar product.

This is because $L_2([0,1])$ is a Hilbert space.

$$\phi(g) = \langle g, v \rangle = \int_0^1 \int_0^1 K(s,t) f(t) dt v(s) ds$$

$$= \int_0^1 \left(\int_0^1 K(s,t) v(s) ds \right) f(t) dt$$

(on change of order of integration by Fubini's theorem 10.5.3)

$$= \int_0^1 (T^* v)(t) f(t) dt$$

$$= \langle T^* v, f \rangle.$$

where, $T^* v(s) = \int_0^1 K(t,s) v(t) dt$

Thus, in the given case, the adjoint operator is also an integral operator, the kernel $K(t,s)$ which is obtained by interchanging the arguments of $K(s,t)$. $K(t,s)$ is called the **transpose** of the kernel $K(s,t)$.

5.6.17   Theorem

Given A, a bounded linear operator mapping a normed linear space E_x into a normed linear space E_y, its **adjoint** A^* is also a bounded linear operator, and $\|A\| = \|A^*\|$.

Let $f_1 = A^* g_1$ and $f_2 = A^* g_2$.

Hence, $g_1(y) = g_1(Ax) = f_1(x)$, $x \in E_x$, $f_1 \in E_x^*$, $y \in E_y$, $g_1 \in E_y^*$. Also $g_2(y) = g_2(Ax) = f_2(x)$.

Now, g_1 and g_2 are linear functionals and hence f_1 and f_2 are linear functionals.

Now, $(g_1 + g_2)(y) = g_1(y) + g_2(y) = f_1(x) + f_2(x) = (f_1 + f_2)(x)$

or $f_1 + f_2 = A^*(g_1 + g_2)$ or $A^*g_1 + A^*g_2 = A^*(g_1 + g_2)$

Thus, A^* is a linear functional.

Moreover, $|A^*g(x)| = |f(x)| = |g(Ax)| \leq ||g|| \, ||A|| \, ||x||$

or, $$||A^*g|| = \sup_{x \neq \theta} \frac{|A^*g(x)|}{||x||} \leq ||g|| \, ||A||$$

Hence, $$\frac{||A^*g||}{||g||} \leq ||A||$$

Therefore, $$||A^*|| = \sup_{g \neq \theta} = \frac{||A^*g||}{||g||} \leq ||A|| \qquad (5.74)$$

Let x_0 be an arbitrary element of E_x. Then, by theorem 5.1.4, there exists a functional $g_0 \in E_y^*$ such that $||g_0|| = 1$ and $g_0(Ax_0) = ||Ax_0||$.

Hence, $||Ax_0|| = g_0(Ax_0) = f_0(x_0) \leq ||f_0|| \, ||x_0||$

$$= ||A^*g_0|| \, ||x_0|| \leq ||A^*|| \, ||g_0|| \, ||x_0||$$

or, $$||A|| = \sup_{x \neq \theta} \frac{||Ax_0||}{||x_0||} \leq ||A^*|| \quad [\text{since } ||g_0|| = 1] \qquad (5.75)$$

It follows from (5.74) and (5.75) that

$$||A|| = ||A^*||$$

5.6.18 *Adjoint operator for an unbound linear operator*

Let A be an unbounded linear operator defined on a subspace L_x dense in E_x with range in the space E_y. The notion of an adjoint to such an unbounded operator can be introduced. Let $g \in E_y^*$ and let

$$g(Ax) = f_0(x), \ x \in L_x$$

Let $x_1, x_2 \in L_x$.

Then $g(A(x_1 + x_2)) = g(Ax_1 + Ax_2)$

$$= g(Ax_1) + g(Ax_2)$$

since g is a linear functional defined on E_y.

$$= f_0(x_1) + f_0(x_2)$$

On the other hand, $g(A(x_1 + x_2)) = f_0(x_1 + x_2)$, showing that f_0 is additive. Similarly, we can show that f_0 is homogeneous. Thus, f_0 is linear. But f_0 is not in general bounded. In case f_0 is bounded, since L_x is everywhere dense in E_x, f_0 can be extended to the entire space E_x.

In case A^* is not defined on the whole space E_y^* which contains θ, it must be defined on some subspace $L_y^* \subset E_y^*$. This will lead to the linear functional $f \in E_x^*$ being set in correspondence to the linear functional $g \in E_y^*$. This operator A^* is also called the **adjoint** of the unbounded linear operator A.

Thus we can write $f_0 = A^*g$, $g \in L_y^*$.

Let $g_1, g_2 \in L_y^*$. Then, for fixed $x \in L_x$.

$$g_1(Ax) = f_{1,0}(x) \left.\right\} $$
$$\text{Similarly,} \quad g_2(Ax) = f_{2,0}(x) \left.\right\} \tag{5.76}$$

Therefore, $(g_1 + g_2)(Ax) = (f_{1,0} + f_{2,0})(x)$ $\tag{5.77}$

Thus, $(g_1 + g_2) \in L_y^*$, showing that L_y^* is a subspace. It follows form (5.74) that $f_{1,0} = A^*g_1$, $f_{2,0} = A^*g_2$. Hence (5.76) gives

$$A^*(g_1 + g_2) = f_{1,0} + f_{2,0} = A^*g_1 + A^*g_2$$

This shows that A^* is a linear operator, but generally **not bounded**.

5.6.19   *The matrix form of operators in space with basis and the adjoint*

Let E be a Banach space with a basis and A a bounded linear operator mapping E into itself.

Let $\{e_i\}$ be a basis in E and $x \in E$ can be written as

$$x = \sum_{i=1}^{\infty} \alpha_i e_i$$

Thus, A being bounded,

$$y = Ax = \sum_{i=1}^{\infty} \alpha_i Ae_i$$

Since Ae_i is again an element of E, it can be represented by

$$Ae_i = \sum_{k=1}^{\infty} p_{ki} e_k$$

Then we can write,

$$y = Ax = \lim_n \sum_{i=1}^{n} \alpha_i Ae_i = \lim_n \sum_{i=1}^{n} \alpha_i \left(\sum_{k=1}^{\infty} p_{ki} e_k \right) \tag{5.78}$$

Thus, $\quad y = \sum_{k=1}^{\infty} \beta_k e_k$ $\tag{5.79}$

where, $\quad \beta_k = \sum_{i=1}^{\infty} p_{ki} \alpha_i$ $\tag{5.80}$

Let $\{\phi_j\}$ be a sequence of functionals biorthogonal to the sequence $\{e_i\}$, i.e.,

$$\phi_j(e_k) = \begin{cases} 1 & \text{if } j = k \\ 0 & \text{if } j \neq k \end{cases} \tag{5.81}$$

Then (5.79) and (5.80) imply,

$$\begin{aligned} \beta_m = \phi_m(y) &= \phi_m \left\{ \lim_n \sum_{i=1}^n \alpha_i \left(\sum_{k=1}^\infty p_{ki} e_k \right) \right\} \\ &= \lim_n \phi_m \left\{ \sum_{i=1}^n \alpha_i \left(\sum_{k=1}^\infty p_{ki} e_k \right) \right\} \\ &= \lim_n \sum_{i=1}^n \alpha_i \sum_{k=1}^n p_{ki} \phi_m(e_k) \\ &= \lim_n \sum_{i=1}^n p_{mi} \alpha_i \end{aligned} \tag{5.82}$$

Equation (5.82) shows that the operator A is uniquely defined by the infinite matrix (p_{ki}). Thus, the components of the element $y = Ax$ are uniquely defined by the components of the element x. Thus a finite matrix gets extended to an infinite dimensional matrix.

5.6.20 Adjoint A^* of an operator A represented by an infinite matrix

Let A^* denote the operator adjoint to A and A^* map E^* into itself. Let $f = A^* g$, i.e., $g(y) = g(Ax) = f(x)$ for every $x \in E$.

Furthermore, let $g = \sum_{i=1}^\infty c_i f_i$ and $f = \sum_{i=1}^\infty d_i f_i$

Then $g(Ax) = g\left\{ A \left(\sum_{i=1}^\infty \alpha_i e_i \right) \right\} = g \left\{ \lim_n \sum_{k=1}^n \left(\sum_{i=1}^\infty p_{ki} \alpha_i \right) e_k \right\}$

$$= \lim_{n \to \infty} \left\{ \sum_{k=1}^n \left(\sum_{i=1}^\infty p_{ki} \alpha_i \right) g(e_k) \right\}$$

$$= \lim_{n \to \infty} \sum_{k=1}^n \left(\sum_{i=1}^\infty p_{ki} \alpha_i \right) c_k = \lim_{n \to \infty} \sum_{i=1}^\infty \left(\sum_{k=1}^n p_{ki} c_k \right) \alpha_i$$

On the other hand,

$$g(Ax) = f(x) = \sum_{i=1}^\infty \alpha_i f_i(x) = \sum_{i=1}^\infty d_i \alpha_i$$

Consequently, $\sum_{i=1}^\infty d_i \alpha_i = \lim_{n \to \infty} \sum_{i=1}^\infty \left(\sum_{k=1}^n p_{ki} c_k \right) \alpha_i \tag{5.83}$

Let $x = e_m$, i.e., $\alpha_m = 1$, $\alpha_i = 0$, for $i \neq m$. Thus (5.83) gives,

$$d_m = \lim_n \sum_{k=1}^{n} p_{km} c_k = \sum_{k=1}^{\infty} p_{km} c_k$$

Thus, $d_m = A^* c_m$ where c_m is the mth component of g. $A^* = (a_{ji})$ is the transpose of $A = (a_{ij})$. Thus, in the case of a matrix with infinite number of elements, the adjoint operator is the transpose of the corresponding matrix.

Such representation of operator and their adjoints hold for instance in the space l_2.

Note 5.6.4. Many equations of mathematical physics are converted into algebraic equations so that numerical methods can be adopted to solve them.

5.6.21 *Representation of sum, product, inverse on adjoints of such operator which admit of infinite matrix representation*

Given A, an operator which admits of infinite matrix representation in a Banach space with a basis, we have seen that the adjoint operator A^* admits of a similar matrix representation.

By routine manipulation we can show that

(i) $(A + B)^* = A^* + B^*$.

(ii) $(AB)^* = B^* A^*$ where A and B are conformable for multiplication.

(iii) $(A^{-1})^* = (A^*)^{-1}$, where A^{-1} exists.

Problems

1. Prove that the dual space of $\mathbb{C}^n$ is $\mathbb{C}^n$.

2. Prove that the dual space of $(\mathbb{C}^n, ||\cdot||_\infty)$ is the space $(\mathbb{C}^n, ||\cdot||_1)$.

3. Prove that the dual space of l_2 is l_2.

4. Show that, although the sequence space l_1 is separable, its dual $(l_1)^*$ is not separable.

5. Show that if E is a normed linear space its conjugate is a Banach space.

6. Show that the space l_p. $1 < p < \infty$ is reflexive but l_1 is not reflexive.

7. If E, a normed linear space is reflexive and $X \subset E$ is a closed subspace, then show that X is reflexive.

8. Show that a Banach space E is reflexive if and only E^* is reflexive.

9. If E is a Banach space and E^* is reflexive, then show that $\phi(E)$ is closed and dense in E^{**}.

10. Let E be a compact metric space. Show that $C(E)$ with the sup norm is reflexive if and only if, E has only a finite number of points.

CHAPTER 6

SPACE OF BOUNDED LINEAR FUNCTIONALS

In the previous chapter, the notion of functionals and their extensions was introduced. We have also talked about the space of functionals or conjugate space and adjoint operator defined on the conjugate space. In this chapter, the notion of the conjugate of a normed linear space and its adjoints has been revisited. The null space and the range space of a bounded linear operator and its transpose (adjoint) are related. Weaker concept of convergence in a normed linear space and its dual (conjugate) are considered. The connection of the notion of reflexivity with weak convergence and with the geometry of the normed linear spaces is explored.

6.1 Conjugates (Duals) and Transposes (Adjoints)

In 5.6 we have seen that the conjugate (dual) space E^* of a Banach space E, as the space of bounded linear functionals mapping the Banach space $E \to \mathbb{R}$. Thus, if $f \in E^*$,

$$||f|| = \sup_{x \neq \theta} \frac{|f(x)|}{||x||}, \quad x \in E.$$

If, $f_1, f_2 \in E^*$, then $f_1 = f_2 \implies f_1(x) = f_2(x) \ \forall \ x \in E$

Again, $f_1(x) = f_2(x) \implies (f_1 - f_2)(x) = 0 \implies ||f_1 - f_2|| = 0 \implies f_1 = f_2$.

On the other hand, the consequence of the Hahn-Banach extension theorem 5.1.4 shows that $x_1 = x_2$ in E if and only if $f(x_1) = f(x_2)$ for all $f \in E^*$. This shows that

$$||x|| = \sup_{f \neq \theta} \frac{|f(x)|}{||f||}, \quad x \in E.$$

in analogy with the definition of $||f||$ above.

This interchangeability between E and E^* explains the nomenclature 'conjugate or dual' for E^*.

6.1.1 Definition: restriction of a mapping

If $F : E_1 \to E_2$, E_1 and E_2 being normed linear spaces, and if $E_0 \subseteq E_1$, then $F|_{E_0}$ defined for all $x \in E_0$ is called the **restriction of F to E_0**.

6.1.2 Theorem

Let E be a normed linear space.

(a) **Let E_0 be a dense subset of E. For $f \in E^*$ let $F(f)$ denote the restriction of f to E_0. Then the map F is a linear isometry from E^* onto E_0^*.**

(b) **IF E^* is separable then so is E.**

Proof: Let $f \in E^*$. Now $F(f)$, being defined on $E_0 \subseteq E$, belongs to E_0^*. $||F(f)|| = ||f||$ and that the map is linear. Then by theorem 5.1.3, $F(f)$ defined on E_0 can be extended to the entire space with preservation of norm. Hence, F is onto (surjective).

(b) [See theorem 5.6.11.]

6.1.3 Theorem

Let $1 \leq p < \infty$ and $\dfrac{1}{p} + \dfrac{1}{q} = 1$. For a fixed $y \in l_q$, let

$$f_y(x) = \sum_{i=1}^{\infty} \xi_i y_i \quad \text{where} \quad x = \{\xi_i\} \in l_p.$$

Then $f_y \in (l_p)^*$ and $||f_y|| = ||y||_q$

The map $f : l_q \to l_p^*$ defined by

$$F(y) = f_y \quad y \in l_q$$

is a linear isometry from l_q into $(l_p)^*$.

If $1 \leq p < \infty$. Then F is surjective (onto).

In fact, if $f \in l_p^*$ and $y = (f(e_1), f(e_2)\ldots)$ then $y \in l_q$ and $f = F(y)$.

Proof: Let $y \in l_q$. For $x \in l_p$, we have

$$\sum_{i=1}^{\infty} |\xi_i y_i| \leq ||x||_p ||y||_q.$$

For $p = 1$ or ∞ the above is true and follows by letting $n \to \infty$ in Holder's inequality (sec. 1.4.3) if $1 \le p < \infty$. Hence f_y is well-defined, linear and $\|f_y\| \le \|y\|_q$. Next, to prove $\|y\|_q \le \|f_y\|$. If $y = \theta$, there is nothing to prove. Assume, therefore, that $y \ne \theta$. The above inequality can be proved by following arguments as in 5.6.3.

If we let $F(y) = f_y$, $y \in l_q$, Then, F is a Linear isometry from l_q into $(l_p)^*$ for $2 \le p < \infty$.

Let $1 \le p < \infty$. To show that F is surjective consider $f \in (l_p)^*$ and and let $y = (f(e_1), f(e_2), \ldots)$.

If, $p = 1$, we show from the expression for y that $y \in l_\infty$. Let $1 \le p < \infty$ and for $n = 1, 2, \ldots$ define $y^n = (y_1, y_2, \ldots y_n, 0 \ldots 0)$.

Thus, $y^n \in l_q$. $\|y^n\|_q \le \|f_{y^n}\|$. Now,

$$\|f_{y^n}\| = \sup \left\{ \left| \sum_{i=1}^{\infty} x_i y_i \right| : x \in l_p, \ \|x\|_p \le 1 \right\}$$

Let us consider $x \in l_p$ with $\|x\|_p \le 1$ and define $x^n = (x_1, x_2, \ldots x_n, 0 \ldots 0)$. Then x^n belongs to l_p, $\|x^n\|_p \le \|x\|_p \le 1$ and

$$f(x^n) = \sum_{i=1}^{n} x_i f(e_i) = \sum_{i=1}^{n} x_i y_i = f_{y^n}(x).$$

Thus, $\|f_{y^n}\| \le \|f\| = \sup\{|f(x)| : x \in l_p, \ \|x\|_p \le 1\}$

so that $\left(\sum_{j=1}^{\infty} |y_j|^q \right)^{\frac{1}{q}} = \lim_{n \to \infty} \|y^n\|_q \le \lim_{n \to \infty} \sup \|f_{y^n}\|$

$$\le \|f\| < \infty, \text{ that is } y \in l_p.$$

Now, let $x \in l_p$. Since $p < \infty$, we see that $x = \lim_{n \to \infty} \sum_{i=1}^{n} x_i e_i$. Hence, by the continuity and the linearity of f,

$$f(x) = \lim_{n \to \infty} f \left(\sum_{i=1}^{n} x_i e_i \right) = \sum_{i=1}^{\infty} x_i f(e_i) = \sum_{i=1}^{\infty} x_i y_i = f_y(x).$$

Thus, $f = f_y$ that is $F(y) = f$ showing that F is surjective.

In what follows we take c_0 as the space of scaler sequences converging to zero and c_{00}, as the space of scalar sequences having only finitely many non-zero terms.

6.1.4   Corollary

Let $1 \le p < \infty$ and $\dfrac{1}{p} + \dfrac{1}{q} = 1$

(i) The dual of $\mathbb{R}^n (\mathbb{C}^n)$ with the norm $\| \cdot \|_p$ is linearly isometric to $\mathbb{R}^n (\mathbb{C}^n)$ with the norm $\| \cdot \|_q$.

(ii) The dual of c_{00} with the norm $|| \cdot ||_p$ is linearly isometric to l_q.

(iii) The dual of c_0 with the norm $|| \ ||_\infty$ is linearly isometric to l_1.

Proof: (i) If we replace the summation $\sum_{i=1}^{\infty} \xi_i y_i$ with the summation $\sum_{i=1}^{n} \xi_i y_i$ in theorem 6.1.3 and follow its argument we get the result.

(ii) If $1 \le p < \infty$. Then c_{00} is a dense subspace of l_p, so that the dual of c_{00} is linearly isometric to l_q by theorems 6.1.2(a) and 6.1.3.

Let $p = \infty$, so that $q = 1$. Consider $y \in l_1$ and define,

$$f_y(x) = \sum_{j=1}^{\infty} x_j y_j, \quad x \in c_{00}.$$

Following 6.1.3 we show that $f_y \in (c_{00})^*$ and $||f_y|| \le ||y||_1$. Next, we show that $||f_y|| = \sum_{j=1}^{\infty} |y_j| = ||y||_1$ and that the map $F : l_1 \to (c_{00})^*$ given by $F(y) = f_y$ is a linear isometry from l_1 into $(c_{00})^*$.

To prove F is surjective, we consider f in $(c_{00})^*$ and let $y = (f(e_1), f(e_2), \ldots)$. Next we define for $n = 1, 2, \ldots$

$$x_j^n = \left\{ \begin{array}{ll} sgn \ y_j & \text{if } 1 \le j \le n \\ 0 & \text{if } j > n \end{array} \right\}$$

so that $||f|| \ge f(x_n) = \sum_{j=1}^{n} x_j^n y_j = \sum_{j=1}^{n} |y_j|, \ n = 1, 2, \ldots$

so that $y \in l_1$. If $x \in c_{00}$ then $x = \sum_{i=1}^{n} x_i e_i$

for some n and hence

$$f(x) = \sum_{j=1}^{n} x_j f(e_j) = \sum_{i=1}^{n} x_i y_i = f_y(x).$$

Thus, $f = f_y$ that is $F(y) = f$, showing that f is surjective.

(iii) Since c_{00} is dense in c_0, we use theorem 6.1.1(a) and (b) above.

Note 6.1.1. Having considered the dual of a normed linear space E, we now turn to a similar concept for a bounded linear operator on E_x, a normed linear space.

Let E_x and E_y be two normed linear spaces and $A \in (E_x \to E_y)$. Define a map $A^* : E_y^* \to E_x^*$ as follows. For $\phi \in E_y^*$ and $x \in E_x$, let $x\phi(y) = \phi(Ax) = f(x)$, where $x \in E_x$, $y \in E_y$ and $f \in E_x^*$. Then we can write

$$f = A^* \phi.$$

A^* is called **adjoint** or **transpose** of A. A^* is linear and bounded [see 5.6.13 to 5.6.16].

6.1.5   Theorem

Let E_x, E_y and E_z be normed linear spaces.

(i) Let $A, B \in (E_x \to E_y)$ and $k \in \mathbb{R}(\mathbb{C})$. Then $(A+B)^* = A^* + B^*$, and $(kA)^* = \bar{k}A^*$.

(ii)Let $A \in (E_x \to E_y)$ and $C \in (E_y \to E_z)$. Then $(CA)^* = A^*C^*$.

(iii) Let $A \in (E_x \to E_y)$. Then $||A^*|| = ||A|| = ||A^{**}||$.

Proof:

(i)For proof of $(A+B)^* = A^* + B^*$ and $(kA)^* = \bar{k}A^*$ see 5.7.13.

(ii) Since A^* maps E_y^* into E_x^* we can find $f \subset E_x^*$ and $\phi \in E_y^*$ such that $\phi(y) = \phi(Ax) = f(x)$. Next, since C^* maps E_z^* into E_y^* we can find $\psi \in E_z^*$ for $\phi \in E_y^*$ such that $\psi(z) = \psi(Cy) = \bar{\phi}(y)$.

Thus $\psi(z) = \psi(CAx) = \overline{f}(x)$.

Thus $\overline{f} = (CA)^*\psi$.

Now $\overline{f} = A^*\bar{\phi} = A^*(C^*\psi)$.

Hence $(CA)^* = A^*C^*$.

(iii) To show that $||A|| = ||A^*||$ see theorem 5.6.17.

Now, $A^{**} = (A^*)^*$. Hence the above result yields $||A^{**}|| = ||A^*||$.

We have $f = A^*\phi$, i.e., $A^* : E_y^* \to E_x^*$ since $\phi \in E_y^*$ and $f \in E_x^*$.

Since A^{**} is the adjoint of A^*, A^{**} maps $E_x^{**} \to E_y^{**}$.

If we write $f(x) = F_x(f)$, then for fixed x, F_x can be treated as a functional defined on E_x^*. Therefore, $F_x \in E_x^{**}$.

Thus, for $F_x \in E_x^{**}$, $\phi \in E_y^*$, we have $A^{**}(F_x)(\phi) = F_x(A^*(\phi))$.

In particular, let $x \in E_x$ and $F_x = \prod_{E_x}(x)$, where $\prod_{E_x}(x)$ is the canonical embedding of E_x into E_x^{**}. Thus, for every $\phi \in E_y^*$, we obtain

$A^{**}(\prod_{E_x}(x))(\phi) = \prod_{E_x}(x)(A^*(\phi)) = A^*(\phi)(x)$
$= \phi(Ax) = \prod_{E_y}(y)(A(x))(\phi).$

Hence $A^{**}\prod_{E_x}(x) = \prod_{E(y)}(y)A$. Schematically,

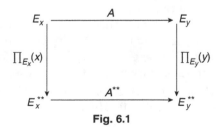

Fig. 6.1

6.1.6   Example

Let $E_x = c_{00} = E_y$, with the norm $||\cdot||_\infty$. Then, by 6.1.4, E_x^* is linearly isometric to l_1 and by 6.1.3. E_x^{**} is linearly isometric to l_∞. The completion of c_{00} (that is the closure of $\prod_{c_{00}}$ in (c_{00}^{**}) is linearly isometric to c_0. Let $A \in (c_{00} \to c_{00})$. Then A^* can be thought of as a norm preserving linear extension of A to l_∞.

We next explore the difference between the null spaces and the range spaces of A, A^* respectively.

6.1.7   Theorem

Let E_x and E_y be normed linear spaces and $A \in (E_x \to E_y)$. Then

(i) $N(A) = \{x \in E_x : f(x) = 0 \text{ for all } f \in R(A^*)\}$.

(ii) $N(A^*) = \{\phi \in E_y^* : \phi(y) = 0 \text{ for all } y \in R(A)\}$.

   In particular, A^* is one-to-one if and only if $R(A)$ dense in E_y.

(iii) $R(A) \subset \{y \in E_y : \phi(y) = 0 \text{ for all } \phi \in N(A^*)\}$, where equality holds if and only if $R(A)$ is closed in E_y.

(iv) $R(A^*) \subset \{f \in E_x^* : f(x) = 0 \text{ for all } x \in N(A)\}$, where equality holds if E_x and E_y are Banach spaces and $R(A)$ is closed in E_y.

In the above, $N(A)$ denotes the null space of A. $R(A)$ denotes the range space of A, $N(A^*)$ and $R(A^*)$ will have similar meanings.

Proof: (i) Let $x \in E_x$. Let $f \in E_x^*$ and $\phi \in E_y^*$.
Then $A^*\phi(x) = f(x) = \phi(Ax)$.

Therefore, $Ax = 0$ if and only $f(x) = 0 \; \forall \; f \in R(A^*)$.

(ii) Let $\phi \in E_x^*$ Then $A^*\phi = 0$ if and only if $\phi(Ax) = A^*\phi(x) = 0$ for every $x \in E_x$.

Now, A^* is one-to-one, that is, $N(A^*) = \{\theta\}$ if and only if $\phi = \theta$ wherever $\phi(y) = 0$ for every $y \in R(A)$. Hence, by theorem 5.1.5, this happens if and only if the closure of $R(A) = E_y$, i.e., $R(A)$, is dense in E_y.

(iii) Let $y \in R(A)$ and $y = Ax$ for some $x \in E_x$. If $\phi \in N(A^*)$ then $\phi(y) = \phi(Ax) = f(x) = A^*\phi(x) = 0$.
Hence $R(A) \subset \{y \in E_y : \phi(y) = 0 \text{ for all } \phi \in N(A^*)\}$.

If equality holds in this inclusion, then $R(A)$ is closed in E_y. Since $R(A) = \cap\{N(\phi) : \phi \in N(A^*)\}$, and each $N(\phi)$ is a closed subspace of E_y. Conversely, let us assume that $R(A)$ is closed in E_y. Let $y_0 \notin R(A)$, then by 5.1.5 there is some $\phi \in E_y^*$ such that $\phi(y_0) \neq 0$ but $\phi(y) = 0$ for every $y \in R(A)$. In particular, $A^*(\phi)(x) = f(x) = \phi(Ax) = 0$ for all $x \in E_x$ i.e., $\phi \in N(A^*)$. This shows that $y_0 \notin \{y \in E_y : \phi(y) = 0 \text{ for all } \phi \in N(A^*)\}$. Thus, equality holds in the inclusion mentioned above.

(d) Let $f \in R(A^*)$ and $f = A^*\phi$ for some $\phi \in E_y^*$. If $x \in N(A)$, then $f(x) = A^*\phi(x) = \phi(Ax) = \phi(0) = 0$. Hence, $R(A^*) \subset \{f \in E_x : f(x) = 0$, for all $x \in N(A)\}$.

Let us assume that $R(A)$ is closed in E_y, and that E_x and E_y are Banach spaces, we next want to show that the above inclusion reduces to an equality. Let $f \in E_x^*$ be such that $f(x) = \phi(Ax) = 0$ wherever $Ax = 0$. We need to find $\phi \in E_y^*$ such that $A^*\phi = f$, that is, $\phi(A(x)) = f(x)$ for every $x \in E_x$. Let us define $\psi : R(A) \to \mathbb{R}(\mathbb{C})$ by $\psi(y) = f(x)$, if $y = Ax$.

Since $f(x) = 0$ for all $x \in N(A)$, $\psi(y_1 + y_2) = f(x_1 + x_2)$, if $y_1 = Ax_1$ and $y_2 = f(x_2)$. Since f is linear,

$$\psi(y_1 + y_2) = f(x_1 + x_2) = f(x_1) + f(x_2) = \psi(y_1) + \psi(y_2).$$
$$\psi(\alpha y) = f(\alpha x) = \alpha f(x), \alpha \in \mathbb{R}(\mathbb{C}).$$

Thus ψ is well defined and linear.

Also, the map $A : E_x \to R(A)$, is linear, bounded and surjective, where E_x is a Banach space and so is the closed subspace $R(A)$ of the Banach space E_y. Hence, by the open mapping theorem [see 7.3], there is some $r > 0$ such that for every $y \in R(A)$, there is some $x \in E_x$ with $Ax = y$ and $||x|| \le \gamma||y||$, so that

$$|\psi(y)| = |f(x)| \le ||f|| \, ||x|| \le \gamma||f|| \, ||y||.$$

This shows that ψ is a continues linear functional on $R(A)$. By the Hahn-Banach extension theorem 5.2.3, there is some $\phi \in E_y^*$ such that $\phi|_{R(A)} = \psi$. Then $A^*(\phi)(x) = \phi(Ax) = \psi(Ax) = f(x)$ for every $x \in E_x$, as desired.

Problems

1. Prove that the dual space of $(\mathbb{C}^n, || \cdot ||_1)$ is isometrically isomorphic to $(\mathbb{C}^n, || \cdot ||_\infty)$.

2. Show that the dual space of $(c_0, || \cdot ||_\infty)$ is $(l_1, || \cdot ||_1)$.

3. Let $|| \cdot ||_1$ and $|| \cdot ||_2$ be two norms on the normed linear space E with $||x||_1 \le K||x||_2$, $\forall x \in E$ and $K > 0$, prove that $(E^*, || \cdot ||_1) \subseteq (E^*, || \cdot ||_2)$.

4. Let E_x and E_y be normed spaces. For $F \in (E_x \to E_y)$, show that

$$||F|| = \sup\{|\phi(F(x))| : x \in E_x, \, ||x|| \le 1, \, \phi \in E_y^*, \, ||\phi|| \le 1\}.$$

5. If S in a linear subspace of a Banach space E, define the annihilator S^0 of S to be the subset $S^0 = \{\phi \in E^* : \phi(s) = 0 \text{ for all } s \in S\}$. If T is a subspace of E^*, define $T^0 = \{x \in E : f(x) = 0 \text{ for all } f \in T\}$. Show that

   (a) S^0 is a closed linear subspace of E^*,
   (b) $S^{00} = \overline{S}$ where $\overline{S}$ is the closure of S,
   (c) If S is a closed subspace of E, then S^* is isomorphic to E^*/S^0.

6. 'c' denotes the vector subspace of l_∞ consisting of all convergent sequences. Define the **limit functional** $\phi : c \to \mathbb{R}$ by $\phi(x) = \phi(x_1, x_2, \ldots) = \lim_{n \to \infty} x_n$ and $\psi : l_\infty \to \mathbb{R}$ by $\psi(x_1, x_2 \ldots) = \limsup_{n \to \infty} x_n$.

   (i) Show that ϕ is a continuous linear functional where 'c' is equipped with the sup norm.

   (ii) Show that ψ is sublinear and $\phi(x) = \psi(x)$ holds for all $x \in c$.

6.2 Conjugates (Duals) of $L_p([a,b])$ and $C([a,b])$

The problem of finding the conjugate (dual) of $L_p([a,b])$ is deferred until Chapter 10.

6.2.1 Conjugate (dual) of $C([a,b])$

Riesz's representation theorem on functionals on $C([a,b])$ has already been discussed. We have seen that bounded linear functional f on $[a,b]$ can be represented by a Riemann-Stieljes integral

$$f(x) = \int_a^b x(t)dw(t) \tag{6.1}$$

where w is a function of bounded variation on $[a,b]$ and has the total variation

$$\mathrm{Var}(w) = ||f|| \tag{6.2}$$

Note 6.2.1. Let $BV([a,b)]$ denote the linear space of $\mathbb{R}(\mathbb{C})$-valued functions of bounded variation on $[a,b]$. For $w \in BV([a,b))$ consider

$$||w|| = |w(a)| + \mathrm{Var}(w).$$

Thus $||\cdot||$ is a norm on $BV([a,b))$. For a fixed $w \in BV([a,b))$, let us define $f_w : C[a,b] \to \mathbb{R}(\mathbb{C})$ by

$$f_w(x) = \int_a^b x dw. \quad x \in C([a,b]).$$

Then $f_w \in C^*([a,b])$ and $||f_w|| \leq ||w||$. However, $||f_w||$ may not be equal to $||w||$. For example, if $z = w + 1$, then $f_z = f_w$, but $||z|| = ||w|| + 1$, so that either $||f_w|| \neq ||w||$ or $||f_z|| \neq ||z||$.

This shows that distinct functions of bounded variation can give rise to the same linear functional on $C([a,b])$. In order to overcome this difficulty a new concept is introduced.

6.2.2 *Definition [normalized function of bounded variation]*

A function w of bounded variation on $[a, b]$ is said to be **normalised** if $w(a) = 0$ and w is right continuous on $]a, b[$. We denote the set of all normalized functions of bounded variation on $[a, b]$ by $NBV([a, b])$. It is a linear space and the total variation gives rise to a norm on it.

6.2.3 *Lemma*

Let $w \in BV([a, b])$. Then there is a unique $y \in NBV([a, b])$ such that

$$\int_a^b x\,dw = \int_a^b x\,dy$$

for all $x \in C([a, b])$. In fact,

$$y(t) = \begin{cases} 0, & \text{if } t = a \\ w(t^+) - w(a), & \text{if } t \in]a, b[\\ w(b) - w(a), & \text{if } t = b \end{cases}.$$

Moreover, $\text{Var}(y) \leq \text{Var}(w)$.

Proof: Let $y : [a, b] \to \mathbb{R}(\mathbb{C})$ be defined as above. Note that the right limit $w(t^+)$ exists for every $t \in]a, b[$, because $\text{Re}w$, and $\text{Im}w$ are real valued functions of bounded variation and hence each of them is a difference of two monotonically increasing functions. This also shows that w has only a countable number of discontinuities in $[a, b]$.

Let $\epsilon > 0$. We show that $\text{Var}(y) \leq \text{Var}(w) + \epsilon$. Consider a partition $a = t_0 < t_1 < t_2 < \cdots t_{n-1} < t_n = b$.

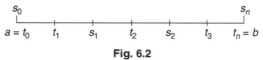

Fig. 6.2

Choose point $s_1, s_2, \ldots, s_{n-1}$ in $]a, b[$, at which w is continuous and which satisfy.

$$t_j < s_j, \ [w(t_j^+) - w(s_j)] < \frac{\epsilon}{2n}, \ j = 1, 2, \ldots, n - 1.$$

Let $s_0 = a$ and $s_n = b$. Then,

$$|y(t_1) - y(t_0)| \leq |w(t_1^+) - w(s_1)| + |w(s_1) - w(s_0)|$$

$$|y(t_j) - y(t_{j-1})| \leq |w(t_j^+) - w(s_j)| + |w(s_j) - w(s_{j-1})|$$

$$+ |w(s_{j-1}) - w(t_{j-1}^+)|, \ j = 2, \ldots (n - 2).$$

$$|y(t_n) - y(t_{n-1})| \leq |w(s_n) - w(s_{n-1})| + |w(s_{n-1}) - w(t_{n-1}^+)|$$

Hence, $\displaystyle\sum_{j=1}^n |y(t_j) - y(t_{j-1})| \leq \sum_{j=1}^n |w(s_j) - w(s_{j-1})| + \frac{\epsilon + (n-2)\epsilon + \epsilon}{2n}$

$$< \sum_{j=1}^n |w(s_j) - w(s_{j-1})| + \epsilon.$$

Since the above is true for every partition P_n of $[a, b]$, $\text{Var}(y) \leq \text{Var}(w) + \epsilon$. As $\epsilon > 0$ is arbitrary, $\text{Var}(y) \leq \text{Var}(w)$. In particular, y is of bounded variation on $[a, b]$. Hence $y \in NBV[a, b]$.

Next, let $x \in C([a, b])$. Apart from the subtraction of the constant $w(a)$, the function y agrees with the function w, except possibly at the points of discontinuities of w. Since these points are countable, they can be avoided while calculating the Riemann-Stieljes sum

$$\sum_{j=1}^{n} x(t_j)[w(t_j) - w(t_{j-1})],$$

which approximates $\displaystyle\int_a^b x dw$, since each sum is equal to $\displaystyle\sum_{j=1}^{n} x(t_j)[y(t_j) - y(t_{j-1})]$ and is approximately equal to $\displaystyle\int_a^b x dy$. Hence $\displaystyle\int_a^b x dw = \int_a^b x dy$.

To prove the uniqueness of y, let $y_0 \in NBV([a, z])$ be such that $\displaystyle\int_a^b x dw = \int_a^b x dy$ for all $x \in C([a, b])$ and $z = y - y_0$. Thus $z(a) = y(a) - y_0(a) = 0 - 0 = 0$.

Also, since $z(b) = z(b) - z(a) = \displaystyle\int_a^b dz = \int_a^b dy - \int_a^b dy_0 = 0$.

Now, let $\xi \in]a, b[$. For a sufficiently small positive h, let

$$x(t) = \begin{cases} 1 & \text{if } a \leq t \leq \xi \\ 1 - \dfrac{t - \xi}{h} & \text{if } \xi < t \leq \xi + h \\ 0 & \text{if } \xi + h < t \leq b. \end{cases}$$

Then $\quad x \in C([a, b])$ and $(x(t)) \leq 1$ for all $t \in [a, b]$.

Since $\quad 0 = \displaystyle\int_a^b x dy - \int_a^b x dy_0 = \int_a^b x dz$

$$= \int_a^\xi dz + \int_\xi^{\xi+h} \left(1 - \frac{t - \xi}{h}\right) dz$$

we have $\quad z(\xi) = \displaystyle\int_a^\xi dz = -\int_\xi^{\xi+h} \left(1 - \frac{t - \xi}{h}\right) dz$.

It follows that $|z(\xi)| \leq \text{Var}_\xi \xi + h$,

where $\text{Var}_\xi \xi + h$ denotes the total variation of z on $[\xi, \xi + h]$. As z is right continuous at ξ, its total variation function $v(t) = \text{Var}_a t$, $t \in [a, b]$ is also right continuous at ξ. Let $\epsilon > 0$, there is some $\delta > 0$ such that for $0 < h < \delta$,

$$|z(\xi)| \leq \text{Var}_\xi \xi + h = v(\xi + h) - v(\xi) < \epsilon.$$

Hence, $z(\xi) = 0$. Thus $z = 0$, that is, $y_0 = y$.

6.2.4 Theorem

Let $E = C([a, b])$. Then E is isometrically isomorphic to the subspace of $BV([a, b])$, consisting of all normalized functions of bounded variation. If y is such a normalized function $(y \in NBV([a, b]))$, the corresponding f is given by

$$f(x) = \int_a^b x(t) dy(t). \tag{6.3}$$

Proof: Formula (6.3) defines a linear mapping $f = Ay$, where y is normalized and $f \in C^*([a, b])$. We evidently have $||f|| \leq \mathrm{Var}(y)$. For a normalized y, $\mathrm{Var}(y)$ is the norm of y because $y(a) = 0$. Now consider any $g \in C^*([a, b])$. The theorem 5.3.3 then tells us that there is a $w \in BV([a, b])$ such that

$$g(x) = \int_a^b x(t) dw(t) \quad \text{and} \quad \mathrm{Var}(w) = ||g||.$$

The integral is not changed if we replace w by the corresponding normalized function y of bounded variation. Then by lemma 6.2.3

$$g(x) = \int_a^b x(t) dw(t) = \int_a^b x(t) dy(t)$$

and $\qquad g = Ty \quad \text{and} \quad ||g|| \leq \mathrm{Var}(y).$

Also $\qquad \mathrm{Var}(y) \leq \mathrm{Var}(w) = ||g||.$

Therefore, $||g|| = \mathrm{Var}(y)$. Since by lemma 6.2.3 there is just one normalized function, corresponding to the functional g a one-to-one correspondence exists between the set of all linear functionals of $C^*([a, b])$ and the set of all elements of $NBV([a, b])$.

It is evident that the sum of functions $y_1, y_2 \in NBV([a, b])$ corresponds to the sum of functionals $g_1, g_2 \in C^*([a, b])$ and the function λy corresponds to the functional λg, in case the functionals g_1, g_2 correspond to the functions $y_1, y_2 \in NBV([a, b])$. It therefore follows that the association between $C^*([a, b])$ and the space of normalized functions of bounded variation $(NBV([a, b])$ is an isomorphism. Furthermore, since

$$||g|| = \mathrm{Var}(y) = ||y||$$

the correspondence is isometric too.

Thus, the dual of a space of continuous functions is a space of normalized functions of bounded variation.

6.2.5 Moment problem of Hausdroff or the little moment problem

Let us consider the discrete analogue of the Laplace transform

$$\mu(s) = \int_0^\infty e^{-su} d\alpha(u) \quad s \in \mathbb{R}(\mathbb{C}) \tag{6.4}$$

where $\alpha : [0, \infty] \to \mathbb{R}(\mathbb{C})$ is of bounded variation on every subinterval of $[0, \infty)$. If we put $t = e^{-u}$ and put $s = n$, a positive integer, the above integral gives rise to the form

$$\mu(n) = \int_0^1 t^n dz(t), \quad n = 0, 1, 2, \qquad (6.5)$$

where $\alpha(u) = -z(e^{-u})$.

The integral (6.4), where z is a function of bounded variation of $[0,1]$, is called the **nth moment** of z. The moments of a distribution of a random variable play an important role in statistics. For example, in the case of a rectangular distribution

$$F(x) = \begin{cases} 0 & x < a \\ \dfrac{x - a}{b - a}, & a \le x \le b \\ L & x > b \end{cases}$$

the frequency density function is given by the following step function [fig. 6.3]

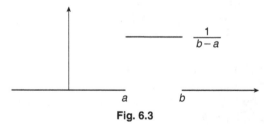

Fig. 6.3

Hence, the nth moment function for a rectangular distribution is

$$\mu(n) = \int_{-\infty}^{\infty} dF(t) = \int_a^n t^n f(t) dt.$$

A sequence of scalars $\mu(n), n = 0, 1, 2, \ldots$ is called a **moments sequence** if there is some $z \in BV([0, 1])$ whose nth moments is $\mu(x), n = 0, 1, 2, \ldots$

For example, if α is a positive integer, then taking $z(t) = \dfrac{t^\alpha}{\alpha}$, $t \in [0, 1]$, we see that

$$\mu(n) = \int_0^1 t^n dz(t) = \int_0^1 t^{n+\alpha-1} dt = \frac{1}{n + \alpha}, \quad n = 0, 1, 2 \ldots$$

Similarly, if $0 < r \le 1$, then $(r^n), n = 0, 1, 2, \ldots$ is a moment sequence since if z is the **characteristic functions** of $[r, 1]$ [see Chapter 10], then

$$\int_0^1 t^n dz(t) = r^n, \quad n = 0, 1, 2, \ldots$$

If $\mu(x)$ is the nth moment of $z \in BV([0,1])$, then

$$|\mu(n)| \leq \mathrm{Var}(z), \ n = 0, 1, 2, \ldots$$

Hence, every moment sequence is bounded. To prove that $\mu(n)$ is convergent [see Limaye [33]]. Thus every scalar sequence need not be a moment sequence. The problem of determining the criteria that a sequence must fulfil in order to become a moment sequence is known as **the moment problem of Hausdroff or the little moment problem**. We next discuss some mathematical preliminaries relevant to the discussion.

6.2.6   The shift operator, the forward difference operator

Let X denote the linear space of all scalar sequences $\mu(n))$, $n = 0, 1, 2, \ldots$ and let $\mathcal{E}: X \to X$ be defined by,
$$\mathcal{E}(\mu(n)) = \mu(n+1), \ \mu \in X, \ n = 0, 1, 2, \ldots$$
$\mathcal{E}$ is called the **shift** operator.

Let I denote the identity operator from X to X.

Define $\Delta = \mathcal{E} - I$. Δ is called then **forward difference** operator.

Thus, for all $\mu \in X$ and $n = 0, 1, 2, \ldots$
$$\Delta(\mu(n)) = \mu(n+1) - \mu(n) \tag{6.6}$$
For $r = 0, 1, 2, \ldots$ we have

$$\Delta^r = (\mathcal{E} - I)^r = \sum_{j=0}^{r} (-1)^{r-j} \binom{r}{j} \mathcal{E}^j,$$

so that $\Delta^r(\mu(n)) = \sum_{j=0}^{r} (-1)^{r-j} \binom{r}{j} \mu(n+j)$

In particular, $\Delta^r(\mu(0)) = \sum_{j=0}^{r} (-1)^{r-j} \binom{r}{j} \mu(j).$ \tag{6.7}

6.2.7   Definition: $P([0,1])$

Let $\mathbf{P}([0,1])$ denote the linear space of all scalar-valued polynomials on $[0,1]$. For $m = 0, 1, 2, \ldots$ let

$$p_m(t) = t^m, t \in [0,1].$$

We next prove the Weierstrass approximations theorem.

6.2.8   The Weierstrass approximations theorem (RALSTON [43])

The Weierstrass approximation theorem asserts that the set of polynomials on $[0,1]$ is dense in $C([0,1])$. Or in other words, $\mathbf{P}([0,1])$ is dense in $C([0,1])$ under the sup norm.

In order to prove the above we need to show that for every continuous function $f \in C([0,1])$ and $\epsilon > 0$ there is a polynomial $p \in \mathbf{P}([0,1])$ such that

$$\max_{x \in [0,1]} \{|f(x) - p(x)| < \epsilon\}.$$

In what follows, we denote by $\binom{n}{k}$, $\dfrac{n!}{k!(n-k)!}$ where n is a positive integer and k an integer such that $0 < k \leq n$. The polynomial $B_n(f)(x)$ defined by

$$B_n(f)(x) = \sum_{k=0}^{n} \binom{n}{k} x^k (1-x)^{n-k} f\left(\frac{k}{n}\right) \tag{6.8}$$

is called the Bernstein polynomial associated with f. We prove our theorem by finding a Bernstein polynomial with the required property. Before we take up the proof, we mention some identities which will be used:

(i) $\displaystyle\sum_{k=0}^{n} \binom{n}{k} x^k (1-x)^{n-k} = [x + (1-x)]^n = 1$ $\tag{6.9}$

(ii) $\displaystyle\sum_{k=0}^{n} \binom{n}{k} x^k (1-x)^{n-k} (k - nx) = 0$ $\tag{6.10}$

(6.10) is obtained by differentiating both sides of (6.9) w.r.t. x and multiplying both sides by $x(1-x)$.

On differentiating (6.10) w.r.t. x, we get

$$\sum_{k=0}^{n} \binom{n}{k} [-nx^k (1-x)^{n-k} + x^{k-1}(1-x)^{n-k-1}(k-nx)^2] = 0$$

Using (6.9), (6.10) reduces to

$$\sum_{k=0}^{n} \binom{n}{k} x^{k-1}(1-x)^{n-k-1}(k-nx)^2 = n \tag{6.11}$$

Multiplying both sides by $x(1-x)$ and dividing by n^2, we obtain,

(iii) $\displaystyle\sum_{k=0}^{n} \binom{n}{k} x^k (1-x)^{n-k} \left(\frac{k}{n} - x\right)^2 = \frac{x(1-x)}{n}$ $\tag{6.12}$

(6.12) is the third identity to be used in proving the theorem.

It then follows from (6.8) and (6.9) that

$$f(x) - B_n(f)(x) = \sum_{k=0}^{n} \binom{n}{k} x^k (1-x)^{n-k} \left[f(x) - f\left(\frac{k}{n}\right)\right] \tag{6.13}$$

or $\quad |f(x) - B_n(f)(x)| \leq \displaystyle\sum_{k=0}^{n} \binom{n}{k} x^k (1-x)^{n-k} \left[f(x) - f\left(\frac{k}{n}\right)\right]$ $\tag{6.14}$

Since f is uniformly continuous on [0,1], we can find a

$$\left. \begin{array}{l} \delta > 0 \quad \text{and} \quad M \text{ s.t.} \quad \left|x - \frac{k}{n}\right| < \delta \;\Rightarrow\; \left|f(x) - f\left(\frac{k}{n}\right)\right| < \frac{\epsilon}{2} \\[2mm] \text{and} \qquad |f(x)| < M \quad \text{for} \quad x \in [0,1]. \end{array} \right\} \tag{6.15}$$

Let us partition the sum on the RHS of (6.13) into two parts, denoted by $\sum'$ and $\sum''$. $\sum'$ stands for the sum for which $\left|x - \frac{k}{n}\right| < \delta$ (x is fixed but arbitrary, and $\sum''$ is the sum of the remaining terms.

Thus, $\displaystyle\sum' - \sum_{\substack{k \\ |x - \frac{x}{n}| < \delta}} \binom{n}{k} x^k (1-x)^{n-k} \left|f(x) - f\left(\frac{k}{n}\right)\right|$

$$< \frac{\epsilon}{2} \sum_{k=0}^{n} \binom{n}{k} x^k (1-x)^{n-k} = \frac{\epsilon}{2}. \tag{6.16}$$

We next show that if n is sufficiently large then $\sum''$ can be made less than $\frac{\epsilon}{2}$ independently of x. Since f is bounded using (6.15),

we get $\displaystyle\sum'' \leq 2M \sum \binom{n}{k} x^k (1-x)^{n-k}$ where the sum is taken for all

k s.t., $\left|x - \dfrac{k}{n}\right| \geq \delta$

(6.12) yields $\delta^2 \displaystyle\sum'' \leq \dfrac{x(1-x)}{n}$

or $\displaystyle\sum''' \leq \dfrac{1}{4\delta^2 n}$ since $\max x(1-x) = \dfrac{1}{4}$ for $x \in [0,1]$

where $\displaystyle\sum''' = \sum_{k,|x-\frac{k}{n}|>\delta} \binom{n}{k} x^k (1-x)^{n-k}$,

taking $n > \dfrac{M}{\delta^2 \epsilon}$, $\displaystyle\sum'' < \dfrac{2M\delta^2 \epsilon}{4M\delta^2} = \dfrac{\epsilon}{2}$.

Hence, $\displaystyle |f(x) - B_n(f)(x)| \leq \sum_{k=0}^{n} \binom{n}{k} x^k (1-x)^{n-k} \left|f(x) - f\left(\frac{k}{n}\right)\right|$

$$< \frac{\epsilon}{2} + \frac{\epsilon}{2} = \epsilon.$$

6.2.9   Definition: $P_{(m)}$

Let us define, for a nonnegative integer m,

$$P_{(m)} = \{p \in \mathbf{P}([0,1]) : p \text{ is of degree} \leq m\}.$$

6.2.10   Lemma $B_n(p) \subset P_{(m)}$ where p is a polynomial of degree $\leq m$

The Bernstein polynomial $B_n(f)$ is given by

$$B_n(f)(x) = \sum_{k=0}^{n} f\left(\frac{k}{n}\right) \binom{n}{k} x^k (1-x)^{n-k}, \; x \in [0,1], \; n = 0,1,2$$

We express $B_n(f)$ as a linear combination of $p_0, p_1, p_2, \ldots$

Since $(1-x)^{n-k} = \sum_{j=0}^{n-k} (-1)^j \binom{n-k}{j} x^j = \sum_{j=0}^{n-k} (-1)^j \binom{n-k}{j} p_j(x)$,

and $p_k p_j = p_{j+k}$, we have

$$B_n(f) = \sum_{k=0}^{n} f\left(\frac{k}{n}\right) \binom{n}{k} \sum_{j=0}^{n-k} (-1)^j \binom{n-k}{j} p_{j+k}.$$

As $\binom{n}{k}\binom{n-k}{j} = \binom{j+k}{k}\binom{n}{j+k}$, we put $j+k = r$

$$B_n(f) = \sum_{r=0}^{n} \left[\sum_{k=0}^{n} (-1)^{r-k} \binom{r}{k} f\left(\frac{k}{n}\right) \right] \binom{n}{r} p_r.$$

In particular, $B_n(f)$ is a polynomial of degree at most n. Also,

$$B_n(p_0)(x) = \sum_{k=0}^{n} \binom{n}{k} x^k (1-x)^{n-k} = [x + (1-x)]^n$$

$$= 1 = p_0(x), \; x \in [0,1].$$

If $n \le m$ then clearly $B_n(p) \in P_{(n)} \subset P_{(m)}$.

Next, fix $n \ge m+1$. Consider the sequence $(\mu_p(k))$, defined by $\mu_p(k) = p\left(\frac{k}{n}\right)$, $k = 0,1,2,\ldots$

Noting the expression for $\Delta^r(\mu)(0)$ and $B_n(f)$ we obtain

$$B_n(p) = \sum_{r=0}^{n} (\Delta^r \mu_P)(0) \binom{n}{r} p_r.$$

Since p is a polynomial of degree at most m, it follows that

$$(\Delta \mu_p)(k) = p\left(\frac{k+1}{n}\right) - p\left(\frac{k}{n}\right)$$

$$= \lambda_0 + \lambda_1 k \cdots + \lambda_m k^{m-1}, \; k = 0,1,2,\ldots$$

for some scalar $\lambda_0, \ldots \lambda_{m-1}$. Proceeding, similarly we conclude that $(\Delta^m \mu_p)(k)$ equals a constant, for $k = 0,1,2,\ldots$ and for each $r \ge m+1$, we have $(\Delta^r \mu_P)(k) = 0$, $k = 01,2,\ldots$

In particular, $(\Delta^r \mu_p)(0) = 0$, for all $r \ge m+1$ so that

$$B_n(p) = \sum_{r=0}^{m} (\Delta^r \mu_p)(0) \binom{n}{r} p_r.$$

Hence, $B_n(p) \in P_{(m)}$.

6.2.11 Lemma

Let h be a linear functional on $\mathbf{P}([0,1])$ with $\mu(n) = h(p_n)$ for $n = 0, 1, 2, \ldots$. If $f_{kl}(x) = x^k(1-x)^l$ for $x \in [0,1]$, then

$$h(f_{kl}) = (-1)^l \Delta^l(\mu)(k), \quad k, l = 0, 1, 2, \ldots$$

Proof: We have $f_{kl} = p_k(1-x)^l$ where $p_k = x^k$

$$= p_k \sum_{i=0}^{l} (-1)^i \binom{l}{i} p_i = \sum_{i=0}^{l} (-1)^i \binom{l}{i} p_{i+k}.$$

Hence, $\quad h(f_{kl}) = \sum_{i=0}^{l} (-1)^i \binom{l}{i} h(p_{i+k}) = \sum_{i=0}^{l} (-1)^i \binom{l}{i} \mu(i+k),$

which equals to $(-1)^l \Delta^l(\mu)(k)$.

We next frame the criterion for a sequence to be a **moment** sequence.

6.2.12 Theorem (Hausdorff, 1921) Limaye [33]

Let $(\mu(n)), n = 0, 1, 2 \ldots$ be a sequence of scalars. Then the following conditions are equivalent,

(i) $(\mu(n))$ is a moment sequence

(ii) For $n = 0, 1, 2, \ldots$ and $k = 0, 1, 2, \ldots, n$, let

$$d_{n,k} = \binom{n}{k}(-1)^{n-k}\Delta^{n-k}(\mu(k)).$$

Then $\displaystyle\sum_{k=0}^{n} |d_{n,k}| \le d$ for all n and some $d > 0$.

(iii) The linear functional $h : \mathbf{P}([0,1]) \to \mathbb{R}(\mathbb{C})$ defined by $h(\lambda_0 p_0 + \lambda_1 p_1 + \cdots + \lambda_n p_n) = \lambda_0 \mu(0) + \cdots + \lambda_n \mu(n)$, is continuous, where $n = 0, 1, 2$ and $\lambda_0, \lambda_1, \ldots, \lambda_n \in \mathbb{R}(\mathbb{C})$.

Further, there is a non-decreasing function on $[0,1]$ whose nth moment is $\mu(n)$ if and only if $d_{n,k} \ge 0$ for all $n = 0, 1, 2, \ldots$ and $k = 0, 1, 2, \ldots, n$. This can only happen if and only if the linear functional h is positive.

Proof: (i) $\Rightarrow$ (ii). Let $z \in BV([0.1])$ be such that the nth moment of z is $\mu(n), n = 0, 1, 2, \ldots$. Then

$$h(p) = \int_0^1 p\,dz, \quad z \in \mathbf{P}([0,1]),$$

define a linear functional h on $\mathbf{P}([0,1])$ such that $h(p_n) = \mu(n)$, for $n = 0, 1, 2, \ldots$. By lemma 6.2.11,

$$d_{n,k} = \binom{n}{k}(-1)^{n-k}\Delta^{n-k}(\mu)(k)$$

$$= \binom{n}{k}h(f_{k,n-k}) \text{ for } n = 0, 1, 2, \ldots$$

and $k = 0, 1, 2, \ldots, n$. Since $f_{k,n-k} \geq 0$ on $[0, 1]$, it follows that

$$|d_{n,k}| \leq \binom{n}{k} \int_0^1 f_{k,n-k} dv_z,$$

where $v_z(x)$ is the total variation of z on $[0, x]$.

But, for $n = 0, 1, 2, \ldots$

$$\sum_{k=0}^n \binom{n}{k} f_{k,n-k} = B_n(1) = 1.$$

Hence, $\quad \displaystyle\sum_{k=0}^n |d_{n,k}| \leq \int_0^1 B_n(1) dv_z = \text{Var } z,$

where Var z is the total variation of z on $[0,1]$.

Note that if z is non-decreasing, then since $f_{k,n-k} \geq 0$ we have $d_{n,k} = \binom{n}{k} \int_0^1 f_{k,n-k} dz \geq 0$ for all $n = 0, 1, 2, \ldots$ and $k = 0, 1, 2, \ldots, n$

(ii) $\Rightarrow$ (iii) For a nonnegative integer m, let h_m denote the restriction of h to $P_{(m)}$. Since h is linear on $P([0, 1])$ and $P_{(m)}$ is a finite dimensional subspace or $P([0, 1])$, it follows that h_m is continuous, since every linear map on a finite dimensional normed linear space is continuous.

Let $p \in \mathbf{P}([0, 1])$. Since

$$B_n(p) = \sum_{k=0}^n p\left(\frac{k}{n}\right) \binom{n}{k} x^k (1 - x)^{n-k}$$

and $\quad h(p_n) = \mu(n)$ for $n = 0, 1, 2, \ldots$, we have

$$h(B_n(p)) = \sum_{k=0}^n p\left(\frac{k}{n}\right) \binom{n}{k} h(x^k(1 - x)^{n-k})$$

$$= \sum_{k=0}^n p\left(\frac{k}{n}\right) \binom{n}{k} (-1)^{n-k} \Delta^{n-k}(\mu)(k).$$

$$= \sum_{k=0}^n p\left(\frac{k}{n}\right) d_{n,k},$$

by lemma 6.2.11. Hence,

$$|h(B_n(p))| \leq \sum_{k=0}^n \left| p\left(\frac{k}{n}\right) \right| |d_{n,k}|$$

$$\leq ||p||_\infty \sum_{k=0}^n |d_{n,k}| \leq d||p||_\infty.$$

Now, let the degree of p be m. Then $B_n(p) \in \mathbf{P}_{(m)}$, for all $n = 0, 1, 2, \ldots$ as proved earlier.

Since $||B_n(p) - p||_\infty \to 0$ as $n \to \infty$ and h_m is continuous,

$$|h(p)| = |h_m(p)| = \lim_{n \to \infty} |h_m(B_n(p))| \le d||p||_\infty.$$

This shows that h is continuous on $\mathbf{P}([0,1])$.

If, $d_{m,k} \ge 0$ for all m and k, and if $p \ge 0$ on $[0,1]$

then $h(p) = \lim_{n \to \infty} h(B_n(p)) = \lim_{n \to \infty} \sum_{k=0}^{n} p\left(\frac{k}{n}\right) d_{n,k} \ge 0,$

i.e., h is a positive functional.

In this case, $\sum_{k=0}^{n} |d_{n,k}| = \sum_{k=0}^{n} d_{n,k} = \int_0^1 B_n(1)dz = \int_0^1 dz = \mu(0).$

(iii) $\Rightarrow$ (ii) Since $\mathbf{P}([0,1])$ is dense in $C([0,1])$ with the sup norm $|| \ ||_\infty$, there is some $F \in C^*([0,1])$ with $F|_{\mathbf{P}([0,1])} = h$ and $||F|| = ||h||$.

By Riesz representations theorem for $C([0,1])$ there is some $z \in NBV([0,1])$ such that

$$F(f) = \int_0^1 f dz, \quad f \in C([0,1]).$$

In particular, for $n = 0, 1, 2, \ldots$

$$\mu(n) = h(p_n) = F(p_n) = \int_0^1 t^n dz(t),$$

that is $\mu(n)$ is the nth moment of z.

If the functional h is positive and $f \in C([0,1])$ with $f \ge 0$ on $[0,1]$, then $B_n(f) \ge 0$ on $[0,1]$ for all n and we have $F(f) = \lim_{n \to \infty} F(B_n(f)) = \lim_{n \to \infty} h(B_n(f)) \ge 0$, i.e., F is a positive functional on $C([0,1])$.

By Riesz representation theorem for $C([0,1])$, we can say that there is a non-decreasing function z such that $F = F_z$. In particular $\mu(n)$ is the nth moment of a non-decreasing function z.

Thus, if $(\mu(n))$ is a moment sequence, then there exists a unique $y \in NBV([0,1])$ such that $\mu(n)$ is the nth moment of y. This follows from lemma 6.2.3 by noting that $\mathbf{P}([0,1])$ is dense in $C([0,1])$ with the sup norm $|| \ ||_\infty$.

Problems

1. For a fixed $x \in [a,b]$, let $F_x \in C^*([a,b])$ be defined by $F_x(f) = f(x)$, $f \in C([a,b])$.

   Let y_x be the function in $NBV([a,b])$ which represents F_x as in theorem 5.3.3. If $x = a$, then y_x is the **characteristic function** of $[a,b]$, and if $a < x \le b$, then y_x is the characteristic function of $[x,b]$.

If $a \leq x_1 < x_2 < \cdots x_{n-1} < x_n \leq b$, and

$$F(f) = k_1 f(x_1) + \cdots + k_n(f(x_n)), \quad f \in C([a, b]),$$

then show that the function in $NBV([a, b])$ corresponding to $F \in C^*([a, b])$ is a step function.

2. Prove the inequality

$$\left| \int_a^b f dg \right| < \max[|f(x)| : x \in [a, b]] \cdot \text{Var } g.$$

3. Show that for any $g \in BV([a, b])$ there is a unique $\bar{g} \in BV([a, b])$, continuous from the right, such that

$$\int_a^b f d\bar{g} = \int_a^b f dg \text{ for all } f \in C([a, b]) \quad \text{and} \quad \text{Var } (\bar{g}) \leq \text{Var } (g).$$

4. Show that a sequence of scalars $\mu(n), n = 0, 1, 2, \ldots$, is a moment sequence if and only if

$$\mu(n) = \mu_1(n) - \mu_2(n) + i\mu_3(n) - i\mu_4(n), \quad \text{where} \quad i = \sqrt{-1}$$

and $\quad (-1)^{n-k} \Delta^{n-k} \mu_j(k) \geq 0$

for all $k = 0, 1, 2, \ldots, n, \ j = 1, 2, 3, 4$ and $n = 0, 1, 2, \ldots,$.

5. Let $y \in NBV([a, b])$ and $\mu(n) = \int_0^1 t^n dy(t)$, for $n = 0, 1, 2, \ldots$

Then show that

$$\text{Var } (y) = \sup\{ \sum_{k=0}^{n} \binom{n}{k} |\Delta^{n-k}(\mu)(k)|; \ n = 0, 1, 2, \ldots \}$$

where Δ is the forward difference operator.

6.3 Weak* and Weak Convergence

6.3.1 *Definition: weak* convergence of functionals*

Let E be a normed linear space. A sequence $\{f_n\}$ of linear functionals in E^* is said to be **weak* convergent** to a linear functional $f_0 \in E^*$, if $f_n(x) \to f_0(x)$ for every $x \in E$.

Thus, for linear functionals the notion of weak convergence is equivalent to pointwise convergence.

6.3.2   Theorem

If a sequence $\{f_n\}$ of functionals weakly converges to itself, then $\{f_n\}$ converges weakly to some linear functional f_0.

For notion of pointwise convergence see 4.5.2. Theorem 4.4.2 asserts that E^* is complete, where E^* is the space conjugate to the normed linear space E. Therefore if $\{f_n\} \in E^*$ is Cauchy, $\{f_n\} \to f_0 \in E^*$. Therefore, for every $x \in E$ $f_n(x) \to f_0(x)$ as $n \to \infty$.

6.3.3   Theorem

Let $\{f_n\}$ be a sequence of bounded linear functionals defined on the Banach space E_x.

A necessary and sufficient condition for $\{f_n\}$ to converge weakly to f as $n \to \infty$ is

(i) $\{\|f_n\|\}$ is bounded

(ii) $f_n(x) \to f(x)$ $\forall$ $x \in M$ where the subspace M is everywhere dense in E_x.

Proof: Let $f_n \to f$ weakly, i.e., $f_n(x) \to f(x)$ $\forall$ $x \in E_x$.

It follows from theorem 4.5.6 that $\{\|f_n\|\}$ is bounded. Since M is a subspace of E_x, condition (ii) is valid.

We next show that the conditions (i) and (ii) are sufficient. Let $\{\|f_n\|\}$ be bounded. Let $L = \sup \|f_n\|$.

Let $x \in E_x$. Since M is everywhere dense in E_x, $\exists$ $x_0 \in M$ s.t. given arbitrary $\epsilon > 0$,

$$\|x - x_0\| < \epsilon/4M.$$

condition (ii) yields, for the above $\epsilon > 0$, $\exists$ $n > n_0$ depending on ϵ s.t.

$$|f_n(x_0) - f(x_0)| < \frac{\epsilon}{2}.$$

The linear functionals f is defined on M. Hence by Hahn-Banach extension theorem (5.1.3), we can extend f from M to the whole of E_x.

Moreover, $\|f\| = \|f\|_M \leq \sup \|f_n\| = L$

Now, $|f_n(x) - f(x)| \leq |f_n(x) - f_n(x_0)| + |f_n(x_0) - f(x_0)|$

$$+ |f(x_0) - f(x)|$$

$$= |f_n(x - x_0)| + \frac{\epsilon}{2} + |f(x - x_0)|$$

$$< M\|x - x_0\| + \frac{\epsilon}{2} + M\|x - x_0\|$$

$$< M\frac{\epsilon}{4M} + \frac{\epsilon}{2} + M \cdot \frac{\epsilon}{4M}.$$

$$= \epsilon \text{ for } n > n_0(t).$$

Hence, $f_n(x) \to f(x)$ $\forall$ $x \in E_x$,

i.e., $f_n \xrightarrow[n \to \infty]{\text{weakly}} f.$

6.3.4   Application to the theory of quadrature formula

Let $x(t) \in C([a, b])$. Then

$$f(x) = \int_a^b x(t)dt \tag{6.17}$$

is a bounded linear functional in $C([a, b])$.

Consider $f_n(x) = \sum_{k=1}^{kn} C_k^{(n)} x(t_k^{(n)}), \quad n = 1, 2, 3, \ldots \tag{6.18}$

$C_k^{(n)}$ are called weights, $t_k^{(n)}$ are the nodes,

$$a \le t_1^{(n)} \le \cdots \le t_{k_{n-1}}^{(n)} \le t_{k_n}^{(n)}$$

(6.18) is a linear functional.

6.3.5   Definition: quadrature formula

Let $C_k^{(n)}$ be so chosen in (6.18), such that $f(x)$ and $f_n(x)$ coincide for all polynomials of a degree less then equal to n, i.e.,

$$f(x) \simeq f_n(x) \qquad \text{if} \quad x(t) = \sum_{p=0}^n a_n t^p \tag{6.19}$$

The relation $f(x) \simeq f_n(x)$ which becomes an equality for all polynomials of degree less then equal to n, is called a quadrature formula. For example, in the case of Gaussian quadrature: $t_1^{(n)} \ne a$ and the last element $\ne b$. $t_k^{(n)}, (k = 1, 2, \ldots n)$ are the n roots of $P_n(t) = \frac{1}{2^n n!}(t^2 - 1)^n = 0$.

Consider the sequence of quadrature formula:

$$f(x) \simeq f_n(x), \ n = 1, 2, 3, \ldots$$

The problem that arises is whether the sequence $\{f_n(x)\}$ converges to the value of $f(x)$ as $n \to \infty$ for any $x(t) \in C([0, 1])$. The theorem below answers this question.

6.3.6   Theorem

The necessary and sufficient condition for the convergence of a sequence of quadrature formula, i.e., in order that

$$\lim_{n \to \infty} \sum_{k=1}^{k_n} C_k^{(n)} x(t_k^{(n)}) = \int_0^1 x(t)d\sigma(t)$$

holds for every continuous function $x(t)$, is that $\sum_{k=1}^{kn} |C_k^{(n)}| \le K = \text{const}$, must be true for every n,

Proof: By definition of the quadrature formula, the functional f_m satisfies

$$f_m(x) = f(x) \text{ for } m \leq n \tag{6.20}$$

for every polynomial $x(t)$ of degree n,

$$f_m(x) = \sum_{k=1}^{k_m} C_k^{(m)} x(t_k^{(m)}). \tag{6.21}$$

Each f_m is bounded since $|x(t^{(m)})| \leq ||x||$ by the definition of the norm.

Consequently, $|f_m(x)| \leq \sum_{k=1}^{k_m} |C_k^{(n)} x(t_k^{(m)})| \leq \left(\sum_{k=1}^{k_m} |C_k^{(n)}| \right) ||x||.$

For later use we show that f_m has the norm,

$$||f_m|| = \sum_{k=1}^{k_m} |C_k^{(m)}|, \tag{6.22}$$

i.e., $||f_m||$ cannot exceed the right-hand side of (6.22) and equality holds if we take an $x_0 \in C([0,1])$ s.t. $|x_0(t)| < 1$ on J and

$$x_0(t_k^{(n)}) = sgn C_k^{(n)} = \begin{cases} 1 & \text{if } C_k^{(n)} \geq 0 \\ -1 & \text{if } C_k^{(n)} < 0 \end{cases}.$$

Since then $||x_0|| = 1$ and

$$f_n(x_0) = \sum_{k=0}^{k_n} C_k^{(n)} sgn C_k^{(n)} = \sum_{k=0}^{k_n} |C_k^{(n)}|.$$

For a given $x \in E_x$, (6.21) yields an approximate value $f_n(x)$ for $f(x)$ in (6.20).

We know that the set **P** of all polynomials with real coefficients is dense in the real space $E_x = C([0,1])$ by the Weierstrass approximation theorem (th 1.4.32). Thus, the sequence of all functionals $\{f_n\}$ converges to the functional f on a set of all polynomials, everywhere dense in $C([0,1])$.

Since $\sum_{k=0}^{k_m} |C_m| \leq k = $ const., it follows from (6.20) $||f_m||$ is bounded. Hence, by theorem 6.3.3, $f_n(x) \to f(x)$ for every continuous function $x(t)$.

6.3.7   Theorem

If all the coefficients $C_k^{(n)}$ of quadrature formulae are positive, then the sequence of quadrature formulae $f(x) \simeq f_n(x)$, $n = 1, 2, \ldots$ is convergent for every continuous function $x(t)$.

In fact, $f_n(x_0) = f(x_0)$ for any n and $x_0(t) \equiv 1$.

Hence, $\|f_n(x_0)\| = \sum_{k=1}^{k_n} |C_k^{(n)}| = \sum_{k=1}^{k_n} C_k^{(n)} = \int_0^1 d\sigma = \sigma(1) - \sigma(0).$

Therefore, the hypothesis of theorem 6.3.6 is satisfied.

6.3.8   Weak convergence of sequence of elements of a space

6.3.9   Definition: Weak convergence of a sequence of elements

Let E be a normed linear space, $\{x_n\}$ a sequence of elements in E, $x \in E$, $\{x_n\}$ is said to **converge weakly to the element** x if for every linear functional $f \in E^*$, $f(x_n) \to f(x)$ as $n \to \infty$ and in symbols we write $x_n \xrightarrow{w} x$. We say that x is the weak limit of the sequence of elements $\{x_n\}$.

6.3.10   Lemma: A sequence cannot converge weakly to two limits

Let $\{x_n\} \subset E$, a normed linear space, converge weakly to $x_0, y_0 \in E$ respectively, $x_0 \neq y_0$. Then, for any linear functional $f \in E^*$,

$$f(x_0) = f(y_0) = \lim_{n \to \infty} f(x_n).$$

f being linear, $f(x_0 - y_0) = 0$. The above is true for any functional belonging to E^*. Hence, $x_0 - y_0 = \theta$, i.e., $x_0 = y_0$. Hence the limit is unique.

It is easy to see that any subsequence $\{x_{n_k}\}$ also converges weakly to x if $x_n \xrightarrow{w} x$.

6.3.11   Definition: strong convergence of sequence of elements

The convergence of a sequence of elements (functions) with respect to the norm of the given space is called **strong convergence**.

6.3.12   Lemma

The strong convergence of a sequence $\{x_n\}$ in a normed linear space E to an element $x \in E$ implies weak convergence.

For any functional $f \in E^*$,

$$|f(x_n) - f(x)| = |f(x_n - x)| \leq \|f\| \, \|x_n - x\| \to 0$$

as $n \to \infty$, since $\|f\|$ is finite, $x_n \to x^*$ strongly $\Rightarrow x_n \xrightarrow{w} x^*$.

Note 6.3.1. The converse is not always true. Let us consider the sequence of elements $\{\sin n\pi t\}$ in $L_2([0,1])$.

Put $x_n(t) = \sin n\pi t$ s.t. $f(x_n) = \int_0^1 \sin n\pi t \alpha(t) dt$ where $\alpha(t)$ is a square integrable function uniquely defined with respect to the functional f. Obviously, $f(x_n)$ is the nth Fourier coefficient of $\alpha(t)$ relative to $\{\sin n\pi t\}$.

Consequently, $f(x_n) \to 0$ as $n \to \infty$ so that $x_n \xrightarrow{w} 0$ as $n \to \infty$.

On the other hand,

$$\|x_n - x_n\|^2 = \int_0^1 (\sin n\pi t - \sin m\pi t)^2 dt$$

$$= \int_0^1 \sin^2 n\pi t dt - 2 \int_0^1 \sin n\pi t \sin m\pi t dt$$

$$+ \int_0^1 \sin^2 m\pi t dt = 1 \text{ if } n \neq m.$$

Thus, $\{x_n\}$ does not converge strongly.

6.3.13   Theorem

In a finite dimensional space, notions of weak and strong convergence are equivalent.

Proof: Let E be a finite dimensional space and $\{x_n\}$ a given sequence such that $x_n \xrightarrow{w} x$. Since E is a finite dimensional there is a finite system of linearly independent elements $e_1, e_2, \ldots, e_m$ s.t. every $x \in E$ can be represents as,

$$x = \xi_1 e_1 + \xi_2 e_2 + \cdots + \xi_m e_m.$$

Let, $x_n = \xi_1^{(n)} e_1 + \xi_2^{(n)} e_2 + \cdots + \xi_m^{(n)} e_m.$

$$x_0 = \xi_1^{(0)} e_1 + \xi_2^{(0)} e_2 + \cdots + \xi_m^{(0)} e_m.$$

Now, consider the functionals f_i such that

$$f_i(e_i) = 1, \quad f_i(e_j) = 0 \ i \neq j.$$

Then $f_i(x_n) = \xi_i^{(n)}, \quad f_i(x_0) = \xi_i^{(0)}.$

But since $f(x_n) \to f(x_0)$, for every functional f,

then also $f_i(x_n) \to f_i(x_0)$,   i.e., $\xi_i(n) \to \xi_i^{(0)}$.

$$\|x_p - x_q\| = \left\| \sum_{i=j}^m (\xi_i^{(p)} - \xi_i^{(q)}) e_i \right\| \leq \sum_{i=1}^m |\xi_i^{(p)} - \xi_i^{(q)}| \, \|e_i\|$$

$\to 0$ as $p, q \to \infty$, showing that in a finite dimensional normed linear space, weak convergence of $\{x_p\} \Longleftrightarrow$ strong convergence or $\{x_p\}$.

6.3.14   Remark

There also exist infinite dimensional spaces in which strong and weak convergence of elements are equivalent.

Let $E = l_1$ of sequences $\{\xi_1, \xi_2, \ldots \xi_n \ldots\}$

s.t. the series $\sum_{i=1}^{\infty} |\xi_i|$ converges.

We note that in l_1, strong convergence of elements implies co-ordinatewise convergence.

6.3.15   Theorem

If the **normed linear** space E is separable, then we can find an equivalent norm, such that the weak convergence $x_n \xrightarrow{w} x$ and $||x_n|| \to ||x_0||$ in the new norm imply the strong convergence of the sequence $\{x_n\}$ to x_0.

Proof: Let E be a normed linear space. Since the space is separable, it has a countably everywhere dense set $\{e_i\}$ where $||e_i|| = 1$.

Let $x_n = \xi_1^{(n)} e_1 + \cdots + \xi_i^{(n)} e_i + \cdots$

$\qquad x_0 = \xi_1^{(0)} e_1 + \cdots + \xi_i^{(0)} e_i + \cdots$

where $x_n \xrightarrow{w} x_0$ as $n \to \infty$.

Let us consider the functionals $f_i \in E^*$ such that

$\qquad f_i(e_i) = 1, \ f_i(e_j) = 0 \ i \neq j.$

Now $f_i(x_n) = \xi_i^{(n)}$ and $f_i(x_0) = \xi_i^{(0)}$

Since $f_i(x_n) \to f_i(x_0)$ as $n \to \infty$,

$\qquad \xi_i^{(n)} \to \xi_i^{(0)}, \ i = 1, 2, 3, \ldots$

If $||x_n|| \to ||x_0||$ as $n \to \infty$

$$\left\| \sum_{i=1}^{\infty} \xi_i^{(n)} e_i \right\| \to \left\| \sum_{i=1}^{\infty} \xi_i^{(0)} e_i \right\|$$

as $n \to \infty$.

Let us introduce in E a new norm $|| \cdot ||_1$ as follows

$$||x_n - x_0||_1 = \left\| \sum_{i=0}^{\infty} (\xi_i^{(n)} - \xi_i^{(0)}) e_i \right\|_1 = \frac{||x_n - x_0||}{1 + ||x_n - x_0||} \qquad (6.23)$$

Since $||x_n - x_0|| \geq 0$

$\qquad ||x_n - x_0||_1 \leq ||x_n - x_0||.$

Again, since $\{||x_n||_1\}$ is convergent and hence bounded, $||x_n||_1 \leq M$ (say).

$\qquad ||x_n - x_0|| \leq (1 - M)^{-1} ||x_n - x_0||_1$

Thus, $||x_n - x_0||_1 \leq ||x_n - x_0||(1 - M)^{-1} ||x_n - x_0||_1$

Thus $|| \cdot ||_1$ and $|| \cdot ||$ are equivalent norms.

(6.23) yields that $||x_n|| \leq \dfrac{M}{1 - M} = L$ (say).

Hence $\left\| \sum_{i=1}^{\infty} \xi_i^{(n)} e_i \right\| \leq L$ (say).

Let $\quad S_m = \sum_{i=1}^{m} \xi_i^{(n)} e_i \quad$ and $\quad S = \sum_{i=1}^{\infty} \xi_i^{(n)} e_i$

Let $\epsilon > 0$ and $S_m - S < \epsilon$ for $m \geq m_0(\epsilon)$.

Now, $\quad x_n - x_0 = \sum_{i=1}^{\infty} (\xi_i^{(n)} - \xi_i^{(0)}) e_i$

Let $\quad \epsilon' = \dfrac{\epsilon}{2L} \quad$ and for $\epsilon > 0$, $n \geq n_0(\epsilon')$, $m \geq m_0(\epsilon)$

then $\quad \left| \sum_m (\xi_i^{(m)} - \xi_i^{(0)}) e_i \right| \leq \sum_n |\xi_i^{(n)} - \xi_i^{(0)}| \, \|e_i\| < \epsilon' 2L = \dfrac{\epsilon}{2L} \cdot 2L = \epsilon$

for $\quad n \geq n_0(\epsilon')$, $m \geq m_0(\epsilon)$.

Thus, $\|x_n - x_0\| \to 0$ as $n \to \infty$, proving strong convergence of $\{x_n\}$.

6.3.16 Theorem

If the sequence $\{x_n\}$ of a normed linear space E converges weakly to x_0, then there is a sequence of linear combinations $\left\{ \sum_{k=1}^{k_n} C_k^{(n)} x_k \right\}$ which converges strongly to x_0.

In other words, x_0 belongs to a closed linear subspace L, spanned by the elements $x_1, x_2, \ldots x_n, \ldots$.

Proof: Let us assume that the theorem is not true, i.e., x_0 does not belong to the closed subspace L. Then, by theorem 5.1.5, there is a linear functional $f \in E^*$, such that $f(x_0) = 1$ and $f(x_n) = 0$, $n = 1, 2, \ldots$. But this means that $f(x_n)$ does not converge to $f(x_0)$, contradicting the hypothesis that $x_n \xrightarrow{w} x_0$.

6.3.17 Theorem

Let A be a bounded linear operator with domain E_x and range in E_y, both normed linear spaces. If the sequence $\{x_n\} \subset E_x$ converges weakly to $x_0 \subset E_x$, then the sequence $\{Ax_n\} \subset E_y$ converges weakly to $Ax_0 \in E_y$.

Proof: Let $\phi \in E_y^*$ be any functional. Then $\phi(Ax_n) = f(x_n)$, $f \in E_x^*$. Analogously $\phi(Ax_0) = f(x_0)$.

Since $x_n \xrightarrow{w} x_0$, $f(x_n) \to f(x_0)$ i.e., $\phi(Ax_n) \to \phi(Ax_0)$. Since ϕ is an arbitrary functional in E_y^*, it follows that $Ax_n \xrightarrow{w} Ax_0$. Thus, every bounded linear operator is not only strongly, but also weakly continuous.

6.3.18 Theorem

If a sequence $\{x_n\}$ in a normed linear space converges weakly to x_0, then the norm of the elements of this sequence is bounded.

We regard x_n $(n = 1, 2, \ldots)$ as the elements of E^{**}, conjugate to E^* then the weak convergence of $\{x_n\}$ to x_0 means that the sequence of functions

$x_n(f)$ converges to $x_0(f)$ for all $f \in E^*$. But by the theorem 4.5.7 (Banach-Steinhaus theorem) the norm $\{||x_n||\}$ is bounded, this completes the proof.

6.3.19 Remark

If x_0 is the weak limit of the sequence $\{x_n\}$, then

$$||x_0|| \leq \lim \inf ||x_n||;$$

Moreover, the existence of this finite inferior limit follows from the preceding theorem.

Proof: Let us assume that $||x_0|| > \lim \inf ||x_n||$. Then there is a number α such that $||x_0|| > \alpha > \lim \inf ||x_n||$. Hence, there is a sequence $\{x_{n_i}\}$ such that $||x_0|| > \alpha > ||x_{n_i}||$. Let us construct a limit functional f_0 such that

$$||f_0|| = 1 \quad \text{and} \quad f_0(x_0) = ||x_0|| > \alpha.$$

Then $\quad f_0(x_{n_i}) \leq ||f_0|| \, ||x_{n_i}|| = ||x_{n_i}|| < \alpha$ for all i.

Consequently, $f_0(x_n)$ does not converge to $f_0(x_0)$, contradicting the hypothesis that $x_n \xrightarrow{w} x_0$.

Note 6.3.2. The following example shows that the inequality $||x_0|| < \lim \inf ||x_n||$ can actually hold.

In the space $L_2([0, 1])$ we consider the function

$$x_n(t) = \sqrt{2} \sin n\pi t. \quad \text{Now,} \quad ||x_n(t)||^2 = \langle x_n(t), x_n(t) \rangle$$

$$= \int_0^1 x_n^2(t) dt = \int_0^1 2 \sin^2 n\pi t \, dt = 1.$$

Thus, $\lim_n ||x_n|| = 1$. On the other hand, for every linear functional f,

$$f(x_n) = \sqrt{2} \int_0^1 g(t) \sin n\pi t \, dt = \sqrt{2} c_n,$$

c_n's are the Fourier coefficients of $g(t) \in L_2([0, 1])$.

Thus, $f(x_n) \to 0$ as $n \to \infty$ for every linear functional f, i.e., $x_n \xrightarrow{w} \theta$ consequently, $x_0 = \theta$.

and $\quad ||x_0|| = 0 < 1 = \lim_n ||x_n||$.

6.3.20 Theorem

In order that a sequence $\{x_n\}$ of a normed linear space E converges weakly to x_0, it is necessary and sufficient that

(i) the sequence $\{||x_n||\}$ is bounded and

(ii) $f(x_n) \to f(x_0)$ for every f of a certain set Ω of linear functionals, linear combination of whose elements are everywhere dense in E^*.

Proof: This theorem is a particular case of theorem 6.3.3. This is because convergence of $\{x_n\} \subset E$ to $x_0 \in E$ is equivalent to the convergence of the linear functionals $\{x_n\} \subset E^{**}$ to $x_o \in E^{**}$.

6.3.21 Weak convergence in certain spaces

(a) Weak convergence in l_p.

6.3.22 Theorem

In order that a sequence $\{x_n\}$, $x_n = \{\xi_i^{(n)}\}$, $\xi_i^{(n)} \in l_p$ converges to $x_0 = \{\xi_i^{(0)}\}$, $\xi_i^{(0)} \in l_p$, it is necessary and sufficient that

(i) the sequence $\{\|x_n\|\}$ be bounded and

(ii) $\xi_i^{(n)} \to \xi_i^{(0)}$ as $n \to \infty$ and for all i (in general, however, non-uniformly).

Proof: We note that the linear combinations of the functionals $f_i = (0, 0, \ldots 1, \ldots 0)$, $i = 1, 2, \ldots$ are everywhere dense in $l_q = l_p^*$. Hence, by the theorem 6.3.20, in order that $x_n \xrightarrow{w} x_0$ it is necessary and sufficient that (i) $\{\|x_n\|\}$ is bounded and $f_i(x_n) = \xi_i^{(n)} \to f_i(x_0) = \xi_i^{(0)}$ for every i.

Thus, weak convergence in l_p is equivalent to coordinate-wise convergence together with the boundedness of norms.

(b) Weak convergence in Hilbert spaces

Let H be a Hilbert space and $x \in H$. Now any linear functional f defined on H can be expressed in the form $f(x) = \langle x, y \rangle$, where $y \in H$ corresponds to x.

Now, $x_n \xrightarrow{w} x_0 \Rightarrow f(x_n) \to f(x_0) \Rightarrow \langle x_n, y \rangle \to \langle x, y \rangle$ for every $y \in H$.

6.3.23 Lemma

In a Hilbert space H, if $x_n \to x$ and $y_n \to y$ strongly as $n \to \infty$ then $\langle x_n, y_n \rangle \to \langle x, y \rangle$ as $n \to \infty$, where $\langle \ , \ \rangle$ denotes a scalar product in H.

Proof: $|\langle x_n, y_n \rangle - \langle x, y \rangle| = |\langle x_n, y_n \rangle - \langle x_n, y \rangle + \langle x_n, y \rangle - \langle x, y \rangle|$

$= |\langle x_n, y_n - y \rangle + \langle x_n - x, y \rangle| \leq \|x_n\| \, \|y_n - y\| + \|x_n - x\| \, \|y\|$

$\to 0$ as $\to \infty$, because $\|x_n\|$ and $\|y\|$ are bounded.

Note 6.3.3. If, however, $x_n \xrightarrow{w} x$, $y_n \xrightarrow{w} y$, then **generally** $\langle x_n, y_n \rangle$ does **not** converge to $\langle x, y \rangle$.

For example, if $x_n = y_n = e_n, \{e_n\}$ an arbitrary orthonormal sequence, then $e_n \xrightarrow{w} \theta$ but

$$\langle e_n, e_n \rangle = \|e_n\|^2 = 1 \text{ does not converge to } \theta$$
$$= \langle 0, 0 \rangle.$$

However, if $x_n \to x$, $y_n \xrightarrow{w} y$, then

$\langle x_n, y_n \rangle \to \langle x, y \rangle$, provided $\|y_n\|$ is totally bounded.

Let $P = \sup \|y_n\|$. Then

$$|\langle x_n, y_n \rangle - \langle x, y \rangle| \leq |\langle x_n - x, y_n \rangle + \langle x, y_n - y \rangle|$$

$$\leq P||x_n - x|| + |\langle x, y_n - y \rangle| \to 0 \text{ as } n \to \infty.$$

Finally, we note that if $x_n \xrightarrow{w} x$ and $||x_n|| \to ||x||$, then $x_n \to x$.

This is because,

$$||x_n - x||^2 = \langle x_n - x, \ x_n - x \rangle = [\langle x_n, x_n \rangle - \langle x, x \rangle]$$
$$+ [\langle x, x \rangle - \langle x, x_n \rangle] + [\langle x, x \rangle - \langle x_n, x \rangle]$$
$$\leq [||x_n||^2 - ||x||^2] + [\langle x, x \rangle - \langle x, x_n \rangle] + [\langle x, x \rangle - \langle x_n, x \rangle]$$
$$\to 0 \text{ as } n \to \infty.$$

Problems

1. Let E be a normed linear space.

   (a) If X is a closed convex subset of E, $\{x_n\}$ is a sequence in X and $x_n \xrightarrow{w} x$ in E, then prove that $x \in X$ (6.3.20).

   (b) Let Y be a closed subspace of E. If $x_n \xrightarrow{w} x$ in E, then show that $x_n + Y \xrightarrow{w} x + Y$ in E/Y.

2. In a Hilbert space H, if $\{x_n\} \xrightarrow{w} x$ and $||x_n|| \to ||x||$ as $n \to \infty$, show that $\{x_n\}$ converges to x strongly.

3. Let $f_m(x) = \sum_{m=1}^{pn} C_{nm} x(t_{n,m})$ be a sequence of quadrature formulae

   for, $f(x) = \int_a^b k(x,t)dt$ on the Banach space $E_x = C([a,b])$.

   ($C_{n,m}$ are the weights and $t_{n,m}$ are the nodes).

   Show that $||f_n|| = \sum_{m=1}^{pn} |C_{n,m}|$.

   Further, if (i) $f_n(x_k) \to f(x_k)$, $k = 0, 1, 2, \ldots$

   and (ii) $C_{n,m} \geq 0 \ \forall \ n, m$,

   show that $f_n(x) \to f(x) \ \forall \ x \in C([a,b])$.

   Hence, show that the sequence of Gaussian quadrature formulae $G_n(x) \to f(x)$ as $n \to \infty$.

4. Show that a sequence $\{x_n\}$ in a normed linear space is norm bounded wherever it is weakly convergent.

5. Given $\{f_n\} \subset E^*$ where E is a normed linear space, show that $\{f_n\}$ is weakly convergent to $f \in E^*$ implies that $||f|| \leq \lim_{n \to \infty} \inf ||f_n||$.

6. In l_1, show that $x_n \xrightarrow{w} x$ iff $||x_n - x|| \to 0$.

7. A space E is called weakly sequentially complete if the existence of $\lim_{n \to \infty} f(x_n)$ for each $f \in E^*$ implies the existence of $x \in E$ such that

$\{x_n\}$ converges weakly to x. Show that the space $C([a,b])$ is not weakly sequentially complete.

8. If $x_n \xrightarrow{w} x_0$ in a normed linear space E, show that $x_0 \in \overline{Y}$, where $Y = \text{span}\{x_n\}$. (Use theorem 5.1.5).

9. Let $\{x_n\}$ be a sequence in a normed linear space E such that $x_n \xrightarrow{w} x$ in E. Prove that there is a sequence $\{y_n\}$ or linear combination of elements of $\{x_n\}$ which converges strongly to x. (Use Hahn-Banach theorem).

10. In the space l_2, we consider a sequence $\{T_n\}$, where $T_n : l_2 \to l_2$ is defined by

$$T_n x = (0, 0, \ldots, 0, \xi_1, \xi_2, \ldots), \quad x = \{x_n\} \in l_2.$$

Show that

   (i) T_n is linear and bounded
   (ii) $\{T_n\}$ is weakly operator convergent to $\mathbf{0}$, but not strongly.

(Note that l_2 is a Hilbert space).

11. Let E be a separable Banach space and $M \subset E^*$ a bounded set. Show that every sequence of elements of M contains a subsequence which is weak* convergent to an element of E^*.

12. Let $E = C([a,b])$ with the sup norm. Fix $t_0 \epsilon(a,b)$. For each positive integer n with $t_0 + \frac{4}{n} < b$, let

$$x_n(t) = \begin{cases} 0 & \text{if } a \le t \le t_0, t_0 + \dfrac{4}{n} \le t \le b \\ n(t - t_0) & \text{if } t_0 \le t \le t_0 + \dfrac{2}{n}, \\ n\left(\dfrac{4}{n} - t + t_0\right) & \text{if } t_0 + \dfrac{2}{n} \le t \le t_0 + \dfrac{4}{n}, \end{cases}$$

Then show that $x_n \xrightarrow{w} \theta$ in E but $x_n(t) \not\to \theta$ in E.

13. Let E_x be a Banach space and E_y be a normed space. Let $\{F_n\}$ be a sequence in $(E_x \to E_y)$ such that for each fixed $x \in E_x, \{F_n(x)\}$ is weakly convergent in E_y. If $F_n(x) \xrightarrow{w} y$ in E_y, let $F(x) = y$. Then show that $F \in (E_x \to E_y)$ and

$$||F|| \le \lim_{n \to \infty} \inf ||F_n|| \le \sup ||F_n|| < \infty, \quad n = 1, 2, \ldots$$

(Use theorem 4.5.7 and example 14).

14. Let E be a normed linear space and $\{x_n\}$ be a sequence in E. Then show that $\{x_n\}$ is weakly convergent to E if and only if (i) $\{x_n\}$ is a bounded sequence in E and (ii) there is some $x \in E$ such that $f(x_n) \to f(x)$ for every f in some subset of E^* whose span is dense in E^*.

6.4   Reflexivity

In 5.6.6 the notion of **canonical or natural embedding** of a normed linear space E into its second conjugate E^{**} was introduced. We have discussed when two normed linear spaces are said to be **reflexive** and some relevant theorems. Since the conjugate spaces E, E^*, E^{**} often appear in discussions on reflexivity of spaces, there may be some relationships between weak convergence and reflexivity. In what follows some results which were not discussed in 5.6 are discussed.

6.4.1   Theorem

Let E be a normed linear space and $\{f_1, \ldots f_n\}$ be a linearly independent subset of E^*. Then there are $e_1, e_2, \ldots, e_n$ in E such that $f_j(e_i) = \delta_{ij}$ for $i, j = 1, 2, \ldots n$.

Proof: We prove by induction on m. If $m = 1$, then since $\{f_1\}$ is linearly independent, let $a_0 \in E$ with $f_1(a_0) \neq 0$. Let $e_1 = \frac{a_0}{f_1(a_0)}$. Hence, $f_1(e_1) = 1$. Next let us assume that the result is true for $m = k$. Let $\{f_1, f_2, \ldots, f_{k+1}\}$ be a linearly independent subset of E^*. Since $\{f_1, f_2, \ldots f_k\}$ is linearly independent, there are $a_1, a_2, \ldots a_k$ in E such that $f_j(a_i) = \delta_{ij}$ for $1 \leq i, j \leq k$. We claim that there is some $a_0 \in E$ such that $f_j(a_0) = 0$ for $1 \leq j \leq k$ but $f_{k+1}(a_0) \neq 0$. For $x \in E$, let $a_0 = f_1(x)a_1 + f_2(x)a_2 + \cdots + f_k(x)a_k$.

Then $(x - a) \in \displaystyle\bigcap_{j=1}^{k} N(f_j)$. If $\displaystyle\bigcap_{j=1}^{k} N(f_j) \subset N(f_{k+1})$,

then $f_{k+1}(x) = f_{k+1}(x - a_0) + f_{k+1}(a_0)$

$$= f_1(x)f_{k+1}(a_1) + \cdots + f_k(x)f_{k+1}(a_k),$$

so that $f_{k+1} = f_{k+1}(a_1)f_1 + \cdots + f_{k+1}(a_k)f_k$

$\in \mathrm{span}\, \{f_1, f_2, \ldots, f_k\}$.

violating the linear independence of $\{f_1, \ldots f_{k+1}\}$. In the above $N(f_j)$ stands for the nullspace of f_j.

Hence our claim is justified. Now let

$$e_{k+1} = \frac{a_0}{f_{k+1}(a_0)} \quad \text{and for } i = 1, 2, \ldots, k,$$

let $e_i = a_i - f_{k+1}(a_i)e_{k+1}$.

Then $f_j(e_i) = f_j(a_i) - f_{k+1}(a_i)f_j(e_{k+1})$.

$$= f_j(a_i) - f_{k+1}(a_i)\, \frac{f_j(a_0)}{f_{k+1}(a_0)}.$$

$$= \begin{cases} 1 & \text{for } j = i = 1, 2, \ldots, k \\ 0 & \text{for } j \neq i \end{cases}$$

$$f_j(e_{k+1}) = f_j \left(\frac{a_0}{f_{k+1}(a_0)} \right) = \frac{f_j(a_0)}{f_{k+1}(a_0)} = 0$$

for $j = 1, 2, \ldots, k$, since $f_j(a_0) = 0$.

Also $\quad f_{k+1}(e_{k+1}) = 1$.

Hence $\quad f_j(e_i) = \delta_{ij}, \; i, j = 1, 2, \ldots k+1$

6.4.2 Theorem (Helley, 1912) [33]

Let E be a normed linear space.

(a) Consider $f_1, f_2, \ldots, f_m$ in E^*, $k_1, k_2, \ldots k_m$ in $\mathbb{R}$ $(\mathbb{C})$ and $\alpha \geq 0$. Then, for every $\epsilon > 0$, there is some $x_\epsilon \in E$ such that $f_j(x_\epsilon) = k_j$ for each $j = 1, 2, \ldots, m$ and $||x_\epsilon|| < \alpha + \epsilon$ if and only if

$$\left| \sum_{j=1}^m h_j k_j \right| \leq \alpha \left\| \sum_{j=1}^m h_j f_j \right\|$$

for all $h_1, h_2, \ldots, h_m$ in $\mathbb{R}$ $(\mathbb{C})$.

(b) Let S be a finite dimensional subspace of E^* and $F_x \in E^{**}$. If $\epsilon > 0$, then there is some $x_\epsilon \in E$ such that

$$F \big|_S = \phi(x_\epsilon) \big|_S \quad \text{and} \quad ||x_\epsilon|| < ||F_x|| + \epsilon.$$

Proof: (a) Suppose that for every $\epsilon > 0$, there is some $x_\epsilon \in E$ such that $f_j(x_\epsilon) = k_j$ for each $j = 1, \ldots, m$ and $||x_\epsilon|| < \alpha + \epsilon$. Let us fix $h_1, h_2, \ldots, h_m$ in $\mathbb{R}$ $(\mathbb{C})$. Then

$$\left| \sum_{j=1}^m h_j k_j \right| = \left| \sum_{j=1}^m h_j f_j(x_\epsilon) \right| = \left| \left(\sum_{j=1}^m h_j f_j \right)(x_\epsilon) \right|$$

$$\leq \left\| \sum_{j=1}^m h_j f_j \right\| ||x_\epsilon|| < (\alpha + \epsilon) \left\| \sum_{j=1}^m h_j f_j \right\|.$$

As this is true for every $\epsilon > 0$, we conclude that

$$\left| \sum_{j=1}^m h_j k_j \right| \leq \alpha \left\| \sum_{j=1}^m h_j f_j \right\|.$$

Conversely, suppose that for all $h_1, h_2, \ldots, h_m$ in $\mathbb{R}$ $(\mathbb{C})$, $\left| \sum_{j=1}^m h_j k_j \right| \leq$

$\alpha \left\| \sum_{j=1}^m h_j f_j \right\|$. It may be noted thats $\{f_1, f_2, \ldots, f_m\}$ can be assumed to be a linearly independent set. If that is not so, let $f_1, f_2, \ldots, f_n$ with $n \leq m$

be a maximal linearly independent subset of $\{f_1, f_2, \ldots f_m\}$. Given $\epsilon > 0$, let $x_\epsilon \in E$ be such that $\|x_\epsilon\| \leq \alpha + \epsilon$ and $f_j(x_\epsilon) = k_j$ for $j = 1, 2, \ldots, n$. If $n < l \leq m$, then $f_l = h_1 f_1 + \cdots + h_n f_n$ for some $h_1, h_2, \ldots, h_n$ in $\mathbb{R}$ ($\mathbb{C}$).

Hence, $\quad f_l(x_\epsilon) = h_1 f_1(x_\epsilon) + \cdots + h_n f_n(x_\epsilon) = h_1 k_1 + \cdots + h_n k_n$.

But, $\quad \left| k_l - \sum_{j=1}^{n} h_j k_j \right| \leq \alpha \left\| f_l - \sum_{j=1}^{n} h_j f_j \right\| = 0,$

so that $f_l(x_\epsilon) = k_l$ as well.

Consider the map $\mathcal{F} : E \to \mathbb{R}^m(\mathbb{C}^m)$ given by $\mathcal{F}(x) = (f_1(x), \ldots, f_m(x))$.

Clearly, $\mathcal{F}$ is a linear map. Next, we show that it is a surjective (onto) mapping. To this end consider $(h_1, h_2, \ldots, h_m) \in \mathbb{R}^m$ (or $\mathbb{C}^m$). Since $\{f_1, f_2, \ldots, f_m\}$ are linearly independent, it follows from theorem 6.4.1 that there exist $e_1, e_2, \ldots, e_m$ in E such that $f_j(e_i) = \delta_{ij}$, $1 \leq i, j \leq m$. If we take $x = h_1 e_1 + \cdots + h_m e_m$, then it follows that $\mathcal{F}(x) = (h_1, h_2, \ldots, h_m)$. We next want to show that $\mathcal{F}$ maps each open subset of E onto an open subset of $\mathbb{R}^m$ (or $\mathbb{C}^m$). Since $\mathcal{F}$ is non-zero we can find a non-zero vector, 'a' in E s.t. $\mathcal{F}(a) = (1, 1, \ldots 1) \in \mathbb{R}^m(\mathbb{C}^m)$. Let $\mathcal{P}$ be an open set in E. Then there exists an open ball $U(x, r) \subset E$ with $x \in E$ and $r \in \mathbb{R}$.

We can now find a scalar k such that

$$x - ka \in U(x, r) \quad \text{where} \quad 0 < |k| < \frac{r}{\|a\|}.$$

Hence, $x - ka \in \mathcal{P}$ with the above choice of k.

Therefore, $\mathcal{F}(x - ka) = \mathcal{F}(x) - k\mathcal{F}(a) = \mathcal{F}(x) - k \in \mathcal{F}(E)$. Thus

$$\left\{ k' \in \mathbb{R}(\mathbb{C}) : \mathcal{F}(x) = |k'| < \frac{r}{\|a\|} \right\} \subset \mathcal{F}(E),$$

showing $\mathcal{F}(E)$ is open in $\mathbb{R}^m$ ($\mathbb{C}^m$).

Thus $\mathcal{F}$ maps each open subset of E onto an open subset of $\mathbb{R}^m$ ($\mathbb{C}^m$).

Let $\epsilon > 0$ and let us consider $U_\epsilon = \{x \in E : \|x\| < \alpha + \epsilon\}$. We want to show that there is some $x \in U_\epsilon$ with $\mathcal{F}(x_\epsilon) = (k_1, k_2, \ldots, k_m)$. If that be not the case, then $(k_1, \ldots, k_m)$ does not belong to the open convex set $\mathcal{F}(U_\epsilon)$. By the Hahn-Banach separation theorem (5.2.10) for $\mathbb{R}^m$ ($\mathbb{C}^m$) there is a continuous linear functional g on $\mathbb{R}^m$ ($\mathbb{C}^m$) such that

$$Re\ g((f_1(x), \ldots, f_m(x))) \leq Re\ g((k_1, \ldots k_m))$$

for all $x \in U_\epsilon$. By 5.4.1, there is some $(h_1, h_2, \ldots, h_m) \in \mathbb{R}^m(\mathbb{C}^m)$ such that

$$g(c_1, c_2, \ldots, c_m) = c_1 h_1 + \cdots + c_m h_m$$

for all $(c_1, c_2, \ldots, c_m) \in \mathbb{R}^m(\mathbb{C}^m)$. Hence,

$$Re[h_1 f_1(x) + \cdots + h_m f_m(x)] \leq Re(h_1 k_1 + \cdots + h_m k_m)$$

for all $x \in U_\epsilon$. If $h_1 f_1(x) + \cdots + h_m f_m(x) = re^{i\theta}$ with $r \geq 0$ and $-\pi < \theta < \pi$, then by considering $xe^{-i\theta}$ in place of x, it follows that

$$|h_1 f_1(x) + \cdots + h_m f_m(x)| \leq Re(h_1 k_1 + \cdots + h_m k)$$

for all $x \in U_\epsilon$. But

$$\sup\left\{ \left| \sum_{j=1}^m h_j f_j(x) \right| : x \in U_\epsilon \right\} = (\alpha + \epsilon) \left\| \sum_{j=1}^m h_j f_j \right\|$$

Hence, $(\alpha + \epsilon) \left\| \sum_{j=1}^m h_j f_j \right\| \leq Re \sum_{j=1}^m h_j k_j \leq \left| \sum_{j=1}^m h_j k_j \right| \leq \alpha \left\| \sum h_j f_j \right\|$

This contradiction shows that there must be some $x_\epsilon \in U_\epsilon$ with $\mathcal{F}(x_\epsilon) = (k_1, k_2, \ldots, k_m)$ as wanted.

(b) Let $\{f_1, f_2, \ldots, f_m\}$ be a basis for the finite dimensional subspace S of E^* and let $k_j = F_x(f_j)$, $j = 1, 2, \ldots, m$. Then for all $h_1, h_2, \ldots, h_m$ in $\mathbb{R}^m$ ($\mathbb{C}^m$), we have

$$\left| \sum_{j=1}^m h_j k_j \right| = \left| \sum_{j=1}^m h_j F_x(f_j) \right| = \left| F_x\left(\sum_{j=1}^m h_j f_j \right) \right| \leq \|F_x\| \left\| \sum_{j=1}^m h_j f_j \right\|.$$

Let $\alpha = \|F_x\|$ in (a) above, we see that for every $\epsilon > 0$, there is some $x_\epsilon \in E$ such that

$$\|x_\epsilon\| \leq \|F_x\| + \epsilon \text{ for } j = 1, 2, \ldots, m$$

$$\phi(x_\epsilon)(f_j) = f_j(x_\epsilon) = k_j = F_x(f_j),$$

i.e., $\phi(x_\epsilon)\big|_S = F_x\big|_S$ as desired.

6.4.3 Remark

(i) It may be noted if we restrict ourselves to a finite dimensional subspace of E, then we are close to **reflexivity**.

The relationship between reflexivity and weak convergence is demonstrated in the following theorem.

6.4.4 Theorem (Eberlein, 1947)

Let E be a normed linear space. Then E is reflexive if and only if every bounded sequence has a weakly convergent subsequence.

Proof: For proof see Limaye [33].

6.4.5 Uniform convexity

We next explore some geometric condition which implies reflexivity. In 2.1.12 we have seen that a closed unit ball of a normed linear space E is a convex set of E. In the case of the strict convexity of E, the mid-point of the segment joining two points on the unit sphere of E does not lie on the unit sphere of E. Next comes a concept in E which implies reflexivity of E.

A normed space E is said to be **uniformly convex** if, for every $\epsilon > 0$, there exists some $\delta > 0$ such that for all x, and y in E with $||x|| \leq 1$, $||y|| \leq 1$ and $||x - y|| \geq \epsilon$, we have $||x + y|| \leq 2(1 - \delta)$.

This idea admits of a geometrical interpretation as follows: given $\epsilon > 0$, there is some $\delta > 0$ such that if x and y are in the closed unit ball of E, and if they are at least ϵ apart, then their mid-point lies at a distance at best δ from the unit sphere. Here δ may depend on ϵ. In what follows the relationship **between a strictly convex** and a **uniformly convex** space is discussed.

6.4.6 Definition

A normed space E is said to be strictly convex if, for $x \neq y$ in E with $||x|| = 1 = ||y||$, we have

$$||x + y|| < 2.$$

6.4.7 Lemma

A **uniformly convex** space is **strictly convex**. This is evident from the definition itself.

6.4.8 Lemma

If E is finite dimensional and **strictly convex**, then E is **uniformly convex**.

Proof: For $\epsilon > 0$, let

$$\Lambda = \{(x,y) \in E \times E : ||x|| \leq 1, \ ||y|| \leq 1, \ ||x - y|| \geq \epsilon\}.$$

Then Λ is a closed and bounded subset of $E \times E$. We next show that Λ is compact, i.e., show that every sequence in Λ has a convergent subsequence, converging in E. Let $u^n = (x^n, y^n)$ in Λ, $(\ ,\ )$ denote the cartesian product. Let $\{e_1, \ldots, e_m\}$ be a basis in E. Then we can write $x^n = \sum_{j=1}^{m} p_j^n e_j$ and $y^n = \sum_{j=1}^{m} q_j^n e_j$ where p_j^n and q_j^n are scalars for $j = 1, 2, \ldots, m$. Since $\{(x^n, y^n)\}$ is a bounded sequence in $E \times E$, $\{p_j^n\}$ and $\{q_j^n\}$ are bounded for $j = 1, 2, \ldots, m$.

By Bolzano–Weierstrass theorem (Cor. 1.6.19), $\{z_j^n\}$ has a convergent subsequence $\{z_j^{n_k}\}$ for each $j = 1, 2, \ldots, m$.

Hence, $\{u^{n_k}\} = \{(x^{n_k}, y^{n_k})\}$ converges to some element $u \in E \times E$ as $k \to \infty$. Since $u^{n_k} \in \Lambda$ and Λ is closed, $u \in \Lambda \subseteq E \times E$. Therefore, if $x^{n_k} \to x$ and $y^{n_k} \to y$ as $k \to \infty$, we have $u = (x, y) \in \Lambda$. Thus Λ is compact.

For, $(x, y) \in \Lambda$, let

$$f(x, y) = 2 - ||x + y||.$$

Now, f is a continuous and strictly positive function on Λ. Hence there is some $\delta > 0$ such that $f(x, y) \geq 2\delta$ for all $(x, y) \in \Lambda$. This implies the uniform convexity of E.

6.4.9 Remark

A strictly convex normed linear space need not in general be uniformly convex. Let $E = c_{00}$, a space of numerical sequences with finite number of non-zero terms. Let

$$\Lambda_n = \{x \in \Lambda : x_j = 0 \text{ for all } j > n\}.$$

For $x \in \Lambda_1$, let $||x|| = |x_1|$.

Let us assume that $||x||$ is defined for all $x \in \Lambda_{n-1}$.

If $x \in \Lambda_n$, then $x = z_{n-1} + x_n e_n$, for some $z_{n-1} \in \Lambda_{n-1}$. Define

$$||x|| = (||z_{n-1}||^n + |x_n|^n)^{1/n}.$$

By making an appeal to induction we can verify that $|| \cdot ||$ is a strictly convex norm on E.

For $n = 1, 2, \ldots$ let

$$x_n = \frac{e_1 + e_n}{2^{1/n}} \quad \text{and} \quad z_n = \frac{-e_1 + e_n}{2^{1/n}},$$

Then $||x_n|| = 1 = |z_n||$

$$||x_n + z_n|| = 2^{(n-1)/n} = ||x_n - z_n||.$$

Thus, $||x_n - z_n|| \geq 1$ for all n. But $||x_n + z_n|| \to 2$ as $n \to \infty$. Hence, E is not uniformly convex.

6.4.10 Remark

It is noted that the normed spaces $l_1, l_\infty, C([a, b])$ are not strictly convex. It was proved by Clarkson [12] that the normed spaces l_p and $L_p([a, b])$ with $1 < p < \infty$ are uniformly convex.

6.4.11 Lemma

Let E be a uniformly convex normed linear space and $\{x_n\}$ be a sequence in E such that $||x_n|| \to 1$ and $||x_n + x_m|| \to 2$ as $n, m \to \infty$. Then $\lim\limits_{n,m \to \infty} ||x_n - x_m|| = 0$. That is, $\{x_n\}$ is a Cauchy sequence.

Proof: If $\{x_n\}$ is not Cauchy in E, given $\epsilon > 0$ and a positive integer n_0, there are $n, m \geq n_0$ with

$$||x_n - x_m|| \geq \epsilon.$$

This implies that for a given $x \in E$ and a positive integer m_0, there is $n, m > m_0$ with

$$||x_n - x|| + ||x_m - x|| \geq ||x_n - x_m|| \geq \epsilon.$$

Taking $n = m > m_0$ we have

$$||x_m - x|| \geq \frac{\epsilon}{2}.$$

Since $||x_n|| \to 1$ as $n \to \infty$ we see that for each $k = 1, 2, \ldots$ there is a positive integer n_k such that

$$||x_{n_k}|| \leq 1 + \frac{1}{k} \text{ for all } n \geq n_k.$$

Choosing $m_1 = n_1$. Then $||x_{m_1}|| \leq 1 + 1 = 2$.

Let $m_0 = \max\{m_1, n_1\}$ and $x = x_{m_1}$

We see that there is some $m_2 > m_0$ with

$$||x_{m_2} - x_{m_1}|| \geq \frac{\epsilon}{2}.$$

We note that $||x_{m_2}|| \leq 1 + \frac{1}{2}$ since $m_2 > m_0 \geq n_2$. Thus we can find a subsequence $\{x_{m_k}\}$ of $\{x_m\}$ such that for $k = 1, 2, \ldots,$

$$||x_{m_{k+1}} - x_{m_k}|| \geq \frac{\epsilon}{2} \quad \text{and} \quad ||x_{m_k}|| \leq 1 + \frac{1}{k}.$$

By the uniform convexity of E, there is a $\delta > 0$ such that $||x+y|| \leq 2(1-\delta)$ wherever x and y are in E, $||x|| \leq 1$, $||y|| \leq 1$ and $||x - y|| \geq \epsilon$.

Let us put $y_k = x_{m_k}$ for $k = 1, 2, \ldots,$ then

$$\left\| \frac{y_k}{1 + \frac{1}{k}} \right\| \leq 1, \ ||y_{k+1}|| \leq 1 + \frac{1}{k+1} \leq 1 + \frac{1}{k} \text{ i.e.,}$$

$$\left\| \frac{y_{k+1}}{1 + \frac{1}{k}} \right\| \leq 1 \quad \text{and} \quad \left\| \frac{y_{k+1} - y_k}{1 + \frac{1}{k}} \right\| \geq \frac{\epsilon}{2} = \epsilon' \text{ (say)}.$$

Hence, $\quad \left\| \dfrac{y_{k+1} + y_k}{1 + \frac{1}{k}} \right\| \leq 2(1 - \delta).$

Thus, $\limsup_{k \to \infty} ||y_{k+1} + y_k|| \leq 2(1 - \delta) < 2,$

i.e., $\limsup_{k \to \infty} ||x_{m_{k+1}} + x_{m_k}|| < 2.$

The above contradicts the fact that

$$||x_m + x_n|| \to 2 \quad \text{as } m, n \to \infty.$$

Hence, $\{x_n\}$ is a Cauchy sequence in E.

6.4.12   Theorem (Milman, 1938) [33]

Let E be a Banach space which is uniformly convex in some equivalent norm. Then E is **reflexive**.

Proof: We first show that a reflexive normed space remains reflexive in an equivalent norm.

From theorem 4.4.2, we can conclude that the space of bounded linear functionals defined on a normed linear space E is complete and hence a Banach space. Thus the dual E^* and in turn the second dual E^{**} of the normed linear space E are Banach spaces. Since E is reflexive, E is isometrically isomorphic to E^{**} and hence E is a Banach space. Also, in any equivalent norm on E, the dual E^* and the second dual E^{**} remains unchanged, so that E remains reflexive.

Hence, we can assume without loss of generality that E is a uniformly convex Banach space in the given norm $|| \ ||$ on E.

Let $F_x \in E^{**}$. Without loss of generality we assume that $||F_x|| = 1$. To show that there is some $x \in E$ with $\phi(x) = F_x$ $\phi : E \to E^{**}$ being a canonical embedding. First, we find a sequence $\{f_n\}$ in E^* such that $||f_n|| = 1$ and $|F_x(f_n)| > 1 - \frac{1}{n}$ for $n = 1, 2, \ldots$.

For a fixed n, let $S_n = \text{span} \{f_1, f_1, \ldots, f_n\}$.

We put $\epsilon_n = \dfrac{1}{n}$ in Helley's theorem (6.4.2) and find $x_n \epsilon E$ such that

$$F_{x_n}\big|_{S_n} = \phi(x_n)\big|_{S_n} \quad \text{and} \quad ||x_n|| < 1 + \frac{1}{n}.$$

Then for $n = 1, 2, \ldots$ and $m = 1, 2, \ldots, n$.

$$F(f_m) = \phi(x_n)(f_m) = f_m(x_n)$$

so that $1 - \dfrac{1}{n} < |F(f_n)| = |f_n(x_n)| \leq ||x_n|| < 1 + \dfrac{1}{n}$

and $\qquad 2 - \dfrac{2}{n} < |2F(f_n)| = |f_n(x_n) + f_n(x_m)|$

$$\leq ||x_m + x_n|| \leq 2 + \frac{1}{n} + \frac{1}{m}.$$

Then we have,

$$\lim_{n \to \infty} ||x_n|| = 1 \quad \text{and} \quad \lim_{n,m \to \infty} ||x_n + x_m|| = 2.$$

By lemma 6.4.11 $\{x_n\}$ is a Cauchy sequence in E. Since E is a Banach space, let $x_n \to x$ in E. Then $||x|| = 1$. Also, since $F(f_m) = f_m(x_n)$ for all $n \geq m$, the continuity or f_m shows that

$$F(f_m) = f_m(x), \ m = 1, 2, \ldots.$$

Let us next consider $f \in E^*$. Replacing S_n by the span of $\{f, f_1, f_2, \ldots, f_n\}$, we find some $z \in E$, such that $||z|| = 1$ and

$$F(f) = f(z) \quad \text{and} \quad F(f_m) = f_m(z).$$

We want to show that $x = z$.

Now, $\quad ||x + z|| \geq |f_m(x + z)| = 2|F(f_m)| > 2 - \dfrac{2}{m},$

for all $m = 1, 2, \ldots$ so that $||x + z|| \geq 2$.

Since $||x|| = 1$, $||z|| = 1$, the strict convexity of E implies that $x = z$ and $F(f) = f(z) = f(x)$, for all $f \in E^*$, i.e., $F = \phi(x)$. Hence, E is reflexive.

6.4.13   Remark

The converse of Milman's theorem is false [see Limaye[33]].

Problems

1. Let E be a reflexive normed linear space. Then show that E is strictly convex (resp. smooth) if and only if E^* is smooth (resp. strictly convex).

    (Hint: A normed linear space E is said to be smooth if, for every $x_0 \in E$ with $||x_0|| = 1$, there is a unique supporting hyperplane [see 4.3.7] for $\overline{B}(\theta, 1)$ at x_0.)

2. [**Weak Schauder basis.** Let E be a normed linear space. A countable subset $\{a_1, a_2, \ldots\}$ of E is called a weak Schauder basis for E if $||a_i|| = 1$ for each i and for every $x \in E$, there are unique

    $$\alpha_i \in \mathbb{R}(\mathbb{C}) \ i = 1, 2, \ldots \text{ such that } \sum_{i=1}^{n} \alpha_i a_i \xrightarrow{w} x \text{ as } n \to \infty.$$

    Weak* Schauder bases. A countable subset $\{f_1, f_2, \ldots\}$ of E^* is called a Weak* Schauder basis if $||f_i|| = 1$ for all i and for every $g \in E^*$, there are unique $\beta_i \in \mathbb{R}(\mathbb{C})$, $i = 1, 2, \ldots$ such that

    $$\sum_{i=1}^{n} \beta_i f_i \xrightarrow{w} g \text{ as } n \to \infty.]$$

    Let E be a reflexive normed linear space and $\{a_1, a_2, \ldots\}$ be a Schauder basis for E with coefficient functionals $\{g_1, g_2, \ldots\}$. If $f_n = g_n/||g_n||$, $n = 1, 2, \ldots$ then show that $\{f_1, f_2, \ldots\}$ is a Schauder basis for E^* with coefficient functionals $(||g_1||F_{a_1}, ||g_2||F_{a_2}, \ldots)$.

3. Let E be a separable normed linear space. Let $\{x_n\}$ be a dense subset of $\{x \in E : ||x|| = 1\}$.

    (a) Then show that there is a sequence $\{f_n\}$ in E^* such that $||f_n|| = 1$ for all n and for every $x \neq \theta$ in E, $f_n(x) \neq 0$ for some n and for $x \in E$ if

    $$||x||_0 = \left(\sum_{n=1}^{\infty} \frac{|f_n(x)|^2}{2^n} \right)^{\frac{1}{2}}$$

    then $|| \ ||_0$ is a norm on E, in which E is strictly convex and $||x||_0 \leq ||x||$ for all $x \in E$.

    (b) There is an equivalent norm on E in which E is strictly convex (Hint: consider $||x||_1 = ||x|| + ||x||_0$).

(c) Show that l_1 is strictly convex but not reflexive in some norm which is equivalent to the norm $||\ ||_1$.

4. Let E be a uniformly convex normed linear space, $x \in E$, and $\{x_n\}$ be a sequence in E.

   (a) If $||x|| = 1$, $||x_n|| \to 1$ and $||x_n + x|| \to 2$ then show that $||x_n - x|| \to 0$.

   (b) Show that $x_n \to x$ in E if and only if $x_n \xrightarrow{w} x$ in E and $\limsup\limits_{n \to \infty} ||x_n|| \le ||x||$.

6.5 Best Approximation in Reflexive Spaces

The problem of best approximation of functions concerns finding a proper combination of known functions so that the said combination is closest to the above function. P.L. Chebyshev [43] was the first to address this problem. Let E be a normed linear space, $x \in E$ is an arbitrary element. We want to approximate x by a finite linear combination of linearly independent elements $x_1, x_2, \ldots x_n \in E$.

6.5.1 Lemma

If $\sum\limits_{i=1}^{n} \alpha_i^2$ increases indefinitely, then $\phi(\alpha_1, \alpha_2, \ldots, \alpha_n) = ||x - \alpha_1 x_1 - \alpha_2 x_2 - \cdots - \alpha_n x_n|| \to \infty$.

Proof: We have

$$\phi(\alpha_1, \alpha_2, \ldots, \alpha_n) \ge ||\alpha_1 x_1 + \cdots + \alpha_n x_n|| - ||x||.$$

The continuous function

$$\psi(\alpha_1, \alpha_2, \ldots, \alpha_n) = ||\alpha_1 x_1 + \alpha_2 x_2 + \cdots + \alpha_n x_n||$$

of the parameters $\alpha_1, \alpha_2, \ldots, \alpha_n$ assumes its minimum 'm' on a unit ball

$$S = \left\{ \alpha_1, \alpha_2, \ldots, \alpha_n \in E_n : \sum_{i=1}^{n} \alpha_i^2 = 1 \right\}$$

in E_n where E_n denotes the n-dimensional Euclidean space.

Since a unit ball in E_n is compact, the continuous function $\psi(x_1, x_2, \ldots x_n)$ assumes its minimum on S. Since $x_1, x_2, \ldots, x_n$ are linearly independent the value of $\psi(x_1, x_2, \ldots, x_n)$ is always positive.

Therfore $m > 0$.

Given an arbitrary $K > 0$,

$$\phi(\alpha_1, \alpha_2, \ldots \alpha_n) \ge ||\alpha_1 x_1 + \alpha_2 x_2 + \cdots + \alpha_n x_n|| - ||x||$$

$$= \sqrt{\sum_{i=1}^{n} \alpha_i^2} \cdot \left\| \frac{\alpha_1 x_1 + \cdots + \alpha_n x_n}{\sqrt{\sum_{i=1}^{n} \alpha_i^2}} \right\| - \|x\|$$

$$\geq \sqrt{\sum_{i=1}^{n} \alpha_i^2} \cdot m - \|x\|,$$

since $\left\{ \dfrac{\alpha_1}{\sum_i \alpha_i^2}, \dfrac{\alpha_2}{\sum_i \alpha_i^2}, \cdots, \dfrac{\alpha_n}{\sum_i \alpha_i^2} \right\}$ lie on a unit ball.

Thus if $\sqrt{\sum_{i=1}^{n} \alpha_i^2} > \dfrac{1}{m}(K + \|x\|), \quad \phi(\alpha_1, \alpha_2, \ldots \alpha_n) > K$

which proves the lemma.

6.5.2   Theorem

There exist real numbers $\alpha_1^{(0)}$, $\alpha_2^{(0)}$, ..., $\alpha_n^{(0)}$, such that $\phi(\alpha_1, \alpha_2, \ldots, \alpha_n) = \|x - \alpha_1 x_1 - \alpha_2 x_2 \ldots \alpha_n x_n\|$ assumes its minimum for $\alpha_1 = \alpha_1^{(0)}, \alpha_2 = \alpha_2^{(0)} \ldots \alpha_n = \alpha_n^{(0)}$.

Proof: If x depends linearly on $x_1, x_2, \ldots, x_n$, then the theorem is true immediately. Let us assume that x does not lie in the subspace spanned by $x_1, x_2, \ldots, x_n$.

We first show that $\phi(\alpha_1, \alpha_2, \ldots, \alpha_n)$ is a continuous function of its arguments.

Now $|\phi(\alpha_1, \alpha_2, \ldots, \alpha_n) - \phi(\beta_1, \beta_2, \ldots, \beta_n)|$

$$= \left\| x - \sum_{i=1}^{n} \alpha_1 x_i \right\| - \left\| x - \sum_{i=1}^{n} \beta_i x_i \right\|$$

$$\leq \left\| \sum_{i=1}^{n} (\alpha_i - \beta_i) x_i \right\| \leq \max_{1 \leq i \leq n} |\alpha_i - \beta_i| \sum_{i=1}^{n} \|x_i\|.$$

If, $S' = \left\{ (\lambda_1, \lambda_2, \ldots \lambda_n) \in E_n : \sum_{i=1}^{n} \lambda_i \leq r^2 \right\},$

then outside S', the previous lemma yields, $\phi(\alpha_1, \alpha_2, \ldots, \alpha_n) \geq \|x\|$.

Now, the ball $S' \subset E_n$ being compact, $\phi(\alpha_1, \alpha_2, \ldots \alpha_n)$ being a continuous function assumes its minimum r at some point $(\alpha_1^{(0)}, \alpha_2^{(0)}, \ldots, \alpha_n^{(0)})$. But $r \leq \phi(0, 0, \ldots, 0) = \|x\|$. Hence, r is the least value of the function $\phi(\alpha_1, \alpha_2, \ldots \alpha_n)$ on the entire space of the points $\alpha_1, \alpha_2, \cdots \alpha_n$, which proves the theorem.

6.5.3 Remark

(i) The linear combination $\sum_{i=1}^{n} \lambda_i^{(0)} x_i$, giving the best approximation of the element x, is in general not unique.

(ii) Let Y be a finite dimensional subspace of $C([0,1])$. Then the best approximation out of Y is unique for every $x \in C([a,b])$ if and only if Y satisfies the Haar condition [see Kreyszig [30]].

(iii) However, there exist certain spaces in which the best approximation is everywhere **uniquely** defined.

6.5.4 Definition: strictly normed

A space E is said to be **strictly normed** if the equality $\|x + y\| = \|x\| + \|y\|$ for $x \neq \theta$, $y \neq \theta$, is possible only when $y = ax$, with $a > 0$.

6.5.5 Theorem

In a strictly normed linear space the best approximation of an arbitrary x in terms of a linear combination of a given finite system of linearly independent elements is unique.

Proof: Let us suppose that there exist two linear combinations $\sum_{i=1}^{n} \alpha_i x_i$ and $\sum_{i=1}^{n} \beta_i x_i$ such that

$$\left\| x - \sum_{i=1}^{n} \alpha_i x_i \right\| = \left\| x - \sum_{i=1}^{n} \beta_i x_i \right\| = d,$$

where $\quad d = \min_{r_i} \left\| x - \sum_{i=1}^{n} r_i x_i \right\| > 0,$

then $\quad \left\| x - \sum_{i=1}^{n} \frac{\alpha_i + \beta_i}{2} x_i \right\| \leq \frac{1}{2} \left\| x - \sum_{i=1}^{n} \alpha_i x_i \right\| + \frac{1}{2} \left\| x - \sum_{i=1}^{n} \beta_i x_i \right\|$

$$= \frac{1}{2} d + \frac{1}{2} d = d.$$

and since $\quad \left\| x - \sum_{i=1}^{n} \frac{\alpha_i + \beta_i}{2} x_i \right\| \geq d,$

we have $\quad \left\| x - \sum_{i=1}^{n} \frac{\alpha_i + \beta_i}{2} x_i \right\| = d,$

Consequently, $\quad \left\| x - \sum_{i=1}^{n} \frac{\alpha_i + \beta_i}{2} x_i \right\| = \left\| \frac{1}{2}\left(x - \sum_{i=1}^{n} \alpha_1 x_i \right) \right\|$

$$+ \left\| \frac{1}{2}\left(x - \sum_{i=0}^{n} \beta_i x_i \right) \right\|.$$

The space being **strictly normed.**

$$x - \sum_{i=1}^{n} \beta_i x_i = a \left\{ x - \sum_{i=1}^{n} \alpha_i x_i \right\}.$$

If $a \neq 1$, then x would be a linear combination of the elements $x_1, x_2, \ldots, x_n$, which is a contradiction. Thus, $a = 1$, but then $\sum_{i=1}^{n} (\alpha_i - \beta_i) x_i = 0$ and $x_1, x_2, \ldots, x_n$ are linearly independents. Hence we get $\alpha_i = \beta_i$, $i = 1, 2, \ldots, n$.

6.5.6 Remark

(i) $L_p([0,1])$ and l_p for $p > 1$ are **strictly normed.**

(ii) $C([0,1])$ is not strictly normed.

Let us take $x(t)$ and $y(t)$ as two non-negative linearly independent functions taking the maximum values at one and the same point of the interval.

Then, $||x + y|| = \max |x(t) + y(t)| = |x(\hat{t}) + y(\hat{t})|$

$$= x(\hat{t}) + y(\hat{t}) = ||x|| + ||y||.$$

But $y \neq ax$. Hence, $C([0,1])$ is not strictly normed.

6.5.7 Lemma

Let E be a reflexive normed linear space and M be a non-empty closed convex subset of E. Then, for every $x \in E$, there is some $y \in M$ such that $||x - y|| = \text{dist}\,(x, M)$, that is there is a best approximation to x from M.

Proof: Let $x \in E$ and $d = \text{dist}\,(x, M)$. If $x \in M$ then the result is trivially satisfied. Let $x \notin M$. Then there is a sequence $\{y_n\}$ in M such that $||x - y_n|| \to d$ as $n \to \infty$. Since x is known and $\{x - y_n\}$ is bounded, $\{y_n\}$ is bounded, and since E is reflexive, $\{y_n\}$ contains a weakly convergent subsequence $\{y_{n_p}\}$, (6.4.4). Now, $\{y_{n_p}\} \subset M$ and M being closed, $\lim_{p \to \infty} \phi_1(y_{n_p}) = \phi(y^1)$, $y^1 \in M$ and ϕ_1 is any linear functional. Therefore, $\lim_{p \to \infty} \phi_1(x - y_{n_p}) = \phi_1(x - y^1)$.

Since $\{y_{n_p}\}$ is a subsequence of $\{y_n\}$,

$$\lim_{p \to \infty} ||x - y_{n_p}|| = \lim_{n \to \infty} ||x - y_n|| = d.$$

Thus, by theorem 6.3.15, $\{x - y_{n_p}\}$ is strongly convergent to $\{x - y^{(1)}\}$. Hence, $\{y_{n_p}\}$ is strongly convergent to $y^{(1)}$. Similarly if $\{y_n\}$ contains another weakly convergent subsequence, then we can find a $y^2 \in M$ s.t. $\{y_{n_q}\}$ is strongly convergent to y^2 s.t. $||x - y^2|| = d$. Since M is convex $y^1, y^2 \in M \Rightarrow [\lambda y^1 + (1 - \lambda) y^2] \in M$, $0 \leq \lambda \leq 1$. Thus $||x - (\lambda y^1 + (1 - \lambda) y^2)|| = d$.

6.5.8   Minimization of functionals

The problem of best approximation is just a particular case of a wider and more comprehensive problem, namely the problem of minimization of a functional.

The classical **Weierstrass existence theorem** tells the following:

(W) The minimum problem

$$F(u) = \min!, \ u \in M \tag{6.25}$$

has a solution provided the functional $F : M \to R$ is continuous on the nonempty compact subset M of the Banach space E.

Unfortunately, this result is not useful for many variational problems because of the following crucial drawback:

In infinite-dimensional Banach spaces, closed balls are not compact.

This is the main difficulty in calculus of variations. To overcome this difficulty the notion of *weak convergence* is introduced.

The basic result runs as follows:

(C) *In a reflexive Banach space, each bounded sequence has a weakly convergent subsequence.* (6.4.4).

If H is a Hilbert space, it is reflexive and the convergence condition (C) is a consequence of the Riesz theorem.

In a reflexive Banach space, the convergence principle (C) implies the following fundamental generalization to the classical Weierstrass theorem (W):

(W*) The minimum problem (6.25) has a solution provided the functional $F : M \to R$ is weakly sequentially lower semicontinuous on the closed ball M of the reflexive Banach space E.

More generally, this is also true if M is a nonempty bounded closed convex set in the reflexive Banach space E. These things will be discussed in Chapter 13.

Problems

1. Prove that a one-to-one continuous linear mapping of one Banach space onto another is a homeomorphism. In particular, if a one-to-one linear mapping A of a Banach space onto itself is continuous, prove that its inverse A^{-1} is automatically continuous.

2. Let $A : E_x \to E_y$ be a linear continuous operator, where E_x and E_y are Banach spaces over $\mathbb{R}$ ($\mathbb{C}$). If the inverse operator $A^{-1} : E_y \to E_x$ exist, then show that it is continuous.

3. Let $A : E_x \to E_y$ be a linear continuous operator, where E_x and E_y are Banach spaces over $\mathbb{R}$ ($\mathbb{C}$). Then show that the following two conditions are equivalent:

(i) Equation $Au = v$, $u \in E_x$, is well-posed that is, by definition, for each given $v \in E_y$, $Au = v$ has a unique solution u, which depends continuously on v.

(ii) For each $v \in E_y$, $Au = v$ has a solutions u, and $Aw = \theta$ implies $w = \theta$.

CHAPTER 7

CLOSED GRAPH THEOREM AND ITS CONSEQUENCES

7.1 Closed Graph Theorem

Bounded linear operators are discussed in chapter 4. But in 4.2.11 we have seen that differential operators defined on a normed linear space are not bounded. But they belong to a class of operator known as closed operators. In what follows, the relationship between closed and bounded linear operators and related concepts is discussed.

Let E_x and E_y be normed linear space, let $T : \mathcal{D}(T) \subset (E_x \to E_y)$ be a linear operator and let $\mathcal{D}(T)$ stand for domain of T.

7.1.1 Definition: graph

The graph of an operator T is denoted by $G(T)$ and defined by

$$G(T) = \{(x, y) : x \in \mathcal{D}(T),\ y = Tx\}. \tag{7.1}$$

7.1.2 Definition: closed linear operator

A linear operator $T : \mathcal{D}(T) \subseteq E_x \to E_y$ is said to be a closed operator if its graph $G(T)$ is closed in the normed space $E_x \times E_y$.

The two algebraic operations of the vector space $E_x \times E_y$ are defined as usual, that is:

For, $x_1, x_2 \in E_x$, $y_1, y_2 \in E_y$,

$$\left.\begin{array}{l} (x_1, y_1) + (x_2, y_2) = (x_1 + x_2, y_1 + y_2) \\[4pt] \text{For, } x \in E_x,\ y \in E_y, \end{array}\right\} \tag{7.2}$$

$$\alpha(x, y) = (\alpha x, \alpha y)$$

where α is a scalar and the norm on $E_x \times E_y$ is defined by

$$||(x, y)|| = ||x|| + ||y||. \tag{7.3}$$

Under what conditions will a closed linear operator be bounded? An answer is given by the following theorem which is known as the **closed graph theorem**.

7.1.3   Theorem: closed graph theorem

Let E_x and E_y be Banach spaces and let $T : \mathcal{D}(T) \to E_y$ be a closed linear operator where $\mathcal{D}(T) \subset E_x$. Then, if $\mathcal{D}(T)$ is closed in E_x, the operator T is bounded.

Proof: We first show that $E_x \times E_y$ with norm defined by (7.3) is complete. Let $\{z_n\}$ be Cauchy in $E_x \times E_y$, where $z_n = (x_n, y_n)$. Then, for every $\epsilon > 0$, there is an $N = N(\epsilon)$ such that

$$||z_n - z_m||_{E_x \times E_y} = ||(x_n, y_n) - (x_m, y_m)||$$
$$= ||x_n - x_m|| + ||y_n - y_m|| < \epsilon \tag{7.4}$$

$\forall\, n, m \geq N(\epsilon)$.

Hence, $\{x_n\}$ and $\{y_n\}$ are Cauchy sequences in E_x and E_y respectively. E_x being complete,

$$x_n \to x \text{ (say)} \in E_x \text{ as } n \to \infty.$$

Similarly, E_y being complete, $y_n \to y$ (say) $\in E_y$.

Hence, $\{z_n\} \to z = (x, y) \in E_x \times E_y$, as $n \to \infty$.

Hence, $E_x \times E_y$ is complete.

By assumption, $G(T)$ is closed in $E_x \times E_y$ and $\mathcal{D}(T)$ is closed in E_x. Hence, $\mathcal{D}(T)$ and $G(T)$ are complete.

We now consider the mapping $P : G(T) \to \mathcal{D}(T)$.

$$(x, Tx) \to x$$

We see that P is linear, because

$$P[(x_1, Tx_1) + (x_2, Tx_2)] = P[(x_1 + x_2, T(x_1 + x_2))]$$
$$= x_1 + x_2 = P(x_1, Tx_1) + P(x_2, Tx_2),$$

where $x_1, x_2 \in E_x$.

P is bounded, because

$$||P[(x, Tx)]|| = ||x|| \leq ||x|| + ||Tx|| = ||(x, Tx)||.$$

P is bijective, the inverse mapping

$$P^{-1} : \mathcal{D}(T) \to G(T) \quad \text{i.e. } x \to (x, Tx).$$

Since $G(T)$ and $\mathcal{D}(T)$ are complete, we can apply the bounded inverse theorem (theorem 7.3) and see that P^{-1} is bounded, say $||(x, Tx)|| \leq b||x||$, for some b and all $x \in \mathcal{D}(T)$.

Therefore, $||x|| + ||Tx|| \leq b||x||$

Hence, $\qquad ||Tx|| \leq ||x|| + ||Tx|| \leq b||x||$

for all $x \in \mathcal{D}(T)$. Thus, T is bounded.

7.1.4 Remark

$G(T)$ is closed if and only if $z \in \overline{G(T)}$ implies $z \in G(T)$. Now, $z \in \overline{G(T)}$ if and only if there are $z_n = (x_n, y_n) \in G(T)$ such that $z_n \to z$, hence $x_n \to x$, $Tx_n \to Tx$.

This leads to the following theorem, where an important criterion for an operator T to be closed is discovered.

7.1.5 Theorem (closed linear operator)

Let $T : \mathcal{D}(T) \subset E_x \to E_y$ be a linear operator, where E_x and E_y are normed linear spaces. Then T is closed if and only if it fulfils the following condition: If $x_n \to x$ where $x_n \in \mathcal{D}(T)$ and $Tx_n \to y$ together imply that $x \in \mathcal{D}(T)$ and $Tx = y$.

7.1.6 Remark

(i) If T is a continuous linear operator, then T is closed.

Since T is continuous, $x_n \to x$ in E_x implies that $Tx_n \to Tx$ in E_y.

(ii) A closed linear operator need not be continuous. For example, let $E_x = E_y = \mathbb{R}$ and $Tx = \frac{1}{x}$ for $x \neq 0$ and $T\theta = 0$. Here, if $x_n \to 0$, then $Tx_n \to 0$, showing that T is closed. But T is not continuous.

(iii) Given T is closed, and that two sequences, $\{x_n\}$ and $\{\bar{x}_n\}$, in the domain converge to the same limit x; if the corresponding sequences $\{Tx_n\}$ and $\{T\bar{x}_n\}$ both converge, then the latter have the same limit.

T being closed, $x_n \to x$ and $Tx_n \to y_1$ imply $x \in \mathcal{D}(T)$ and $Tx = y_1$.

Since $\{\bar{x}_n\} \to x$, T being closed, $\bar{x}_n \to x$ and $T\bar{x}_n \to y_2$ imply that $x \in \mathcal{D}(T)$ and $y_2 = Tx$.

Thus, $\{Tx_n\}$ and $\{T\bar{x}_n\}$ have the same limit.

7.1.7 Example (differential operator)

We refer to example 4.2.11.

We have seen that the operator A given by $Ax(t) = x'(t)$ where $E_x = C([0, 1])$ and $\mathcal{D}(A) \subset E_x$ is the subspace of functions having continuous derivatives, is not bounded. We show now that A is a closed operator. Let $x_n \in \mathcal{D}(A)$ be such that

$$x_n \to x \quad \text{and} \quad Ax_n = x'_n \to y.$$

Since convergence in the norm of $C([0, 1])$ is uniform convergence on $[0,1]$, from $x'_n \to y$ we have

$$\int_0^t y(\tau)d\tau = \int_0^t \lim_{n\to\infty} x_n'(\tau)d\tau = \lim_{n\to\infty} \int_0^t x_n'(\tau)d\tau = x(t) - x(0),$$

i.e., $x(t) = x(0) + \int_0^t y(\tau)d\tau.$

This shows that $x \in \mathcal{D}(A)$ and $x' = y$. The theorem 7.1.5 now implies that A is closed.

Note 7.1.1. Here, $\mathcal{D}(A)$ is not closed in $E_x = C([0,1])$, for otherwise A would be **bounded** by the **closed graph theorem**.

7.1.8   Theorem

Closedness does not imply boundedness of a linear operator. Conversely, boundedness does not imply closedness.

Proof: The first statement is shown to be true by examples 7.1.6 (ii) and 7.1.7. The second statement is demonstrated by the following example. Let $T : \mathcal{D}(T) \to \mathcal{D}(T) \subset E_x$ be the identity operator on $\mathcal{D}(T)$, where $\mathcal{D}(T)$ is a proper dense subspace of a normed linear space E_x. It is evident that T is linear and bounded. However, we show that T is not closed. Let us take $x \in E_x - \mathcal{D}(T)$ and a sequence $\{x_n\}$ in $\mathcal{D}(T)$ which converges to x.

7.1.9   Lemma (closed operator)

Since a broad class of operators in mathematical and theoretical physics are differential operators and hence unbounded operators, it is important to determine the domain and extensions of such operators. The following lemma will be an aid in investigation in this direction.

Let $T : \mathcal{D}(T) \to E_y$ be a bounded linear operator with domain $\mathcal{D}(T) \subseteq E_x$, where E_x and E_y are normed linear spaces. Then:

(a) If $\mathcal{D}(T)$ is a closed subset of E_x, then T is closed.

(b) IF T is closed and E_y is complete, then $\mathcal{D}(T)$ is a closed subset of E_x.

Proof: (a) If $\{x_n\}$ is in $\mathcal{D}(T)$ and converges, say, $x_n \to x$ and is such that $\{Tx_n\}$ also converges, then $x \in \overline{\mathcal{D}(T)} = \mathcal{D}(T)$, since $\mathcal{D}(T)$ is closed.

$Tx_n \to Tx$ since T is continuous.

Hence, T is closed by theorem 7.1.5.

(b) For $x \in \overline{\mathcal{D}(T)}$ there is a sequence $\{x_n\}$ in $\mathcal{D}(T)$ such that $x_n \to x$. Since T is bounded,

$$\|Tx_n - Tx_m\| = \|T(x_n - x_m)\| \le \|T\| \, \|x_n - x_m\|.$$

This shows that $\{Tx_n\}$ is Cauchy, $\{Tx_n\}$ converges, say, $Tx_n \to y \epsilon E_y$ because E_y is complete. Since T is closed, $x \in \mathcal{D}(T)$ by theorem 7.1.5 and $Tx = y$. Hence, $\mathcal{D}(T)$ is closed because $x \in \overline{\mathcal{D}(T)}$ was arbitrary.

7.1.10   Projection mapping

We next discuss the partition of a Banach space into two subspaces and the related question of the existence of operators, which are projections onto subspaces. These ideas help very much in the analysis of the structure of a linear transformation. We provide here an illustration of the use of closed graph theorem.

7.1.11   Definition: direct sum

A vector space E is said to be the direct sum of two of its subspaces M and N, i.e.,

$$E = M \oplus N \tag{7.5}$$

if every $x \in E$ has a unique decomposition

$$x = y + z, \tag{7.6}$$

with $y \in M$ and $z \in N$.

Thus, if $E = M \oplus N$ then $M \cap N = \{\theta\}$.

7.1.12   Definition: projection

A linear map P from a linear space E to itself is called a projection if $P^2 = P$.

If P is a projection then $(I - P)$ is also a projection.

For
$$(I - P)^2 = (I - P)(I - P) = I - P + P^2 - P$$
$$= I - P + P - P = I - P.$$

Moreover, $P(I - P) = P - P^2 = P - P = \mathbf{0}$.

7.1.13   Lemma

If a normed linear space E is the direct sum of two subspaces M and N and if P is a projection of E onto M, then

(i) $Px = x$ if and only if $x \in M$;

(ii) $Px = \theta$ if and only if $x \in N$.

Proof: If $x \in E$, then $x = y + z$ where $y \in M$ and $z \in N$.

Since P is a projection of E onto M

$$Px = y$$

if $Px = x$, $y = x$ and $z = 0$. Similarly, if $x \in M$, $Px = x$.

If $Px = \theta$ then $y = \theta$ and hence $x \in N$. Similarly, if $x \in N$, $Px = \theta$.

If $R(P)$ and $N(P)$ denote respectively the range space and null space of P, then

$$R(P) = N(I - P), \quad N(P) = R(I - P).$$

Therefore, $E = R(P) + N(I - P)$ and $R(P) \cap N(P) = \{\theta\}$ for every projection P defined on E.

The closedness and the continuity of a projection can be determined by the closedness of its range space and null space respectively.

7.1.14 Theorem

Let E be a normed linear space and $P : E \to E$ be a projection. Then P is a closed map, if and only if, the subspaces $R(P)$ and $N(P)$ are closed in E. In that case, P is in fact, continuous if E is a Banach space.

Proof: Let P be a closed map, $y_n \in R(P)$, $z_n \in N(P)$. Further, let $y_n \to y$, $z_n \to z$ in E. Then $Py_n = y_n \to y$, $Pz_n = \theta \to \theta$ in E so that $Py = y$ and $Pz = \theta$. Then $y \in R(P)$ and $z \in N(P)$. The above shows that $R(P)$ and $N(P)$ are closed subspaces in E.

Conversely, let $R(P)$ and $N(P)$ be closed in E. Let $x_n \to x$ and $Px_n \to y$ in E. Since $R(P)$ is closed and $Px_n \in R(P)$ we see that $y \in R(P)$. Also, since $N(P)$ is closed and $x_n - Px_n \in N(P)$, we see that $x - y \in N(P)$. Thus, $x - y = z$ with $z \in N(P)$. Thus, $x = y + z$, with $y \in R(P)$ and $z = x - y \in N(P)$. Hence, $Px = y$, showing that P is a closed mapping.

If E is a Banach space and $R(P)$ and $N(P)$ are closed, then by the closed graph theorem (theorem 7.1.3) the closed mapping P is in fact continuous.

7.1.15 Remark

(i) Let E be a normed linear space and M a subspace of E. Then there exists a projection P defined on E such that $R(P) = M$. Let $\{f_i\}$ be a (Hamel basis) for M. Let $\{f_i\}$ be extended to a basis $\{h_i\}$ such that $\{h_i\} = \{f_i\} \cup \{g_i\}$ for the space E. Let $N = \text{span } \{g_i\}$, then $E = M + N$ and $M \cap N = \{\Phi\}$. The above shows that there is a projection of E onto M along N.

(ii) A question that arises is that, given E is a normed linear space and M a closed subspace of E, does there exist a closed projection defined on E such that $R(P) = M$? By theorem 7.1.14, such a projection exists if and only if, there is a closed subspace N of E such that $E = M + N$ and $M \cap N = \{\Phi\}$. In such a case, N is called a **closed complement** of M in E.

7.2 Open Mapping Theorem

7.2.1 Definition: open mapping

Let E_x and E_y be two Banach spaces and T be a linear operator mapping $E_x \to E_y$. Then, $T : \mathcal{D}(T) \to E_y$ with $\mathcal{D}(T) \subseteq E_x$ is called an open mapping if for every open set in $\mathcal{D}(T)$ the image is an open set in E_y.

Note 7.2.1. A continuous mapping, $T : E_x \to E_y$ has the property that for every open set in E_y the inverse image is an open set. This does not imply that T maps open sets in E_x into open sets in E_y. For example, the mapping $\mathbb{R} \to \mathbb{R}$ given by $t \to \sin t$ is continuous but maps $]0, 2\pi[$ onto $[-1, 1]$.

7.2.2 Theorem: open mapping theorem

A bounded linear operator T mapping a Banach space E_x onto all of a Banach space E_y is an open mapping.

Before proving the above theorem, the following lemma will be proved.

7.2.3 Lemma (open unit ball)

A bounded linear operator T from a Banach space E_x onto all of a Banach space E_y has the property that the image $T(B(0,1))$ of the unit ball $B(0,1) \subseteq E_x$ contains an open ball about $0 \in E_y$.

Proof: The proof comprises three parts:

(i) The closure of the image of the open ball $B\left(0, \frac{1}{2}\right)$ contains an open ball B^*.

(ii) $\overline{T(B_n)}$ contains an open ball V_n about $0 \in E_y$ where $B_n = B(0, 2^{-n}) \subset E_x$.

(iii) $T(B(0,1))$ contains an open ball about $0 \in E_y$.

(i) Given a set $A \subseteq E_x$ we shall write [see figs 7.1(a) and 7.1(b)]

$$\alpha A = \{x \in E_x : x = \alpha a, \ \alpha \text{ a scalar}, \ a \in A\} \tag{7.7}$$

$$A + g = \{x \in E_x : x = a + g, \ a \in A, \ g \in E_x\} \tag{7.8}$$

and similarly for subsets of E_y.

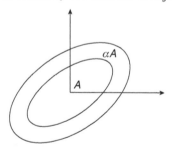 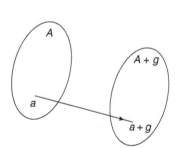

Fig. 7.1(a) Illustration of formula (7.7)   **Fig. 7.1(b) Illustration of formula (7.8)**

We consider the open ball $B_1 = B\left(0, \frac{1}{2}\right) \subseteq E_x$.

Any fixed $x \in E_x$ is in kB_1 with real k sufficiently large ($k > 2\|x\|$). Hence, $E_x = \bigcup_{k=1}^{\infty} kB_1$.

Since T is surjective and linear,

$$E_y = T(E_x) = T\left(\bigcup_{k=1}^{\infty} kB_1\right) = \bigcup_{k=1}^{\infty} kT(B_1) = \bigcup_{k=1}^{\infty} \overline{kT(B_1)}. \tag{7.9}$$

Since E_y is complete and $\bigcup\limits_{k=1}^{\infty} kT(B_1)$ is equal to E_y, (7.9) holds. Since E_y is complete, it is a set of the second category by Baire's category theorem (theorem 1.4.20). Hence $E_y = \bigcup\limits_{k=1}^{\infty} \overline{kT(B_1)}$ cannot be expressed as the countable union of nowhere dense sets. Hence, at least one ball $\overline{kT(B_1)}$ must contain an open ball. This means that $\overline{T(B_1)}$ must contain an open ball $B^* = B(y_0, \epsilon) \subset \overline{T(B_1)}$. It therefore follows from (7.8) that

$$B^* - y_0 = B(0, \epsilon) \subset \overline{T(B_1)} - y_0. \tag{7.10}$$

We show now $B^* - y_0 = \overline{T(B(0,1))}$.

This we do by showing that

$$\overline{T(B_1)} - y_0 \subseteq \overline{T(B(0,1))}. \tag{7.11}$$

Let $y \in \overline{T(B_1)} - y_0$. Then $y + y_0 \in \overline{T(B_1)}$ and we remember that $y_0 \in \overline{T(B_1)}$ too.

Since $y + y_0 \in \overline{T(B_1)}$, there exists $w_n \in B_1$ such that $u_n = Tw_n \subseteq T(B_1)$ such that $u_n \to y + y_0$. Similarly, we can find $z_n \in B_1$ such that

$$v_n \to Tz_n \subseteq T(B_1) \quad \text{and} \quad v_n \to y_0.$$

Since $w_n, z_n \in B_1$ and B_1 has radius $\frac{1}{2}$, we obtain $\|w_n - z_n\| \leq \|w_n\| + \|z_n\| < 1$.

Hence $\quad w_n - z_n \in B(0, 1)$.

Now, $\quad T(w_n - z_n) = Tw_n - Tz_n \to y$.

Thus, $\quad y \in \overline{T(B(0,1))}$. Thus if $y \in \overline{T(B_1)} - y_0$,

then $y \in \overline{T(B(0,1))}$ and y is arbitrary, it follows that

$$\overline{T(B_1)} - y_0 \subseteq \overline{T(B(0,1))}.$$

Hence, (7.11) is proved. Using (7.9) we see that (7.8) yields

$$B^* - y_0 \subseteq B(0, \epsilon) \subseteq \overline{T(B_1)} - y_0 \subseteq \overline{TB(0,1)}. \tag{7.12}$$

Let $B_n = B(0, 2^{-n}) \subseteq E_x$. Since T is linear, $\overline{T(B_n)} = 2^{-n}\overline{T(B(0,1))}$.

If follows from (7.12), $V_n = 2^{-n}B(0, \epsilon) = B(0, \epsilon/2^n) \subseteq \overline{T(B_n)}$. $\tag{7.13}$

(iii) We finally prove that $V_1 = B\left(0, \dfrac{1}{2}\epsilon\right) \subseteq T(B(0,1))$.

For that, let $y \in V_1$, (7.13) yields $V_1 = B\left(0, \dfrac{1}{2}\epsilon\right) \subseteq \overline{T(B_1)}$. Hence $y \in \overline{T(B_1)}$. Since $\overline{T(B_1)}$ is a closed set, there is a $v \in T(B_1)$ close to y such that,

$$||v - y|| < \frac{\epsilon}{4}.$$

Now, $v \in T(B_1)$ implies that there is a $x_1 \in B_1$ such that $v = Tx_1$.

Hence $\quad ||y - Tx_1|| < \frac{\epsilon}{4}.$

From the above and (7.13), putting $n = 2$, we see that

$$y - Tx_1 \in V_2 = \overline{T(B_2)}.$$

We can again find a $x_2 \in B_2$ such that

$$||(y - Tx_1) - Tx_2|| < \frac{\epsilon}{8}.$$

Hence $\quad y - Tx_1 - Tx_2 \in V_3 \subseteq \overline{T(B_3)}$ and so on.

Proceeding in the above manner we get at the nth stage, an $x_n \in B_n$ such that

$$\left\| y - \sum_{k=1}^{n} Tx_k \right\| < \frac{\epsilon}{2^{n+1}}, \quad (n = 1, 2, \ldots) \tag{7.14}$$

writing $z_n = x_1 + x_2 + \cdots + x_n$, and since $x_k \in B_k$, so that $||x_k|| < 1/2^k$ we have, for $n > m$,

$$||z_n - z_m|| \leq \sum_{k=m+1}^{n} ||x_k|| < \sum_{k=m+1}^{n} \frac{1}{2^k}$$

$$= \frac{1}{2^{m+1}} \left(1 + \frac{1}{2} + \frac{1}{2^2} + \cdots + \frac{1}{2^{n-m-1}} \right) \rightarrow 0 \text{ as } m \rightarrow \infty.$$

Hence, $\{z_n\}$ is a Cauchy sequence and E_x being complete $\{z_n\} \rightarrow z \in E_x$. Also $z \in B(0, 1)$ since $B(0, 1)$ has radius 1 and

$$\sum_{k=1}^{\infty} ||x_k|| < \sum_{k=1}^{\infty} \frac{1}{2^k} = 1. \tag{7.15}$$

Since T is continuous, $Tz_n \rightarrow Tz$ and (7.14) shows that $Tz = y$. Hence $y \in T(B(0, 1))$.

Proof of theorem 7.2.2

We have to prove that for every open set $A \subseteq E_x$ the image $T(A)$ is open in E_y.

Let $y = Tz \in T(A)$, where $z \in A$. Since A is open, it contains an open ball with centre z. Hence $A - z$ contains an open ball with centre 0. Let the radius of the ball be r and $k = \frac{1}{r}$. Then $k(A - z)$ has an open ball $B(0, 1)$. Then, by lemma 7.2.3. $T(k(A - z)) = k[T(A) - Tz]$ contains an open ball about 0 and therefore $T(A) - Tz$ contains a ball about 0. Hence, $T(A)$ contains an open ball about $Tz = y$. Since $y \in T(A)$ is arbitrary, $T(A)$ is open.

7.3   Bounded Inverse Theorem

Let E_x and E_y be Banach spaces. Hence, if the linear operator T is bijective (i.e., injective and surjective) then T^{-1} is continuous and thus bounded. Since T is open, by the open mapping theorem 7.2.2, given $A \subseteq E_x$ is open, $T(A)$ is open. Again T is bijective, i.e., injective and surjective. Therefore, T^{-1} exists. For every open set A in the range of T^{-1}, the domain of T^{-1} contains an open set A. Hence by theorem 1.6.4, T^{-1} is continuous and linear and hence bounded (4.2.4).

7.3.1   Remark

(i) The inverse of a bijective closed mapping from a complete metric space to a complete metric space is closed.

(ii) The inverse of a bijective, linear, continuous mapping from a Banach space to a Banach space is linear and continuous.

(iii) If the normed linear spaces are not complete then the above ((i) and (ii)) may not be true.

Let $E_x = c_{00}$ with $\| \ \|_1$ and $E_y = c_{00}$ with $\| \cdot \|_\infty$. If $P(x) = x$ for $x \in E_x$, then $P : E_x \to E_y$ is bijective, linear and continuous. But P^{-1} is not continuous since, for $x_n = (\underbrace{1,1,1,\ldots,1}_{n},0,0,0\ldots)$ we have $\|x_n\|_\infty = 1$ and $\|P^{-1}(x_n)\| = \|x_n\|_1 = n$ for all $n = 1,2,\ldots$.

7.3.2   Definition: stronger norm, comparable norm

Given E a normed linear space, $\| \cdot \|$ on E is said to be stronger than the norm $\| \cdot \|'$ if for every $x \in E$ and every $\epsilon > 0$, there is some $\delta > 0$ such $B_{\|\cdot\|}(x,\delta) \subseteq B_{\|\cdot\|'}(x,\epsilon)$. Here, $B_{\|\cdot\|}(x,\delta)$ denotes an open ball in E w.r.t. $\| \cdot \|$. Similarly $B_{\|\cdot\|'}(x,\epsilon)$ denotes an open ball in E w.r.t. $\| \cdot \|'$. In other words, $\| \cdot \|$ is **stronger** than $\| \cdot \|'$ if and only if every open subset of E with respect to $\| \cdot \|'$ is also an open subset with respect to $\| \cdot \|$.

The norms $\| \cdot \|$ and $\| \cdot \|'$ are said to be **comparable** if one of them is stronger than the other. For definition of two equivalent norms, $\| \cdot \|$ and $\| \cdot \|'$, see 2.3.5.

7.3.3   Theorem

Let $\| \cdot \|$ and $\| \cdot \|'$ be norms on a linear space E. Then the norm $\| \cdot \|$ is stronger than $\| \cdot \|'$ if and only if there is some $\alpha > 0$ such that $\|x\|' \leq \alpha\|x\|$ for all $x \in E$.

Proof: Let $\| \cdot \|$ be stronger than $\| \cdot \|'$, then there is some $r > 0$ such that

$$\theta \in \{x \in E : \|x\| < r\} \subset \{x \in E : \|x\|' < 1\}.$$

Let $\theta \neq x \in E$ and $\epsilon > 0$. Since $\left\| \dfrac{rx}{(1+\epsilon)\|x\|} \right\| < r,$

then $\left\|\dfrac{rx}{(1+\epsilon)\|x\|}\right\|' < 1$, i.e., $\|x\|' < \dfrac{(1+\epsilon)}{r}\|x\|.$

Since $\epsilon > 0$ is arbitrary, $\|x\|' \le \dfrac{1}{r}\|x\|.$

or $\qquad \|x\|' \le \alpha\|x\|$ with $\alpha = \dfrac{1}{r}.$

Conversely, let $\|x\|' \le \alpha\|x\|$ for all $x \in E$.

Let $\{x_n\}$ be a sequence in E such that $\|x_n - x\| \to 0$.

Since $\|x_n - x\|' \le \alpha\|x_n - x\| \to 0$.

Hence, the $\|\cdot\|$ is stronger than the norm $\|\cdot\|'$.

7.3.4 Two-norm theorem

Let E be a Banach space in the norm $\|\cdot\|$. Then a norm $\|\cdot\|'$ of the linear space E is equivalent to the norm $\|\cdot\|$ if and only if, E is also a Banach space in the norm $\|\cdot\|'$ and the norm $\|\cdot\|'$ is comparable to the norm $\|\cdot\|$.

Proof: If the norms $\|\cdot\|$ and $\|\cdot\|'$ are equivalent, then clearly they are comparable.

Therefore, $\alpha_1\|x\| \le \|x\|' \le \alpha_2\|x\|$, $\alpha_1, \alpha_2 \ge 0$ for all $x \in E$. Let $\{x_n\}$ be a Cauchy sequence in the Banach space E with norm $\|\cdot\|$.

Then, if $x_n \to x$ in E with norm $\|\cdot\|$,

$$\|x_n - x\|' \le \alpha_2\|x_n - x\|$$

and hence $\{x_n\} \to x$ in the norm $\|\cdot\|'$.

Therefore, E is a Banach space with respect to the norm $\|\cdot\|'$.

Conversely, let us suppose that E is a Banach space in the norm $\|\cdot\|'$ and that the norm $\|\cdot\|'$ is comparable to the norm $\|\cdot\|$.

Let us suppose without loss of generality that $\|\cdot\|$ is stronger than $\|\cdot\|'$. Then, by theorem 7.3.3, we can find a $\alpha > 0$ such that $\|x\|' \le \alpha\|x\|$, for all $x \in E$. Let $\overline{E}$ denote the linear space with the norm $\|\cdot\|'$ and let us consider the identity map $I : E \to \overline{E}$. Clearly I is bijective, linear and continuous. By the bounded inverse theorem 7.3, $I^{-1} : \overline{E} \to E$ is also continuous, that is, $\|x\| \le \beta\|x\|'$ for all $x \in E$ and some $\beta > 0$.

Letting $\alpha' = \dfrac{1}{\beta}$ we have,

$$\alpha'\|x\| \le \|x\|' \le \alpha\|x\|.$$

Therefore, it follows from 2.3.5 $\|\cdot\|$ and $\|\cdot\|'$ are equivalent.

7.3.5 Remark

(i) The result above shows that two comparable complete norms on a normed linear space are equivalent.

Problem [7.1, 7.2 and 7.3]

1. Given that E is a Banach space, $\mathcal{D}(T) \subseteq E$ is closed, and the linear operator T is bounded, show that T is closed.

2. If T is a linear transformation from a Banach E_x into a Banach space E_y, find a necessary and sufficient condition that a subspace G of $E_x \times E_y$ is a graph of T.

3. Given E_x, E_y and E_z, three normed linear spaces respectively:

   (i) If $F : E_x \to E_y$ is continuous and $G : E_y \to E_z$ is closed, then show that $G \cdot F : E_x \to E_z$ is closed.

   (ii) If $F : E_x \to E_y$ is continuous and $G : E_x \to E_y$ is closed, then show that $F + G : E_x \to E_y$ is closed.

4. Let E_x and E_y be normed linear spaces and $T : E_x \to E_y$ be linear. Let $\hat{T} : E_x/N(T) \to E_y$ be defined by $\hat{T}(x + N(T)) = T(x), x \in E_x$. Show that T is closed if and only if $N(T)$ is closed in E_x and $\hat{T}$ is closed.

5. Let E_x and E_y be normed linear spaces and $A : E_x \to E_y$ be linear, such that the range $R(A)$ of A is finite dimensional. Then show that A is continuous if and only if the null space $N(A)$ of A is closed in E_x.

   In particular, show that a linear functional f on E_x is continuous if and only if $N(A)$ is closed in E_x.

6. Let E_x be a normed linear space and $f : E_x \to R$ be linear. Then show that f is closed if and only if f is continuous.

   (Hint: Problems 7.4 and 7.5).

7. Let E_x and E_y be Banach spaces and let $T : E_x \to E_y$ be a closed linear operator, then show that

   (i) if C is compact in E_x, $T(C)$ is closed in E_y and

   (ii) if K is compact in E_y, $T^{-1}(K)$ is closed in E_x.

8. Give an example of a discontinuous operator A from a Banach space E_x to a normed linear space E_y, such that A has a closed graph.

9. Show that the null space $N(A)$ of a closed linear operator $A : E_x \to E_y, E_x, E_y$ being normed linear spaces, is a closed subspace of E_x.

10. Let E_x and E_y be normed linear spaces. If $A_1 : E_x \to E_y$ is a closed linear operator and $A_2 \in (E_x \to E_y)$, show that $A_1 + A_2$ is a closed linear operator.

11. Show that $A : \mathbb{R}^2 \to \mathbb{R}$, defined by $(x_1, x_2) \to \{x_1\}$ is open. Is the mapping $\mathbb{R}^2 \to \mathbb{R}^2$ given by $(x_1, x_2) \to (x_1, 0)$ an open mapping?

12. Let $A : c_{00} \to c_{00}$ be defined by,

$$y = Ax = \left(\xi_1, \frac{1}{2} \xi_2, \frac{1}{3} \xi_3, \dots \right)^T$$

where $x = \{\xi_i\}$. Show that A is linear and bounded but A^{-1} is unbounded.

13. Let E_x and E_y be Banach spaces and $A : E_x \to E_y$ be an injective bounded linear operator. Show that $A^{-1} : \mathcal{R}(A) \to E_x$ is bounded if and only if, $\mathcal{R}(A)$ is closed in E_y.

14. Let $A : E_x \to E_y$, be a bounded linear operator where E_x and E_y are Banach spaces. If A is bijective, show that there are positive real numbers α and β such that $\alpha \|x\| \le \|Ax\| \le \beta \|x\|$ for all $x \in E_x$.

15. Prove that the closed graph theorem can be deduced from the open mapping theorem.

    (Hint: $E_x \times E_y$ is a Banach space and the map $(x, A(x)) \to x \in E_x$ is one-to-one and onto, $A : E_x \to E_y$).

16. Let E_x and E_y be Banach spaces and $A \in (E_x \to E_y)$ be surjective. Let $y_n \to y$ in E_y. If $Ax = y$, show that there is a sequence $\{x_n\}$ in E_x, such that $Ax_n = y_n$ for each n and $x_n \to x$ in E_x.

17. Show that the uniform bounded principle for functionals [see 4.5.5] can be deduced from the closed graph theorem.

18. Let E_x and E_y be Banach spaces and E_z be a normed linear space. Let $A_1 \in (E_x \to E_z)$ and $A_2 \in (E_y \to E_z)$. Suppose that for every $x \in E_x$ there is a unique $y \in E_y$ such that $Bx = A_2 y$, and define $A_1 x = y$. Then show that $A_1 \in (E_x \to E_y)$.

19. Let E_x and E_y be Banach spaces and $A \in (E_x \to E_y)$. Show that $R(A)$ is linearly homeomorphic to $E_x / N(A)$ if and only if $R(A)$ is closed in E_y.

    (Hint: Two metric spaces are said to be **homeomorphic** to each other if there is a **homeomorphism** from E_x onto E_y. Use theorem 7.1.3.)

20. Let E_x denote the sequence space $l_p (1 \le p \le \infty)$. Let $\| \cdot \|'$ be a complete norm on E_x such that if $\|x_n - x\|' \to 0$ then $x_j^n \to x_j$ for every $j = 1, 2, \dots$. Show that $\| \cdot \|$ is equivalent to the usual norm $\| \ \|_p$ on E_x.

21. Let $\| \cdot \|'$ be a complete norm on $C([a, b])$ such that if $\|x_n - x\|' \to 0$ then $x_n(t) \to x(t)$ for every $t \in [a, b]$. Show that $\| \cdot \|'$ is equivalent to any norm on $C([a, b])$.

22. Give an example of a bounded linear operator mapping $C^1([0, 1]) \to C^1([0, 1])$ which is not a closed operator.

7.4   Applications of the Open Mapping Theorem

Recall the definition of a Schauder basis in 4.8.3. Let E be a normed linear space. A denumerable subset $\{e_1, e_2, \ldots\}$ of E is called a Schauder basis for E if $\|e_n\| = 1$ for each n and if for every $x \in E$, there are unique scalars $\alpha_1, \alpha_2, \ldots \alpha_n \cdots$ in $\mathbb{R}$ ($\mathbb{C}$) such that $x = \sum\limits_{i=1}^{\infty} \alpha_i e_i$.

In case $\{e_1, e_2, \ldots\}$ is a Schauder basis for E then for $n = 1, 2, \ldots$, let us define functionals $f_n : E \to \mathbb{R}(\mathbb{C})$ by $f_n(x) = \alpha_n(x)$ for $x = \sum\limits_{n=1}^{\infty} \alpha_n e_n \in E$. f_n is well-defined and linear on E. It is called the nth coefficient functional on E.

7.4.1   Theorem

The functionals $f_n = \alpha_n(x)$ for a given $x \in E$ are **bounded**.

We consider the vector space E' of all sequence $(\alpha_1, \alpha_2, \ldots, \alpha_n, \ldots)$ for which $\sum\limits_{n=1}^{\infty} \alpha_n e_n$ converges in E. The norm $\|y\| = \sup\limits_{n} \left\| \sum\limits_{i=1}^{n} \alpha_i e_i \right\|$ converts E' into a normed vector space. We show that E' is a Banach space. Let $y_m = \{\alpha_n^{(m)}\}$, $m = 1, 2, \ldots$ be a Cauchy sequence in E'. Let $\epsilon > 0$, then there is a N, such that $m, p > N$ implies

$$\|y_m - y_p\| = \sup_{n} \left\| \sum_{i=1}^{n} (\alpha_i^{(m)} - \alpha_i^{(p)}) e_i \right\| < \epsilon.$$

But this implies $\|\alpha_n^{(m)} - \alpha_n^{(p)}\| < 2\epsilon$ for every n. Hence for every n, $\lim\limits_{m \to \infty} \alpha_n^{(m)} = \alpha_n$ exists. It remains to be shown that

$$y = (\alpha_1, \alpha_2, \ldots, \alpha_n, \ldots) \in E' \quad \text{and} \quad \lim_{n \to \infty} y_n = y.$$

Now, $y_m = \{\alpha_1^{(m)}, \alpha_2^{(m)}, \ldots \alpha_n^{(m)} \ldots\} \in E'$.

Since in E' convergence implies coordinatewise convergence and since

$$\lim_{n \to \infty} \alpha_n^m = \alpha_n, \quad \lim_{n \to \infty} y_m = (\alpha_1, \alpha_2, \ldots \alpha_n, \ldots).$$

Now, $\|y - y_m\| = \sup\limits_{n} \left\| \sum\limits_{i=1}^{n} (\alpha_i - \alpha_i^m) e_i \right\| \to 0$ as $m \to \infty$.

Hence $y = (\alpha_1, \alpha_2, \ldots \alpha_n, \ldots) \in E'$ and $\{y_m\}$ being Cauchy, E' is a Banach space.

Let us next consider a mapping $P : E' \to E$ for which $y = (\alpha_1, \alpha_2, \ldots) \in E'$ such that $P_y = \sum\limits_{n=1}^{\infty} \alpha_n e_n \in E$. If $z = (\beta_1, \beta_2, \ldots) \in E'$ such that

$Pz = \sum_{n=1}^{\infty} \beta_n e_n \in E$, then $P(y+z) = \sum_{n=1}^{\infty} (\alpha_n + \beta_n) e_n = Py + Pz$ showing P is linear. Since $\{e_i\}$ are linearly independent $Py = \theta \Longleftrightarrow y = \theta$. Hence P is one-to-one. Now $\{e_n\}$ being a Schauder basis in E, every element in E is representable in the form $\sum_{n=1}^{\infty} r_n e_n$ where $(r_1, r, \ldots r_n \ldots) \in E'$. Hence P is onto. P is bounded since

$$\sup_n \left\| \sum_{i=1}^{n} \alpha_i e_i \right\| \geq \lim_n \left\| \sum_{i=1}^{n} \alpha_i e_i \right\|.$$

By the open mapping theorem, the inverse P^{-1} of P is bounded. Now,

$$|\alpha_n(x)| = |\alpha_n| = \|\alpha_n e_n\| = \left\| \sum_{i=1}^{n} \alpha_i e_i - \sum_{i=1}^{n-1} \alpha_i e_i \right\| \leq 2 \sup_n \left\| \sum_{i=1}^{n} \alpha_i e_i \right\|$$

$$= 2\|y\| = 2\|P^{-1}x\| \leq 2\|P^{-1}\| \, \|x\|.$$

This proves the boundedness of $\alpha_n(x)$.

CHAPTER 8

COMPACT OPERATORS ON NORMED LINEAR SPACES

This chapter focusses on a natural and useful generalisation of bounded linear operators having a finite dimensional range. The concept of a compact linear operator is introduced in section 8.1. Compact linear operators often appear in applications. They play a crucial role in the theory of integral equations and in various problems of mathematical physics. The relation of compactness with weak convergence and reflexivity is highlighted. The spectral properties of a compact linear operator are studied in section 8.2. The notion of the Fredholm alternative and the relevant theorems are provided in section 8.3. Section 8.4 shows how to construct a finite rank approximations of a compact operator. A reduction of the finite rank problem to a finite dimensional problem is also given.

8.1 Compact Linear Operators

8.1.1 Definition: compact linear operator

A linear operator mapping a normed linear space E_x onto a normed linear space E_y is said to be **compact** if it maps a bounded set of (E_x) into a compact set of (E_y).

8.1.2 Remark

(i) A linear map A from a normed linear space E_x into a normed linear space E_y is continuous if and only if it sends the open unit ball $B(0,1)$ in

282

E_x to a bounded subset of E_y.

(ii) A compact linear operator A is stronger than a bounded linear operator in the sense that $A(B(0,1))$ is a compact subset of E_y given $B(0,1)$ an open unit ball.

(iii) A **compact** linear operator is also known as a **completely continuous operator** in view of a result we shall prove in 8.1.14(a).

8.1.3 Remark

(i) A compact linear operator is **continuous**, but the converse is not always true. For example, if E_x is an infinite dimensional normed linear space, then the identity map I on E_x is clearly linear and continuous, but it is not compact. See example 1.6.16.

8.1.4 Lemma

Let E_x and E_y be normed linear spaces.

Further, A_1, A_2 map $B(0,1)$ into $A_1(B(0,1))$ and $A_2(B(0,1))$ which are respectively compact.

Then (i) $(A_1 + A_2)(B(0,1))$ is compact.

(ii) $A_1 A_2(B(0,1))$ is compact.

Proof: Let $||x_n|| \leq 1$ and $\{x_n\}$ is a Cauchy sequence. Since $A_1(B(0,1))$ is compact and is hence sequentially compact (1.6), $\{A_1 x_n\}$ contains a convergent subsequence $\{A_1 x_{n_p}\}$.

A_2 being compact, we can similarly argue that $\{A_2 x_{n_q}\}$ is convergent.

Let $\{x_{n_r}\}$ be a subsequence of both $\{x_{n_p}\}$ and $\{x_{n_q}\}$.

Then $\{x_{n_r}\}$ is Cauchy.

Moreover, $(A_1 + A_2)(x_{n_r})$ is convergent.

Hence $(A_1 + A_2)$ is compact.

(ii) Let $||x_n|| \leq 1$ and $\{x_n\}$ is convergent in E_x. $\{A_2 x_n\}$ being compact and sequentially compact (1.6.17) $\{A_2 x_n\}$ contains a convergent subsequence $\{A_2 x_{n_p}\} \subseteq E_y$. Hence $\{A_2 x_{n_p}\}$ is bounded. A_1 being compact $(A_1 A_2)(x_{n_p})$ is a compact sequence. Hence, $A_1 A_2$ maps bounded sequence $\{x_n\}$ into a compact sequence. Hence $A_1 A_2$ is compact.

8.1.5 Examples

(1) Let $E_x = E_y = C([0,1])$ and let

$$Ax = y(t) = \int_0^1 K(t,s)x(s)ds.$$

The kernel $K(t,s)$ is continuous on $0 \leq t, s \leq 1$. We want to show that A is compact. Let $\{x(t)\}$ be a bounded set of functions of $C([0,1]), ||x|| \leq \alpha$, let $K(t,s)$ satisfy $L = \max_{t,s} |K(t,s)|$. Then $y(t)$ satisfies

$$|y(t)| \leq \left| \int_0^1 |K(t,s)|x(s)ds \right| \leq L\alpha,$$

showing that $y(t)$ is uniformly bounded. Furthermore, we show that the functions $y(t)$ are uniformly continuous.

Given $\epsilon > 0$, we can find a $\delta > 0$ on account of the uniform continuity of $K(t, s)$, such that

$$|K(t_2, s) - K(t_1, s)| < \epsilon/\alpha$$

for $|t_2 - t_1| < \delta$ and every $s \in [0, 1]$.

Therefore, $\quad |y(t_2) - y(t_1)| = \left| \int_0^1 (K(t_2, x) - K(t_1, s))x(s)ds \right|$

$$< \max_s |K(t_2, s) - K(t_1, s)| \cdot ||x(s)||$$

$$< \frac{\epsilon}{\alpha} \cdot \alpha = \epsilon \text{ wherever } |t_2 - t_1| < \delta$$

for all $y(t)$. Hence $\{y(t)\}$ is uniformly continuous.

By Arzela-Ascoli's theorem (1.6.23) the set of functions $\{y(t)\}$ is compact in the sense of the metric of the space $C([0, 1])$. Hence the operator A is compact.

8.1.6   *Lemma*

If a sequence $\{x_n\}$ is *weakly convergent* to x_0 and compact, then the sequence is strongly convergent to x_0.

Proof: Let us assume by way of contradiction that $\{x_n\}$ is not strongly convergent. Then, given $\epsilon > 0$, we can find an increasing sequence of indices $n_1, n_2, \ldots, n_k \ldots$ such that $||x_{n_k} - x_0|| \geq \epsilon$. Since the sequence $\{x_{n_i}\}$ is compact, it contains a convergent subsequence $\{x_{n_{i_j}}\}$. Thus, let $\{x_{n_{i_j}}\}$ converge strongly to u_0.

Moreover, $x_{n_{i_j}} \xrightarrow[j \to \infty]{w} u_0$. Since at the same time $x_{n_{i_j}} \xrightarrow[j \to \infty]{w} x_0$, $u_0 = x_0$. Thus on one hand, $||x_{n_{i_j}} - x_0|| \geq \epsilon$, whereas on the other $||x_{n_{i_j}} - x_0|| \to 0$, a contradiction to our hypothesis. Hence the lemma.

8.1.7   *Theorem*

A compact operator A maps a weakly convergent sequence into a strongly convergent sequence.

Let the sequence $\{x_n\}$ converge weakly to x_0. Then the norms of the elements of this sequence are bounded (theorem 6.3.3). Thus A maps a bounded sequence $\{x_n\}$ into a compact sequence $\{Ax_n\}$. Let $y_n = Ax_n$. Since a compact linear operator is bounded, and since $\{x_n\}$ is weakly convergent to x_0, by theorem 6.3.17 Ax_n converges weakly to $Ax_0 = y_0$. Given A is compact and $\{x_n\}$ is bounded, $\{y_n\}$ where $\{y_n\} = \{Ax_n\}$ is compact. Now, as because $\{y_n\} \xrightarrow[n \to \infty]{w} y_0$ and $\{y_n\}$ is compact by lemma 8.1.6.

$$y_n \to y_0.$$

8.1.8   Theorem

Let A be a linear compact operator mapping an infinite dimensional space E into itself and let B be an arbitrary bounded linear operator acting in the same space. Then AB and BA are compact.

Proof: See lemma 8.1.4.

Note 8.1.1. In case, A is a compact linear operator mapping a linear space $E \to E$ and admits of an inverse A^{-1}, then $A \cdot A^{-1} = I$. Since I is not compact, A^{-1} is not bounded.

8.1.9   Theorem

If a sequence $\{A_n\}$ of compact linear operators mapping a normed linear space E_x into a Banach space E_y converges strongly to the operator A, that is if $||A_n - A|| \to 0$, then A is also a compact operator.

Proof: Let M be a bounded set in E_x and α a constant such that $||x|| \le \alpha$ for every $x \in M$. For given $\epsilon > 0$, there is an index n_0 such that $||A_n - A|| < \epsilon/\alpha$, for $n \ge n_0(\epsilon)$. Let $A(M) = L$ and $A_{n_0}(M) = N$. We assert that the set $A_{n_0}(M) = N$ is a finite ϵ-net of L. Let us take for every $y \in L$ one of the pre-images $x \in M$ and put $y_0 = A_{n_0} x \in N$, to receive $||y - y_0|| = ||Ax - A_{n_0} x|| \le ||A - A_{n_0}|| \, ||x|| < \epsilon/\alpha \cdot \alpha = \epsilon$. On the other hand, since A_{n_0} is compact and M is bounded, the set N is compact. It follows then L for every $\epsilon > 0$ has a compact ϵ-net and is therefore itself compact (theorem 1.6.18). Thus, the operator A maps an arbitrary bounded set into a set whose closure is compact set and hence the operator A is compact.

8.1.10   Example

1. If $E_x = E_y = L_2([0,1])$, then the operator, $Ax - y = \int_0^1 K(t,s)x(s)ds$ with $\int_0^1 \int_0^1 K^2(t,s)dtds < \infty$, is compact.

Proof: Let us assume first that $K(t,s)$ is a continuous kernel. Let M be a bounded set in $L_2([0,1])$ and let

$$\int_0^1 x^2(t)dt \le \alpha^2 \text{ for all } x(t) \in M.$$

Consider the set of functions

$$y(t) = \int_0^1 K(t,s)x(s)dx, \quad x(s) \in M.$$

It is to be shown that the functions $y(t)$ are uniformly bounded and equicontinuous (theorem 1.6.22). This implies the compactness of the set $\{y(t)\}$ in the sense of uniform convergence and also in the sense of convergence in the mean square. By Cauchy-Bunyakovsky-Schwartz inequality (1.4.3) we have

$$|y(t)| = \left| \int_0^1 K(t,s)x(s)ds \right| \leq \left(\int_0^1 K^2(t,s)ds \right)^{\frac{1}{2}} \left(\int_0^1 x^2(s)ds \right)^{\frac{1}{2}} \leq L\alpha$$

where $L = \max_{t,s} |K(t,s)|$. Consequently the functions $y(t)$ are uniformly bounded. Furthermore,

$$|y(t_2) - y(t_1)| \leq \left(\int_0^1 [K(t_2,s) - K(t_1,s)]^2 ds \right)^{\frac{1}{2}} \left(\int_0^1 x^2(s)ds \right)^{\frac{1}{2}} < \epsilon$$

for $|t_2 - t_1| < \delta$ where δ is chosen such that

$$|K(t_2,s) - K(t_1,s)| < \epsilon/\alpha$$

for $|t_2 - t_1| < \delta$.

The estimate $|y(t_2) - y(t_1)| < \epsilon$ does not depend on the positions of t_1, t_2 on $[0,1]$ and also does not depend on the choice of $y(t) \in M$. Hence, the functions $y(t)$ are equicontinuous.

Thus, in the case of a continuous kernel the operator is compact.

Next let us assume $K(t,s)$ to be an arbitrary square-integrable kernel. We select a sequence of continuous kernels $\{K_n(t,s)\}$ which converges in the mean to $K(t,s)$, i.e., a sequence such that

$$\int_0^1 \int_0^1 (K(t,s) - K_n(t,s))^2 dt ds \to 0 \text{ as } n \to \infty$$

Set $\quad A_n x = \int_0^1 K_n(t,s)x(s)ds$.

Then,

$$\|Ax - A_n x\| = \left\{ \int_0^1 \left[\int_0^1 K(t,s)x(s)ds - \int_0^1 K_n(t,s)x(s)dx \right]^2 dt \right\}^{\frac{1}{2}}$$

$$\leq \left\{ \int_0^1 \left[\int_0^1 [K(t,s) - K_n(t,s)]^2 ds \int_0^1 x^2(s)ds \right] dt \right\}^{\frac{1}{2}}$$

$$= \left\{ \int_0^1 [K(t,s) - K_n(t,s)]^2 ds dt \right\}^{\frac{1}{2}} \|x\|$$

Hence,

$$\|A - A_n\| = \sup_{x \neq 0} \frac{\|Ax - A_n x\|}{\|x\|} \leq \left\{ \int_0^1 \int_0^1 [K(t,s) - K_n(t,s)]^2 dt ds \right\}^{\frac{1}{2}}$$

Since $K_n(t,s) \to K(t,s)$ as $n \to \infty$

$\|A - A_n\| \to 0$ as $n \to \infty$. Since all the A_n are compact, A is also compact by theorem 8.1.9.

8.1.11   Remark

The limit of a weakly convergent sequence $\{A_n\}$ of compact operators is not necessarily compact.

Let us consider an infinite dimensional Banach space E with a basis $\{e_i\}$. Then every $x \in E$ can be written in the form

$$x = \sum_{i=1}^{\infty} \xi_i e_i.$$

Let $S_n x = \displaystyle\sum_{i=1}^{n} \xi_i e_i$ where $S_n x$ is a projection of x to a finite dimensional space.

Let us consider the unit ball $B(0,1) = \{x : x \in E, ||x|| \leq 1\}$.

Then $S_n(B(0,1))$ is closed and bounded in the n-dimensional space E_n and hence compact.

Thus, S_n is compact.

As $n \to \infty$, $S_n x \xrightarrow{w} x$ or $S_n \xrightarrow{w} I$, where the identity operator I is not compact.

8.1.12   Theorem (Schauder, 1930) [49]

Let E_x and E_y be normed linear spaces and $A \in (E_x \to E_y)$. If A is compact then A^* is a compact linear operator mapping E_y^* into E_x^*. The converse holds if E_y is a Banach space.

Proof: Let A be a compact linear operator mapping E_x into E_y. Let us consider a bounded sequence $\{\phi_n\}$ in E_y^*. For $y_1, y_2 \in E_y$

$$|\phi_n(y_1) - \phi_n(y_2)| \leq ||\phi_n||\, ||y_1 - y_2|| \leq \alpha ||y_1 - y_2||.$$

Let $L = \overline{A(B(0,1))}$. Then $\{\phi_{n|L} : n = 1, 2, \ldots\}$ is a set of uniformly bounded equicontinuous functions on the compact metric space L. By the Arzela-Ascoli's theorem (1.6.23) $\{\phi_{n|L}\}$ has a subsequence $\{\phi_{n_j|L}\}$ which converges uniformly on L.

For $i, j = 1, 2, \ldots$, we have

$$\begin{aligned}
||A^*(\phi_{n_i}) - A^*(\phi_{n_j})|| &= \sup\{|A^*(\phi_{n_i} - \phi_{n_j})(x)| : ||x|| \leq 1\} \\
&= \sup\{|(\phi_{n_i} - \phi_{n_j})(Ax)| : ||x|| \leq 1\} \\
&\leq \sup\{|\phi_{n_i}(y) - \phi_{n_j}(y)| : y \in L\}.
\end{aligned}$$

Since the sequence $\{\phi_{n_j|L}\}$ is uniformly Cauchy on L; we see that $(A^*(\phi_{n_j}))$ is a Cauchy sequence in E_x^*. It must converge in E_x^* since E_x^* is a complete normed linear space and hence a Banach space (theorem 4.4.2). We have thus shown that $(A^*(\phi_n))$ has a convergent subsequence. Thus A^* maps a bounded sequence in E_y^* into a convergent subsequence in E_x^* and is hence a compact operator.

Conversely, let us assume that E_y is a Banach space and $A \in (E_x \to E_y)$ and A^* is a compact operator mapping E_y^* into E_x^*. Then we can show that A^{**} is a compact operator mapping E_x^{**} into E_y^{**} by following arguments put forward as in above. Now let us consider the canonical embedding $\phi_{E_x} : E_x \to E_x^{**}$ and $\phi_{E_y} : E_y \to E_y^{**}$ introduced in sec. 5.6.6. Since $A^{**}\phi_{E_x} = \phi_{E_y}A$ by sec. 6.1.5 we see that $\phi_{E_y}A(B(0,1)) = \{A^{**}\phi_{E_x}(x) : x \in B(0,1) \subseteq E_x\}$, is contained in $\{A^{**}(f^*) : f^* \in E_x^{**},\ \|f^{**}\| < 1\}$.

This last set is totally bounded in E_y^{**} since A^{**} is a compact map. As a result, $\phi_{E_y}(A(B(0,1))$ is a totally bounded subset of E_y^{**}. Since ϕ_{E_y} is an isometry, $A(B(0,1))$ is a totally bounded subset of E_y. As E_y is a Banach space, and $A(B(0,1))$ is a totally bounded subset of it, then its closure $\overline{A(B(0,1))}$ is complete and totally bounded. Hence by theorem 1.6.18, $\overline{A(B(0,1))}$ is compact. Hence, A is a compact operator mapping E_x into E_y.

8.1.13   Theorem

Let E_x and E_y be normed linear spaces and $A : E_x \to E_y$ be linear. If A is continuous and the range of A is finite dimensional then A is compact and $R(A)$ is closed in E_y.

Conversely, if E_x and E_y are Banach spaces, A is compact and $R(A)$ is closed in E_y, then A is continuous and its range is of finite dimensions.

Proof: Since A is linear and continuous, it is bounded by theorem 4.2.4. Since $R(A)$ is a finite dimensional subspace of E_y, it is closed [see theorem 2.3.4]. Thus if $\{x_n\}$ is a bounded sequence in E_x, $\{Ax_n\}$ is a bounded and closed subset of the finite dimensional space $R(A)$. We next show that A is compact. By the th. 2.3.1 every finite dimensional normed linear space of a given dimension n is isomorphic to the n-dimensional Euclidean space $\mathbb{R}^n$. By Heine-Borel theorem a closed and bounded subset of $\mathbb{R}^n$ is compact. Therefore A is relatively compact.

Conversely, let us assume that E_x and E_y are Banach spaces, A is compact such that $R(A)$ is closed in E_y. Then A is continuous. Also, $R(A)$ is a Banach space and $A : E_x \to R(A)$ is onto. Then by the open mapping theorem (7.2.2), $A(B(0,1))$ is open. Hence there is some $\delta > 0$ such that

$$X = \{y \in R(A) : \|y\| \leq \delta\} \subset A(B(0,1)).$$

Since $R(A)$ is closed, we have

$$\{y \in R(A) : \|y\| \leq \delta\} = \overline{X} \subseteq \overline{A(B(0,1))} \subseteq R(A).$$

As $\overline{A(B(0,1))}$ is compact, we find the closed ball of radius δ about zero in the normed linear space $R(A)$ is compact. We can next show using 2.3.8 that $R(A)$ is finite dimensional.

8.1.14   Remark

An operator A on a linear space E is said to be of *finite rank* if the range of A is finite dimensional.

8.1.15   Theorem

Let E_x and E_y be normed linear spaces and $A : E_x \to E_y$ be linear.

(a) Let A be a compact operator mapping E_x into E_y. If $x_n \xrightarrow{w} x$ in E_x, then $Ax_n \to Ax$ in E_y.

(b) Let E_x be reflexive and $Ax_n \to Ax$ in E_y wherever $x_n \xrightarrow{w} x$ in E_x. Then A is a compact linear operator mapping $E_x \to E_y$.

Proof: (a) Let $x_n \xrightarrow{w} x$ in E_x. By theorem 6.3.3, $\{x_n\}$ is a bounded sequence in E_x. Let us suppose by way of argument, that $Ax_n \not\to Ax$. Then, given $\epsilon > 0$, there is a subsequence $\{x_{n_i}\}$ such that $||Ax_{n_i} - Ax|| \geq \epsilon$ for all $i = 1, 2, \dots$. Since A is compact and $\{x_{n_i}\}$ is a bounded sequence, there is a subsequence $\{x_{n_{i_j}}\}$ of $\{x_{n_i}\}$ such that $Ax_{n_{i_j}}$ converges as $j \to \infty$, to some element y in E_y. Then $||y - Ax|| \geq \epsilon$, so that $y \neq Ax$. On the other hand, if $f \in E_y^*$ then $f \circ A \in E_x^*$ and since $x_n \xrightarrow{w} x$ in E_x, we have
$$f(Ax) = \lim_{j \to \infty} f(Ax_{n_{i_j}}) = f \lim_{j \to \infty} (Ax_{n_{i_j}}) = f(y)$$

Thus $f(y - Ax) = 0$ for every $f \in E_y^*$. Then by 5.1.4 we must have $y = Ax$. This contradiction proves that $A(x_n) \to Ax$ in E_y.

(b) Let $\{x_n\}$ be a bounded sequence in E_x. Since E_x is reflexive, Eberlein's theorem (6.4.4) shows that $\{x_n\}$ has a weak convergent subsequence $\{x_{n_i}\}$. Let $x_{n_i} \xrightarrow{w} x$ in E_x. Then by our hyperthesis $Ax_{n_i} \to Ax$ in E_y. Thus, for every bounded sequence $\{x_n\}$ in E_x, Ax_n contains a subsequence which converges to E_y. Hence, A is a compact map by 8.1.1.

8.1.16   Remark

(a) The requirement of reflexivity of E_x cannot be dropped from 8.1.15(b). For example, if A denotes the identity operator from l_1 to l_1, then Schur's lemma [see Limaye [33]] shows that $Ax_n \to Ax$ wherever $x_n \xrightarrow{w} x$ in l_1. However, the identity operator is not compact.

8.1.17   Theorem

The range of a compact operator A is separable.

Proof: Let K_n be the image of the ball $\{x : ||x|| \leq n\}$. Since A is compact, $\overline{K}_n$ is compact and therefore, also a separable set [see 1.6.19]. Let L_n be a countable everywhere dense set in $\overline{K}_n$. Since $K = \bigcup_{n=1}^{\infty} K_n$ is the range of A, $L = \bigcup_{n=1}^{\infty} L_n$ is a countable, everywhere dense set in K.

8.2   Spectrum of a Compact Operator

In this section we develop the Riesz-Schauder theory of the spectrum of a compact operator on a normed linear space E_x over $\mathbb{R}$ ($\mathbb{C}$). We show that this spectrum resembles the spectrum of a finite matrix except for the number 0. We begin the study by referring to some preliminary results (4.7.17–4.7.20).

8.2.1   Theorem

Let E_x be a normed linear space, A is a compact linear operator mapping E_x into E_x and $0 \neq k \in \mathbb{R}(\mathbb{C})$. If $\{x_n\}$ is a bounded sequence in E_x such that $Ax_n - kx_n \to y$ in E_x, then there is a subsequence $\{x_{n_i}\}$ of $\{x_n\}$ such that $x_{n_i} \to x$ in E_x and $Ax - kx = y$.

Proof: Since $\{x_n\}$ is bounded and A is a compact operator, $\{x_n\}$ has a subsequence $\{x_{n_i}\}$ such that $\{A(x_{n_i})\}$ converges to some z in E_x, then

$$kx_{n_i} = kx_{n_i} - Ax_{n_i} + Ax_{n_i} \to -y + z,$$

so that $x_{n_i} \to (z - y)/k = x$ (say). Also, since A is continuous, $Ax - kx = \lim_{i \to \infty} \{Ax_{n_i} - kx_{n_i}\} = z - \{z - y\} = y$.

8.2.2   Remark

The above result shows that if A is a compact linear operator mapping E_x into E_x, and if $\{x_n\}$ is a bounded sequence of approximate solution of $Ax - kx = y$, then a subsequence of $\{x_n\}$ converges to an exact solution of the above equation. The following result, which is based on Riesz lemma (2.3.7), is instrumental in analysing the **spectrum of a compact operator**.

8.2.3   Lemma

Let E_x be a normed linear space and $A : E_x \to E_x$.

(a) Let $0 \neq k \in \mathbb{R}(C)$ and E_y be a proper closed subspace of E_x such that $(A - kI)E_x \subseteq E_y$. Then there is some $x \in E_x$ such that $||x|| = 1$ and for all $y \in E_y$,

$$||Ax - Ay|| \geq \frac{|k|}{2}.$$

(b) Let A be a compact linear operator mapping $E_x \to E_x$ and $k_0, k_1, \ldots$, be scalars with $|k_n| \geq \delta$ for some $\delta > 0$ and $n = 0, 1, 2, \ldots$.

Let $E_0, E_1, \ldots, E^0, E^1, \ldots$ be closed subspaces of E_x, such that for $n = 0, 1, 2, \ldots$,

$$E_{n+1} \subseteq E_n, \ (A - k_n I)(E_n) \subseteq E^{n+1},$$
$$E^n \subseteq E^{n+1}, \ (A - k_{n+1} I)(E^{n+1}) \subseteq E^n.$$

Then there are non-negative integers p and q such that

$$E_{p+1} = E_p \quad \text{and} \quad E^{q+1} = E^q.$$

Proof: First, we note that $A(E_y) \subseteq E_y$, since

$$Ay = [Ay - ky] + ky \in E_y \text{ for all } y \in E_y.$$

Now by the Riesz lemma (2.3.7), there is some $x \in E_x$, such that $||x|| = 1$ and dist $(x, E_y) \geq \dfrac{1}{2}$.

Let us consider $y \in E_y$. Since $Ax - kx \in E_y$ and $Ay \in E_y$, we have

$$||Ax - Ay|| = ||kx - [kx - Ax + Ay]|| = |k| \left\| x - \frac{1}{k}[kx - Ax + Ay] \right\|$$

$$\geq |k| \text{ dist } (x, E_y) \geq \frac{1}{2}.$$

(b) Let us suppose now that E_{p+1} is a proper closed subspace of E_p for each $p = 0, 1, 2$. By (a) above we can find an $y_p \in E_p$, such that $||y_p|| = 1$ and for all $y \in E_{p+1}$,

$$||Ay_p - Ay_{p+1}|| \geq \frac{|k|}{2} \geq \frac{\delta}{2}, \ p = 0, 1, 2, \ldots.$$

If follows that $\{y_p\}$ is a bounded sequence in E_x and

$$||Ay_p - Ay_r|| \geq \frac{\delta}{2}, \ p, r = 0, 1, \text{ with } p \neq r.$$

The above shows that $\{A_{y_p}\}$ cannot have a convergent subsequence. But this contradicts the fact that A is compact. Hence there is some nonnegative integer p such that $E_p = E_{p+1}$.

It can similarly be proved that there is some nonnegative integer q such that $E^{q+1} = E^q$.

8.2.4 Definitions: $\rho(A), \delta(A), \sigma_e(A), \sigma_a(A)$

In view of the discussion in 4.7.17–4.7.20, we write the following definitions:

(i) **Resolvent set:** $\rho(A) : \{\lambda \in \mathbb{R}(\mathbb{C}) : A - \lambda I \text{ is invertible}\}$.

(ii) **Spectrum $\sigma(A)$** $: \{\lambda \in \mathbb{R}(\mathbb{C}) : A - \lambda I \text{ does not have an inverse}\}$. A scalar belonging to $\sigma(A)$ is known as **spectral value** of A.

(iii) **Eigenspectrum $\sigma_e(A)$ of A** consists of all λ in $\mathbb{R}$ ($\mathbb{C}$), such that A is not injective or one-to-one. Thus, $\lambda \in \sigma_e(A)$ if and only if there is some non-zero x in E_x such that $Ax = \lambda x$. λ is called an **eigenvalue** of A and x is called the corresponding **eigenvector** of A. The subspace $N(A - \lambda I)$ is known as the **eigenspace** of A, corresponding to the eigenvalue λ.

(iv) **The approximate eigenspectrum $\sigma_a(A)$** consists of all λ in $\mathbb{R}$ ($\mathbb{C}$), such that $(A - \lambda I)$ is not bounded below. Thus, $\lambda \in \sigma_a(A)$ if and only if, there is a sequence in E_x such that $||x_n|| = 1$ for each n and $||Ax_n - \lambda x_n|| \to 0$ as $n \to \infty$. Then λ is called an **approximate**

eigenvalue of A. If $\lambda \in \sigma_e(A)$ and x is a corresponding eigenvector, then letting $x_n = x/||x||$ for all n, we conclude that $\lambda \in \sigma_a(A)$. Hence,

$$\sigma_e(A) \subset \sigma_a(A) \subset \sigma(A).$$

(iv) An operator A on a linear space E_x is said to be of **finite rank** if the range of A is finite dimensional.

8.2.5 Theorem

Let E_x be a normed linear space and A be a compact linear operator, mapping E_x into E_y.

(a) Every non-zero spectral value of A is an eigenvalue of A, so that

$$\{\lambda : \lambda \in \sigma_e(A), \lambda \neq 0\} = \{\lambda : \lambda \in \sigma(A), \lambda \neq 0\}.$$

(b) If E_x is infinite dimensional, then $0 \in \sigma_a(A)$

(c) $\sigma_a(A) = \sigma(A)$.

Proof: (a) Let $0 \neq \lambda \in \mathbb{R}(\mathbb{C})$. If λ is not an eigenvalue, then $A - \lambda I$ is one-to-one. We prove that λ is not a spectral value of A, i.e., $(A - \lambda I)$ is invertible. We first show that $(A - \lambda I)$ is bounded below. Otherwise, we can find a sequence $\{x_n\}$ in E_x, such that $||x_n|| = 1$ for each n and $||(A - kI)(x_n)|| \to 0$ as $n \to \infty$. Then, by theorem 8.2.1, there is a subsequence $\{x_{n_i}\}$ of $\{x_n\}$ such that $x_{n_i} \to x$ in E_x and $Ax - \lambda x = 0$. Since $A - \lambda I$ is one-to-one, we have $x = \theta$. But $||x|| = \lim_{i \to \infty} ||x_{n_i}|| = 1$. This leads to a contradiction. Thus, $A - \lambda I$ is bounded below.

Next we show that $A - \lambda I$ is onto, i.e., $R(A - \lambda I) = E_x$. First, we show that $R(A - \lambda I)$ is a closed subspace of E_x. Let $(Ax_n - \lambda x_n)$ be a sequence in $R(A - \lambda I)$ which converges to some element $y \in E_x$. Then $((A - \lambda I)x_n)$ is a bounded sequence in E_x and since $(A - \lambda I)$ is bounded below, i.e., $||(A - \lambda I)x|| \geq m||x||$, we see that $\{x_n\}$ is also a bounded sequence in E_x. By theorem 8.2.1, there is a subsequence $\{x_{n_i}\}$ of (x_n), such that $x_{n_i} \to x$ in E_x and $Ax - \lambda x = y$. Thus, $y \in R(A - \lambda I)$ showing that the range of $(A - \lambda I)$ is closed in E_x.

Now, let $E_n = R((A - \lambda I)^n)$ for $n = 0, 1, 2, \ldots$. Then we show by induction that each E_n is closed in E_x. For $n = 0$, $E_0 = E_x$, for $n = 1$, E_1 is closed. For $n \geq 2$ $(A - \lambda I)^n$

$$= A^n - \lambda A^{n-1} + \cdots + {}^nC_r(-1)^r A^{n-r}\lambda^r + \cdots$$
$$+ (-1)^{n-1}A\lambda^{n-1} + (-1)^n\lambda^n I.$$

$$= P_n(A) - \lambda_n I$$

where $\lambda_n = -(-1)^n\lambda^n$ and $P_n(A)$ is a nth degree polynomial in A.

Then, by lemma 8.1.4., $P_n(A)$ is a compact operator and clearly $\lambda_n \neq 0$. Further, since $(A - \lambda I)$ is one-to-one, $(A - \lambda I)^n$ is also one-to-one.

If we replace A with $P_n(A)$ and λ with λ_n and follow the arguments put forward above, we conclude that $R(P_n(A) - \lambda_n I) = E_n$ is a closed subspace of E_x.

Since $E_{n+1} \subseteq E_n$ and $E_{n+1} = (A - \lambda I)(E_n)$ and part (b) of lemma 8.2.3 shows that that there is a non-negative integer p with $E_{p+1} = E_p$. If $p = 0$ then $E_1 = E_0$. If $p > 0$, we want to show that $E_p = E_{p-1}$.

Let $y \in E_{p-1}$, that is, $y = (A - \lambda I)^{p-1}x$ for some $x \in E_x$. Then $(A - \lambda I)y = (A - \lambda I)^p x \in E_p = E_{p+1}$, so that there is some $x \in E_x$ with $(A - \lambda I)y = (A - \lambda I)^{p+1}x$. Since $(A - \lambda I)(y - (A - \lambda I)^p x) = \theta$ and since $(A - \lambda I)$ is one-to-one, it follows that $y - (A - \lambda I)^p x = \theta$, i.e., $y = (A - \lambda I)^p x \in E_p$. Thus, $E_p = E_{p-1}$. Proceeding as in above, if $p > 1$, we see that $E_{p+1} = E_p = E_{p-1} = E_{p-2} = \cdots = E_1 = E_0$. But $E_1 = R(A - \lambda I)$ and $E_0 = E_x$. Hence $A - \lambda I$ is one-to-one.

Being bounded below and onto, $(A - \lambda I)$ has an inverse. Hence, every non-zero spectral value of A is an eigenvalue of A. Since $\sigma_e(A) \subset \sigma(A)$ always, the proof (a) is complete.

(b) Let E_x be infinite dimensional. Let us consider an infinite linearly independent set $\{e_1, e_2, \ldots\}$ of E_x and let $E^n = \text{span } \{e_1, e_2, \ldots e_n\}$, $n = 1, 2, \ldots$. Then E^n is a proper subspace of E^{n+1}. E^n is of finite dimension and is closed by theorem 2.3.4. By the Ricsz lemma (theorem 2.3.7), there is some element $a_{n+1} \in E^{n+1}$ such that $\|a_{n+1}\| = 1$, dist $(a_{n+1}, E^n) \geq \frac{1}{2}$. Let us assume that A is bounded below i.e., $\|Ax\| \geq m\|x\|$ for all $x \in E_x$ and some $m > 0$. Then for all $p, q = 1, 2, \ldots$, and $p \neq q$, we have,

$$\|Aa_p - Aa_q\| \geq m\|a_p - a_q\| \geq \frac{m}{2},$$

so that $\{Aa_p\}$ cannot have a convergent subsequence, which contradicts the fact that A is compact.

Hence, A is not bounded below. Hence $0 \in \sigma_a(A)$.

(c) If E_x is finite dimensional and $D(A) = E_x$, then the operator A can be represented by a matrix, (a_{ij}); then $A - \lambda I$ is also represented by a matrix and $\sigma(A)$ is composed of those scalars λ which are the roots of the equation

$$\begin{vmatrix} a_{11} - \lambda & a_{12} & & a_{1n} \\ & & & \\ a_{n1} & a_{n2} & & a_{nn} - \lambda \end{vmatrix} = 0 \text{ [see Taylor, [55]]}$$

Hence $\sigma_a(A) = \sigma(A)$.

If E_x is infinite dimensional, then $0 \in \sigma_a(A)$ by (b) above. Also, since $\sigma_e(A) \subseteq \sigma_a(A) \subseteq \sigma(A)$ always, if follows from (a) above that $\sigma_a(A) = \sigma(A)$.

8.2.6 Lemma

Let E_x be a linear space, $A : E_x \to E_x$ linear, $Ax_n = \lambda_n x_n$, for some $\theta \neq x_n \in E_x$ and $\lambda_n \in \mathbb{R}(\mathbb{C})$, $n = 0, 1, 2, \ldots$.

(a) Let $\lambda_n \neq \lambda_m$ wherever $n \neq m$. Then $\{x_1, x_2, \ldots\}$ is a linearly independent subset of E_x.

(b) Let E_x be a normed linear space. A is a compact linear operator mapping $E_x \to E_x$ and the set $\{x_1, x_2, \ldots\}$ is linearly independent and infinite. Then $\lambda_n \to 0$ as $n \to \infty$.

Proof: (a) Since $x_1 \neq \theta$, the set $\{x_1\}$ is linearly independent. Let $n = 2, 3, \ldots$ and assume that $\{x_1, x_2, \ldots, x_n\}$ is linearly independent. Let, if possible, $x_{n+1} = \alpha_1 x_1 + \alpha_2 x_2 + \cdots + \alpha_n x_n$ for some $\alpha_1, \alpha_2, \ldots \alpha_n$ in $\mathbb{R}$ $(\mathbb{C})$.

Then, $\lambda_{n+1} x_{n+1} = \alpha_1 \lambda_{n+1} x_1 + \alpha_2 \lambda_{n+1} x_2 + \cdots + \alpha_n \lambda_{n+1} x_n$ and also

$$\lambda_{n+1} x_{n+1} = A(x_{n+1}) = \sum_{i=1}^{n} \alpha_i A x_i = \alpha_1 \lambda_1 x_1 + \alpha_2 \lambda_2 x_2 + \cdots + \alpha_n \lambda_n x_n.$$

Thus we get on subtraction,

$$\alpha_1(\lambda_{n+1} - \lambda_1)x_1 + \alpha_2(\lambda_{n+1} - \lambda_2)x_2 + \cdots + \alpha_n(\lambda_{n+1} - \lambda_n)x_n = \theta.$$

Since $x_2, \ldots x_n$ are linearly independent, $\alpha_j(\lambda_j - \lambda_{n+1}) = 0$ for each j. As $x_{n+1} \neq \theta$, we see that $\alpha_j \neq 0$ for some j, $1 \leq j \leq n$, so that $\lambda_{n+1} = \lambda_j$.

But this is impossible. thus the set $\{x_1, x_2, \ldots x_{n+1}\}$ is linearly independent. Using mathematical induction we conclude that $\{x_1, x_2, \ldots\}$ are linearly independent.

(b) For $n = 1, 2, \ldots$, let $E^n = \text{span } \{x_1, x_2, \ldots, x_n\}$. Since x_{n+1} does not belong to E^n, E^n is a proper subspace of E^{n+1}. Also, E^n is closed in E_x by th. 2.3.4, and $(A - \lambda_{n+1} I)(E^{n+1}) \subset E^n$ since $(A - \lambda_{n+1} I)x_{n+1} = \theta$. If $\lambda_n \not\to 0$ as $n \to \infty$, we can assume by passing to a subsequence that $|\lambda_n| \geq \delta > 0$ for all $n = 1, 2, \ldots$. Now 8.2.3(b) yields that $E^{q+1} = E^q$ for some positive integer q which contradicts the fact that E^q is a proper subspace of E^{q+1}. Hence $\lambda_n \to 0$ as $n \to \infty$.

8.2.7 Theorem

Let E_x be a normed linear space and A be a compact linear operator mapping E_x into E_x.

(a) The eigenspectrum and the spectrum of A are countable sets and have '0' as the only possible limiting point. In particular, if $\{\lambda_1, \lambda_2, \ldots\}$ is an infinite set of eigenvalues of A, then $\lambda_n \to 0$ as $n \to \infty$.

(b) Every eigenspace of A corresponding to a non-zero eigenvalue of A is finite dimensional.

Proof: Since $\{\lambda : \lambda \in \sigma(A), \lambda \neq 0\} = \{\lambda : \lambda \in \sigma_e(A), \lambda \neq 0\}$ by 8.2.5(a), we have to show that the set $\sigma_e(A)$ is countable and 0 is the only possible limit point of it.

For $\delta > 0$, let

$$L_\delta = \{\lambda \in \sigma_e(a) : |\lambda| \geq \delta\}.$$

Suppose that L_δ is an infinite set for some $\delta > 0$. Let $\lambda_n \in L_\delta$ for $n = 1, 2, \ldots$ with $\lambda_n \neq \lambda_m$ wherever $n \neq m$. If x_n is an eigenvector of A corresponding to the eigenvalue λ_n, then by theorem 8.2.6(a), the set $\{x_1, x_2, \ldots\}$ is linearly independent, and consequently $\lambda_n \to 0$ as $n \to \infty$ by 8.2.6(b). But this is impossible since $|\lambda_n| \geq \delta$ for each n. Hence L_δ is a finite set for $\delta > 0$. Since $\sigma_e(A) = \bigcup_{n=1}^{\infty} L_{1/n}$ it follows that $\sigma_e(A)$ is a countable set and that $\sigma_e(A)$ has no limit points except possibly the number 0.

Furthermore, $\sigma_e(A)$ is a bounded subset of $\mathbb{R}$ ($\mathbb{C}$) since $|\lambda| \leq \|A\|$ for every $\lambda \in \sigma_e(A)$. If $\{\lambda_1, \lambda_2, \ldots\}$ is an infinite subset of $\sigma_e(A)$, then it must have a limit point by the Bolzano-Weierstrass theorem for $\mathbb{R}$ ($\mathbb{C}$) (theorem 1.6.19). As the only possible limit point is 0, we see that $\lambda_n \to 0$ as $n \to \infty$.

(b) Let $0 \neq \lambda \in \sigma_e(A)$. Suppose that the set of eigenvectors corresponding to an eigenvalue λ forms an infinite set $\{x_1, x_2, \ldots\}$. Let λ take the values $\lambda_1, \lambda_2, \ldots$ corresponding to the eigenvectors $x_1, x_2, \ldots$. Then, by Bolzano-Weierstrass theorem (th. 1.6.19) the set of eigenvalues $\{\lambda_1, \lambda_2, \ldots\}$ have a limit point in $\mathbb{R}$ ($\mathbb{C}$). As the only possible limit point is zero, we see that $\lambda_n \to 0$ as $n \to \infty$. But this is impossible since $\sigma_e(A) \ni \lambda \neq 0$.

Thus the eigenspace of A corresponding to λ is finite dimensional.

We next consider the spectrum of the transpose of a compact operator.

8.2.8 Theorem

Let E_x be a normed linear space and $A \in (E_x \to E_x)$. Then $\sigma(A^*) \subseteq \sigma(A)$.

If E_x is a Banach space, then

$$\sigma(A) = \sigma_a(A) \cup \sigma_e(A^*) = \sigma(A^*).$$

Proof: Let $\lambda \in \mathbb{R}(\mathbb{C})$ be such that $(A - \lambda I)$ is invertible, i.e., $(A - \lambda I)$ has a bounded inverse. If $(A - \lambda I)B = I = B(A - \lambda I)$ for some bounded linear operator B mapping $E_x \to E_x$, then by 6.1.5(ii) $B^*(A^* - \overline{\lambda}I) = I = (A^* - \overline{\lambda}I)B^*$, where A^*, B^* stand for adjoints of A and B respectively. Hence $\sigma(A^*) \subseteq \sigma(A)$.

Let E_x be a Banach space. By 8.2.4(iii) $\lambda \in \sigma(A)$ if and only if either $A - \lambda I$ is not bounded below or $R(A - \lambda I)$ is not dense in E_x. As because $(A - \lambda I)$ is not bounded below $\lambda \in \sigma_a(A)$.

Let $f \in E_x^*$. Then $(A^* - \overline{\lambda}I)f = 0$ if and only if $f((A - \lambda I)x) = (A^* - \overline{\lambda}I)f(x) = 0$ for every $x \in E_x$.

Now, $(A^* - \overline{\lambda}I)$ is one-to-one, i.e., $N(A^* - \overline{\lambda}I) = \{\theta\}$ if and only if $f = \theta$ wherever $f(y) = 0$ for every $y \in R(A - \lambda I)$. This happens if and only if the closure of $R(A - \lambda I)$ is E_y i.e., $R(A - \lambda I)$ is dense in E_y. Hence $\lambda \notin \sigma_e(A^*)$. Thus $\sigma(A) = \sigma_a(A) \cup \sigma_e(A^*)$.

Finally, to conclude $\sigma(A) = \sigma(A^*)$, it will suffice to show that $\sigma_a(A) \subseteq \sigma(A^*)$. Let $\lambda \notin \sigma(A^*)$, that is $A^* - \overline{\lambda}I$ is invertible. If $x \in E_x$ then by 5.1.4, there is some $f \in E_x^*$, such that $f(x) = ||x||$, $||f|| = 1$ so that

$$||x|| = |f(x)| = |(A^* - \overline{\lambda}I)(A^* - \overline{\lambda}I)^{-1}(f)(x)|$$
$$= |(A^* - \overline{\lambda}I)^{-1}(f)(A - \lambda I)(x)|$$
$$\leq ||(A^* - \overline{\lambda}I)^{-1}|| \; ||A(x) - \lambda x||.$$

Thus, $(A - \lambda I)$ is bounded below, that is $\lambda \notin \sigma_a(A)$.

If A is a compact operator we get some interesting results.

8.2.9   Theorem

Let E_x be a normed linear space and A be a compact operator mapping E_x into E_x. Then
   (a) dim $N(A^* - \overline{\lambda}I) =$ dim $N(A - \lambda I) < \infty$ for $0 \neq \lambda \in \mathbb{R}(\mathbb{C})$,
   (b) $\{\lambda : \lambda \in \sigma_e(A^*), \lambda \neq 0\} = \{\lambda : \lambda \in \sigma_e(A), \lambda \neq 0\}$,
   (c) $\sigma(A^*) = \sigma(A)$.

Proof: (a) By theorem 8.1.12, A^* is a compact linear operator mapping E_x into E_x. Then, theorem 8.2.7 yields that the dimension r of $N(A - \lambda I)$ and the dimension s of $N(A^* - \overline{\lambda}I)$ are both finite.

First we show that $s \leq r$.

If $r = 0$, that is $\lambda \notin \sigma_e(A)$, then, by theorem 8.2.5(a), we see that $\lambda \notin \sigma(A)$. Since $\sigma(A^*) \subseteq \sigma(A)$ by theorem 8.2.8 we have $\lambda \notin \sigma(A^*)$. In particular $(A^* - \overline{\lambda}I)$ is one-to-one, i.e., $s = 0$.

Next, let $r > 1$. Consider a basis $\{e_1, e_2, \ldots, e_r\}$ of $N(A - \lambda I)$. Then from 4.8.3 we can find $f_1, \ldots, f_r$ in E_x^* such that $f_j(e_i) = \delta_{i,j}, i, j = 1, 2, \ldots, r$.

Let, if possible, $\{\phi_1, \phi_2, \ldots, \phi_{r+1}\}$ be a linearly independent subset of $N(A^* - \overline{\lambda}I)$ containing $(r + 1)$ elements. By 4.8.2 there are $y_1, y_2, \ldots, y_{r+1}$ in E_x such that

$$\phi_j(y_i) = \delta_{ij}, i, j = 1, 2, \ldots, r + 1.$$

Consider the map $B : E_x \to E_x$ given by

$$B(x) = \sum_{i=1}^{r} f_i(x)y_i, \quad x \in E_x.$$

Since $f_i \in E_x^*$, B is a a a bounded linear operator mapping $E_x \to E_x$ and B is of finite rank. Therefore B is a compact operator on E_x by theorem 8.1.13.

Since A is also compact, lemma 8.1.4 shows that $A - B$ is a compact operator. We show that $A - B - \lambda I$ is one-to-one but not onto and obtain a contradiction.

We note that $\phi_j \in N(A^* - \overline{\lambda}I)$ and hence,

$$\phi_j(A - B - \lambda I)(x) = (A^* - \overline{\lambda}I)(\phi_j)(x) - \phi_j\left(\sum_{i=1}^{r} f_i(x)y_i\right)$$

$$= 0 - \sum_{i=1}^{r} f_i(x)\phi_j(y_i).$$

$$= \begin{cases} -f_i(x) & \text{if } 1 \leq j \leq r \\ 0 & \text{if } j = r + 1 \end{cases}.$$

Now, let $x \in E_x$ satisfy $(A - B - \lambda I)x = \theta$. Then it follows that $-f_j(x) = \phi_j(A - B - \lambda I)x = \phi_j(0) = 0$, for $1 \leq j \leq r$ and in turn, $B(x) = \theta$. Hence $(A - \lambda I)x = \theta$ i.e., $x \in N(A - \lambda I)$. Since $\{e_1, e_2, \ldots, e_r\}$ is a basis of $N(A - \lambda I)$, we have $x = \alpha_1 e_1 + \cdots + \alpha_r e_r$ for some $\alpha_1, \alpha_2, \ldots, \alpha_r$ in $\mathbb{R}$ ($\mathbb{C}$). But

$$0 = f_j(x) = f_j(\alpha_1 e_1 + \alpha_2 e_2 + \cdots + \alpha_r e_r) = \alpha_j, j = 1, 2, \ldots, r$$

so that $x = 0 \cdot e_1 + 0 \cdot e_2 + \cdots + 0 \cdot e_r = \theta$.

Thus $A - B - \lambda I$ is one-to-one because

$$(A - B - \lambda I)x = \theta \implies x = \theta.$$

Next we assert that $y_{r+1} \notin R(A - B - \lambda I)$. For if $y_{r+1} = (A - B - \lambda I)x$ for some $x \in E_x$, then

$$1 = \phi_{r+1}(y_{r+1}) = \phi_{r+1}((A - B - \lambda I)x) = 0,$$

as we have noted above. Hence $(A - B - \lambda I)$ is not onto.

Thus a linearly independent subset of $N(A^* - \overline{\lambda}I)$ can have at most r elements, i.e., $s \leq r$.

To obtain $r \leq s$ we proceed as follows. Let t denote the dimension of $N(A^{**} - \lambda I)$. Considering the compact operator A^* in place of A, we find that $t \leq s$. If Π_{E_x} denotes the canonical embedding of E_x into E_x^{**} considered in sec. 5.6.6, then by theorem 6.1.5. $A^{**}\Pi_{E_x} = \Pi_{E_x}A$. Hence $\Pi_{E_x}(N(A - \lambda I)) \subseteq N(A^{**} - \lambda I)$, so that $r \leq t$. Thus $r \leq t \leq s$. Hence $r = s$.

(b) Let $0 \neq \lambda \in \mathbb{R}(\mathbb{C})$. Part (a) shows that $N(A - \lambda I) \neq \{\theta\}$ if and only if $N(A^* - \overline{\lambda}I) \neq \{\theta\}$, that is, $\lambda \in \sigma_e(A)$ if and only if $\lambda \in \sigma_e(A^*)$.

(c) Since A and A^* are compact operators, we have by theorem 8.2.5

$$\{\lambda : \lambda \in \sigma(A), \lambda \neq 0\} = \{\lambda : \lambda \in \sigma_e(A), \lambda \neq 0\}$$

$$\{\lambda : \lambda \in \sigma(A^*), \lambda \neq 0\} = \{\lambda : \lambda \in \sigma_e(A^*), \lambda \neq 0\}$$

It follows from (b) above that

$$\{\lambda : \lambda \in \sigma(A^*), \lambda \neq 0\} = \{\lambda : \lambda \in \sigma(A), \lambda \neq 0\}$$

If E_x is finite dimensional, then $\det(A - \lambda I) = \det(A^* - \bar{\lambda}I)$. Hence $0 \in \sigma(A^*)$ if and only if $0 \in \sigma(A)$. If E_x is infinite dimensional then E_x^* is infinite dimensional and hence $0 \in \sigma_a(A)$ as well as $0 \in \sigma_a(A^*)$ by theorem 8.2.5(b). Thus, in both cases, $\sigma(A^*) = \sigma(A)$.

If follows from the above that the spectrum of an infinite matrix is very much like the spectrum of a finite matrix, except for the number zero.

8.2.10 *Examples*

1. Let $E_x = l_p$, $1 \leq p \leq \infty$ and $Ax = \left\{ \dfrac{\xi_1}{1}, \dfrac{\xi_2}{2}, \dfrac{\xi_3}{3} \cdots \right\}$ where $x = \{\xi_1, \xi_2, \xi_3, \ldots\} \in l_p$.

Let $A_n = \left\{ \dfrac{1}{1}, \dfrac{1}{2}, \cdots, \dfrac{1}{n}, 0 \cdots 0 \right\}$. Since A_n is finite, A_n is a linear compact operator [see theorem 8.1.13].

Furthermore, $\|(A - A_n)x\|_p^p = \displaystyle\sum_{i=n+1}^{\infty} |\eta_i|^p = \displaystyle\sum_{i=n+1}^{\infty} \dfrac{1}{i^p} |\xi_i|^p$

$$\leq \dfrac{1}{(n+1)^p} \sum_{i=n+1}^{\infty} |\xi_i|^p \leq \dfrac{\|x\|^p}{(n+1)^p}, \ p > 1.$$

Hence $\|(A - A_n)\| = \sup \dfrac{\|(A - A_n)x\|}{\|x\|} \leq \dfrac{1}{n+1}, \ p > 1.$

Hence, $A_n \to A$ as $n \to \infty$ and A_n is compact, by theorem 8.1.9, A is also a compact operator. A is clearly one-to-one, 0 is not an eigenvalue of A, but since A is not bounded below, 0 is a spectral value of A. Also, $\lambda_n = \frac{1}{n}$ is an eigenvalue of A and $\lambda_n \to 0$ as $n \to \infty$.

2. The eigenspace of a compact operator corresponding to the eigenvalue 0 can be infinite dimensional. The easiest example is the zero operator on an infinite dimensional normed linear space.

3. $\lambda = 0$ can be an eigenvalue of a compact operator A, but $\lambda = 0$ may not be an eigenvalue of the transpose A^* and vice versa.

4. Let $E_x = l_p$ and A denote the compact operator on E_x defined by

$$Ax = \left(x_3, \frac{x_4}{3}, \frac{x_5}{4}, \cdots \right)^T \text{ for } x = (x_1, x_2, \ldots)^T \in l_p$$

i.e., $\begin{pmatrix} 0 & 0 & 1 & 0 & \cdots \\ 0 & 0 & 0 & \frac{1}{3} & \cdots \\ 0 & 0 & 0 & \frac{1}{4} & \cdots \end{pmatrix} \begin{pmatrix} x_1 \\ x_2 \\ x_3 \\ x_4 \\ \vdots \end{pmatrix} = \begin{pmatrix} x_3 \\ \frac{x_4}{3} \\ \frac{x_5}{4} \\ \vdots \end{pmatrix}$

Hence A^* can be identified with B on l_p, $\dfrac{1}{p} + \dfrac{1}{q} = 1$, so that

$$B = \begin{pmatrix} 0 & 0 & 0 \cdots \\ 0 & 0 & 0 \cdots \\ 1 & 0 & 0 \cdots \\ 0 & \frac{1}{3} & 0 \cdots \\ \vdots & \vdots & \frac{1}{4} \cdots \end{pmatrix}$$

Hence $Bx = \left(0, 0, x_1, \dfrac{x_2}{3}, \dfrac{x_3}{4} \cdots \right)^T$ for $x = (x_1, x_2, x_3, \ldots) \in l^q$.

Since $A \begin{pmatrix} 1 \\ 0 \\ \vdots \\ 0 \end{pmatrix} = 0 \begin{pmatrix} 1 \\ 0 \\ \vdots \\ 0 \end{pmatrix}$ we see that 0 is an eigenvalue of A. But

since B is one-to-one, 0 is not an eigenvalue of B. Also since $B^* = A$, we see that not only the compact operator B does not have an eigenvalue 0, its adjoint B^* does not have an eigenvalue 0 too.

(c) Let $E_x = C([0,1]), 1 \le p \le \infty$. For $x \in E_x$, let

$$Ax(s) = (1-s) \int_0^s tx(t)dx(t) + s \int_s^1 (1-t)x(t)dx(t), \ s \in [0,1] \quad (8.1)$$

Since the kernel is continuous A is a compact operator mapping E_x into E_x [see example 8.1.10].

The above is a Fredholm integral operator with a continuous kernel given by,

$$K(x,t) = \begin{cases} (1-s)t & \text{if } 0 \le t \le s \le 1 \\ s(1-t) & \text{if } 0 \le s \le t \le 1 \end{cases} \quad (8.2)$$

Let $x \in E_x$ and $0 \ne \lambda \in \mathbb{R}(\mathbb{C})$ be such that

$$Ax = \lambda x.$$

Then for all $s \in [a, b]$

$$\lambda x(s) = (1-s) \int_0^s tx(t)dx(t) + s \int_s^1 (1-t)x(t)dx(t). \quad (8.3)$$

Putting $s = 0$ and $s = 1$, we note that $x(0) = 0 = x(1)$. Since $tx(t)$ and $(1-t)x(t)$ are integrable functions of $t \in [0,1]$, it follows that the right-hand side of the equation given above is an absolutely continuous function of $x \in [0,1]$. Hence x is (absolutely) continuous on $[0,1]$. This implies that $tx(t)$ and $(1-t)x(t)$ are continuous functions of t on $[0,1]$. Thus the right hand is, in fact, a continuously differentiable function of s and we have for all $s \in [0,1]$.

$$\lambda x'(s) = (1-s)sx(s) - \int_0^s tx(t)dx(t) - s(1-s)x(s) + \int_s^1 (1-t)x(t)dx(t)$$

$$= -\int_0^s tx(t)dx(t) + \int_s^1 (1-t)x(t)dx(t).$$

This shows that x' is a continuously differentiable function, and for all $s \in [0,1]$, we have,

$$\lambda x''(s) = -sx(s) - (1-s)x(s) = -x(s).$$

Thus, the differential equation $\lambda x'' + x = 0$ has a non-zero solution, satisfying $x(0) = 1 = x(1)$ if and only if $\lambda = 1/n^2\pi^2, n = 1, 2, \ldots$ and in such a case its most general solution is given by $x(s) = c\sin n\pi s$, $s \in [0,1]$, where $c \in \mathbb{R}(\mathbb{C})$.

Let $\lambda_n = \frac{1}{n^2\pi^2}$, $n = 1, 2, \ldots$ and $x_n(s) = \sin n\pi s$ for $s \in [0,1]$. Thus each λ_n is an eigenvalue of A and the corresponding eigenspace $N(A - \lambda_n I) =$ span $\{x_n\}$ is one dimensional.

Next, let 0 be not an eigenvalue of A. For, if $Ax = \theta$ for some $x \in E_x$, then by differentiating the expression for $Ax(s)$ with respect to s two times, we see that $x(s) = 0$ for all $s \in [0,1]$. On the other hand, since A is compact and E_x is infinite dimensional, 0 is an approximate eigenvalue of A by theorem 8.2.5. Thus,

$$\sigma_e(A) = \left(\frac{1}{\pi^2}, \frac{1}{2^2\pi^2}, \cdots\right) \quad \text{and} \quad \sigma_a(A) = \sigma(A) = \left\{0, \frac{1}{\pi^2}, \frac{1}{2^2\pi^2}, \cdots\right\}.$$

8.2.11   Problems [8.1 and 8.2]

1. Show that the zero operator on any normed linear space is compact.

2. If A_1 and A_2 are two compact linear operators mapping a normed linear space E_x into a normed linear space E_y, show that $A_1 + A_2$ is a compact linear operator.

3. If E_x is finite dimensional and A is linear mapping E_x into E_y, then show that A is compact.

4. If $A \in (E_x \to E_y)$ and E_y is finite dimensional, then show that A is compact.

5. If $A, B \in (E_x \to E_x)$ and A is compact, then show that AB and BA are compact.

6. Let E_x be a Banach space and $P \in (E_x \to E_x)$ be a projection. Then show that P is a compact linear operator if and only if P is of finite rank.

7. Given E_x is an infinite dimensional normed linear space and A a compact linear operator mapping E_x into E_x, show that $\lambda I - A$ is not a compact operator where λ is a non-zero scalar.

8. Let $A = (a_{ij})$ be an infinite matrix with $a_{ij} \in \mathbb{R}(\mathbb{C})$ $i, j \in N$. If $x \in l_p$ and $Ax \in l_r$ where $Ax = \left(\sum\limits_{j=1}^{\infty} a_{ij} x_j \right)$, show that in the following cases $A : l_p \to l_r$ is a compact operator;

(i) $1 \le p < \infty$, $1 \le r < \infty$ and $\sum\limits_{j=1}^{\infty} |a_{ij}| \to 0$ as $i \to \infty$

(ii) $1 \le p < \infty$, $1 \le r < \infty$ and $\sum\limits_{i=1}^{\infty} \left(\sum\limits_{j=1}^{\infty} |u_{ij}|^r \right) < \infty$.

9. Let $E_x = C([a, b])$ with $|| \cdot ||_\infty$ and $A : E_x \to E_x$ be defined by $Ax(s) = \int_a^b K(s, t) x(t) dt$, $x \in E_x$ where $K(\cdot, \cdot) \in C([a, b]) \times C([a, b])$. Let $\{A_n\}$ be the Nyström approximation of A corresponding to a convergent quadrature formula with nodes $t_{1,n}, t_{2,n}, \ldots, t_{n,n}$ in $[a, b]$ and weights $w_{1,n}, w_{2,n}, \ldots w_{n,n}$ in $\mathbb{R}$ $(\mathbb{C})$ i.e., $A_n x(s) = \sum\limits_{j=1}^{n} K(t_{j,n}, t) x(t_{j,n}) w_{j,n}, x \in E_x$, $n \in N$, where nodes and weights are such that $\sum\limits_{j=1}^{n} x(t_{j,n}) w_{j,n} \to \int_a^b x(t) dt$ as $n \to \infty$ for every $x \in C([a, b])$. Then show that (i) $||Ax - A_n x|| \to 0$ for every $x \in C([a, b])$. (ii) $||(A_n - A)A|| \to 0$ and $||(A_n - A)A_n|| \to 0$ as $n \to \infty$.

For quadrature formula see 6.3.5. In order to solve the integral equation numerically $Ax = y$, $x, y \in C([a, b])$ the given equation is approximately reduced to a system of algebraic equations by using 'quadrature'.

(Hint: (i) Show that for each $u \in C([a, b])$, $\{(A_n u(s))\}$ converges to $(Au(s))$. $\{A_n u : n \in \mathbb{N}\}$ is equicontinuous, and hence (ii). Use the result in (i) and the fact that $\{Au : ||u||_\infty \le 1\}$ and $\{A_n u : ||u||_\infty \le 1, n \in \mathbb{N}\}$ are equicontiuous.)

8.3   Fredholm Alternative

In this section, linear equations with compact operators will be considered. F. Riesz has shown that such equations admit the applications of basic consequences from the Fredholm theory of linear integral equations.

8.3.1   *A linear equation with compact operator and its adjoint*

Let A be a compact operator which maps a Banach space E into itself. Consider the equation

$$Au - u = v \tag{8.4}$$

or, $Pu = v$ $\tag{8.5}$

where $P = A - I$. Together with equation (8.5), consider

$$A^*f - f = g \tag{8.6}$$

or, $P^*f = g$ $\tag{8.7}$

where A^* is the adjoint operator of A and acts into the space E^*. By theorem 8.1.12, A^* is a compact operator.

8.3.2   Lemma

Let N be a subspace of the null space of the operator P, that is, a collection of elements u such that $Pu = \theta$. Then N is a finite-dimensional subspace of E.

Proof: Let M be an arbitrary bounded set in N. For every $u \in N$, $Au = u$, that is, the operator A leaves the element of the subspace N invariant and in particular, carries the set M into itself. The subspace N of E is then said to be **invariant** with respect to A.

As A is a compact operator, A carries M into a compact set. Consequently, every bounded set $M \subseteq N$ is compact, implying by theorem 2.3 that N is finite dimensional.

8.3.3   Remark

The elements of the subspace N are eigenvectors of the operator A corresponding to the eigenvalue $\lambda_0 = 1$. The above conclusion remains valid if λ_0 is replaced by any non-zero eigenvalue.

Thus a compact linear operator can have only a finite number of linearly independent eigenvectors corresponding to the same non-zero eigenvalue.

8.3.4   Lemma

Let $L = P(E)$, that is, L be a collection of elements $v \in E$ representable in the form $Au - u = v$. Then L is a subspace.

To prove L is linear we note that if $Au_1 - u_1 = v_1$ and $Au_2 - u_2 = v_2$, then $\alpha_1 v_1 + \alpha_2 v_2 = A(\alpha_1 u_1 + \alpha_2 u_2) - (\alpha_1 u_1 + \alpha_2 u_2), \alpha_1, \alpha_2 \in \mathbb{R}(C)$. Thus, $v_1, v_2 \in L \Rightarrow \alpha_1 v_1 + \alpha_2 v_2 \in L$. We next prove that L is closed. We first show that there is a constant m depending only on $A - I$ such that wherever the equation $Pu = v$ is solvable, at least one of the solutions satisfies the inequality

$$m||u|| \leq ||v||, \quad m > 0 \tag{8.8}$$

Let u_0 be a solution of $Pu = v$. Then every other solution of $Pu = v$ is expressible in the form $u = u_0 + w$ where w is a solution of the homogenous equation

$$Pu = \theta \tag{8.9}$$

Let us consider $F(w) = ||u_0 + w||$, a bounded below continuous functional. Let $d = \inf. F(w)$ and $\{w_n\} \subseteq N$ be the **minimizing sequence**, that is,

$$F(w_n) = ||u_0 + w_n|| \to d \qquad (8.10)$$

The sequence $\{||u_0 + w_n||\}$ has a limit and is hence bounded. However, the sequence $\{||w_n||\}$ is also bounded, since,

$$||w_n|| = ||(u_0 + w_n) - u_0|| \le ||u_0 + w_n|| + ||u_0||$$

Thus $\{w_n\}$ is a bounded sequence in a finite-dimensional space N and hence, by **Bolzano-Weierstrass theorem** (theorem 1.6.19) has a convergent subsequence. Hence, we can find a subsequence $\{w_{n_p}\}$ such that

$$w_{n_p} \to w_0. \text{ Then } F(w_{n_p}) \to F(w_0). \qquad (8.11)$$

From (8.10) and (8.11) it follows that

$$F(w_0) = ||u_0 + w_0|| = d.$$

Therefore, the equation $Pu = v$ always has the solution $\widetilde{u} = u_0 + w_0$ with the minimal norm. In order to show that (8.8) holds for $\widetilde{u}$, we consider the ratio $||\widetilde{u}||/||v||$, and let us assume that the ratio is not bounded. Then there exist sequences v_n and $\widetilde{u}_n$ such that

$$\frac{||\widetilde{u}_n||}{||v_n||} \to \infty.$$

Since, λv_n, evidently, corresponds to the minimal solution $\lambda \widetilde{u}_n$, we can assume, without loss of generality, that $||\widetilde{u}_n|| = 1$; then $||v_n|| \to 0$.

Since the sequence $\{\widetilde{u}_n\}$ is bounded and A is compact, the sequence $\{A\widetilde{u}_n\}$ is compact and consequently contains a convergent subsequence. Again, without loss of generality, let us assume that

$$A\widetilde{u}_n \to \widetilde{u}_0. \qquad (8.12)$$

However, since $\widetilde{u}_n = A\widetilde{u}_n - v_n$,

$$\widetilde{u}_n \to \widetilde{u}_o \text{ since } v_n \to \theta.$$

and consequently,

$$A\widetilde{u}_n \to A\widetilde{u}_0 \qquad (8.13)$$

From (8.12) and (8.13) it follows that

$$A\widetilde{u}_0 = \widetilde{u}_0, \text{ that is, } \widetilde{u}_0 \in N.$$

However, because of the minimality of the norm of the solution $\widetilde{u}$, it follows that $||\widetilde{u}_n - \widetilde{u}_0|| \ge ||\widetilde{u}|| = 1$, contradicting the convergence of $\{\widetilde{u}_n\}$ to $\widetilde{u}_0$.

Thus $||\tilde{u}||/||v||$ is bounded and if $m = \inf \{||v||/||\tilde{u}||\}$, the inequality (8.8) is proved.

Now, suppose we are given a sequence $v_n \in L$ convergent to v_0. We can assume that for some subsequence

$$||v_{n_p+1} - v_{n_p}|| < \frac{1}{2^{n_p+1}}, \quad \text{where} \quad ||v_{n_{p+1}} - v_{n_p}|| < \frac{1}{2^{n_p}}.$$

Let u_{n_0} be a minimal solution of the equation $Pu = v_{n_1}$ and $u_{n_p}, p = 1, 2, \ldots$, a minimal solution of the equation $Pu = v_{n_p+1} - v_{n_p}$.

Then $m||u_{n_p}|| \le ||v_{n_p+1} - v_{n_p}|| < \frac{1}{2^{n_p}}.$

This estimate yields that $\sum\limits_{p=1}^{\infty} u_{n_p}$ converges and if $\tilde{u}$ is the sum of the series, then

$$P\tilde{u} = P\left(\lim_{k\to\infty} \sum_{p=0}^{k} u_{n_p}\right) = \lim_{k\to\infty} \sum_{p=0}^{k} P u_{n_p}$$

$$= \lim_{k}\left[P u_{n_0} + \sum_{p=1}^{k}(v_{n_p+1} - v_{n_p})\right]$$

$$= \lim_{k} v_{n_k+1} = v_0.$$

exhibiting $v_0 \in L$. Hence, L is closed.

8.3.5   Theorem

The equation (8.4) is solvable for given $v \in E$, a Banach space, if and only if $f(v) = 0$ for every linear functional f, such that

$$A^*f - f = \theta \tag{8.14}$$

Proof: Suppose that the equation $Au - u = v$ is solvable, that is, v is expressible in the form $v = Au_0 - u_0$, for some $u_0 \in E$. Let f be any linear functional satisfying $A^*f - f = \theta$. Then

$$f(v) = f(Au_0 - u_0) = f(Au_0) - f(u_0) = A^*f(u_0) - f(u_0) = (A^*f - f)(u_0) = 0.$$

Next we have to show that $v \in L = P(E)$ satisfies the hypothesis of the theorem. Let us suppose $v \notin L$. Since L is closed, v lies at a distance $d > 0$, from L and by theorem 5.1.5, there exists a linear functional f_0 such that $f_0(v) = 1$, and $f_0(z) = 0$ for every $z \in L$. Hence $f_0(Au - u) = (A^*f_0 - f_0)(u) = 0$ for all $u \in E$, that is, $A^*f_0 - f_0 = 0$, a contradiction, because on the one hand by construction $f_0(v) = 1$, where on the other hand $f_0(v) = 0$. Hence $y \in L$, proving the sufficiency.

8.3.6   Remark

An equation $Pu = v$ with the property that it has a solution u if $f(v) = 0$ for every f, satisfying $P^*f = \theta$, is said to be **normally solvable**. The essence of the theorem 8.3.5 is that

$L = P(E)$ is closed, is a sufficient condition for $Pu = v$ to be normally solvable.

8.3.7   Corollary

If a conjugate homogeneous equation $A^*f - f = 0$ has only a trivial solution, then the equation $Au - u = v$ has a solution for any right-hand side.

8.3.8   Theorem

In order that equation (8.6) be solvable for $g \in E^*$ given, it is necessary and sufficient that $g(u) = 0$ for every $u \in E$, such that

$$Au - u = \theta. \tag{8.15}$$

Proof: To prove that the condition is necessary, we note that

$$g(u) = (A^*f - f)u = f(Au - u) = 0 \tag{8.16}$$

For proving that the condition is sufficient we proceed as follows. Let us define the function $f_0(v)$ on the subspace L by means of the equality $f_0(v) = g(u)$, u being one of the pre-images of the element v (i.e., $P^{-1}v$) under the mapping P. The functional f_0 satisfying hypothesis of the theorem is uniquely defined. For if u' is another pre-image of the same element v, there

$$Au - u = Au' - u' \quad \text{i.e.,} \quad A(u - u') - (u - u') = 0,$$

where $g(u - u') = 0$, i.e., $g(u) = g(u')$.

If u_1 and u_2 are solutions of (8.16), we have

$$g(u_1 + u_2) = (A^*f - f)(u_1 + u_2) = f(A(u_1 + u_2))$$
$$- (u_1 + u_2)) = f((Au_1 - u_1) + (Au_2 - u_2)) = 0$$

Since $g \in E^*$, $g(u_1 + u_2) = g(u_1) + g(u_2)$, $f((Au_1 - u_1) + (Au_2 - u_2))$
$$= f(Au_1 - u_1) + f(Au_2 - u_2)$$

This shows that f is additive and homogeneous. To prove the boundedness of f we proceed as follows. We can show, as in lemma 8.3.4, that the inequality $m||u|| \leq ||v||$ is satisfied for at least one of the pre-images u of the element v.

Therefore, $|f_0(v)| = |g(u)| \leq ||g|| \, ||u|| \leq \frac{1}{m}||g|| \, ||v||$ and the boundedness of f_0 is proved. We can extend f_0 by the Hahn-Banach theorem 5.1.3 to the entire space E to obtain a linear functional f, such that

$$f(Au - u) = f(v) = f_0(v) = g(u), \quad \text{or} \quad (A^*f - f)u = g(u).$$

such that f is a solution of (8.6).

8.3.9   Corollary

If the equation $Au - u = \theta$ has only a null solution $u = \theta$, then the equation $A^*f - f = g$ is solvable only when $g = \theta$ on the RHS.

We next want to show that the homogeneous and non-homogeneous equations having solutions in the identical space are also closely related.

8.3.10   Theorem

In order that the equation

$$Au - u = v \tag{8.4}$$

be solvable for every v, where A is a compact operator mapping a Banach space E into itself, it is necessary and sufficient that the corresponding homogeneous equation

$$Au - u = \theta \tag{8.15}$$

has only a trivial solution $u = \theta$. In this case, the solution of equation (8.4) is uniquely defined, and the operator $T = A - I$ has a bounded inverse.

Proof: Let us suppose that the condition is necessary. Let us denote by N_K the null space of the operator T^K. It is clear that $T^K u = \theta \Rightarrow T^{K+1}u = \theta$, that is, $N_K \subset N_{K+1}$.

Let the equation $Au - u = v$ be solvable for every v, and let us assume that the homogeneous equation $Au - u = \theta$ has a non-trivial solution u_1. Let u_2 be a solution of the equation $Au - u = u_1$, and in general let u_{k+1} be a solution of the equation $Au - u = u_k$, $k = 1, 2, 3, \ldots$. We have, $Tu_k = u_{k-1}$, $T^2 u_k = u_{k-2}, \ldots, T^{k-1}u_k = u_1 \neq \theta$. Wherever $T^k u_k = Tu_1 = \theta$. Hence, $u_k \in N_k$ and $u_k \notin N_{k-1}$, that is, each subspace N_{k-1} is a proper subspace of N_k. Then, by Riesz lemma 2.3.7 there is in the subspace N_k, an element v_k with norm 1, such that $||v_k - u|| \geq \frac{1}{2}$ for every $u \in N_{k-1}$. Consider the sequence $\{Av_k\}$, which is compact since $||v_k|| = 1$ (i.e., $\{v_k\}$ is bounded) and A is a compact operator. On the other hand, let v_p and v_q be two such elements with $p > q$.

Since $T^{p-1}(v_q - Tv_p + Tv_q) = T^{p-1}v_q - T^p v_p + T^p v_q = \theta$ noting that $p - 1 \geq q$, then $v_q - Tv_p + Tv_q \in N_{p-1}$ and hence

$$||Av_p - Av_q|| = ||v_p - (v_q - Tv_p + Tv_q)|| \geq \frac{1}{2}.$$

Thus a contradiction arises from the assumption that equation (8.4) has in the presence of a trivial solution of the equation $Tu = \theta$, a nontrivial solution. This proves the necessary part. Next to show that the condition is sufficient.

Suppose that the equation $Tu = \theta$ has only a trivial solution. Then, by corollary 8.3.9, the equation

$$A^*f - f = g \tag{8.6}$$

is solvable for any right side. Since A^* is also a compact operator and E^* a Banach space, we can apply the necessary part of the theorem just proved to equation (8.6). Hence the equation

$$A^*f - f = \theta \tag{8.14}$$

has only a trivial solution. However then equation (8.4) by corollary to theorem 8.3.6 has a solution for every v, and it is proved that the condition is sufficient.

Since by hypothesis of the theorem, equation (8.4) has a unique solution, then the inverse T^{-1} to T (i.e., $(A - I)$) exists and $T^{-1} = (A - I)^{-1}$.

Because of uniqueness property, the unique solution is at the same time minimal, and hence

$$m\|(A - I)^{-1}v\| \leq \|v\|.$$

8.3.11 Theorem

Let us consider the pair of equations,

$$Au - u = \theta \tag{8.15}$$

$$\text{and} \quad A^*f - f = \theta \tag{8.14}$$

where A and A^* are compact operators mapping respectively the Banach space E into itself and the Banach space E^* into itself. Then the above pair of equations have the same number of linearly independent solutions.

Proof: Let $u_1, u_2, \ldots, u_n$ be a basis of the subspace N of solutions of equation (8.15). Similarly, let $f_1, f_2, \ldots, f_m$ be a basis of the subspace of solutions of equation (8.14).

Let us construct a system of functionals $\phi_1, \phi_2, \ldots, \phi_n$ orthogonal to $u_1, u_2, \ldots u_n$, that is, such that

$$\phi_i(u_j) = \delta_{ij}, \quad i, j = 1, 2, \ldots n.$$

Let us also construct a system of elements $w_1, w_2, \ldots, w_m$ biorthogonal to $f_1, f_2, \ldots, f_m$.

Let us assume $n < m$. We consider the operator V given by

$$Vu = Au + \sum_{i=1}^{n} \phi_i(u)w_i.$$

Since A is a compact operator and the right-hand side of Vu contains a finite number of terms, V is a compact operator. We next want to show that the equation $Vu - u = \theta$ has only a trivial solution.

Let u_0 be a solution of $V_u - u = \theta$.

Then $\qquad f_k(Vu_0 - u_0) = 0, \quad \text{or,} \quad f_k\left(Au_0 - u_0 + \sum_{i=1}^{n} \phi_i(u_0)w_i\right) = 0$

or, $\qquad A^* f_k u_0 - f_k u_0 + \sum_{i=1}^{n} \phi_i(u_0)f_k(w_i) = 0$

or, $\qquad (A^* f_k - f_k)u_0 + \sum_{i=1}^{n} \phi_i(u_0)f_k(w_i) = 0.$

Since $\{f_i\}$ and $\{w_i\}$ are biorthogonal to each other, we have from the above equation,

$$(A^* f_k - f_k)u_0 + \phi_k(u_0) = 0.$$

Since f_k is a basis of the subspace of solutions of equation (8.14), $A^* f_k - f_k = \theta$.

Hence, $\qquad \phi_k(u_0) = 0, \ k = 1, 2, \ldots (n < m).$

Hence we have $Vu_0 = Au_0 \quad$ or, $\quad Au_0 - u_0 = Vu_0 - u_0 = \theta.$

Since $u_0 \in N$ and $\{u_i\}$ is a basis of N,

$$u_0 = \sum_{i=1}^{n} \xi_i u_i.$$

However, $\quad \phi_j(u_0) = \sum_{i=1}^{n} \xi_i \phi_j(u_i) = \xi_j.$

Since $\qquad \phi_j(u_0) = 0, \ j = 1, 2, \ldots, n, \ \xi_j = 0.$

Hence, $\qquad u_0 = \theta.$

Since the equation $Vu - u = \theta$ has only a trivial solution, the equation $Vu - u = v$ is solvable for any v and in particular for $v = w_{n+1}$. Let u' be a solution of this equation. Then we can write

$$f_{n+1}(w_{n+1}) = f_{n+1}\left(Au' - u' + \sum_{i=1}^{n} \phi_i(u')\right)(w_i)$$

$$= (A^* f_{n+1} - f_{n+1})u' + \sum_{i=1}^{n} \phi_i(u')f_{n+1}(w_i) = 0$$

where on the other hand $f_{n+1}(w_{n+1}) = 1$. The contradiction obtained proves the inequality $n < m$ to be impossible.

Let us assume, conversely, that $m < n$. Consider in the space E^*, the operator

$$V^*f = A^*f + \sum_{i=1}^{m} f(w_i)\phi_i. \tag{8.16}$$

This operator is adjoint to the operator V.

It is to be shown that the equation $V^*f - f = \theta$ has only a trivial solution.

For all $k = 1, 2, \ldots, n$.

Taking note of the biorthogonality of $\{\phi_i\}$ and $\{u_i\}$

$$(V^*f - f)u_k = (A^*f - f)u_k + \sum_{i=1}^{m} f(w_i)\phi_i(u_k)$$

$$= f(Au_k - u_k) + f(w_k)$$

$$= f(w_k) \tag{8.17}$$

since $\{u_k\}$ is one of the bases of the subspace of solutions of (8.15). Thus, if f_0 is a solution of the the equation $V^*f - f = \theta$ then from (8.17) it follows that $f_0(w_k) = 0$, $k = 1, 2, \ldots, m$.

Hence (8.16) yields $V^*f_0 = A^*f_0$.

Hence $\quad 0 = V^*f_0 - f_0 = A^*f_0 - f_0$,

i.e., f_0 is a solution of $A^*f - f = 0$.

However, $\quad f_0 = \sum_{i=1}^{m} \beta_i f_i = \sum_{i=1}^{m} f_0(w_i)f_i = \theta$,

since $\quad f_0(w_i) = 0$, $i = 1, 2, 3, \ldots$.

Since V^* is a compact operator, by theorem (8.3.10) the equation $V^*f - f = g$ has a solution for any g, particularly for, $g = \phi_{m+1}$. Therefore if f' is a solution of the above equation we have

$$V^*f' - f' = \phi_{m+1}$$

Therefore, $\phi_{m+1}(u_{m+1}) = V^*f'(u_{m+1}) - f'(u_{m+1})$

$$= (A^*f' - f')u_{m+1} + \sum_{i=1}^{m} f'(w_i)\phi_i(u_{m+1})$$

$$= f'(Au_{m+1} - u_{m+1}) = 0.$$

On the other hand, we have by construction $\phi_{m+1}(u_{m+1}) = 1$.

The contradiction obtained proves the inequality $m < n$ to be impossible. Thus $m = n$.

In what follows, we observe that if we combine the theorems 8.3.5, 8.3.8, 8.3.10 and 8.3.11 we obtain a theorem which generalizes the famous Fredholm theorem for linear integral equations to any linear equation with compact operator.

8.3.12 Theorem

Let us consider the equations

$$Au - u = v \tag{8.4}$$

and $\quad A^*f - f = g \tag{8.5}$

where A and A^* are compact operators mapping respectively Banach spaces E and E^* into itself. Then equations (8.4) and (8.5) have a solution for any element on the right side and, in this case, the homogeneous equations

$$Au - u = \theta \tag{8.13}$$

$$A^*f - f = \theta \tag{8.14}$$

have only a trivial solution or the homogeneous equations have the same finite number of linearly independent solutions $u_1, u_2, \ldots, u_n; f_1, f_2, \ldots, f_n$. In that case equation (8.4) will have a solution, if and only if,

$$f_i(v) = 0 \ (g(u_i) = 0), \ i = 1, 2, \ldots, n$$

The general solution of equation (8.4), then, takes the form,

$$u = u_0 + \sum_{i=1}^{n} a_i u_i,$$

u_0 is any solution of equation (8.4) and $a_1, a_2, \ldots a_n$ are arbitrary constants. Correspondingly, the general solution of equation (8.5), has the form

$$f = f_0 + \sum_{i=1}^{n} b_i f_i,$$

f_0 any solution of equation (8.5) and $b_1, b_2, \ldots, b_n$ are arbitrary constants.

We next consider an equation containing a parameter:

$$Au - \lambda u = v, \ \lambda \neq 0. \tag{8.18}$$

Since the equation can be expressed in the form

$$\left(\frac{1}{\lambda}\right) Au - u = \left(\frac{1}{\lambda}\right) v \quad \text{and} \quad \left(\frac{1}{\lambda}\right) A$$

is compact (completely continuous) together with A, the theorem proved for equation (8.4) remains valid for equation (8.15).

Theorem 8.3.10 implies that for a given $\lambda \neq 0$, either the equation $Au - \lambda u = v$ is solvable for any element on the right-hand side, or the homogeneous equation $Au - \lambda u = \theta$ has a non-trivial solution. Hence, every value of the parameter $\lambda \neq 0$ is either regular or is an eigenvalue and the operator A has no other non-zero point spectrum except the eigenvalues.

8.3.13   Theorem

If A is a compact operator, then its spectrum consists of finite or countable point sets. All eigenvalues are located in the interval $[-||A||, \, ||A||]$ and in the case of a countable spectrum, these have only one limit point $\lambda = 0$.

Proof: Let us consider the operator $T_\lambda = A - \lambda I$.

Now, for $\lambda \neq 0$, $T_\lambda = -\lambda \left(I - \frac{1}{\lambda} A\right)$ and by theorem 4.7.12, the operator $\left(I - \frac{1}{\lambda} A\right)$ and hence T_λ has an inverse when $\left(\frac{1}{|\lambda|}\right) ||A|| < 1$, i.e., the spectrum of the operator A lies on $[-||A||, ||A||]$. Let $0 < m < ||A||$. For a conclusive proof it will suffice to exhibit that there can exist only a finite number of eigenvalues λ, such that $|\lambda| \geq m$. If that be not true, it is possible to select a sequence $\lambda_1, \lambda_2, \dots, \lambda_n$ of distinct eigenvalues, and also $|\lambda_i| \geq m$. Let $u_1, u_2, \dots, u_n$ be a sequence of eigenvectors corresponding to these eigenvalues, such that

$$Au_n = \lambda_n u_n.$$

It is required to show that the elements $u_1, u_2, \dots, u_k$ for every k are linearly independent. For $k = 1$, this is trivial. Suppose that $u_1, u_2, \dots, u_k$ are linearly independent.

Let us assume that

$$u_{k+1} = \sum_{i=1}^{k} c_i u_i \tag{8.19}$$

then we have

$$\lambda_{k+1} u_{k+1} = Au_{k+1} = \sum_{i=1}^{k} c_i Au_i = \sum_{i=1}^{k} \lambda_i c_i u_i \tag{8.20}$$

From (8.19) and (8.20) it follows (since $\lambda_{k+1} \neq 0$) that

$$\sum_{i=1}^{k} \left(1 - \frac{\lambda_i}{\lambda_{k+1}}\right) c_i u_i = 0.$$

However this is impossible since $1 - \dfrac{\lambda_i}{\lambda_{k+1}} \neq 0$

and $u_1, u_2, \dots, u_{k+1}$ are linearly independent. Hence the distinct eigenvalues are finite in number. For proof in the case of a countable spectrum see theorem 8.2.7.

8.4   Approximate Solutions

In the last section we have seen that, given A, a linear compact operator mapping a Banach space E into itself and that if the homogeneous equation

$u - Au = \theta$ has only a trivial solution, then the equation $u - Au = v$ has a unique solution.

In this section we consider the question of finding approximate solution to the unique solution. We consider here operators with finite rank, i.e., operators having finite dimensional range. The process of finding such an approximate solution has a deep relevance. In numerical analysis, in case we cannot find the solution of an equation in a closed form, we find an approximation to such operator equation, so that the approximations can be reduced to finite dimensional equations. To make the analysis complete, it is imperative in this case that the approximate operator equations have a unique solution and this solution tends to the exact solution of the original equation in the limit. Thus, if A is a bounded linear operator mapping E into E and if $\widetilde{A}$ is approximate to A, $v_0 \in E$ is approximate to v, thus the element $u_0 \in E$ satisfying $u_0 - \widetilde{A}u_0 = v_0$ is a close approximation to u satisfying $u - Au = v$.

8.4.1 *Theorem*

Let E be a Banach space and A be a compact operator on E, such that $x = \theta$ is the only solution of $x - Ax = \theta$. Then $(I - A)$ is invertible. Let $\widetilde{A} \in (E_x \to E_x)$ satisfy

$$\epsilon = ||(A - \widetilde{A})(I - A)^{-1}|| < 1.$$

Then for given $v, v_0 \in E_x$, there are unique $u, u_0 \in E$ such that

$$u - Au = v, \quad u_0 - Au_0 = v_0$$

and $||u - u_0|| \leq \dfrac{||(I - A)^{-1}||}{1 - \epsilon}(\epsilon||v|| + ||v - v_0||).$

Proof: Since A is compact and $I - A$ is one-to-one, it follows from theorem 8.2.5(a) that $(I - A)$ is invertible. As E is a Banach space and

$$||[(I - A) - (I - \widetilde{A})](I - A)^{-1}|| = \epsilon < 1,$$

it follows from theorem 4.7.12 that $(I - \widetilde{A})$ is invertible and

$$||(I - \widetilde{A})^{-1}|| \leq \frac{||(I - A)^{-1}||}{1 - \epsilon}, \quad ||(I - A)^{-1} - (I - \widetilde{A})^{-1}|| \leq \frac{\epsilon||(I - A)^{-1}||}{1 - \epsilon}.$$

Let $v, v_0 \in E$, since $I - A$ and $I - \widetilde{A}$ are invertible, there are unique $u, u_0 \in E$ such that

$$u - Au = v \quad \text{and} \quad u_0 - \widetilde{A}u_0 = v_0.$$

Also, $u - u_0 = (I - A)^{-1}v - (I - \widetilde{A})^{-1}v_0$

$$= [(I - A)^{-1} - (I - \widetilde{A})^{-1}]v + (I - \widetilde{A})^{-1}(v - v_0).$$

Hence, $||u - u_0|| \leq \dfrac{\epsilon||(I - A)^{-1}||}{1 - \epsilon}||v|| + \dfrac{||(I - A)^{-1}||}{1 - \epsilon}||v - v_0||.$

We would next show how the operator $\widetilde{A}$ can be constructed. We would also show that if $\widetilde{A}$ is an operator of finite rank, then the solution of the equation $u_0 - \widetilde{A}u_0 = v_0$ can be reduced to the solution of a finite system of linear equations which can be solved by standard methods. Next, when the operator A is compact, we can find several ways of constructing a bounded linear operator $\widetilde{A}$ of finite rank such that $||A - \widetilde{A}||$ is arbitrarily small.

8.4.2 Theorem

Let $\widetilde{A}$ be an operator of finite rank on a normed linear space E over $\mathbb{R}(\mathbb{C})$ given by

$$\widetilde{A}u = f_1(u)u_1 + \cdots + f_m(u)u_m, \ u \in E$$

where $u_1, u_2, \ldots, u_m$ are in E, and $f_1, f_2, \ldots, f_m$ are linear functionals on E.

Let $\quad M = \begin{bmatrix} f_1(u_1) & \cdots & f_1(u_m) \\ \vdots & & \\ f_m(u_1) & \cdots & f_m(u_m) \end{bmatrix}$

(a) Consider $v_o \in E$ and let $\widetilde{v} = (f_1(v_0), f_2(v_0), \ldots, f_m(v_0))^T$

Then $\quad u_0 - \widetilde{A}u_0 = v_0 \quad$ and $\quad \widetilde{u} = (f_1(u_0), \ldots, f_m(u_0))^T$,

if and only if $\widetilde{u} - M\widetilde{u} = \widetilde{v}$

and $\quad u_0 = v_0 + \widetilde{u}^1 u_1 + \widetilde{u}^2 u_2 + \cdots + \widetilde{u}^m u_m$

where $\widetilde{u}^i, i = 1, \ldots, m$ is the i-th component of $\overline{u}$.

(b) Let $0 \neq \lambda \in \mathbb{R}(\mathbb{C})$. Them λ is an eigenvalue of $\widetilde{A}$ if and only if λ is an eigenvalue of M.

Furthermore, if $\widetilde{u}$ (resp., u_0) is an eigenvector of M (resp., $\widetilde{A}$) corresponding to λ, then

$$u_0 = \widetilde{u}^1 u_1 + \widetilde{u}^2 u_2 + \cdots + \widetilde{u}^m u_m.$$

(resp., $\widetilde{u} = (f_1(u_0), \ldots, f_m(u_0))^T$ is an eigenvector of $\widetilde{A}$ (resp., M) corresponding to λ,

Proof: Let $u_0 - \widetilde{A}u_0 = v_0$ and $\widetilde{u} = (f_1(u_0), \ldots, f_m(u_0))^T$

Then for $i = 1, 2, \ldots, m$,

$$\begin{aligned}
(M\widetilde{u})^{(i)} &= f_i(u_1)f_1(u_0) + \cdots + f_i(u_m)f_m(u_0) \\
&= f_i(f_1(u_0)u_1 + \cdots + f_m(u_0)u_m) \\
&= f_i(\widetilde{A}u_0) = f_i(u_0 - v_0) = f_i(u_0) - f_i(v_0) \\
&= \widetilde{u}^{(i)} - \widetilde{v}^{(i)}.
\end{aligned}$$

Hence $\widetilde{u} - M\widetilde{u} = \widetilde{v}$.

Also, $u_0 = v_0 + \widetilde{A}u_0 = v_0 + f_1(u_0)u_1 + \cdots + f_m(u_0)u_m$
$$= v_0 + \widetilde{u}^1 u_1 + \cdots + \widetilde{u}^m u_m.$$

Conversely, let $\widetilde{u} - M\widetilde{u} = \widetilde{v}$ and $u_0 = v_0 + \widetilde{u}^1 u_1 + \cdots + \widetilde{u}^m u_m$.

Then, $\widetilde{A}u_0 = f_1(u_0)u_1 + \cdots + f_m(u_0)u_m$

$$= \left[f_1(v_0) + \sum_{j=1}^{m} \widetilde{u}^j f_1(u_j) \right] u_1 + \cdots + \left[f_m(v_0) + \sum_{j=1}^{m} \widetilde{u}^j f_m(u_j) \right] u_m$$

$$= [\widetilde{v}^1 + (M\widetilde{u})^1]u_1 + \cdots + [\widetilde{v}^m + (M\widetilde{u})^m]u_m$$

$$= \widetilde{u}^1 u_1 + \cdots + \widetilde{u}^m u_m = u_0 - v_0.$$

Also for $i = 1, 2, \ldots, m$

$$\widetilde{u}^{(i)} = \widetilde{v}^{(i)} + (M\widetilde{u})^{(i)} = \widetilde{v}^{(i)} + f_i(u_1)\widetilde{u}^{(1)} + f_i(u_2)\widetilde{u}^{(2)}$$
$$+ \cdots + f_i(u_m)\widetilde{u}^{(m)}$$

$$= \widetilde{v}^{(i)} + f_i(\widetilde{u}^{(1)}u_1 + \widetilde{u}^{(2)}u_2 + \cdots + \widetilde{u}^{(m)}u_m)$$

$$= \widetilde{v}^{(i)} + f_i(u_0 - v_0) = f_i(u_0),$$

i.e., $\widetilde{u} = (f_1(u_0), \ldots, f_m(u_0))^T.$

(b) Since $\lambda \neq 0$, let $\mu = \frac{1}{\lambda}$. Replacing $\widetilde{A}$ by $\mu\widetilde{A}$ and letting $v_0 = \theta$ in (a) above, we see that

$$u_0 - \mu\widetilde{A}u_0 = \theta \quad \text{and} \quad u_0 = \widetilde{u}^{(1)}u_1 + \cdots + \widetilde{u}^{(m)}u_m.$$

Hence $\widetilde{A}u_0 = \lambda u_0$ with $u_0 \neq \theta$ if and only if $M\widetilde{u} = \lambda\widetilde{u}$ with $\widetilde{u} \neq \theta$. Thus, λ is an eigenvalue of $\widetilde{A}$ if and only if λ is an eigenvalue of M. Also, the eigenvectors of $\widetilde{A}$ and M corresponding to λ are related by $\widetilde{u} = (f_1(u_0), \ldots, f_m(u_0))$ and

$$u_0 = \widetilde{u}^{(1)}u_1 + \widetilde{u}^{(2)}u_2 + \cdots + \widetilde{u}^{(m)}u_m.$$

Letting $\lambda = 1$ in (b) above, we see that $\widetilde{A}u = u$ has a non-zero solution in E if and only if $Mu = u$ has a non-zero solution in $\mathbb{R}^m$ ($\mathbb{C}^m$). Also, for a given $v_0 \in E$, the general solution of $u - \widetilde{A}u = v_0$ is given by $u = v_0 + \widetilde{u}^{(1)}u_1 + \cdots + \widetilde{u}^{(m)}u_m$, where $\widetilde{u} = (\widetilde{u}^{(1)}, \widetilde{u}^{(2)}, \ldots, \widetilde{u}^{(m)})^T$ is the general solution of $u - Mu = (f_1(v_0), \ldots, f_m(v_0))^T$. Thus, the problem of solving the operator equation $u - \widetilde{A}u = v_0$ is reduced to solving the matrix equation

$$u - Mu = \widetilde{v} \quad \text{where} \quad \widetilde{v} = (f_1(v_0), \ldots, f_m(v_0))^T.$$

We next describe some methods of approximating a compact operator by bounded operators of finite rank. First, we describe some methods related to projections.

8.4.3   Theorem

Let E be a Banach space and A be a compact operator on E. For $n = 1, 2, \ldots$, let $P_n \in (E \to E)$ be a projection of finite rank and

$$A_n^P = P_n A, \; A_n^S = A P_n, \; A_n^G = P_n A P_n.$$

If $P_n u \to u$ in E for every $u \in E$, then $||A_n^P - A|| \to 0$. If, in addition, $P_n^T u^T \to u^T$ in E^*, for $u^T \in E^*$, then $||A_n^S - A|| \to 0$ and $||A_n^G - A|| \to 0$.

Proof: Let $P_n u \to u$ in E for every u in E. Then it follows that $A_n^P u \to Au$ in E for every $u \in E$. Since A is a compact linear operator mapping $E \to E$, the set $G = \{Au : u \in E, \; ||u|| \leq 1\}$ is totally bounded. As E is a Banach space, we show below that $\{P_n v\}$ converges to v uniformly on E. Since G is totally bounded, given $\epsilon > 0$, there are $v_1, \ldots v_m$ in G such that

$$G \subseteq B(v_1, \epsilon) \cup B(v_2, \epsilon) \cup \cdots \cup B(v_m, \epsilon).$$

Now, $P_n v_j \to v_j$ as $n \to \infty$ for each $j = 1, 2, \ldots, m$. Find n_0 such that $||P_n v_j - v_j|| < \epsilon$, for all $n \geq n_0$ and $j = 1, 2, \ldots, m$. Let $v \in G$, and chose v_j in E such that $||v - v_j|| < \epsilon$. Then, for all $n \geq n_0$, we have,

$$||P_n v - v|| \leq ||P_n(v - v_j)|| + ||(P_n - I)v_j|| + ||I(v_j - v)||.$$
$$\leq (||P_n|| + ||I||)||v - v_j|| + ||P_n v_j - v_j|| \leq 3\epsilon.$$

Thus $P_n v$ converges to v uniformly on G. Hence,

$$||A_n^P - A|| = ||(P_n - I)A|| = \sup_{||v|| \leq 1} ||(P_n - I)Av|| \to 0 \text{ as } n \to \infty.$$

Let us next assume that $P_n^* u^T \to u^T$ in E^* for every $u^T \in E^*$ as well. By theorem 8.1.12, A^* is a compact operator on E and

$$(A_n^S)^* = (A P_n)^* = P_n^* A^*.$$

Replacing A by A^* and P_n by P_n^* and recalling theorem 6.1.5 (ii), we see that

$$||A_n^S - A|| = ||(A_n^S - A)^*|| = ||P_n^* A^* - A^*|| \to 0, \text{ as before.}$$

Also, $\;||A_n^G - A|| = ||P_n A P_n - P_n A + P_n A - A||$

$$\leq ||P_n(A P_n - A)|| + ||P_n A - A||$$
$$\leq ||P_n|| \, ||A_n^S - A|| + ||A_n^P - A||.$$

which tends to zero as $n \to \infty$, since the sequence $\{||P_n||\}$ is bounded by theorem 4.5.7.

8.4.4   Remark

Definitions A_n^P, A_n^S, A_n^G

(i) A_n^P is called the projection of A on the n-dimensional subspace and is expressed as $A_n^P = P_n A$.

(ii) $A_n^S = AP_n$ is called the Sloan projection in the name of the mathematician Sloan.

(iii) $A_n^G = P_n AP_n$ is called the Galerkin projection in the name of the mathematician Galerkin.

8.4.5   Example of projections

We next describe several ways of constructing bounded projections P_n of finite rank such that $P_n x \to x$ as $n \to \infty$.

1. Truncation of Schauder expansion

Let E be a Banach space with a Schauder basis $\{e_1, e_2, \ldots\}$. Let $f_1, f_2, \ldots$ be the corresponding coefficient functionals. For $n = 1, 2, \ldots$ define

$$P_n u = \sum_{k=1}^{n} f_n(u) e_k \quad u \in E.$$

Now each $f_k \in E^*$ and hence each $P_n \in (E \to E)$. Now, $P_n^2 = P_n$ and each P_n is of finite rank.

The very definition of the Schauder basis implies that $P_n u \to u$ in E for every $u \in E$. Hence $\|A - A_n^P\| \to 0$ if A is a compact operator mapping $E \to E$.

2. Projection of an element in a Hilbert space

Let H be a separable Hilbert space and $\{u_1, u_2, \ldots\}$ be an orthonormal basis for H [see 3.8.8]. Then for $n = 1, 2, \ldots,$

$$P_n u = \sum_{k=1}^{n} <u, u_k> u_k, \quad u \in H,$$

where $< \cdot >$ is the inner product on H. Note that each P_n is obtained by truncating the Fourier expansion of $u \in H$ [see 3.8.6]. Since H^* can be identified with H (Note 5.6.1) and P_n^* can be identified with P_n we obtain $\|A - A_n^S\| \to 0$ and $\|A - A_n^G\| \to 0$, in addition to $\|A - A_n^P\| \to 0$ as $n \to \infty$.

Piece-wise linear interpolations

Let $E = C([a, b])$ with the sup norm. For $n = 1, 2, \ldots,$ consider n nodes $t_1^{(n)}, t_2^{(n)} \cdots t_n^{(n)}$ in $[a, b]$: i.e., $a = t_0^{(n)} \le t_1^n < \cdots < t_n^{(n)} \le t_{n+1}^{(n)} = b$.

For $j = 1, 2, \ldots, n$, let $u_j^{(n)} \in C([a, b])$ be such that

(i) $u_j^{(n)}(t_i^{(n)}) = \delta_{ij}, \ i = 1, \ldots n$

(ii) $u_1^{(n)}(a) = 1, \ u_j^{(n)}(a) = 0$ for $j = 2, \ldots, n$

$u_n^{(n)}(b) = 1, \ u_j^{(n)}(b) = 0$ for $j = 1, 2, \ldots, n - 1,$

(iii) $u_j^{(n)}$ is linear on each of the subintervals $[t_k^{(n)}, t_{k+1}^{(n)}], \ k = 0, 1, 2, \ldots, n.$

The functions $u_1^{(n)}, u_2^{(n)}, \ldots, u_n^{(n)}$ are known as the **hat functions** because of the shapes of their graphs. Let $t \in [a, b]$. Then $u_j^{(n)}(t) \geq 0$ for all $j = 1, 2, \ldots, n$. If $t \in [t_k^{(n)}, t_{k+1}^{(n)}]$, then $u_k^{(n)}(t) + u_{k+1}^{(n)}(t) = 1$ and $u_j^{(n)}(t) = 0$ for all $j \neq k, \ k + 1$. Thus $u_1^{(n)}(t) + u_2^{(n)}(t) + \cdots + u_n^{(n)}(t) = 1$.

For $x \in C([a, b])$, define

$$P_n(x) = \sum_{j=1}^{n} x(t_j^{(n)}) u_j^{(n)}.$$

Then P_n is called a **piecewise linear interpolatory projection**. Let $h_n = \max\{t_j^{(n+1)} - t_j^{(n)}; \ j = 0, 1, 2, \ldots, n\}$ denote the mesh of the partition of $[a, b]$ by the given nodes. We show that $P_n x \to x$ in $C([a, b])$, provided $h_n \to 0$ as $n \to \infty$. Let us fix $x \in C([a, b])$ and let $\epsilon > 0$. By the uniform continuity of x on $[a, b]$, there is some $\delta > 0$ such that $|x(s) - x(t)| < \epsilon$ wherever $|s - t| < \delta$. Let us choose N such that $h_n < \delta$ for all $n \geq N$. Consider $n \geq n_0$ and $t \in [a, b]$.

If $u_j^{(n)}(t) \neq 0$, then $t \in [t_{j-1}^{(n)}, t_{j-1}^{(n)}]$, so that

$$|t_j^{(n)} - t| \leq h_n < \delta \quad \text{and} \quad |x(t_j^{(n)}) - x(t)| < \epsilon.$$

Hence, $|P_n x(t) - x(t)| = \left| \sum_{j=1}^{n} (x(t_j^{(n)}) - x(t)) u_j^{(n)}(t) \right|$

$$\leq \sum_{j=1}^{n} |x(t_j^{(n)}) - x(t)| u_j^{(n)}(t)$$

$$\leq \sum_{j=1}^{n} \epsilon u_j^{(n)}(t) = \epsilon.$$

Thus $\|P_n x(t) - x(t)\|_\infty \to 0$, provided $h_n \to 0$.

3. Linear integral equation with degenerate kernels

Let E denote either $C([a, b])$ or $L_2([a, b])$. We consider the integral equation

$$\int_0^1 k(t, s) x(s) ds = y(t). \tag{8.21}$$

The kernel $k(t, s)$ is said to be **degenerate** if

$$k(t, s) = \sum a_i(t) b_i(s), \ t, s \in [a, b]. \tag{8.22}$$

Thus the kernel can be expressed as the sum of the products of functions, one depending exclusively on t and the other depending exclusively on s. a_i and b_i belong to $E, i = 1, 2, \ldots, m$.

Thus, the equation (8.21) can be reduced to the form

$$\int_a^b k(t,s)x(s)ds = \int_a^b \sum_{i=1}^m a_i(t)b_i(s)x(s)ds = y(t)$$

Hence, $\sum_{i=1}^m a_i(t)\left\{\int_a^b b_i(s)x(s)ds\right\} = y(t)$ for $t \in [a,b]$.

We write $\tilde{A}x(t) = \sum_{i=1}^m \left\{\int_a^b b_i(s)x(s)ds\right\} a_i(t) = y(t)$.

We note that $\tilde{A} \in (E \to E)$. Also $\tilde{A}$ is of finite rank because $R(\tilde{A}) \subseteq$ span $\{a_1(t),\ldots,a_m(t)\}$.

8.4.6   Theorem

Let $E = C([a,b])$ (resp. $L_2([a,b])$ and $K(t,s) \in C([a,b] \times [a,b])$. Let $(k_n(\ ,\ ))$ be a sequence of degenerate kernels in $C([a,b] \times [a,b])$ (respectively, $L_2([a,b] \times [a,b])$ such that $||k - k_n||_\infty \to 0$. (resp. $||k - k_n||_2 \to 0$). If A and A_n^D are the Fredholm integral operators with kernels $k(\cdot,\cdot)$ and $k_n(\cdot,\cdot)$ respectively, then $||A - A_n|| \to 0$, where $||\cdot||$ denotes the operator norm in $(E \to E)$.

Proof: Let $E = C([a,b])$.

Then $||(A - A_n^D)x||_\infty = \left\|\int_a^b (k - k_n)(t,s)x(s)ds\right\|_\infty$

$$\leq (b-a)||(k - k_n)||_\infty ||x||_\infty.$$

Hence, $||(A - A_n^D)||_\infty \leq (b-a)||k - k_n||_\infty \to 0$ as $n \to \infty$.

Next, let $E = L_2([a,b])$.

$$||(A - A_n^D)x||_2 = \left\|\int_a^b (k - k_n)(t,s)x(s)ds\right\|_2$$

$$\leq \left(\int_a^b [k(t,s) - k_n(t,s)]^2 ds\right)^{\frac{1}{2}} \left(\int_a^b x^2(s)ds\right)^{\frac{1}{2}}$$

$$= ||k - k_n||_2 ||x||_2 \quad \text{[see example 8.1.10]}$$

Hence, $||A - A_n^D||_2 \leq ||k - k_n||_2 \to 0$ as $n \to \infty$.

4.   Truncation of a Fourier expansions

Let $k(\ ,\ ) \in L_2([a,b] \times [a,b])$ and $\{e_1, e_2, \ldots\}$ be an orthonormal basis for $L_2([a,b])$. For $i,j = 1,2,\ldots$ let $w_{i,j}(t,s) = e_i(t)\overline{e_j(s)}, t,s \in [a,b]$. Then $\{w_{i,j}; i,j = 1,2,\ldots\}$ is an orthonormal basis for $L_2([a,b] \times [a,b])$. Then by 3.8.6,

$$k = \sum_{i,j}(k, w_{i,j})w_{i,j}.$$

For $n, 1, 2, \ldots$ and $s, t \in [a, b]$. Let

$$k_n(t, s) = \sum_{i,j=1}^{n} \langle k, w_{i,j} \rangle w_{i,j}(t, s) = \sum_{i,j=1}^{n} \langle k, w_{i,j} \rangle e_i(t)\overline{e_j(s)},$$

where $\quad < k, w_{i,j} >= \int_a^b \int_a^b k(t, s)\overline{e_i(t)}e_j(s)dtds \quad i, j = 1, 2, \ldots$.

Thus $k_n(\cdot)$ is a degenerate kernel and $\|k - k_n\|_2 \to 0$ as $n \to \infty$.

8.4.7 Examples

Let us consider the infinite dimensional homogeneous system of equations

$$x_i - \sum_{j=1}^{\infty} a_{i,j}x_j = 0, \; i = 1, 2, \ldots \infty. \tag{8.23}$$

Let $\{x_i\}$ be a denumerable set and let

$$a_{i,j} \in \mathbb{R}(\mathbb{C}) \quad \text{where} \quad \sum_{i,j=1}^{\infty} |a_{i,j}|^2 < \infty.$$

Let the only square-summable solution of (8.23) be zero.

Let $\quad A = \begin{bmatrix} a_{11} & a_{12} & \cdots & a_{1n} & \cdots \\ \vdots & & & & \\ a_{n1} & a_{n2} & \cdots & a_{nn} & \cdots \\ \cdots & \cdots & \cdots & \cdots & \cdots \\ \cdots & \cdots & \cdots & \cdots & \cdots \\ \cdots & \cdots & \cdots & \cdots & \cdots \end{bmatrix} \tag{8.24}$

Let $\quad X = \begin{bmatrix} x_1 \\ x_2 \\ \vdots \\ x_n \\ \vdots \end{bmatrix} \in l_2, \quad Y = \begin{bmatrix} y_1 \\ y_2 \\ \vdots \\ y_n \\ \vdots \end{bmatrix} \in l_2.$

Consider the infinite dimensional equation

$$X - AX = Y, \tag{8.25}$$

If we now truncate the equation to a n-dimensional subspace then (8.25) reduces to

$$X_n - A_n X_n = Y_n \tag{8.26}$$

where $\quad A_n = \begin{bmatrix} a_{11} & a_{12} & \cdots & a_{1n} \\ \vdots & & & \\ a_{n1} & a_{n2} & \cdots & a_{nn} \end{bmatrix} \quad X_n = \begin{bmatrix} x_1 \\ \vdots \\ x_n \end{bmatrix}.$

$$Y_n = \begin{bmatrix} y_1 \\ y_2 \\ \vdots \\ y_n \end{bmatrix}$$

Since the homogeneous equation (8.23) has the only solution as zero, the equation (8.25) has a unique solution. If we let $X_n^{(i)} = 0$, for $i = n+1, \ldots$ then, the sequence $\{X_n\}$ converges in l_2 to the unique solution

$$\hat{X} = \begin{pmatrix} \hat{x}_1 \\ \hat{x}_2 \\ \vdots \\ \hat{x}_n \\ \vdots \end{pmatrix} \text{ of the denumerable system given by}$$

$$x_i - \sum_{j=1}^{\infty} a_{ij} x_j = y_i, \quad i = 1, 2, \ldots \infty. \tag{8.27}$$

In fact, $\quad \|X - X_n\| \le \dfrac{\|(I-A)^{-1}\|}{1 - \epsilon_n}(\epsilon_n \|Y\|_2 + \|Y - Y_n\|_2)$

provided $\quad \epsilon_n = \|A - A_n\| < \dfrac{1}{\|(I-A)^{-1}\|}$,

where $\quad AX = \left(\displaystyle\sum_{j=1}^{\infty} a_{ij} x_j \right), \quad X \in l_2, \, i = 1, 2, \ldots, n, \ldots, \infty$

$$A_n X = \left(\begin{cases} \displaystyle\sum_{j=1}^{n} a_{ij} x_j & \text{if } i = 1, 2, \ldots, n \\ 0 & \text{if } i > n \end{cases} \right.$$

These results follow from theorem 8.4.1 and theorem 8.4.3 if we note that A is a compact operator and if

$$P_n X = (x_1, \ldots, x_n, 0, \ldots, 0)^T, \quad X \in E,$$

then $A_n = P_n A P_n$. Since P_n is obtained by truncating the Fourier series of $X \in l_2$, we see that $P_n X \to X$ for every X in l_2 and $P_n^* X^T \to X^T$ for every X^T in $(l_2)^*$. Hence $\|A - A_n^G\|_2 \to 0$.

We note that $(x_1, \ldots x_n)^T$ is a solution of the system (8.26) if and only if $(x_1, \ldots x_n, 0 \ldots 0)^T$ is a solution of the system

$$X - A_n X = (y_1, \ldots, y_n, 0, 0, \ldots)^T, \quad X \in l_2.$$

Problems [8.3 and 8.4]

1. Let $E_x = C([0,1])$ and define $A : E_x \to E_x$ by $Ax(t) = u(t)x(t)$ where $u(t) \in E_x$ is fixed. Find $\sigma(A)$ and show that it is closed (Hint:

$\lambda(t) = u(t)$ which is a continuous function defined on a compact set $[0,1]$.)

2. Let A be a compact linear operator mapping E_x into E_x and $\lambda \neq 0$. Then show that $A - \lambda I$ is injective if and only if it is surjective.

3. If $\lambda = \lambda_i$ for some i, show that the range of $\lambda - T$ consists of all vectors orthogonal to the eigenspace corresponding to λ_i, where T is self-adjoint and compact. For such a vector show that the general solution of $(\lambda - T)u = f$ is

$$u = \frac{1}{\lambda} f + \frac{1}{\lambda} \sum_{\lambda_j \neq \lambda} \lambda_j \frac{< f, u_j >}{\lambda - \lambda_j} + g,$$

where g is an arbitrary element of the eigenspace corresponding to λ_i [see Taylor [55], ch. VI].

4. If A maps a normed linear space E_x into E_y and A is compact, then show $\mathcal{R}(A)$ is separable.

(Hint: $\mathcal{R}(A) = \bigcup_{n=1}^{\infty} A(Z_n)$ where $Z_n = \{x : ||x|| \leq n\}$).

Since A is compact, it may be seen that every infinite subset of $A(Z_n)$ has a limit point in E_y. Consequently, for each positive integer m there is a finite set of points in $A(Z_n)$ such that the balls of radius $\frac{1}{m}$ with center at these points cover $A(Z_n)$.)

5. Show that the operator C, such that

$$Cu = -\frac{d^2 u}{dx^2},$$

subject to the boundary conditions

$$u'(0) = u'(1) = 0,$$

is positive but not positive-definite.

(Hint: If $\langle u, v \rangle = \int_0^1 u(x)v(x)dx$, show that $\langle Cu, u \rangle \geq 0$. If $u = 1$ then $\langle Cu, u \rangle = 0$ [see 9.5].)

6. Find the eigenvalues and eigenvectors of the operator C given in problem 5.

(Hint: Take $u(x) = \sum C_n \cos n\pi x$

where $\int_0^1 \cos n\pi x \cos m\pi x = 0$ for $n \neq m$.)

7. Suppose E_x and E_y are infinite dimensional normed linear spaces. Show that if $A : E_x \to E_x$ is a surjective linear operator, then A is not a compact operator.

8. Let $P_0 = (1/ij)$, $i, j = 1, 2, \ldots$ and $v_0 \in l_2$. Then, show that the unique $u_0 \in l_2$ satisfying $u_0 - P_0 u_0 = v_0$ is given by

$$u_0 = v_0 + \left[\frac{6}{6 - \pi^2} \sum_{j=1}^{\infty} \frac{v_{0,j}}{j} \right] \left(1, \frac{1}{2}, \frac{1}{3}, \cdots \right)^T$$

$$\left(\text{Hint: } P_0 u_0 = \left(\sum_{j=1}^{\infty} \frac{u_{0,j}}{j} \right) \left(1, \frac{1}{2}, \frac{1}{3}, \cdots \right)^T \text{ for } u_0 \in l_2 \right)$$

9. Let $E_x = C([0, 1])$, $k(\cdot, \cdot) \in C([0, 1] \times [0, 1])$ and P be the Fredholm integral operator with kernel $k(,)$.

(a) If $|\mu| < 1/\|k\|_\infty$, then for every $v \in C([0, 1])$, show that there is a unique $u \in C([0, 1])$ such that $u - \mu P u = v$. Further, let

$$u_n(s) = v(s) + \sum_{j=1}^{n} \mu^j \left[\int_0^1 k^{(j)}(s, t) v(t) d\sigma(t) \right], s \in [0, 1],$$

where $k^{(j)}(\cdot, \cdot)$ is the jth iterated kernel then, show that

$$\|u_n - u\| \le \|v\|_\infty \frac{|\mu|^{n+1} \|k\|_\infty^{n+1}}{1 - |\mu| \, \|k\|_\infty}$$

(b) If $k(s, t) = 0$, for all $s \le t$ and $0 \ne \mu \in \mathbb{R}(\mathbb{C})$ then show that

$$\|u_n - u\|_\infty \le \|v\|_\infty \sum_{j=n+1}^{\infty} \frac{|\mu|^j \|k\|_\infty}{j!}.$$

CHAPTER 9

ELEMENTS OF SPECTRAL THEORY OF SELF-ADJOINT OPERATORS IN HILBERT SPACES

A Hilbert space has some special properties: it is self-conjugate as well as an inner product space. Besides adjoint operators, this has given rise to self-adjoint operators which have immense applications in analysis and theoretical physics. This chapter is devoted to a study of self-adjoint bounded linear operators.

9.1 Adjoint Operators

In 5.6.14 we defined **adjoint operators** in a Banach space. In a Hilbert space we can use inner product to obtain an adjoint to a given linear operator. Let H be a Hilbert space and A a bounded linear operator defined on H with range in the same space. Let us consider the functional

$$f_y(x) = \langle Ax, y \rangle, \ y \in H \tag{9.1}$$

As a linear functional in the Hilbert space, $f_y(x)$ can always be written in the form $f_y(x) = \langle x, y^* \rangle$ where y^* is some element in H. Let us suppose that there exists another element $z^* \in H$ s.t. $f_y(x) = \langle x, z^* \rangle$. Then we have $\langle x, y^* - z^* \rangle = 0$. Since x is arbitrary and $x \perp (y^* - z^*)$ we have $y^* = z^*$. Thus $y^* \in H$ can be uniquely associated to the element y, identifying the functional f_y. Thus we can find a correspondence between y and y^* given by $y^* = A^*y$. A^* is an operator defined on H with range in H. Then operator A^* is associated with A by

$$\langle Ax, y \rangle = \langle x, A^*y \rangle, \tag{9.2}$$

and is called the **adjoint operator** of A. If A^* is not unique, let us suppose, $\langle Ax, y \rangle = \langle x, A^*y \rangle = \langle x, A_1^*y \rangle$.

Hence $\langle x, (A^* - A_1^*)y \rangle = 0$ i.e., $x \perp (A^* - A_1^*)y$. Since x is arbitrary, $(A^* - A_1^*)y = 0$. Again the result is true for any y implying that $A^* = A_1^*$.

It can be easily seen that the definition of adjoint operator derived here formally coincides with the definition given in 5.6.14, for the case of Banach spaces. It can be easily proved that the theorems on adjoint operators for Banach spaces developed in 5.6 remain valid in complex Hilbert space too.

Note 9.1.1. In 5.6.18, we defined the adjoint of an unbounded linear operator in a space E_x. Let H be a Hilbert space. Let A be a linear operator (unbounded) with domain $\mathcal{D}(A)$ everywhere dense in H. If the scalar product $\langle Ax, y \rangle$ for a given fixed y and every $x \in \mathcal{D}(A)$ can be represented in the form

$$\langle Ax, y \rangle = \langle x, y^* \rangle,$$

then it can be seen that y belongs to the domain $\mathcal{D}_{A^*}$ of the operator, adjoint of A. The adjoint operator A^* itself is thus defined by

$$A^*y = y^*.$$

It can be argued as in above, that y^* is unique and A^* is a linear operator. Here, $y \in \mathcal{D}(A^*)$.

9.1.1 Lemma

Given a complex Hilbert space H, the operator A^* adjoint to a bounded linear operator A is bounded and $||A|| = ||A^*||$.

Proof: $||Ax||^2 = \langle Ax, Ax \rangle = \langle x, A^*Ax \rangle \leq ||x|| \, ||A^*Ax||$

Or, $\sup_{x \neq \theta} \dfrac{||Ax||}{||x||} \leq \sup_{Ax \neq \theta} \dfrac{||A^*Ax||}{||Ax||}$. Hence, $||A|| \leq ||A^*||$.

Similarly, considering $||A^*x||^2$ we can show that $||A^*|| \leq ||A||$.

Hence, $||A|| = ||A^*||$, showing that A^* is bounded.

9.1.2 Lemma

In H, $A^{**} = A$.

Proof: The operator adjoint ot A^* is denoted by A^{**}.

We have $\langle A^*x, y \rangle = \overline{\langle y, A^*x \rangle} = \overline{\langle Ay, x \rangle} = \langle x, Ay \rangle, \ \forall \, x, y \in H$

Therefore, $\langle A^*x, y \rangle = \langle x, A^{**}y \rangle = \langle x, Ay \rangle$ showing that $A^{**} = A$.

9.1.3 Remark

(i) $A^{***} = A^*$.

(ii) For A and B linear operators $(A + B)^* = A^* + B^*$.

(iii) $(\lambda A)^* = \bar{\lambda} A^*, \lambda \in \mathbb{C}$

(iv) $(AB)^* = B^* A^*$.

(v) If A has an inverse A^{-1} then A^* has an inverse and $(A^*)^{-1} = (A^{-1})^*$.

Proof: (i) $A^{***} = (A^{**})^* = A^*$ using lemma 9.1.2.

(ii) For $x, y \in H$ $\langle (A + B)^* x, y \rangle = \langle x, (A + B)y \rangle = \langle x, Ay \rangle + \langle x, By \rangle$
$$= \langle A^* x, y \rangle + \langle B^* x, y \rangle.$$

or, $\langle [(A + B)^* - A^* - B^*]x, y \rangle = 0$.

Since x and y are arbitrary, $(A + B)^* = A^* + B^*$.

(iii) $\langle \lambda A x, y \rangle = \langle \lambda\ Ax, y \rangle = \langle x, \bar{\lambda} A^* y \rangle = \langle x, (\lambda A)^* y \rangle\ \forall\ x, y \in H$.

Hence, $(\lambda A)^* = \bar{\lambda} A^*$.

(iv) $\langle AB x, y \rangle = \langle Bx, A^* y \rangle = \langle x, B^* A^* y \rangle = \langle x, (AB)^* y \rangle$

where $x, y \in H$.

Hence, $(AB)^* = B^* A^*$.

(v) Let A mapping H into H have an inverse A^{-1}.

$$\langle A^{-1} x, y \rangle = \langle x, (A^{-1})^* y \rangle = \langle \overline{y, A^{-1} x} \rangle = \langle \overline{AA^{-1} y, A^{-1} x} \rangle$$
$$= \langle \overline{A^{-1} y, A^* A^{-1} x} \rangle = \langle \overline{y, (A^{-1})^* A^* A^{-1} x} \rangle$$
$$= \langle (A^{-1})^* A^* A^{-1} x, y \rangle.$$

Thus, $(A^{-1})^* A^* = I$ \hfill (9.3)

, Again $\langle A^* (A^{-1})^* x, y \rangle = \langle (A^{-1})^* x, Ay \rangle = \langle x, A^{-1} Ay \rangle = \langle x, y \rangle$.

Thus $A^* (A^{-1})^* = I$. \hfill (9.4)

Hence it follows from (9.3) and (9.4) that $(A^*)^{-1} = (A^{-1})^*$.

9.2 Self-Adjoint Operators

9.2.1 Self-adjoint operators

A bounded linear operator A is said to be **self-adjoint** if it is equal to its adjoint, i.e., $A = A^*$. **Self-adjoint operators** on a Hilbert space H are also called **Hermitian.**

Note 9.2.1. A linear (**not necessarily bounded**) operator A with domain $\mathcal{D}(A)$ dense in H is said to be **symmetric**, if for all $x, y \in \mathcal{D}(A)$, the equality

$$\langle Ax, y \rangle = \langle x, Ay \rangle$$

holds. If A is unbounded, it follows from Note 9.1.1 that

$$y \in \mathcal{D}(A) \implies y \in \mathcal{D}(A^*).$$

Hence $\mathcal{D}(A) \subseteq \mathcal{D}(A^*)$. In other words, $A \subseteq A^*$ or A^* is an extension of A. For A bounded, $\mathcal{D}(A) = \mathcal{D}(A^*) = \mathcal{H}$. For, $A = A^*$ and $\mathcal{D}(A)$ dense in H, A is called **self-adjoint**.

9.2.2   Examples

1. In an n-dimensional complex Euclidean space, a linear operator A can be identified with the matrix (a_{ij}) with complex numbers as elements. The operator adjoint to $A = (a_{ij})$ is $A^* = (\overline{a_{ji}})$. A self-adjoint operator is a Hermitian matrix if $a_{ij} = \overline{a_{ji}}$.

If (a_{ij}) is real then a Hermitian matrix becomes a symmetric matrix.

2. **Adjoint operator corresponding to a Fredholm operator in $L_2([0,1])$.**

If $Tf = g(s) = \int_0^1 k(s,t)f(t)dt$ (5.6.15), the kernel of the adjoint operator T^* in complex $L_2([0,1])$ is $\overline{k(t,s)}$.

T is self-adjoint if $k(s,t) = \overline{k(t,s)}$.

3. In $L_2([0,1])$ let the operator A be given by $Ax = tx(t) \in L_2([0,1])$ with every function $x(t) \in L_2([0,1])$. It can be seen that A is self-adjoint.

9.2.3   Remark

Given A, a self-adjoint operator, then

(i) λA is self-adjoint where λ is real.

(ii) $(A + B)$ is self-adjoint if A and B are respectively self-adjoint.

(iii) AB is self-adjoint if A and B are respectively self-adjoint and $AB = BA$.

(iv) If $A_n \to A$ in the sense of norm convergence in the space of operators and all A_n are self-adjoints, then A is also self-adjoint.

Proof: For (i)–(iii) see 9.1.3 (ii)–(iv).

(iv) Let $A_n \to A$ as $n \to \infty$ in the space of bounded linear operators.

Using 5.6.16, $||A_n - A|| = ||A_n^* - A^*||$.

Since $A_n \to A$, as $n \to \infty$, $\lim_{n\to\infty} A_n^* = A^*$. Since A_n is self-adjoint, $A_n = A$

Thus, A_n tends to both A and A^* as $n \to \infty$.

Hence, $A = A^*$.

9.2.4   Definition: bilinear hermitian form

A functional is said to be of **bilinear form** if it is a functional of two vectors and is linear in both the vectors.

Bilinear Hermitian Form

Let us consider $\langle Ax, y \rangle$ where A is **self-adjoint**.
Now, $\langle A(\alpha x_1 + \beta x_2), y \rangle = \alpha \langle Ax_1, y \rangle + \beta \langle Ax_2, y \rangle$

Moreover, $\langle Ax, y \rangle = \langle x, Ay \rangle$

Thus, $\langle Ax, y \rangle$ is a bilinear functional.

If A is self-adjoint, then we denote $\langle Ax, y \rangle$ by $A(x, y)$. Thus, $A(x, y) = \overline{A(y, x)}$.

This form is bounded in the sense, that $|A(x, y)| \le C_A \|x\| \, \|y\|$ where C_A is some constant.

9.2.5   Lemma

Thus every self-adjoint operator A generates some bounded bilinear hermitian form

$$A(x, y) = \langle Ax, y \rangle = \langle x, Ay \rangle.$$

Conversely if a bounded linear Hermitian form $A(x, y)$ is given, then it generates some self-adjoint operator A, satisfying the equality

$$A(x, y) = \langle Ax, y \rangle. \tag{9.5}$$

Proof: The first part follows from the definition in 9.2.4. Let us consider the bilinear hermitian form given by (9.5). Let us keep y fixed in $A(x, y)$ and obtain a linear functional of x. Consequently, $A(x, y) = \langle x, y^* \rangle$, y^* is a uniquely defined element. Thus we get an operator A, defined by $Ay = y^*$, and such that $\langle x, Ay \rangle = A(x, y)$.

Now $\langle x, A(y_1 + y_2) \rangle = A(x, y_1 + y_2) = A(x, y_1) + A(x, y_2) = \langle x, Ay_1 \rangle + \langle x, Ay_2 \rangle$

Thus A is linear. Moreover, $|\langle x, Ay \rangle| = |A(x, y)| \le C_A \|x\| \, \|y\|$.

Putting $x = Ay$, we get from the above, $|\langle Ay, Ay \rangle| \le C_A \|Ay\| \, \|y\|$ or $\|Ay\| \le C_A \|y\|$.

Hence $\|A\| \le C_A$, showing that A is bounded. To prove self-adjointness of A, we note that for $x, y \in H$, we have, $\langle x, Ay \rangle = \overline{A(y, x)} = \overline{(y, Ax)} = \langle Ax, y \rangle$, implying that $A = A^*$ and $A(x, y) = \langle Ax, y \rangle$.

9.3   Quadratic Form

9.3.1   Definition: quadratic form

A bilinear hermitian form $A(x, y)$ given by (9.5) is said to be a **quadratic form** if $x = y$ so that $A(x, x)$ is a quadratic form.

Further, (i) $A(x, x)$ is real since $A(x, y) = \overline{A(y, x)}$.

(ii) $A(\alpha x + \beta y, \ \alpha x + \beta y) = \alpha \overline{\alpha} A(x, x) + \alpha \overline{\beta} A(x, y)$

$$+ \overline{\alpha} \beta A(y, x) + \beta \overline{\beta} A(y, y)$$

9.3.2   Lemma

Every bilinear hermitian form $A(x, y)$ can be uniquely defined by a quadratic hermitian form.

The bilinear form $A(x, y)$ is defined by

$$A(x, y) = \frac{1}{4}\{[A(x_1, x_1) - A(x_2, x_2)] + i[A(x_3, x_3) - A(x_4, x_4)]\}$$

where $x_1 = x + y$, $x_2 = x - y$ and $x_3 = x + iy$, $x_4 = x - iy$.

The quadratic form $A(x, x)$ is bounded, that is, $|A(x, x)| \leq C_A ||x||^2$, if and only if the corresponding bilinear hermitian form is bounded. Moreover, $||A|| = \max(|m|, |M|) = \sup_{||x||=1} |(Ax, x)|$ where m and M are defined below.

Proof: Let $m = \inf_{||x||=1} \langle Ax, x \rangle$ and $M = \sup_{||x||=1} \langle Ax, x \rangle$. The numbers m and M are called the **greatest lower bound** and the **least upper bounds** respectively of the self-adjoint operator A.

Let $||x|| = 1$. Then,

$$|\langle Ax, x \rangle| \leq ||Ax|| \cdot ||x|| \leq ||A||||x||^2 = ||A||$$

and, consequently, $C_A = \sup_{||x||=1} |\langle Ax, x \rangle| \leq ||A||$.   (9.6)

On the other hand, for every $y \in H$, we have, $\langle Ay, y \rangle \leq C_A ||y||^2$.

Let z be any element in H, different from zero.

We put $t = \left(\dfrac{||Az||}{||z||} \right)^{\frac{1}{2}}$ and $u = \dfrac{1}{t} Az,$

we get $||Az||^2 = \langle Az, Az \rangle = \langle A(tz), u \rangle$

$$= \frac{1}{4}\{\langle A(tz + u), (tz + u) \rangle - \langle A(tz - u), (tz - u) \rangle\}$$

$$\leq \frac{1}{4} C_A[||tz + u||^2 + ||tz - u||^2]$$

$$= \frac{1}{2} C_A[||tz||^2 + ||u||^2]$$

$$= \frac{1}{2} C_A \left[t^2 ||z||^2 + \left(\frac{1}{t^2} \right) ||Az||^2 \right]$$

$$= \frac{1}{2} C_A[||Az|| \ ||z|| + ||z|| \ ||Az||] = C_A ||Az|| \ ||z||.$$

Hence, $||Az|| \leq C_A ||z||.$

Therefore, $||A|| = \sup_{z \neq \theta} \dfrac{||Az||}{||z||} \leq C_A = \sup_{||x||=1} (Ax, x)$ (9.7)

It follows from (9.6) and (9.7) that

$$||A|| = \max\{|m|, |M|\} = \sup_{||x||=1} \langle Ax, x \rangle.$$

9.4 Unitary Operators, Projection Operators

In this section we study some well-behaved bounded linear operators in a Hilbert space which commute with their adjoints.

9.4.1 Definitions: normal, unitary operators

(i) **Normal operator:** Let A be a bounded linear operator mapping a Hilbert space H into itself. A is called a **normal** operator if $A^*A = AA^*$.

(ii) **Unitary operator:** The bounded linear operator mapping a Hilbert space into itself is called unitary if $AA^* = I = A^*A$. Hence A has an inverse and $A^{-1} = A^*$.

9.4.2 Example 1

Let $H = \mathbb{R}^2(\mathbb{C}^2)$ and $x = (x_1, x_2)^T$

$$Ax = \begin{pmatrix} x_1 - x_2 \\ x_1 + x_2 \end{pmatrix}, \quad A^*x = \begin{pmatrix} x_1 + x_2 \\ -x_1 + x_2 \end{pmatrix}.$$

$$A^*Ax = \begin{pmatrix} 2x_1 \\ 2x_2 \end{pmatrix} = AA^*x.$$

Thus A is a normal operator.

9.4.3 Example 2

Let
$$A = \begin{pmatrix} \cos\theta & -\sin\theta \\ \sin\theta & \cos\theta \end{pmatrix}$$

Then
$$A^* = \begin{pmatrix} \cos\theta & \sin\theta \\ -\sin\theta & \cos\theta \end{pmatrix}$$

Therefore, $AA^* = A^*A = \begin{pmatrix} 1 & 0 \\ 0 & 1 \end{pmatrix}.$

Thus A is unitary.

9.4.4 Remark

(i) If A is **unitary** or **self-adjoint** then A is **normal**.

(ii) The converse is not always true.

(iii) The operator A in example 9.4.2 although normal is not unitary.

(iv) The operator A in example 9.4.3 is unitary and necessarily normal.

9.4.5 Remark

If B is a normal operator and C is a bounded operator, such that $C^*C = I$, then operator $A = CBC^*$ is normal.

For $A^* = CB^*C^*$.   Now, $AA^* = CBC^* \cdot CB^*C^* = CBB^*C^*$

Again $A^*A = CB^*C^* \cdot CBC^* = CB^*BC^*$.   Hence $AA^* = A^*A$.

9.4.6   Example

Let $H = l_2$ and $x = (x_1, x_2, \ldots)^T$ in H, let $Cx = (0, x_1, x_2, \ldots)^T$.

Then $C^*x = (x_2, x_3, \ldots)^T$ for $x \in H$.

Hence $C^*Cx = (x_1, x_2, \ldots)^T$

$\quad\quad\quad CC^*x = (0, x_2, x_3, \ldots)^T$ for all $x \in H$

where $C = \begin{pmatrix} 0 & 0 & \cdots & 0 \\ 1 & 0 & \cdots & 0 \\ \cdots & \cdots & 1 & \cdots \\ 0 & 0 & \cdots & 1 \end{pmatrix}$, $C^* = \begin{pmatrix} 0 & 1 & \cdots & 0 \\ \vdots & 0 & 1 & \vdots \\ 0 & 0 & \cdots & 1 \end{pmatrix}$

Thus $C^*C = I$ but $CC^* \neq I$.

9.4.7   Definition

Given a linear operator A and a unitary operator U, the operator $B = UAU^{-1} = UAU^*$ is called an operator **unitarily equivalent** to A.

9.4.8   Projection Operator

Let H be a Hilbert space and L a closed subspace of H. Then, by **orthogonal projection theorem** for every $x \in H$, $y \in L$, and $z \in L^\perp$, x can be uniquely represented by $x = y + z$.

Then $Px = y$ and $(I - P)x = z$.

This motivates us to define the **projection operator**, see 3.6.2.

9.4.9   Theorem

P *is a self-adjoint operator with its norm equal to one and* P *satisfies* $P^2 = P$.

We first show that P is a linear operator.

Let, $x_1 = y_1 + z_1$, $y_1 \in L$, $z_1 \in L^\perp$ and $x_2 = y_2 + z_2$, $y_2 \in L$, $z_2 \in L^\perp$.
Now, $y_1 = Px_1$, $y_2 = Px_2$.

Since $\alpha x_1 + \beta x_2 = [\alpha y_1 + \beta y_2] + [\alpha z_1 + \beta z_2]$, therefore,
$P(\alpha x_1 + \beta y_2) = \alpha y_1 + \beta y_2 = \alpha Px_1 + \beta Px_2$, $\alpha, \beta \in \mathbb{R}(\mathbb{C})$.

Hence P is linear.

Since $y \perp z$, we have $||x||^2 = ||y + z||^2 = \langle y + z, \ y + z \rangle$

$\quad\quad\quad\quad\quad\quad\quad = \langle y, y \rangle + \langle z, y \rangle + \langle y, z \rangle + \langle z, z \rangle$

$\quad\quad\quad\quad\quad\quad\quad = ||y||^2 + ||z||^2$

Thus, $||y||^2 = ||Px||^2 \leq ||x||^2$,   i.e. $||Px|| \leq ||x||$ for every x.
Hence, $||P|| \leq 1$.

Since for $x \in L$, $Px = x$ and consequently $||Px|| = ||x||$, it follows that $||P|| = 1$.

Next we want to show that P is a self-adjoint operator. Let $x_1 = y_1 + z_1$ and $x_2 = y_2 + z_2$

We have $Px_1 = y_1$, $Px_2 = y_2$.

Therefore, $\langle Px_1, x_2 \rangle = \langle y_1, Px_2 \rangle = \langle y_1, y_2 \rangle$.

Similarly, $\langle x_1, Px_2 \rangle = \langle Px_1, y_2 \rangle = \langle y_1, y_2 \rangle$.

Consequently, $\langle Px_1, x_2 \rangle = \langle x_1, Px_2 \rangle$.

Hence, P is self-adjoint.

Since, $x = y + z$, with $y \in L$ and $z \perp L$, $Px \in L$ for every $x \in H$.

Hence, $P^2 x = P(Px) = Px$ for every $x \in H$.

Hence, projection P in a Hilbert space satisfies $P^2 = P$.

9.4.10   Theorem

Every self-adjoint operator P satisfying $P^2 = P$ is an orthogonal projection on some subspace L of the Hilbert space H.

Proof: Let L have the element y where $y = Px$, x being any element of H. Now, if $y_1 = Px_1 \in L$ and $y_2 = Px_2 \in L$ for $x_1, x_2 \in H$, then $y_1 + y_2 = Px_1 + Px_2 = P(x_1 + x_2) \in L$. Similarly, $\alpha y = \alpha Px = P(\alpha x) \in L$ for $\alpha \in \mathbb{R}(\mathbb{C})$. Hence, L is a linear subspace. Now, let $y_n \to y_0 \cdot y_n \in L$ since $y_n = Px_n$ for every $x_n \in H$. Now $Py_n = P^2 x_n = Px_n = y_n$.

Since P is continuous, $y_n \to y_0 \Rightarrow Py_n \to Py_0$. However, since $Py_n = y_n$, $y_n \to Py_0$. Consequently, $y_0 = Py_0$, i.e., $y_0 \in L$, showing that L is closed. Finally, $x - Px \perp Px$ since P is self-adjoint and $P^2 = P$, as is shown below $\langle x - Px, Px \rangle = \langle Px - P^2 x, x \rangle = 0$.

Thus, it follows from the definition of L that P is the projection of H onto this subspace. Moreover, corresponding to an element $x \in H$, $Px \in L$ is unique. For if otherwise, let $P_1 x$ be another projection on L. But that violates the orthogonal projection theorem [see 3.5].

9.4.11   Remark

(i) L mentioned above consists of all elements of the form $Px = x$, $x \in L$.

(ii) By the orthogonal projection theorem, we can write $x = y + z$, where $y \in L$, $z \in L^{\perp}$ and $x \in H$. If we write $y = Px$, then $z = (I - P)x$. Thus $(I - P)$ is a projection on $L^{\perp}$.

Moreover, $(I - P)^2 = I - 2P + P^2 = I - 2P + P = I - P$.

$$\langle (I - P)x, y \rangle = \langle x, (I - P)^* y \rangle = \langle x, y \rangle - \langle x, Py \rangle$$
$$= \langle x, (I - P)y \rangle \text{ for all } x, y \in H.$$

Thus $(I - P)$ is a projection operator.

9.4.12   Theorem

For the projections P_1 and P_2 to be orthogonal, it is necessary and sufficient that the corresponding subspace L_1 and L_2 are orthogonal.

Let $y_1 = P_1 x$ and $y_2 = P_2 x, x \in H$. Let P_1 be orthogonal to P_2. Then $\langle y_1, y_2 \rangle = \langle P_1 x, P_2 x \rangle = \langle x, P_1 P_2 x \rangle = 0$, since P_1 is orthogonal to P_2.

Since y_1 is any element of L_1 and y_2 is any element of L_2, we conclude that $L_1 \perp L_2$. Similarly let $L_1 \perp L_2$. Then for $y, \in L_1$ and $y_2 \in L_2$ we have $\langle y_1, y_2 \rangle = 0$ or $\langle P_1 x, P_2 x \rangle = \langle x, P_1 P_2 x \rangle = 0$ showing that $P_1 \perp P_2$.

9.4.13   Lemma

The necessary and sufficient condition that the sum of two projection operators P_{L_1} and P_{L_2} be a projection operator is that P_{L_1} and P_{L_2} must be mutually orthogonal. In this case $P_{L_1} + P_{L_2} = P_{L_1 + L_2}$.

Proof: Let $P_{L_1} + P_{L_2}$ be a projection operator P.

Then $(P_{L_1} + P_{L_2})^2 = P_{L_1} + P_{L_2}$.

Therefore $P_{L_1}^2 + P_{L_1} P_{L_2} + P_{L_2} P_{L_1} + P_{L_2}^2 = P_{L_1} + P_{L_2}$.

Hence $P_{L_1} P_{L_2} + P_{L_2} P_{L_1} = \mathbf{0}$.

Multiplying LHS of the above equation by P_{L_1}, we have,

$$P_{L_1}^2 P_{L_2} + P_{L_1} P_{L_2} P_{L_1} = 0 \quad \text{or} \quad P_{L_1} P_{L_2} + P_{L_1} P_{L_2} P_{L_1} = 0.$$

Multiplying RHS by P_{L_1} we have,

$$P_{L}, P_{L_2} P_{L_1} + P_{L_1} P_{L_2} P_{L_1}^2 = 0 \quad \text{or} \quad P_{L_1} P_{L_2} P_{L_1} = 0.$$

Multiplying LHS by $P_{L_1}^{-1}$ we have $P_{L_2} P_{L_1} = 0$ i.e. $P_{L_2} \perp P_{L_1}$.

Next, let us suppose that $P_{L_1} P_{L_2} = 0$.

Then $(P_{L_1} + P_{L_2})^2 = (P_{L_1} + P_{L_2})(P_{L_1} + P_{L_2})$

$$= P_{L_1}^2 + P_{L_1} P_{L_2} + P_{L_2} P_{L_1} + P_{L_2}^2$$

$$= P_{L_1}^2 + P_{L_2}^2 = P_{L_1} + P_{L_2}.$$

Thus $P_{L_1} + P_{L_2}$ is a projection operator.

Since $P_{L_1} \perp P_{L_2}, \; L_1 \text{ is } \perp L_2$ (9.8)

If $x \in H$, thus $Px = P_{L_1} x + P_{L_2} x = x_1 + x_2$ with $x_1 + x_2 \in L_1 + L_2$.

Further, if $x = x_1 + x_2$ is an element in $L_1 + L_2$ thus

$$x = x_1 + x_2 = P_{L_1}(x_1 + x_2) + P_{L_2}(x_1 + x_2)$$

$$= P_{L_1 + L_2}(x_1 + x_2),$$

since $P_{L_1} x_2 = 0$ (9.9)

and $P_{L_2} x_1 = 0$.

It follows from (9.8) and (9.9) that $Px = x = P_{L_1 + L_2} x$.

Hence P is a projection.

9.4.14   Lemma

The necessary and sufficient condition for the product of two projections P_{L_1} and P_{L_2} to be a projection is that the projection operator i.e. $P_{L_1}P_{L_2} = P_{L_2}P_{L_1}$. In this case $P_{L_1}P_{L_2} = P_{L_1 \cap L_2}$.

Proof: Since $P = P_{L_1}P_{L_2}$ is self-adjoint, we have

$$P_{L_1}P_{L_2} = (P_{L_1}P_{L_2})^* = P_{L_2}^* P_{L_1}^* = P_{L_2}P_{L_1},$$

taking note that P_{L_1} and P_{L_2} are self-adjoint.

Hence P_{L_1} commutes with P_{L_2}.

Conversely, if $P_{L_1}P_{L_2} = P_{L_2}P_{L_1}$, then

$$P^* = (P_{L_1}P_{L_2})^* = P_{L_2}^* P_{L_1}^* = P_{L_2}P_{L_1} = P_{L_1}P_{L_2} = P.$$

Thus P is self-adjoint.

Furthermore, $(P_{L_1}P_{L_2})^2 = P_{L_1}P_{L_2}P_{L_1}P_{L_2} = P_{L_1}P_{L_1}P_{L_2}P_{L_2}$
$$= P_{L_1}^2 P_{L_2}^2 = P_{L_1}P_{L_2}.$$

Hence $P = P_{L_1}P_{L_2}$ is a projection.

Let $x \in H$ be arbitrary. Then, $Px = P_{L_1}P_{L_2}x = P_{L_2}P_{L_1}x$.

Thus Px belongs to L_1 and L_2, that is to $L_1 \cap L_2$.

Now, let $y \in L_1 \cap L_2$. Then $Py = P_{L_1}(P_{L_2}y) = P_{L_1}y = y$.

Thus P is a projection on $L_1 \cap L_2$. This proves the lemma.

9.4.15   Definition

The projection P_2 is said to be a part of the projection P_1, if $P_1 P_2 = P_2$.

9.4.16   Remark

(i) $P_1 P_2 = P_2 \Rightarrow (P_1 P_2)^* = P_2^* \Rightarrow P_2 P_1 = P_2$.

(ii) P_{L_2} is a part of P_{L_1} if and only if L_2 is a subspace of L_1.

9.4.17   Theorem

The necessary and sufficient condition for a projection operator P_{L_2} to be a part of the projection operator P_{L_1} is the inequality $||P_{L_2}x|| \le ||P_{L_1}x||$ being satisfied for all $x \in H$.

Proof: $P_{L_2}P_{L_1}x = P_{L_2}x$ yields

$$||P_{L_2}x|| = ||P_{L_2}P_{L_1}x|| \le ||P_{L_2}|| \, ||P_{L_1}x|| \le ||P_{L_1}x|| \qquad (9.10)$$

Conversely if (9.10) be true, then for every $x \in L_2$,

$$||P_{L_1}x|| \ge ||P_{L_2}x|| = ||x|| \quad \text{and since} \quad ||P_{L_1}x|| \le ||x||$$

we have $||P_{L_1}x|| = ||x||$.

Therefore, $||P_{H \ominus L_1}x||^2 = ||x||^2 - ||P_{L_1}x||^2 = 0$ and hence, $x \in L_1$.

Therefore, $P_{L_1}x \in L_2$ for every $x \in H$, which implies that

$$P_{L_1}P_{L_2}x = P_{L_2}x \quad \text{i.e.} \quad P_{L_1}P_{L_2} = P_{L_2}.$$

9.4.18   Theorem

The difference $P_1 - P_2$ of two projections is a projection operator, if and only if P_2 is a part of P_1. In this case, $L_{P_1-P_2}$ is the orthogonal complement of L_{P_2} in L_{P_1}.

Proof: If $P_1 - P_2$ is a projection operator, then so is $I - (P_1 - P_2) = (I - P_1) + P_2$.

Then, by lemma 9.4.13, $(I - P_1)$ and P_2 are mutually orthogonal, i.e., $(I - P_1)P_2 = 0$ i.e. $P_1P_2 = P_2$ showing that P_2 is a part of P_1.

Conversely, let P_2 be a part of P_1 i.e. $P_1P_2 = P_2$ or $(I - P_1)P_2 = \mathbf{0}$, i.e., $I - P_1$ is orthogonal to P_2.

Therefore, by lemma 9.4.13, $(I - P_1) + P_2$ is a projection operator and $I - [(I - P_1) + P_2] = P_1 - P_2$ is aslo a projection operator. The condition $P_1P_2 = P_2$ implies that $P_1 - P_2$ and P_2 are orthogonal. Then, because of lemma 9.4.13, $L_{P_1} = L_{P_1-P_2} + L_{P_2}$.

9.5   Positive Operators, Square Roots of a Positive Operator

9.5.1   Definition: A non-negative operator, a positive operator

A self-adjoint operator, A in H over $\mathbb{R}(\mathbb{C})$ is said to be non-negative if $\langle Ax, x \rangle \geq 0$ for all $x \in H$ and is denoted by $A \geq 0$. A self-adjoint operator A in H over $\mathbb{R}(\mathbb{C})$ is said to be positive if $\langle Ax, x \rangle \geq 0$ for all $x \in H$ and $\langle Ax, x \rangle \neq 0$ for at least one x and is written as $A > 0$.

9.5.2   Definition: stronger and smaller operator

If A and B are self-adjoint operators and $A - B$ is positive, i.e., $A - B > 0$, then A is said to be *greater* then B or B is *smaller* then A and expressed as $A > B$.

9.5.3   Remark

The relation $\geq$ on the set of self-adjoint operators on H is a partial order. The relation is

(i) reflexive, i.e., $A \geq A$

(ii) transitive, i.e., $A \geq B$ and $B \geq C \Rightarrow A \geq C$

(iii) antisymmetric, i.e., $A \geq B$ and $B \geq A \Rightarrow A = B$.

(iv) $A \geq B$, $C \geq D \Rightarrow A + C \geq B + D$.

(v) For any A, AA^* and A^*A are non-negative.

For $x \in H$, $\langle AA^*x, x \rangle = \langle A^*x, A^*x \rangle = ||A^*x||^2 \geq 0$

$\langle A^*Ax, x \rangle = \langle Ax, Ax \rangle = ||Ax||^2 \geq 0$.

(vi) If $A \geq 0$ and A^{-1} exists, then $A^{-1} > 0$. $A \geq 0 \Rightarrow \langle Ax, x \rangle \geq 0$.
Now, A^{-1} exists $\Rightarrow Ax = \theta \Rightarrow x = \theta$.

Hence, $\langle A^{-1}x, x \rangle = 0 \Rightarrow x = \theta$, for if x is non-null, $A^{-1}x$ is non-null
and $< A^{-1}x, x > 0$. Thus, $\langle A^{-1}x, x \rangle > 0$ for non-null x.

(vii) If A and B are positive operators and the composition AB exists, then
AB may not be a positive operator. For example, let $H = \mathbb{R}^2(\mathbb{C}^2)$, and

$$A(x_1, x_2)^T = (x_1 + x_2,\ x_1 + 2x_2)$$
$$B(x_1, x_2)^T = (x_1 + x_2,\ x_1 + x_2).$$
$$AB(x_1, x_2)^T = (2x_1 + 2x_2,\ 3x_1 + 3x_2)$$
$$BA(x_1, x_2)^T = (2x_1 + 3x_2, 2x_1 + 3x_2)$$

for all $(x_1, x_2) \in \mathbb{R} \times \mathbb{R}(\mathbb{C} \times \mathbb{C})$

AB is not a positive operator since AB is not self-adjoint.

9.5.4   Example

Let us consider the symmetric operator B

$$Bu = -\frac{d^2u}{dx^2}$$

the functions $u(x)$ being subject to the boundary conditions $u(0) = u(1) = 0$, the field Ω being the segment $0 < x < 1$.

$\mathcal{D}(\mathcal{B}) = \{u(x) : u(x) \in C^2(0, 1), u(0) = u(1) = 0\}$. Take $H = L_2([0, 1])$.
Then, for all $u, v \in \mathcal{D}(\mathcal{B})$,

$$\langle Bu, v \rangle = \int_0^1 \frac{d^2x}{dx^2} v(x) dx = -\int_0^1 \frac{d^2v}{dx^2} u(x) dx = \langle Bv, u \rangle$$

Hence, B is symmetric. Therefore,

$$\langle Bu, u \rangle = \int_0^1 \left(\frac{du}{dx}\right)^2 dx - \left[u\frac{du}{dx}\right]_{x=0}^{x=1} = \int_0^1 \left(\frac{du}{dx}\right)^2 dx \geq 0.$$

9.5.5   Theorem

If two positive self-adjoint operators A and B commute, then their product is also a positive operator.

Proof: Let us put

$$A_1 = \frac{A}{||A||}, \quad A_2 = A_1 - A_1^2, \ldots A_{n+1} = A_n - A_n^2, \ldots \text{ and show that}$$

$0 \leq A_n \leq I$ for every n.   (9.11)

The above is true for $n = 1$. Let us suppose that (9.11) is true for $n = k$.

Then $\langle A_k^2(I - A_k)x, x \rangle = \langle (I - A_k)A_kx, A_kx \rangle \geq 0$ since $(I - A_k)$ is a positive operator. Hence $A_k^2(I - A_k) \geq 0$.

Analogously $A_k(I - A_k)^2 \geq 0$.

Hence, $A_{k+1} = A_k^2(I - A_k) + A_k(I - A_k)^2 \geq 0$

and $I - A_{k+1} = (I - A_k) + A_k^2 \geq 0$.

Consequently, (9.11) holds for $n = k + 1$.

Moreover, $A_1 = A_1^2 + A_2 = A_1^2 + A_2^2 + A_3 = \cdots$

$$= A_1^2 + A_2^2 + \cdots + A_n^2 + A_{n+1},$$

whence $\displaystyle\sum_{k=1}^{n} A_k^2 = A_1 - A_{n+1} \leq A_1,$ since $A_{n+1} \geq 0,$

that is, $\displaystyle\sum_{k=1}^{n}\langle A_kx, A_kx \rangle \leq \langle A_1x, x \rangle.$

Consequently, the series $\displaystyle\sum_{k=1}^{\infty}||A_kx||^2$ converges and $||A_kx|| \to 0$ as $k \to \infty$.

Hence, $\displaystyle\left(\sum_{k=1}^{n} A_k^2\right)x = A_1x - A_{n+1}x \to A_1x$ as $n \to \infty$ (9.12)

since B commutes with A and hence with A_1.

$$BA_2 = B(A_1 - A_1^2) = BA_1 - BA_1^2$$
$$= A_1B - A_1^2B = (A_1 - A_1^2)B,$$

i.e., B commutes with A_2.

Let B commute with A_k, $k = 1, 2, \ldots, n$.

$$BA_{n+1} = B(A_n - A_n^2) = A_nB - A_nBA_n$$
$$= (A_n - A_n^2)B = A_{n+1}B.$$

Hence B commutes with A_k, $k = 1, 2, \ldots, n, \ldots$.

$$\langle ABx, x \rangle = ||A||\langle BA_1x, x \rangle = ||A|| \lim_{n \to \infty}\left[\sum_{k=1}^{n}\langle BA_k^2x, x \rangle\right].$$

$$= ||A|| \lim_{n \to \infty}\sum_{k=1}^{n}\langle BA_kx, A_kx \rangle \geq 0$$

Using (9.12) $\langle ABx, x \rangle = ||A||\langle BA_1x, x \rangle = \langle BAx, x \rangle.$

9.5.6   Theorem

If $\{A_n\}$ is a monotone increasing sequence of mutually commuting self-adjoint operators, bounded above by a self-adjoint operator B commuting *with all the*

$$A_n : A_1 \leq A_2 \leq \cdots \leq A_n \leq \cdots \leq B \tag{9.13}$$

then the sequence $\{A_n\}$ converges pointwise to a self-adjoint operator A and $A \leq B$.

Proof: Let $C_n = B - A_n$. Since B and A_n for all n are self-adjoint operator $\{C_n\}$ is a sequence of self-adjoint operators. $\langle C_n x, x \rangle = \langle (B - A_n)x, x \rangle \geq 0$ because of (9.13). Hence, C_n is a non-negative operator.

Moreover, $C_n C_m = (B - A_n)(B - A_m) = B^2 - BA_m - A_n B + A_n A_m$

$$= B^2 - A_m B - BA_n + A_m A_n = (B - A_m)(B - A_n)$$

since B commutes with A_n for all n and $A_n A_m$ commute. Hence, $C_n C_m = C_m C_n$. Moreover, $\{C_n\}$ forms a monotonic decreasing sequence. Consequently, for $m < n$, the operator $(C_m - C_n)C_m$ and $C_n(C_m - C_n)$ are also positive. Moreover,

$$\langle C_m^2 x, x \rangle \geq \langle C_m C_n x, x \rangle \geq \langle C_n^2 x, x \rangle \geq 0$$

This implies that the monotonic decreasing non-negative numerical sequence $\{\langle C_n^2 x, x \rangle\}$ has a limit. Hence, it follows from the above inequality, $\{\langle C_m C_n x, x \rangle\}$ also tends to the same limit as $n, m \to \infty$. Therefore,

$$||C_m x - C_n x||^2 = \langle (C_m - C_n)^2 x, x) \rangle = \langle C_m^2 x, x \rangle - 2\langle C_m C_n x, x \rangle$$
$$+ \langle C_n^2 x, x \rangle \to 0 \text{ as } n, m \to \infty.$$

Thus, the sequence $\{C_n x\}$ and thereby also $\{A_n x\}$ converges to some limit Ax for arbitrary x, that is $Ax = \lim_{n \to \infty} A_n x$. Hence A is a self-adjoint operator, satisfying $A \leq B$.

9.5.7 Remark

If, in theorem 9.5.6, the inequality (9.13) is replaced by $A_1 \geq A_2 \geq \cdots \geq A_n \geq \cdots B$ and the other conditions remain unchanged, then the conclusion of theorem 9.5.6 remains unchanged, except that $A \geq B$.

9.5.8 Square roots of non-negative operators

9.5.9 Definition: square root

The self-adjoint operator B is called a **square root** of the non-negative operator A, if $B^2 = A$.

9.5.10 Theorem

There exists a unique positive square root B of every positive self-adjoint operator A; it commutes with every operator commuting with A.

Proof: Without loss of generality, it can be assumed that $A \leq I$. Let us put $B_0 = 0$, and

$$B_{n+1} = B_n + \frac{1}{2}(A - B_n^2), \quad n = 0, 1, 2, \ldots \tag{9.14}$$

Suppose that B_k is self-adjoint, positive and commutes with every operator commuting with A, for $k = 1, 2, \ldots n$.

Then $\langle B_{n+1}x, x \rangle = \langle [B_n + \frac{1}{2}(A - B_n^2)]x, x \rangle$

$$= \langle B_n x, x \rangle + \frac{1}{2}\langle Ax, x \rangle - \frac{1}{2}(B_n^2 x, x)$$

$$= \langle x, B_n x \rangle + \frac{1}{2}\langle x, Ax \rangle - \frac{1}{2}\langle x, B_n^2 x \rangle$$

$$= \langle x, B_{n+1}x \rangle.$$

$$\langle B_{n+1}x, x \rangle = \langle B_n x, x \rangle + \frac{1}{2}\langle (A - B_n^2)x, x \rangle \tag{9.15}$$

Now, $(I - B_{n+1}) = \frac{1}{2}(I - B_n)^2 + \frac{1}{2}(I - A) \tag{9.16}$

and $B_{n+1} - B_n = \frac{1}{2}[(I - B_{n-1}) + (I - B_n)](B_n - B_{n-1}) \tag{9.17}$

Now, $B_0 = 0, B_1 = \frac{1}{2}A \leq I$. Also $B_1 > 0$.

Therefore, it follows from (9.16) and (9.17) that $B_n \leq I$ and $B_n \leq B_{n+1}$ for all n. Thus it follows from (9.15) that B_{n+1} is self-adjoint, positive and commutes with every operator commuting with A. Thus $\{B_n\}$ is a monotonic increasing sequence bounded above. This sequence converges in limit to some self-adjoint positive operator B. Taking limits in (9.14) we have,

$$B = B + \frac{1}{2}(A - B^2) \quad \text{that is } B^2 = A.$$

Finally, B commutes with every operator that commutes with A. This is because that B_n possesses the above property. Thus B is the positive square root of A. If B is not unique let B_1 be another square root of A. Then $B^2 - B_1^2 = 0$.

Therefore, $\langle (B^2 - B_1^2)x, y \rangle = 0$ or $\langle (B + B_1)(B - B_1)x, y \rangle = 0$.

Let us take y such that $(B - B_1)x = y$.

Then $0 = \langle (B + B_1)y, y \rangle = \langle By, y \rangle + \langle B_1 y, y \rangle$.

Since B and B_1 are positive, $\langle By, y \rangle = \langle B_1 y, y \rangle = 0$.

However, since the roots are positive, we have $B = C^2$ where C is a self-adjoint operator. Since

$$\|Cy\|^2 = \langle C^2 y, y \rangle = \langle By, y \rangle = 0, \quad \text{hence } Cy = 0.$$

Consequently, $By = C(Cy) = 0$ and analogously $B_1 y = 0$.

However, then, $\|B_1 x - Bx\|^2 = \langle (B - B_1)^2 x, x \rangle = \langle (B - B_1)y, x \rangle = 0$, that is, $Bx = B_1 x$ for every $x \in H$ and the uniqueness of the square root is proved.

9.5.11 Example

Let $H = L_2([0, 1])$. Let the operator A be defined by $Ax(t) = tx(t)$, $x(t) \in L_2([0, 1])$. Then, $||Ax||^2 = \langle Ax, Ax \rangle = \int_0^1 t^2 x^2(t) dt \leq ||x||^2$. Hence, A is bounded.

$$\langle Ax, x \rangle^2 = \int_0^1 tx(t) \cdot x(t) dt = \int_0^1 tx^2(t) dt = \int_0^1 (\sqrt{t}x(t))(\sqrt{t}x(t)) dt = (Bx, Bx)$$

where $Bx(t) = +\sqrt{t}x(t)$.

Problems

1. Suppose A is linear and maps a complex Hilbert space H into itself. Then, if $\langle Ax^2, x \rangle \geq 0$ and $\langle Ax, x \rangle = 0$ for each $x \in H$, show that $A = 0$.

2. Let $\{u_1, u_2, \ldots\}$ be an orthonormal system in a Hilbert space H, $T \in (H \to H)$ and $a_{i,j} = \langle Tu_j, u_i \rangle$, $i, j = 1, 2, \ldots$ Then show that the matrix $\{a_{i,j}\}$ defines a bounded linear operator Q on H with respect to $u_1, u_2, u_3 \cdots$. Show further that $Q = PTP$, where

$$Px = \sum_j \langle x, u_j \rangle u_j, \ x \in H.$$

If $u_1, u_2, u_3 \ldots$ constitute an orthonormal basis for H, then prove that $Q = T$.

3. Let P and Q denote Fredholm integral operators on $H = L_2([a, b])$ with kernels $p(\cdot, \cdot)$ and $q(\cdot, \cdot)$ in $L_2([a, b] \times [a, b])$, respectively. Then show that $P = Q$ if and only if $p(\cdot, \cdot)$ and $q(\cdot, \cdot)$ are equal almost everywhere on $[a, b] \times [a, b]$. Further, show that PQ is a Fredholm integral operator with kernel

$$p \circ q(s, t) = \int_a^b p(s, u)q(u, t) d\mu(u), (s, t) \in [a, b] \times [a, b]$$

and that $||PQ|| \leq ||p \circ q(s, t)||_2 \leq ||p||_2 ||q||_2$

(Hint: To find PQ use Fubini's theorem (sec. 10.5).)

4. Consider the shift operator A and a multiplication operator B on l_2 such that

$$Ax(n) = \begin{cases} 0 & \text{if } n = 0 \\ x(n-1) & \text{if } n \geq 1. \end{cases}$$

$$Bx(n) = (n+1)^{-1}x(n) \ \text{if } n \geq 0.$$

Put $C = AB$. Show that C is a compact operator which has no eigenvalue and whose spectrum consists of exactly one point.

5. Let $\lambda_1 \leq \lambda_2 \leq \cdots \lambda_n$ be the first n consecutive eigenvalues of a self-adjoint **coercive** operator 'A' mapping a Hilbert space H into itself and let $u_1, u_2, \ldots, u_n$ be the corresponding orthonormal eigenfunctions. Let there exist a function $u = u_{n+1} \neq 0$ which maximizes the functional

$$\frac{\langle Au, u \rangle}{\langle u, u \rangle}, \quad u \in D_A \subseteq H,$$

under the supplementary conditions, $\langle u, u_1 \rangle = 0$, $\langle u, u_2 \rangle = 0 \cdots \langle u, u_n \rangle = 0$.

Then show that u_{n+1} is the eigenfunction corresponding to the eigenvalue

$$\lambda_{n+1} = \frac{\langle Au_{n+1}, u_{n+1} \rangle}{\langle u_{n+1}, u_{n+1} \rangle}.$$

Show further that $\lambda_n \leq \lambda_{n+1}$.

(Hint: A symmetric non-negative operator 'A' is said to be coercive if there exists a non-negative number α such that $\langle Au, u \rangle \geq \alpha \langle u, u \rangle \; \forall \, u \in D(A)$, $\alpha > 1$.)

6. Let $D(A)$ be the subspace of a Hilbert space $H^1([0,1])$ of functions $u(x)$ with continuous first derivatives on [0,1] with $u(0) = u(1) = 0$ and $A = -\dfrac{d^2}{dx^2}$.

Find the adjoint A^* and show that A is symmetric.

(Hint: $\langle u, v \rangle = \int_0^1 u(x)v(x)dx$ for all $u, v \in H^2$.)

7. (*Elastic bending of a clamped beam*) Let

$$Au = \frac{d^2}{dx^2}\left(b(x)\frac{d^2u}{dx^2}\right) + ku = f(x), \quad k > 0, 0 < x < L.$$

subject to the boundary conditions,

$$u = \frac{du}{dx} = 0 \quad \text{at } x = 0 \quad \text{and} \quad x = L$$

Show that the operator A is symmetric on its domain.

8. For $x \in L_2[0, \infty[$, consider

$$U_1(x)u = \sqrt{\frac{2}{\pi}} \frac{d}{du} \int_0^\infty \frac{\sin us}{s} x(s)d\mu(s) \quad \text{[see ch. 10]}$$

$$U_2(x)u = \sqrt{\frac{2}{\pi}} \frac{d}{du} \int_0^\infty \frac{1 - \cos us}{s} x(s)d\mu(s)$$

Show that

(i) $U_1(x)u$ and $U_2(x)u$ are well-defined for almost all $u \in [0, \infty)$

(ii) $U_1(x), U_2(x) \in L_2[0, \infty[$

(iii) The mappings U_1 and U_2 are bounded operators on $L_2[0, \infty[$, which are self-adjoint and unitary.

9. Let $A \in (H \to H)$ be self-adjoint. Then show that

(i) $A^2 \geq 0$ and $A \leq ||A||I$.

(ii) if $A^2 \leq A$ then $0 \leq A \leq I$.

9.6   Spectrum of Self-Adjoint Operators

Let us consider the operator $A_\lambda = A - \lambda I$, where A is self-adjoint and λ a complex number.

In sec. 4.7.17 we defined a **resolvent operator and regular** values of an operator. By theorem 4.7.13, if $||(1/\lambda)A|| < 1$ (that is, if $|\lambda| > ||A||$), then λ is a **regular** value of A and consequently, then entire spectrum of A lies inside and on the boundary of the disk $|\lambda| \leq ||A||$. This is true for arbitrary linear operators acting into a Banach space. For a self-adjoint operator defined on a Hilbert space, the plane comprising the spectrum of the operator is indicated more precisely below.

9.6.1   *Lemma*

Let A be a self-adjoint linear operator in a Hilbert space over $\mathbb{R}(\mathbb{C})$. Then all of its eigenvalues are real.

Let $x \neq \theta$ be an eigenvector of A and λ the corresponding eigenvalue.

Then, $Ax = \lambda x$.

Pre-multiplying both sides with x^*

we have $x^* A x = \lambda x^* x$.   (9.18)

Taking adjoint of both sides we have,

$$x^* A^* x = \overline{\lambda} x^* x. \tag{9.19}$$

From (9.18) and (9.19) it follows that

$$x^* A x = \lambda x^* x = \overline{\lambda} x^* x, \text{ showing that } \lambda = \overline{\lambda} \text{ i.e. } \lambda \text{ is real.}$$

9.6.2   *Lemma*

Eigenvectors belonging to different eigenvalues of a self-adjoint operator in a Hilbert space H over $\mathbb{R}(\mathbb{C})$ are orthogonal.

Let x_1, x_2 be two eigenvectors of a self-adjoint operator corresponding to different eigenvalues λ_1 and λ_2.

Then we have

$$Ax_1 = \lambda_1 x_1, \tag{9.20}$$

$$Ax_2 = \lambda x_2 \tag{9.21}$$

Premultiplying (9.20) by x_2^* and (9.21) with x_1^* we have,

$$x_2^* A x_1 = \lambda_1 x_2^* x_1$$

$$x_1^* A x_2 = \lambda_2 x_1^* x_2$$

Therefore, $\overline{x_2^* A x_1} = x_1^* A x_2 = \lambda_1 \overline{x_1^* x_2} = \lambda_2 x_2^* x_1$.

Since A is self-adjoint λ_1, λ_2 are real.

Therefore, $(\lambda_1 - \lambda_2)x_1^* x_2 = 0$.

Since $\lambda_1 \neq \lambda_2$, $x_1^* x_2 = 0$ i.e. $x_1 \perp x_2$.

9.6.3   Theorem

For the point λ to be a regular value of the self-adjoint operator A, it is necessary and sufficient that there is a positive constants C, such that

$$||(A - \lambda I)x|| = ||A_\lambda x|| = ||Ax - \lambda x|| \geq C||x|| \tag{9.22}$$

for every $x \in H$ over $\mathbb{R}(\mathbb{C})$.

Proof: Suppose that $R_\lambda = A_\lambda^{-1}$ is bounded and $||R_\lambda|| = K$. For every $x \in H$, we have $||x|| = ||R_\lambda A_\lambda x|| \leq K||A_\lambda x||$ whence $||A_\lambda x|| \geq (1/K)||x||$, proving the conditions is necessary.

We next want to show that the condition is sufficient. Let $y = Ax - \lambda x$ and x run through H. Then y runs through some linear subspace L. By (9.22) there is a one-to-one correspondence between x and y. For if x_1 and x_2 correspond to the same element y,

we have $A(x_1 - x_2) - \lambda(x_1 - x_2) = 0$,

whence $||x_1 - x_2|| \leq \left(\dfrac{1}{C}\right) ||A_\lambda(x_1 - x_2)| = 0 \tag{9.23}$

We next show that L is everywhere dense in H. If it were not so, then there would exist a non-null element $x_0 \in H$ such that $\langle x_0, y \rangle = 0$ for every $y \in H$. Hence $\langle x_0, Ax - \lambda x \rangle = 0$ for every $x \in H$. In other words, $\langle (A - \overline{\lambda})x_0, x \rangle = 0$, A being self-adjoint. Hence $\langle Ax_0 - \overline{\lambda}x_0, x \rangle = 0$ for non-zero x_0 and for every $x \in H$. It then follows that

$$\langle Ax_0 - \overline{\lambda}x_0, x \rangle = 0.$$

The above equality is impossible, either for complex λ because the eigenvalues of a self-adjoint operator A are real. If λ is real, i.e., $\lambda = \overline{\lambda}$, then we have from (9.23)

$$||x_0|| \le (1/C)||Ax_0 - \lambda x_0|| = 0.$$

Next, let $\{y_n\} \subset L$, $y_n = A_\lambda x_n$ and $\{y_n\} \to y_0$.

By (9.22) $||x_n - x_m|| \le \left(\dfrac{1}{C}\right) ||A_\lambda x_n - A_\lambda x_m|| = \left(\dfrac{1}{C}\right) ||y_n - y_m||.$

$\{y_n\}$ is a Cauchy sequence and hence $||y_n - y_m|| \to 0$ as $n, m \to \infty$. However, then $||x_n - x_m|| \to 0$ as $n, m \to \infty$. Since H is a complete space, there exists a limit for $\{x_n\} : x = \lim\limits_r x_n$. Moreover,

$$A_\lambda x = \lim_n A_\lambda x_n = \lim_n y_n = y \quad \text{i.e. } y \in L.$$

Thus L is a closed subspace everywhere dense in H, i.e., $L = H$. In addition, since the correspondence $y = A_\lambda x$ is one-to-one, there exists an inverse operator $x = A_\lambda^{-1} y = R_\lambda y$ defined on the entire H. Inequality (9.22) yields

$$||R_\lambda y|| = ||x|| \le \left(\frac{1}{C}\right) ||A_\lambda x|| = \left(\frac{1}{C}\right) ||y||,$$

i.e. R_λ is a bounded operator and $||R_\lambda|| \le \dfrac{1}{C}$.

9.6.4   Corollary

The point λ belongs to the spectrum of a self-adjoint operator A if and only if there exists a sequence $\{x_n\}$ such that

$$||Ax_n - \lambda x_n|| \le C_n ||x_n||, \ C_n \to 0 \ \text{ as } n \to \infty. \tag{9.24}$$

If we take $||x_n|| = 1$, then (9.24) yields

$$||Ax_n - \lambda x_n|| \to 0, \ ||x_n|| = 1 \tag{9.25}$$

9.6.5   Theorem

The spectrum of a self-adjoint operator A lies entirely on a segment $[m, M]$ of the real axis, where

$$M = \sup_{||x||=1} \langle Ax, x \rangle \quad \text{and} \quad m = \inf_{||x||=1} \langle Ax, x \rangle.$$

Proof: Since A is self-adjoint

$\langle Ax, x \rangle = \langle x, Ax \rangle = \overline{\langle Ax, x \rangle}$ i.e. $\langle Ax, x \rangle$ is real.

Also, $\dfrac{|\langle Ax, x \rangle|}{||x||^2} \le \dfrac{||Ax|| \, ||x||}{||x||^2} \le ||A||$

Then, $C_A = \sup\limits_{||x||=1} \langle Ax, x \rangle| \le ||A||.$

On the other hand, for every $y \in H$ it follows from Lemma 9.3.2 that $\langle Ay, y \rangle \le C_A ||y||^2.$

Let $\quad m = \inf\limits_{x \neq \theta} \dfrac{\langle Ax, x \rangle}{||x||^2}, \ M = \sup\limits_{x \neq \theta} \dfrac{\langle Ax, x \rangle}{||x||^2}$ $\qquad\qquad$ (9.26)

i.e., $\quad m \leq \dfrac{\langle Ax, x \rangle}{||x||^2} \leq M$

Let λ_1 be any eigenvalue of A and x_1 the corresponding eigenvector. Then $Ax_1 = \lambda_1 x_1$ and $m \leq \lambda_1 \leq M$.

Now if $\quad y = A_\lambda x = Ax - \lambda x, \quad$ then $\langle y, x \rangle = \langle Ax, x \rangle - \lambda \langle x, x \rangle$;

$$\langle x, y \rangle = \langle \overline{y, x} \rangle = \langle Ax, x \rangle - \overline{\lambda} \langle x, x \rangle.$$

Hence, $\langle x, y \rangle - \langle y, x \rangle = (\lambda - \overline{\lambda})\langle x, x \rangle = 2i\beta||x||^2$, where $\lambda = \alpha + i\beta$

or, $\quad 2|\beta|\ ||x||^2 = |\langle x, y \rangle - \langle y, x \rangle| \leq |\langle x, y \rangle| + |\langle y, x \rangle| \leq 2||x||\ ||y||$

and therefore, $||y|| \geq |\beta|||x||$, that is, $||A_\lambda x|| \geq |\beta|\ ||x||$ $\qquad\qquad$ (9.27)

Since $\beta \neq 0$, it follows from theorem 9.6.3, that $\lambda = \alpha + i\beta$ with $\beta \neq 0$ is a regular value of the self-adjoint operator A.

In view of the above result we can say that the spectrum can lie on the real axis. We next want to show that if λ lie outside $[m, M]$ on the real line then it is a regular value.

For example, if $\lambda > M$, then $\lambda = M + k$ with $k > 0$.
We have $\langle A_\lambda x, x \rangle = \langle Ax, x \rangle - \lambda \langle x, x \rangle \leq M\langle x, x \rangle - \lambda \langle x, x \rangle = -k||x||^2$

where $\qquad\qquad |\langle A_\lambda x, x \rangle| \geq k||x||^2$

On the other hand, $\ |\langle A_\lambda x, x \rangle| \leq ||A_\lambda x||\ ||x||$. Thus $||A_\lambda x|| \geq k||x||$ showing that λ is regular. Similar arguments can be put forward if $\lambda \leq m$.

9.6.6   Theorem

M and m belong to the point spectrum.

Proof: If A is replaced by $A_\mu = A - \mu I$, then the spectrum is shifted by μ to the left and M and m change to $M - \mu$ and $m - \mu$ respectively. Thus without loss of generality it can be assumed that $0 \leq m \leq M$. Then $M = ||A||$ [see lemma 9.3.2].

We next want to show that M is in the point spectrum. Since $M = ||A||$, we can consider a sequence $\{x_n\}$ with $||x_n|| = 1$ such that

$$\langle Ax_n, x_n \rangle = M - \epsilon_n, \ \epsilon_n \to 0 \ \text{ as } n \to \infty.$$

Further, $\quad ||Ax_n|| \leq ||A||\ ||x_n|| = ||A|| = M$.

Therefore, $||Ax_n - Mx_n||^2 = \langle Ax_n - Mx_n, \ Ax_n - Mx_n \rangle$

$$= \langle Ax_n, Ax_n \rangle - 2M\langle Ax_n, x_n \rangle + M^2||x_n||^2.$$

$$= ||Ax_n||^2 - 2M(M - \epsilon_n) + M^2 \leq M^2 - 2M(M - \epsilon_n) + M^2.$$

$$= 2M\epsilon_n.$$

Hence, $||Ax_n - Mx_n|| = \sqrt{2M\epsilon_n}.$

Therefore, $||Ax_m - Mx_n|| \to 0$ as $n \to \infty$ and $||x_n|| = 1.$

Using corollary 9.6.4, we can conclude from the above that M belongs to the spectrum. Similarly, we can prove that m belongs to the spectrum.

9.6.7   Examples

1.   If A is the identity operator I, then the spectrum consists of the single eigenvalue 1 for which the corresponding eigenspace $H_1 = H$. $R_\lambda = \left[\dfrac{1}{(\lambda - 1)}\right] I$ is a bounded operator for $\lambda \neq 1$.

2. The operator $A : L_2([0,1]) \to L_2([0,1])$ is defined by $Ax = tx(t)$, $0 \leq t \leq 1$.

Example 9.5.11 shows that A is a non-negative operator. Here $m = 0$ and $M \leq 1$. Let us show that all the points of the segment $[0,1]$ belong to the spectrum of A, implying that $M = 1$.

Let $0 \leq \lambda \leq 1$ and $\epsilon > 0$. Let us consider the interval $[\lambda, \lambda + \epsilon]$ or $[\lambda - \epsilon, \lambda]$ lying in $[0, t]$.

$$x_\epsilon(t) = \begin{cases} \dfrac{1}{\sqrt{\epsilon}} & \text{for } t \in [\lambda, \lambda + \epsilon] \\ 0 & \text{for } t \notin [\lambda, \lambda + \epsilon] \end{cases}$$

Since $\displaystyle\int_0^1 x_\epsilon^2(t)dt = \int_\lambda^{\lambda+\epsilon} \dfrac{1}{\epsilon}dt = 1.$

Hence, $x_\epsilon(t) \in L_2([0,1]),\ ||x_\epsilon|| = 1.$

Furthermore, $A_\lambda x_\epsilon(t) = (t - \lambda)x_\epsilon(t).$

Therefore, $||A_\lambda x_\epsilon(t)||^2 = \dfrac{1}{\epsilon}\displaystyle\int_\lambda^{\lambda+\epsilon}(t - \lambda)^2 dt = \dfrac{\epsilon^2}{3}.$

We have $||A_\lambda x_\epsilon|| \to 0$ as $\epsilon \to 0$. Consequently, for λ, $0 \leq \lambda \leq 1$ is in the point spectrum.

At the same time, the operator has no eigenvalues. In fact, $A_\lambda x(t) = (t - \lambda)x(t)$.

If $A_\lambda x(t) = 0$, then $(t - \lambda)x(t) \to 0$ almost everywhere on $[0, 1]$ and thus $x(t)$ is also equal to zero, almost everywhere.

9.7   Invariant Subspaces

9.7.1   Definition: invariant subspace

A subspace L of H is called **invariant** under an operator A, if $x \in L \Rightarrow Ax \in L$.

9.7.2   Example

Let λ be the eigenvalue of A, and N_λ the collection of eigenvectors corresponding to this eigenvalue which includes zero as well.

Since $Ax = \lambda x, x \in N_\lambda \Rightarrow Ax \in N_\lambda$. Hence N_λ is an **invariant** subspace.

9.7.3   Remark

If the subspace L is invariant under A, we say that L **reduces** the operator A.

9.7.4   Lemma

For self-adjoint A, the invariance of L implies the invariance of its orthogonal complements, $M = H - L$.

Let $x \in M$, implying $\langle x, y \rangle = 0$ for every $y \in L$. However, $Ay \in L$ for $y \in L$, and $\langle x, Ay \rangle = 0$, i.e., $\langle Ax, y \rangle = 0$ for every $y \in L$. Hence $x \perp L$ and $Ax \perp L$ implies M is invariant under A. Moreover, $M = H - L$. Let G_λ denote the range of the operator A_λ, i.e., the collection of all elements of the form $y = Ax - \lambda x$, λ an eigenvalue. We want to show that

$$H = \overline{G_\lambda} + N_\lambda. \text{ Let } y \in G_\lambda,\ u \in N_\lambda,$$

then $\langle y, u \rangle = \langle Ax - \lambda x, u \rangle = \langle x, Au - \lambda u \rangle = \langle x, 0 \rangle = 0.$

Consequently, $G_\lambda \perp N_\lambda$. If $y \in \overline{G_\lambda}$ and $y \notin G_\lambda$,

then $y = \lim_n y_n, \quad \text{where } y_n \in G_\lambda \cdot \langle y_n, u \rangle = 0$

$\Rightarrow$ $\langle y, u \rangle = \lim_n \langle y_n, u \rangle = 0.$

Consequently, $\overline{G_\lambda} \perp N_\lambda$.

Now, let $\langle y, u \rangle = 0$ for every $y \in G_\lambda$. For any $x \in H$,

$$0 = \langle Ax - \lambda x, u \rangle = \langle x, Au - \lambda u \rangle \Rightarrow Au = \lambda u$$

since x is arbitrary. Therefore, $u \in N_\lambda$.

Consequently, $N_\lambda = H - G_\lambda = H - \overline{G_\lambda}$.

9.7.5   Lemma

$\overline{G_\lambda}$ *is an invariant subspace under a self-adjoint operator A where G_λ stands for the range of the operator A_λ.*

Proof: Let N denote the orthogonal sum of all the subspaces N_λ, i.e., a closed linear span of all the eigenvectors of the operator A. If H is separable, then it is possible to construct in every N_λ a finite or countably orthonormal system of eigenvectors which span N_λ for a particular λ. Since the eigenvectors of distinct members of N_λ are orthogonal, by combining these systems, we obtain an orthogonal system of eigenvectors $\{e_n\}$, contained completely in the span N.

The operator A defines in the invariant subspace L an operator A_L in

$(L \to L)$; namely $A_L x = Ax$ for $x \in L$. It can be easily seen that A_L is also a self-adjoint operator.

9.7.6 Lemma

If the invariant subspace L and M are orthogonal complements of each other, then the spectrum of A is the set-theoretic union of the spectra of operators A_L and A_M.

Proof: Let λ belong to the point spectrum of A_L (or A_M). Then, there is a sequence of elements $\{x_n\} \subseteq L$(or M) such that $\|x_n\| = 1$, $\|A_{L,\lambda} x_n\| \to 0(\|A_{M,\lambda} x_n\| \to 0)$. However, $\|A_{L,\lambda} x_n\| = \|A_\lambda x_n\|$ $(\|A_{M,\lambda} x_n\| \to \|A_\lambda x_n\|)$. Hence, λ belongs to the spectrum of A.

Now, let λ belong to the spectrum of neither A_L nor A_M. Then, there is a positive number C, such that $\|A_\lambda y\| = \|A_{L,\lambda} y\| \geq C\|y\|$, $\|A_{M,\lambda} z\| \geq C\|z\|$, for any $y \in L$ and $z \in M$. However, every $x \in H$ has the form $x = y + z$ with $y \in L$ and $z \in M$, and $\|x\|^2 = \|y\|^2 + \|z\|^2$. Hence,

$$\|A_\lambda x\| = \|A_\lambda y + A_\lambda z\| = (\|A_\lambda y\|^2 + \|A_\lambda z\|^2)^{\frac{1}{2}}$$

$$\geq C(\|y\|^2 + \|z\|^2)^{\frac{1}{2}} = C\|x\|.$$

Thus λ is not in the point spectrum of A.

9.8 Continuous Spectra and Point Spectra

It has already been shown that a Hilbert space H can be represented as the orthogonal sum of two spaces, N, a closed linear hull of the set of all eigenvectors of a self-adjoint operator A, and its orthogonal complement G.

Thus $H = N \oplus G$.

9.8.1 Definition: discrete or point spectrum

The spectrum of A_N is called **discrete or point spectrum** if N is the closed linear hull of all eigenvectors of a self-adjoint operator A.

9.8.2 Definition: continuous spectrum

The spectrum of the operator A_G is called **continuous spectrum** of A if G is the orthogonal complement of N in H.

9.8.3 Remark

(i) If $N = H$, then A has no continuous spectrum and A has a **pure point spectrum.**

This happen in the case of compact operators.

(ii) If $H = G$, then A has no eigenvalues and the operator A has a **purely continuous spectrum.** The operator in example 2 of section 9.6.7 has a purely continuous spectrum.

9.8.4   Spectral radius

Let A be a bounded linear operator mapping a Banach space E_x into itself. The **spectral radius** of A is denoted by $r_\sigma(A)$ and is defined as

$$r_\sigma(A) = \text{Sup } \{|\lambda|, \ \lambda \in \sigma(A)\}.$$

Thus, all the eigenvalues of the operator A lie within the disc with origin as centre and $r_\sigma(A)$ as radius.

9.8.5   Remark

Knowledge of spectral radius is very useful in numerical analysis.

We next find the value of the spectral radius in terms of the norm of the operator A.

9.8.6   Theorem

Let E_x be a complex Banach space and let $A \in (E_x \to E_x)$. Then $r_\sigma(A) = \lim\limits_{n\to\infty} ||A^n||^{\frac{1}{n}}$.

Proof: Note that for any $0 \neq \lambda \in \mathbb{R}(\mathbb{C})$, we have the factorization

$$A^n - \lambda^n I = (A - \lambda I)p(A) = p(A)(A - \lambda I)$$

where $p(A)$ is a polynomial in A. If follows from the above that if $A^n - \lambda^n I$ has a bounded universe in E_x, then $A - \lambda I$ has bounded inverse in E_x.

Therefore, $\lambda^n \in \rho(A^n) \Rightarrow \lambda \in \rho(A)$.

and so, $\lambda \in \sigma(A) \Rightarrow \lambda^n \in \sigma(A^n).$ (9.28)

Hence, if $\lambda \in \sigma(A)$, then $|\lambda|^n \leq ||A^n||$ (by (9.28) and lemma 9.3.2).

$\Rightarrow$ $|\lambda| \leq ||A^n||^{\frac{1}{n}}$ for $\lambda \in \sigma(A).$

Hence, $r_\sigma(A) = \text{Sup } \{|\lambda| : \lambda \in \sigma(A) \leq ||A^n||^{\frac{1}{n}}\}.$

This gives $r_\sigma(A) \leq \lim\limits_{n\to\infty} \text{ inf } ||A^n||^{\frac{1}{n}}$ (9.29)

Further, in view of theorems 4.7.21, the resolvent operator is represented by

$$R_\lambda(A) = -\lambda^{-1} \sum \lambda^{-k} A^k, \ |\lambda| \geq ||A||.$$

Also, we have $Ax = \lambda x$ where λ is an eigenvalue and x the corresponding eigenvector. Therefore

$$|\lambda| \ ||x|| = ||Ax|| \leq ||A|| \ ||x||$$

or, $|\lambda| \leq ||A||$ for any eigenvalue λ.

Hence, $r_\sigma(A) \leq ||A||$. Also $R_\sigma(A)$ is analytic at every point $\lambda \in \sigma(A)$. Let $x \in E_x$ and $f \in E_x^*$. Then the function

$$g(\lambda) = f(R_\lambda(A)x) = -\lambda^{-1} \sum_{n=0}^{\infty} f(\lambda^{-n} A^n x).$$

is analytic for $|\lambda| > r_\sigma(A)$. Hence the singularities of the function g all be in the disc $\{\lambda : |\lambda| \leq r_\sigma(A)\}$. Therefore, the series $\sum\limits_{n=1}^{\infty} f(\lambda^{-n} A^n x)$ forms a bounded sequence. Since this is true for every $f \in E_x^*$, an application of uniform boundedness principle (theorem 4.5.6) shows that the elements $\lambda^{-n} A^n$ form a bounded sequence in $(E_x \to E_x)$. Thus,

$$||\lambda^{-n} A^n|| \leq M < \infty$$

for some positive constant M (depending on λ).

Hence, $||A^n||^{\frac{1}{n}} \leq M^{\frac{1}{n}} |\lambda| \Rightarrow \lim\limits_{n \to \infty} \text{Sup} \, ||A^n||^{\frac{1}{n}} \leq |\lambda|$.

Since λ is arbitrary with $|\lambda| \leq r_\sigma(A)$, it follows that

$$\lim_{n \to \infty} \text{Sup} \, ||A^n||^{\frac{1}{n}} \leq r_\sigma(A). \tag{9.30}$$

It follows from (9.29) and (9.30) that $\lim\limits_{n \to \infty} ||A^n||^{\frac{1}{n}} = r_\sigma(A)$.

9.8.7 Remark

The above result was proved by I. Gelfand [19].

9.8.8 Operator with a pure point spectrum

9.8.9 Theorem

Let A be a self-adjoint operator in a complex Hilbert space and let A have a pure point spectrum.

Then the resolvent operator $R_\lambda = (A - \lambda I)^{-1}$ can be expressed as

$$\sum_n \frac{1}{\lambda_n - \lambda} P_n.$$

Proof: In this case, $N = H$ and therefore there exists a closed orthonormal system of eigenvectors $\{e_n\}$, such that

$$A e_n = \lambda_n e_n \tag{9.31}$$

where λ_n is the corresponding eigenvalue.

Every $x \in H$ can be written as

$$x = \sum_{n=1}^{\infty} c_n e_n \tag{9.32}$$

where the Fourier coefficients c_n are given by

$$c_n = \langle x, e_n \rangle \tag{9.33}$$

The projection operator P_n is given by

$$P_n x = \langle x, e_n \rangle e_n = c_n e_n, \tag{9.34}$$

P_n denotes the projection along e_n.

The series (9.32) can be written as,

$$x = Ix = \sum_n P_n x \text{ or in the form } I = \sum_n P_n \tag{9.35}$$

We know $P_n P_m = 0, \ m \neq n$ $\qquad\qquad\qquad\qquad\qquad\qquad$ (9.36)

By (9.31) and (9.35) $\quad Ax = \sum_n c_n A e_n = \sum_n \lambda_n P_n x$ $\qquad\qquad$ (9.37)

We can write A in the operator form. Then, (9.36) yields,

$$A = \sum_n \lambda_n P_n \tag{9.38}$$

Thus, $\qquad \langle Ax, x \rangle = \langle \sum_n \lambda_n c_n e_n, \sum_n c_m e_m \rangle = \sum_n \lambda_n c_n^2 \qquad$ (9.39)

Thus the quadratic form $\langle Ax, x \rangle$ can be reduced to a sum of squares. Using (9.37), (9.39) can be written as,

$$\langle Ax, x \rangle = \sum_n \lambda_n \langle P_n x, x \rangle \tag{9.40}$$

If λ does not belong to the closed set $\{\lambda_n\}$ of eigenvalues, then there is a $d > 0$ such that $|\lambda - \lambda_n| > d$.

We have $\quad A_\lambda x = (A - \lambda I)x = \sum_n (\lambda_n - \lambda) P_n x$. Since A_λ has an inverse and P_n commutes with A_λ^{-1}, we have

$$x = \sum_n (\lambda_n - \lambda) P_n A_\lambda^{-1} x = \sum_n P_n x.$$

Premultiplying with P_m we have

$$P_m x = (\lambda_m - \lambda) P_m A_\lambda^{-1} x.$$

Hence $\quad R_\lambda x = A_\lambda^{-1} x = \sum_n \frac{1}{\lambda_n - \lambda} P_n x.$ $\qquad\qquad$ (9.41)

Since $\quad P_n x = c_n e_n, \ R_\lambda x = \sum_n \frac{c_n}{\lambda_n - \lambda} e_n$ $\qquad\qquad$ (9.42)

Since $\quad \left| \dfrac{c_n}{\lambda_n - \lambda} \right| \leq \left| \dfrac{c_n}{d} \right|,$

$$\|R_\lambda x\| \leq \frac{1}{d} \left(\sum_n c_n^2 \right)^{\frac{1}{2}} = \frac{\|x\|}{d} \quad \text{or} \quad \|R_\lambda\| \leq \frac{1}{d}.$$

Consequently, λ does not belong to the spectrum. Now it is possible to write (9.42) in the form

$$R_\lambda = \sum_n \frac{1}{\lambda_n - \lambda} P_n. \tag{9.43}$$

9.8.10 Remark

For n dimensional symmetric (hermitian) matrices we have similar expressions for R_λ, with the only difference that for n-dimensional matrices the sum is finite.

Hilbert demonstrated that the class of operators with a pure point spectrum is the class of compact operators.

Problems

1. $A : H \to H$ be a coercive operator (see 9.5, problem 5).

   (i) Show that $[u, v] = \langle Au, v \rangle$, $\forall\, u, v \in \mathcal{D}(A)$ defines a scalar product in H. If $[u_n, u] \to 0$ as $n \to \infty$, u_n is said to tend u in **energy** and the above scalar product is called **energy product**.

   (ii) If $\{\phi_n\}$ be a eigenfunction of the operator A and λ_n the corresponding eigenvalue, show that the solution u_0 of the equation $Au = f$ can be written in the form,

   $$u_0 = \sum_{n=1}^{\infty} \frac{\langle f, \phi_n \rangle}{\lambda_n} \phi_n.$$

   (Hint: Note that $\langle A\phi_n, \phi_n \rangle = [\phi_n, \phi_n] = \lambda_n$ and $\langle A\phi_n, \phi_m \rangle = [\phi_n, \phi_m] = 0$, for $n \neq m$, $\{\phi_n\}$ is a system of functions which is orthogonal in *energy* and is complete in energy.)

2. Let A be a compact operator on H. If $\{u_n\}$ is an infinite dimensional orthonormal sequence in H then show that $Au_n \to 0$ as $n \to \infty$. In particular, if a sequence of matrices $\{a_{i,j}\}$ defines a compact operator on l_2,

   and $\qquad \Pi_j = \sum_{i=1}^{\infty} |a_{i,j}|^2 \quad$ and $\quad \Delta^i = \sum_{j=1}^{\infty} |a_{i,j}|^2$,

   show that $\Pi_j \to 0$ as $j \to \infty$ and $\Delta^i \to 0$ as $i \to \infty$.

3. Let $A \in (H \to H)$ where H is a Hilbert space. Then show that

   (i) A is normal if and only if $\|Ax\| = \|A^*x\|$ for every $x \in H$.

   (ii) if A is normal then $N(A) = N(A^*) = R(A)^\perp$.

4. Let $P \in (H \to H)$ be normal, H being a Hilbert space over $\mathbb{R}(\mathbb{C})$.

   (i) Let X be the set of eigenvectors of P and Y the closure of the span of X. Then show that Y and $Y^\perp$ are closed invariant subspaces for P.

   (Hint: Show that $\sigma(P)$ is a closed and bounded subset of $\mathbb{R}(\mathbb{C})$ and that $\sigma(P) = \sigma_e(P) \cup \{\mu : \bar{\mu} \in \sigma_a(P^*)\}$.)

5. Let $A \in (H \to H)$ be self-adjoint, where H is a Hilbert space over $\mathbb{C}$. Then show that its *Cayley transform*

   (i) $T(A) = (A - iI)(A + iI)^{-1}$ is unitary and $1 \notin \sigma(T(A))$.

6. Let a be a non-zero vector, v be a unit vector, and $\alpha = ||a||$. Define

$$\mu^2 = 2\alpha(\alpha - v^T a) \quad \text{and} \quad u = \left(\frac{1}{\mu}\right)(a - \alpha v).$$

   (i) Show that u is a unit vector and that $(I - 2u\, u^T)a = \alpha v$.

   (ii) If v_1 and v_2 are vectors and σ is a constant, show that $\det (I - \sigma v_1 v_2^T) = 1 - \sigma v_1^T v_2$, and that $\det (I - 2u\, u^T) = -1$.

7. Let $x(t) \in C([a, b])$ and $K(s, t) \in C([a, b] \times [a, b])$.

   If $\quad Ax(s) = \int_a^s K(s, t)x(t)dt$, show that $||A^n|| \leq \dfrac{K^n(b - a)^n}{n!}$

   where $\quad K = \max_{a \leq s, t \leq b} |K(s, t)|$.

   (Hint: For finding A^2 use Fubini's theorem (sec. 10.5).)

8. Let A denote a Fredholm integral operator on $L_2([a, b])$ with kernel $K(\cdot, \cdot) \in L_2([a, b] \times [a, b])$. Then show that

   (i) A is self-adjoint if and only if $\overline{K(t, s)} = K(s, t)$ for almost all (s, t) in $[a, b] \times [a, b]$.

   (ii) A is normal if and only if

$$\int_a^b |\overline{K(u, s)} K(u, t) d\mu(u) = \int_a^b K(s, u)\overline{K(t, u)} d\mu(u)$$

   for almost all (s, t) in $[a, b] \times [a, b]$.

   (Hint: Use Fubini's theorem (sec. 10.5).)

9. Fix $m \in \mathbb{R}(\mathbb{C})$. For $(x_1, x_2) \in \mathbb{R}^2(\mathbb{C}^2)$, define

$$A(x_1, x_2) = (mx_1 + x_2, mx_2).$$

   Then show that $||A|| = \left[\dfrac{2|m|^2 + 1 + \sqrt{4|m|^2 + 1}}{2}\right]^{\frac{1}{2}}$ while $\sigma(A) = \{m\}$, so that $r_\sigma(A) = |m| < ||A||$.

10. Let $A \in (H \to H)$ be normal. Let X be a set of eigenvectors of A, and let Y denote the closure of the span of X. Then show that Y and $Y^\perp$ are closed invariant subspaces of A.

11. Let $A \in (H \to H)$, H a Hilbert space.

    Then show that

    (i) $\lambda \in \omega(A)$ if and only if $\overline{\lambda} \in \omega(A^*)$

(ii) $\sigma_e(A) \subset w(A)$ and $\sigma(A)$ is contained in the closure of $w(A)$.

$w(A)$ is defined as $w(A) = \{\langle Ax, x \rangle : x \in H, \ ||x|| = 1\}$. $w(A)$ is known as **the numerical range** of A.

CHAPTER 10

MEASURE AND INTEGRATION IN L_p SPACES

In this chapter we discuss the theory of Lebesgue measure and p-integrable functions on $\mathbb{R}$. Spaces of these functions provide some of the most concrete and useful examples of many theorems in functional analysis. This theory will be utilized to study some elements of Fourier series and Fourier integrals. Before we introduce the Lebesgue theory in a proper fashion, we point out some of the lacuna of the Riemann theory which prompted a new line of thinking.

10.1   The Lebesgue Measure on $\mathbb{R}$

Before we introduce 'Lebesgue measure' and associated concepts we present some examples.

10.1.1   *Examples*

1. Let S be the set of continuous functions defined on a closed interval $[a, b]$. Let

$$\rho(x, y) = \int_a^b |x(t) - y(t)| dt. \tag{10.1}$$

(X, ρ) is a metric space. But it is **not** complete [see example in note 1.4.11].

$$\text{Let } x_n(t) = \begin{cases} 0 & \text{if } a \leq t \leq c - \dfrac{1}{n} \\ nt - nc + 1 & \text{if } c - \dfrac{1}{n} \leq t \leq c \\ 1 & \text{if } c \leq t \leq b \end{cases}$$

$\{x_n\}$ is a Cauchy sequence. Let $x_n \longrightarrow x$ as $n \to \infty$, then $x_n \longrightarrow x$ as $n \to \infty \Rightarrow x(t) = 0$, $t \in [a, c]$ and $x(t) = 1$ for $t \in (c, b]$, (see note 1.4.11).

The above example shows that $\int_a^b |x(t) - y(t)| dt$ can be used as a metric of a wider class of functions, namely the class of absolutely integrable functions.

2. Consider a sequence $\{f_n(t)\}$ of functions defined by

$$f_n(t) = \lim_{m \to \infty} [\cos(\pi n!)t]^{2m} \tag{10.2}$$

Thus, $f_n(t) = 1$ if $t = \dfrac{k}{n!}$, $k = 0, 1, 2, \ldots, n!$

$= 0$ otherwise

Define, $\rho(f_n, f_m) = \displaystyle\int_0^1 |f_n(t) - f_m(t)| dt$.

Now, for any value of n, $\exists$ some common point at which the functional values of f_n, f_m take the value 1 and hence their difference is zero. On the other hand, at the remaining points $\{(n! + 1) - (m! + 1)\}$ the value $f_n(t) - f_m(t)$ is equal to 1 or -1. But the number of such types of functions is finite and hence $f_n - f_m \neq 0$ only for a finite number of points. Thus $\rho(f_n, f_m) = 0$.

Hence, $\{f_n\}$ is a Cauchy sequence. Hence, $\{f_n(t)\}$ tends to a function $f(t)$ s.t.

$$\left.\begin{array}{l} f(t) = 1, \text{ at all rational points in } 0 \le t \le 1 \\ = 0, \text{ irrational points in } (0, 1) \end{array}\right\} \tag{10.3}$$

Therefore, if we consider the integration in the Riemann sense, the integral $\int_0^1 |f(t)| dt$ does not exist. Hence the space is not complete. We would show later that if the integration is taken in the Lebesgue sense, then the integral exists.

3. Let us define a sequence $\{\Omega_n\}$ of sets as follows:

$$\Omega_0 = [0, 1]$$

$\Omega_1 = \Omega_0$ with middle open interval of length $\frac{1}{4}$ removed.

$\Omega_2 = \Omega_1$ with middle open interval of the component intervals of Ω_1 removed, each of length $\frac{1}{4^2}$.

Then by induction we have already defined Ω_n so as to consist of 2^n disjoint closed intervals of equal length. Let $\Omega_{n+1} = \Omega_n$ with middle open intervals of the component intervals of Ω_n removed, each of length $\frac{1}{4^{n+1}}$.

For each $n = 1, 2, \ldots$, the sum of the lengths of the component open intervals of Ω_n is given by

$$m(\Omega_n) = 1 - \sum_{i=0}^{n-1} \frac{1}{4^{i+1}} 2^i = \frac{1}{2} + \frac{1}{2^{n+1}}. \qquad (10.4)$$

For every $n = 1, 2, \ldots$, let x_n be defined by

$$x_n(t) = \begin{cases} 1 & \text{if } t \in \Omega_n \\ 0 & \text{if } t \notin \Omega_n \end{cases} \qquad (10.5)$$

It may be seen that $\{x_n\}$ is a Cauchy sequence. Let $m > n$. Then

$$\begin{aligned} \rho(x_m, x_n) &= \int_0^1 |x_m(t) - x_n(t)| dt \\ &= m(\Omega_n) - m(\Omega_m) \\ &= \frac{1}{2^{n+1}} - \frac{1}{2^{m+1}} \end{aligned}$$

We shall show that $\{x_n\}$ does not converge to any Riemann integrable function. Let us suppose that there exists a Riemann integrable function x and that

$$\lim_{n \to \infty} \int_0^1 |x(t) - x_n(t)| dt = 0. \qquad (10.6)$$

Let J_1 be the open interval removed in forming Ω_1, J_1, J_2, J_3 the open intervals removed in forming Ω_2 etc.

For each $l = 1, 2, \ldots$ there is an N so that $n > N$ implies $x_n(t) = 0$, $t \in J_l$.

It follows x is equivalent to a function which is identically zero on

$$V = \bigcup_{l=1}^{\infty} J_l$$

But the lower Riemann integral of such a function is zero. Since x is integrable,

$$\int_0^1 x(t) dt = 0.$$

But (10.4) yields that

$$\int_0^1 x_n(t) dt > \frac{1}{2}$$

This contradicts (10.6).

Thus, the space of absolutely integrable functions when integration is used in the Riemann sense is not complete. This may be regarded as a major defect of the Riemann integration. The definition of Lebesgue integration overcomes this defect and other defects of the Riemann integration.

10.1.2 Remark

In example 2 (10.1.1) it may seen that $\{f_n\} \to 1$ at all rational points in $[0, 1]$ and $\longrightarrow 0$ at all irrational points in $[0, 1]$.

It is known that the set of rational points in $[0, 1]$ can be put into one-to-one correspondence with the set of positive integers, i.e., the set of natural numbers [see Simmons [53]]. Hence the set of rational numbers in $[0, 1]$ forms a **countable** set. Thus, the set of rational numbers in $[0, 1]$ can be written as a sequence $\{r_1, r_2, r_3, \ldots\}$.

Let ϵ be any positive real number. Suppose we put an open interval of width ϵ about the first rational number r_1, an interval of width $\epsilon/2$ about r_2 and so on. About r_n we put an open-interval of width $\epsilon/2^{n-1}$. Then we have an open interval of some positive width about every rational number in $[0, 1]$. The sum of the widths of these open intervals is $\epsilon + \frac{\epsilon}{2} + \frac{\epsilon}{2^2} + \cdots + \frac{\epsilon}{2^n} + \cdots = 2\epsilon$. We conclude from all this that **all** rational numbers in $[0, 1]$ can be covered with open intervals, the sum of whose length is an arbitrarily small positive number.

We say that the **Lebesgue measure** of the R of rational numbers in $[0, 1]$ is $l \cdot m \cdot (R) = 0$. This means that the greatest lower bound of the total lengths of a set of open intervals covering the rational number is zero.

The Lebesgue measure of the entire interval $[0, 1]$ is $l \cdot m \cdot [0, 1] = 1$. This is because the greatest lower bound of the total length of any set of open intervals covering the whole set $[0, 1]$ is 1.

Now if we remove the rational numbers in $[0, 1]$ from $[0, 1]$ we are left with the set of irrational numbers the Lebesgue measure of which is 1.

Thus, if we delete from $[0, 1]$ the set M of rational numbers, whose Lebesgue measure is zero, we can find $L \int_0^1 f(t)dt$ in example 2 above, i.e.,

$$\int_{[0,1]-M} f(t)dt = 1 \cdot \quad L \int_0^1 f(t)dt$$

denotes the integration in the Lebesgue sense. The above discussion may be treated as a prelude to a more formal treatment ahead.

10.1.3 The Lebesgue outer measure of a set $E \subset R$

The Lebesgue outer measure of a set $E \subseteq \mathbb{R}$ is denoted by $m^*(E)$ and is defined as

$$m^*(E) = g \cdot l \cdot b \left\{ \sum_{n=1}^{\infty} l(I_n) : E \subset \bigcup_{n=1}^{\infty} I_n \right\},$$

where I_n is an open interval in $\mathbb{R}$ and $l(I_n)$ denotes the length of the interval I_n.

10.1.4   Simple results

(i) $m^*(\phi) = 0$

(ii) $m^*(A) \geq 0$ for all $A \subset \mathbb{R}$

(iii) $m^*(A_1) \leq m^*(A_2)$ for $A_1 \subseteq A_2 \subset \mathbb{R}$

(iv) $m^*\left(\sum\limits_{n=1}^{\infty} A_n\right) \leq \sum\limits_{n=1}^{\infty} m^*(A_n)$ for all subsets $A_1, A_2, \ldots, A_n \ldots \subset \mathbb{R}$

(v) $m^*(I) = l(I)$ for any interval $I \subseteq \mathbb{R}$

(vi) Even when $A_1, A_2, \ldots, A_n \ldots$ are pairwise disjoint subsets of $\mathbb{R}$, we may **not** have

$$m^*\left(\bigcup_{n=1}^{\infty} A_n\right) = \sum_{n=1}^{\infty} m^*(A_n).$$

10.1.5   Definition: Lebesgue measurable set, Lebesgue measure of such a set

A set $S \subseteq \mathbb{R}$ is said to be **Lebesgue Measurable** if

$$m^*(A) = m^*(A \cap S) + m^*(A \cap S^C) \text{ for every } A \subseteq \mathbb{R}$$

Since we have always $m^*(A) \leq m^*(A \cap S) + m^*(A \cap S^C)$ we see that S is measurable (if and only if) for each A we have

$$m^*(A) \geq m^*(A \cap S) + m^*(A \cap S^C).$$

10.1.6   Remark

(i) Since the definition of measurability is symmetric in S and S^C, we have S^C measurable whenever S is.

(ii) Φ and the set $\mathbb{R}$ of all real numbers are measurable.

10.1.7   Lemma

If $m^*(S) = 0$ then S is measurable.

Proof: Let A be any set. Then $A \cap S \subset S$ and so $m^*(A \cap S) \leq m^*(S) = 0$. Also $A \supseteq A \cap S^C$.

Hence, $m^*(A) \geq m^*(A \cap S^C) = m^*(A \cap S) + m^*(A \cap S^C)$.

But $m^*(A) \leq m^*(A \cap S) + m^*(A \cap S^C)$.

Hence S is measurable.

10.1.8   Lemma

If S_1 and S_2 are measurable, so is $S_1 \cup S_2$.

Proof: Let A be any set. Since S_2 is measurable, we have

$$m^*(A \cap S_1^C) = m^*(A \cap S_1^C \cap S_2) + m^*(A \cap S_1^C \cap S_2^C)$$

Since $A \cap (S_1 \cup S_2) = (A \cap S_1) \cup (A \cup S_2 \cap S_1^C)$, we have

$$m^*(A \cap (S_1 \cup S_2)) \leq m^*(A \cap S_1) + m^*(A \cap S_2 \cap S_1^C).$$

Thus, $m^*(A \cap (S_1 \cup S_2)) + m^*(A \cap S_1^C \cup S_2^C)$

$$\leq m^*(A \cap S_1) + m^*(A \cap S_2 \cap S_1^C) + m^*(A \cap S_1^C \cap S_2^C)$$

$$= m^*(A \cap S_1) + m^*(A \cap S_1^C) = m^* A$$

Since $(S_1 \cup S_2)^C = S_1^C \cap S_2^C$. Hence, $S \cup S_2$ is measurable, since the above equality is valid for every set $A \subseteq \mathbb{R}$, where S^C denotes the complement of S in $\mathbb{R}$. If S is measurable then $m^*(S)$ is called the **Lebesgue measure** of S and is denoted simply by $m(S)$.

10.1.9   Remark

(i) Φ and $\mathbb{R}$ are measurable subsets.

(ii) The complements and countable union of measurable sets are measurable.

10.1.10   Lemma

Let A be any set and $S_1, S_2, \ldots, S_n$, a finite sequence of disjoint measurable sets. Then

$$m^* \left(A \cap \left(\bigcup_{i=1}^{n} S_i \right) \right) = \sum_{i=1}^{n} m^*(A \cup S_i)$$

Proof: The lemma can be proved by making an appeal to induction on n.

It is true for $n = 1$. Let us next assume that the lemma is true for $m = n - 1$ sets S_i. Since S_i are disjoint sets, we have

$$A \cap \left(\bigcup_{i=1}^{n} S_i \right) \cap S_n = A \cap S_n \qquad (10.7)$$

$$A \cap \left(\bigcup_{i=1}^{n} S_i \right) \cap S_n^C = A \cap \left(\bigcup_{i=1}^{n-1} S_i \right) \qquad (10.8)$$

Hence the measurability of S_n implies

$$m^* \left(A \cap \left(\bigcup_{i=1}^{n} S_i \right) \right) = m^* \left(A \cap \left(\bigcup_{i=1}^{n} S_i \right) \cap S_n \right)$$

$$+ m^* \left(A \cap \left(\bigcup_{i=1}^{n} S_i \right) \cap S_n^C \right)$$

Using (10.7) and (10.8) we have,

$$m^* \left(A \cap \left(\bigcup_{i=1}^{n} S_i \right) \right) = m^*(A \cap S_n) + m^* \left(A \cap \left(\bigcup_{i=1}^{n-1} S_i \right) \right)$$

$$= m^*(A \cap S_n) + \sum_{i=1}^{n-1} m^*(A \cap S_i) \qquad (10.9)$$

(10.9) is true since by assumption the lemma is true for $m = n - 1$. Thus the lemma is true for $m = n$ and the induction is complete.

10.1.11   Remark [see Royden [47]]

It can be proved that the Lebesgue measure m^* is **countably additive** on measurable sets, i.e., if $S_1, S_2, \ldots$ are pairwise disjoint measurable sets, then

$$m^* \left(\bigcup_{n=1}^{\infty} S_n \right) = \sum_{n=1}^{\infty} m^*(S_n).$$

10.2   Measurable and Simple Functions

10.2.1   Definition: Lebesgue measurable function

An extended real-valued function f on $\mathbb{R}$ is said to be **Lebesgue measurable** if $f^{-1}(S)$ is a measurable subset for every open subset S of $\mathbb{R}$ and if the subsets $f^{-1}(\infty)$ and $f^{-1}(-\infty)$ of $\mathbb{R}$ are measurable.

10.2.2   Definition: complex-valued   Lebesgue   measurable function

A complex-valued function f on $\mathbb{R}$ is said to be **Lebesgue measurable** if the real and imaginary part $\mathcal{R}ef$ and $\mathcal{I}mf$ are both measurable.

10.2.3   Lemma

Let f be an extended real-valued function whose domain is measurable. Then the following statements are equivalent:

(i) For each real number α the set $\{x : f(x) > \alpha\}$ is measurable.

(ii) For each real number α the set $\{x : f(x) \geq \alpha\}$ is measurable.

(iii) For each real number α the set $\{x : f(x) < \alpha\}$ is measurable.

(iv) For each real number α the set $\{x : f(x) \leq \alpha\}$ is measurable.

   If (i)–(iv) are true, then

(v) For each extended real number α the set $\{x : f(x) = \alpha\}$ is measurable.

Proof: Let the domain of f be D, which is measurable.

(i) $\Longrightarrow$ (iv) $\{x : f(x) \leq \alpha\} = D \sim \{x : f(x) > \alpha\}$

and the difference of two measurable sets is measurable. Hence (i) $\Longrightarrow$ (iv). Similarly (iv) $\Longrightarrow$ (i). This is because

$$\{x : f(x) > \alpha\} = D \sim \{x : f(x) \leq \alpha\}.$$

Next to show that (ii) $\Longrightarrow$ (iii) since

$$\{x : f(x) < \alpha\} = D \sim \{x : f(x) \geq \alpha\}.$$

(iii) $\Longrightarrow$ (ii) by arguments similar as in above.
(ii) $\Longrightarrow$ (i) since

$$\{x : f(x) > \alpha\} = \bigcup_{n=1}^{\infty} \left\{x : f(x) \geq \alpha + \frac{1}{n}\right\},$$

and the union of a sequence of measurable sets is measurable. Hence (ii) $\Longrightarrow$ (i).
(i) $\Longrightarrow$ (ii) since

$$\{x : f(x) \geq \alpha\} = \bigcap_{n=1}^{\infty} \left\{x : f(x) > \alpha - \frac{1}{n}\right\}$$

and the intersection of a sequence of measurable sets is measurable. Hence, (i) $\Longrightarrow$ (ii).

Thus the first four statements are equivalent.

If α is a real number,

$$\{x : f(x) = \alpha\} = \{x : f(x) \geq \alpha\} \cap \{x : f(x) \leq \alpha\}$$

and so (ii) and (iv) $\Longrightarrow$ (v) for α real. Since

$$\{x : f(x) = \infty\} = \bigcap_{n=1}^{\infty} \{x : f(x) \geq n\}$$

(ii) $\Longrightarrow$ (v) for $\alpha = \infty$. Similarly, (iv) $\Longrightarrow$ (v) for $\alpha = -\infty$, and we have (ii) and (iv) $\Longrightarrow$ (v).

10.2.4   *Remark*

(i) It may be noted that an extended real valued function f is (Lebesgue) measurable if its domain is measurable and if it satisfies one of the first four statements of the lemma 10.2.3.

(ii) A **continuous** function (with a measurable domain) is measurable, because the preimage of any open set in $\mathbb{R}$ is an open set.

(iii) Each **step** function is measurable.

10.2.5   Lemma

Let K be a constant and f_1 and f_2 be two measurable real-valued functions defined on the same domain. Then the functions $f_1 + K$, Kf_1, $f_1 + f_2$, $f_2 - f_1$ and $f_1 f_2$ are measurable.

Proof: Let $\{x : f_1(x) + K < \alpha\} = \{x : f_1(x) < \alpha - K\}$.

Therefore by condition (iii) of lemma 10.2.3, since f_1 is measurable, $f_1 + K$ is measurable.

If $f_1(x) + f_2(x) < \alpha$, then $f_1(x) < \alpha - f_2(x)$ and by the corollary to the axiom of Archimedes [see Royden [47]] we have a rational number between two real numbers. Hence there is a rational number p such that $f_1(x) < p < \alpha - f_2(x)$.

Hence, $\{x : f_1(x) + f_2(x) < \alpha\} = \cup \{x : f_1(x) < p\} \cap \{x : f_2(x) < \alpha - p\}$.

Since the rational numbers are countable, this set is measurable and so $f_1 + f_2$ is measurable.

Since $-f_2 = (-1)f_2$ is measurable, when f_2 is measurable $f_1 - f_2$ is measurable.

Now, $\{x : f^2(x) > \alpha\} = \{x : f(x) > \sqrt{\alpha}\} \cup \{x : f(x) < -\sqrt{\alpha}\}$ for $\alpha > 0$ and if $\alpha < 0$

$$\{x : f^2(x) > \alpha\} = D,$$

where D is the domain of f. Hence $f^2(x)$ is measurable. Moreover,

$$f_1 f_2 = \frac{1}{2}[(f_1 + f_2)^2 - f_1^2 - f_2^2].$$

Given f_1, f_2 measurable functions, $(f_1 + f_2)^2$, f_1^2 and f_2^2 are respectively measurable functions.

Hence, $f_1 f_2$ is a measurable function.

10.2.6   Remark

Given f_1 and f_2 are measurable,

(i) $\max\{f_1, f_2\}$ is measurable

(ii) $\min\{f_1, f_2\}$ is measurable

(iii) $|f_1|, |f_2|$ are measurable.

10.2.7   Theorem

Let $\{f_n\}$ be a sequence of measurable functions (with the same domain of definition). Then the functions $\sup\{f_1, \ldots, f_n\}$ and $\inf\{f_1, f_2, \ldots, f_n\}$, $\sup_n f_n$, $\inf_n f_n$, $\inf_n \sup_{k \geq n} f_k$, $\sup_n \inf_{k \geq n} f_k$ are all measurable.

Proof: If q is defined by $q(x) = \sup\{f_1(x), f_2(x), \ldots, f_n(x)\}$ then $\{x : q(x) > \alpha\} = \bigcup_{n=1}^{\infty} \{x : f_n(x) > \alpha\}$. Since f_i for each i is measurable, g is measurable.

Similarly, if $p(x)$ is defined by $p(x) = \sup f_n(x)$ then $\{x : p(x) > \alpha\} = \bigcup_{n=1}^{\infty} \{x : f_n(x) > \alpha\}$ and so $p(x)$ is measurable. Similar arguments can be put forward for inf.

10.2.8 Remark

If $\{f_n\}$ is a sequence of measurable functions such that $f_n^{(x)} \longrightarrow f(x)$ for each $x \in \mathbb{R}$ then f is measurable.

10.2.9 Almost everywhere (a.e.)

If f and g are measurable functions, f is said to be equal to g **almost everywhere** (abbreviated as a.e.) on a measurable set S, if

$$m\{x \in S : f(x) \neq g(x)\} = 0.$$

10.2.10 Characteristic function of a set E, simple function

If we refer to (10.3) in example 2 of 10.1 we see that

$$\left. \begin{cases} f(x) = 1 & \text{at all rational points in } [0,1] \\ f(x) = 0 & \text{at all irrational points in } [0,1] \end{cases} \right\}, \quad R\overline{\int}_0^1 f(x)dx = 1 \text{ and}$$

$R\underline{\int}_0^1 f(x)dx = 0.$

Thus the integral $\int_0^1 f(x)dx = 0$ is not Riemann integrable. This has led to the introduction of a function which is 1 on a measurable set and is zero elsewhere. Such a function is integrable and has as its integral the measure of the set.

Definition: characteristic function of E

The function χ_E defined by

$$\chi_E(x) = \begin{cases} 1 & x \in E \\ 0 & x \notin E \end{cases}$$

is called the *characteristic function* of E.

The characteristic function is measurable if and only if E is measurable.

Definition: simple function

A **simple function** is a scalar-valued function on $\mathbb{R}$ whose range is finite. If $a_1, a_2, \ldots, a_n$ are the distinct values of such a function ϕ, then

$$\phi(x) = \sum_{i=1}^{n} a_i \chi_{E_i}(x) \tag{10.10}$$

is called a simple function if the sets E_i are measurable, E_i is given by

$$E_i = \{x \in \mathbb{R} : \phi(x) = a_i\}$$

10.2.11   Remark

(i) The representation for ϕ is not unique.

(ii) The function ϕ is simple if and only if it is measurable and assumes only a finite number of values.

(iii) The representation (10.10) is called the canonical representation and it is characterised by the fact that E_i are disjoint and the a_i distinct and non-zero.

10.2.12   Example

Let $f : \mathbb{R} \longrightarrow [0, \infty[$. Consider the simple function for $n = 1, 2, \ldots$

$$\phi_n(x) = \begin{cases} \frac{(i-1)}{2^n} & \text{if } \frac{(i-1)}{2^n} \leq f(x) < \frac{i}{2^n} \text{ for } i = 1, 2, \ldots, n2^n \\ n & \text{if } f(x) \geq n \end{cases}$$

Then $0 \leq \phi_1(x) \leq \phi_2(x) \cdots \leq f(x)$ and $\phi_n(x) \longrightarrow f(x)$ for each $x \in \mathbb{R}$.

If f is bounded, the sequence $\{\phi_n\}$ converges to f uniformly on $\mathbb{R}$. If $f : \mathbb{R} \longrightarrow [-\infty, \infty]$, then by considering $f = f^+ - f^-$, where $f^+ = \max\{f, 0\}$ and $f^- = \min\{f, 0\}$, we see that $f = f^+$ where $f(x) \geq 0$ and $f = f^-$ when $f(x) = 0$.

Thus, there exists a sequence of simple functions which converges to f at every point of $\mathbb{R}$.

It may be noted that if f is measurable, each of the simple functions is measurable.

10.2.13   The Lebesgue integral ϕ

If ϕ vanishes outside a set of finite measure, we define the integral of ϕ by

$$\int \phi(x)dm(x) = \sum_{i=1}^{n} a_i\mu(E_i) \tag{10.11}$$

when ϕ has the canonical representation $\phi = \sum_{i=1}^{n} a_i\chi_{E_i}$.

10.2.14   Lemma

Let $\phi = \sum_{i=1}^{n} a_i\chi_{E_i}$, with $E_i \cap E_j = \emptyset$ for $i \neq j$. Suppose each set E_i is a measurable set of finite measure. Then

$$\int \phi dm = \sum_{i=1}^{n} a_i m(E_i).$$

Proof: The set $A_a = \{x : \phi(x) = a\} = \bigcup_{a_i=a} E_i$

Hence $am(A_a) = \sum_{a_i=a} a_i m(E_i)$ by the additivity of m, and hence

$$\int \phi(x)dm(x) = \sum a\, m(A_a) = \sum_{i=1}^{n} a_i\, m(E_i).$$

10.2.15   Theorem

Let ϕ and ψ be simple functions, which vanish outside a set of finite measure. Then

$$\int (\alpha\phi + \beta\psi)dm = \alpha \int \phi dm + \beta \int \psi dm$$

and if $\phi \geq \psi$ a.e. then

$$\int \phi dm \geq \int \psi dm.$$

Proof: Let $\{E_i\}$ and $\{E_i'\}$ be the set occurring in canonical representations of ϕ and ψ. Let E_0 and E_0' be the sets where ϕ and ψ are zero. Then the set F_k obtained by taking the intersections of $E_i \cap E_i'$ are members of a finite disjoint collection of measurable sets and we may write

$$\phi = \sum_{i=1}^{n} a_i \chi_{F_i} \qquad \psi = \sum_{i=1}^{n} b_i \chi_{F_i}$$

and so $\alpha\phi + \beta\psi = \sum (\alpha a_i + \beta b_i)\chi_{F_i}$

Hence, using lemma 10.2.13

$$\int (\alpha\phi + \beta\psi)dm = a \int \phi dm + b \int \psi dm$$

Again $\phi \geq \psi$ a.e. $\implies (\phi - \psi) \geq 0$ a.e.

$$\implies \int (\phi - \psi)dm \geq 0$$

since the integral of a simple function which is greater than or equal to zero a.e. is non-negative. Hence, the first part of the theorem yields

$$\int \phi dm \geq \int \psi dm.$$

10.2.16   The Lebesgue integral of a bounded function over a set of finite measure

Let f be a bounded real-valued function and E a measurable set of finite measure. Keeping in mind the case of Riemann integral, we consider for **simple functions** ϕ and ψ the numbers

$$\inf_{\psi \geq f} \int_E \psi \tag{10.12}$$

and $\quad \sup_{\phi \le f} \int_E \phi.$ $\hspace{6cm}$ (10.13)

It can be proved that if f is bounded on a measurable set E with $m(E)$ finite, then the integrals (10.12) and (10.13) will be equal where ϕ and ψ are simple functions if and only if f is measurable [see Royden [47]].

10.2.17   Definition: Lebesgue integral of f

If f is a bounded measurable function defined on a measurable set E with $m(E)$ finite, the Lebesgue integral of f over E is defined as

$$\int_E f(x)dm = \inf_{\psi \ge f} \int_E \psi(x)dx \hspace{4cm} (10.14)$$

for all simple functions $\psi (\ge f)$.

Note 10.2.1. If f is a bounded function defined on $[0,1]$ and f is Riemann integrable on $[0,1]$, then it is measurable and

$$R \int_0^1 f(x)dx = \int_0^1 f(x)dm.$$

10.2.18   Definition

If f is a complex-valued measurable function over $\mathbb{R}$, then we define

$$\int_\mathbb{R} fdm = \int_\mathbb{R} R_e fdm + i \int_\mathbb{R} I_m fdm$$

whenever $\int_\mathbb{R} \mathcal{R}e fdm$ and $\int_\mathbb{R} \mathcal{I}m fdm$ are well-defined.

10.2.19   Definition

If f is a measurable function on $\mathbb{R}$ and $\int_\mathbb{R} |f|dm < \infty$, we say f is an **integrable function** on $\mathbb{R}$.

In what follows we state without proof some important convergence theorems.

10.2.20   Theorem

Let $\{f_n\}$ be a sequence of measurable functions on a measurable subset E of $\mathbb{R}$.

(a) **Monotone convergence theorem:** If $0 \le f_1(x) \le f_2(x) \le \cdots$ and $f_n(x) \longrightarrow f(x)$ for all $x \in E$, then

$$\int_E f_n dm \longrightarrow \int_E fdm. \hspace{3cm} (10.15)$$

(b) **Dominated convergence theorem:** If $|f_n(x)| \leq g(x)$ for all $n = 1, 2, \dots$ and $x \in E$, where g is an integrable function on E and if $f_n(x) \xrightarrow[n \to \infty]{} f(x)$ for all $x \in E$ then f_n, f are integrable on E and

$$\int_E f_n(x) dm \longrightarrow \int_E f(x) dm \qquad (10.16)$$

If in particular, $m(E) < \infty$ and $|f_n(x)| \leq K$ for all $n = 1, 2, \dots$, $x \in E$ and some $K > 0$, then the result in 10.2.20(b) is known as the **bounded convergence theorem.**

Note 10.2.2. If f_1 and f_2 are integrable functions on E, then

$$\int_E (f_1 + f_2) dm = \int_E f_1 dm + \int_E f_2 dm.$$

Proof: The above result is true where f_1 and f_2 are simple measurable functions defined on E [see theorem 10.2.15]. We write $f_1 = f_1^+ - f_1^-$, where $f_1^+ = \max\{f_1, 0\}$ and $f_1^- = \min\{f_1, 0\}$. Similarly we take $f_2 = f_2^+ - f_2^-$, f_2^+ and f_2^- will have similar meanings as those of f_1^+, f_1^-. It may be noted that f_1^+, f_1^-, f_2^+, f_2^- are nonnegative functions.

We now approximate f_1^+, f_1^-, f_2^+, f_2^- by non-decreasing sequence of simple measurable functions and applying the monotone convergence theorem (10.2.20(a)).

10.3   Calculus with the Lebesgue Measure

Let $E = [a, b]$, a finite closed interval in $\mathbb{R}$. We first recapitulate a few definitions pertaining to the Riemann integral. Let f be a bounded real-valued function defined on the interval $[a, b]$ and let $a = \xi_0 < \xi < \cdots < \xi_n = b$ be a subdivision of $[a, b]$. Then for each subdivision we can define the sums

$$S = \sum_{i=1}^{n} (\xi_i - \xi_{i-1}) M_i \quad \text{and} \quad s = \sum_{i=1}^{n} (\xi_i - \xi_{i-1}) m_i$$

where $M_i = \sup_{\xi_{i-1} < x \leq \xi_i} f(x)$, $m_i = \inf_{\xi_{i-1} < x \leq \xi_i} f(x)$.

We then define the upper Riemann integral of f by

$$\int_a^b f(x) dx = \inf S \qquad (10.17)$$

with the infimum taken over all possible subdivisions of $[a, b]$. Similarly, we define the lower integral

$$\int_{\underline{a}}^b f(x) dx = \sup s \qquad (10.18)$$

The upper integral is always at least at large as the lower integral and if the two are equal we say f is *Riemann integrable* and call the common value, the Riemann integral of f. We shall denote it by

$$R \int_a^b f(x)dx \tag{10.19}$$

For the definition of the Lebesgue integral see 10.2.16.

10.3.1   *Remark*

Consider a $\mathbb{R}(\mathbb{C})$-valued bounded function f on $[a,b]$. f is Riemann integrable on $[a,b]$ if and only if the set of discontinuities of f on $[a,b]$ is of (Lebesgue) measure zero. In that case, f is Lebesgue integrable on $[a,b]$ and integral $\int_a^b f(x)dx$ is equal to the integral $\int_a^b f(x)dm$ [see Rudin [48]].

10.4   The Fundamental Theorem for Riemann Integration

A $\mathbb{R}(\mathbb{C})$-valued function F is differentiable on $[a,b]$ and its derivative f is continuous on $[a,b]$ if and only if

$$F(x) = F(a) + \int_a^x f(s)ds \quad a \leq x \leq b$$

for some continuous function f on $[a,b]$. In that case $F'(x) = f(x)$ for all $x \in [a,b]$. For a proof see Rudin [48].

10.4.1   *Absolutely continuous function*

A $\mathbb{R}(\mathbb{C})$-valued function F on $[a,b]$ is said to be **absolutely continuous** on $[a,b]$ if for every $\epsilon > 0$, there is some $\delta > 0$ such that

$$\sum_{i=1}^n |F(x_i) - F(y_i)| < \epsilon$$

whenever $a \leq y_1 < x_1 < \cdots < y_n < x_n \leq b$ and $\sum_{i=1}^n (x_i - y_i) < \delta$.

10.4.2   *Remark*

(i) Every absolutely continuous function is uniformly continuous on $[a,b]$.

(ii) If F is differentiable on $[a,b]$ and its derivative F' is bounded on $[a,b]$, then F is absolutely continuous by the mean value theorem.

10.5   The Fundamental Theory for Lebesgue Integration

A $\mathbb{R}(\mathbb{C})$-valued function F is absolutely continuous on $[a, b]$ if and only if

$$F(x) = F(a) + \int_a^x f \, dm \quad a \le x \le b$$

for some (Lebesgue) integrable function; f on $[a, b]$. In that case $F'(x) = f(x)$ for almost all $x \in [a, b]$ [see Royden [47]].

10.5.1   Total variation, bounded variation

Let $f : [a, b] \longrightarrow \mathbb{R}(C)$ be a function. Then the (total) **variation Var** (f) of f over $[a, b]$ is defined as

$$\text{Var } (f) = \sup \left[\sum_{i=1}^n |f(t_i) - f(t_{i-1})| \right] : P = [t_0, t_1, \ldots, t_n]$$

where $a = t_0$ and $b = t_n$ is a partition of $[a, b]$.

The supremum is taken over all partitions of $[a, b]$. If $\text{Var } (f) < \infty$ holds, f is said to be a function of **bounded variation.**

10.5.2   Remark

(i) An absolutely continuous function on $[a, b]$ is of bounded variation on $[a, b]$.

(ii) If f is of bounded variation on $[a, b]$, then $f'(x)$ exists for almost all $x \in [a, b]$ and f' is (Lebesgue) integrable on $[a, b]$ [see Royden [47]].

(iii) A function of bounded variation on $[a, b]$ need not be continuous on $[a, b]$.

For example, the characteristic function of the set $\left[0, \frac{1}{3}\right]$ is of bounded variation on $[0, 1]$, but it is not continuous on $[0, 1]$

Although our discussion is confined to Lebesgue measure on $\mathbb{R}$, we sometimes need Lebesgue measure on $\mathbb{R}^2$ to apply some results. The Lebesgue measure on $\mathbb{R}^2$ generalizes the idea of area of a rectangle, while the Lebesgue measure on $\mathbb{R}$ generalizes the idea of length of an interval.

10.5.3   Theorem (Fubini and Tonelli) [see Limaye [33]]

Let $m \times m$ denote the Lebesgue measure on $\mathbb{R}^2$ and $k(\cdot, \cdot)$ be a $\mathbb{R}(\mathbb{C})$-valued measurable function on $[a, b] \times [c, d]$. If either $\int \int_{[a,b] \times [c,d]} |K(s, t)| d(m \times m)(s, t) < \infty$ or if $K(s, t) \ge 0$ for all $(s, t) \in [a, b] \times [c, d]$, then $\int_c^d K(s, t) dm(t)$ exists for almost every $s \in [a, b]$ and $\int_a^b K(s, t) dm(s)$ exists for almost every $t \in [c, d]$.

The functions defined by these integrals are integrable on $[a, b]$ and $[c, d]$ respectively.

Moreover, $\displaystyle\int\int_{[a,b]\times[c,d]} K(s,t)d(m \times m)(s,t)$

$$= \int_a^b \left[\int_c^d K(s,t)dm(t) \right] dm(s)$$

$$= \int_c^d \left[\int_a^b K(s,t)dm(s) \right] dm(t).$$

10.6 L_p Spaces and Completeness

We first recapitulate the two important inequalities, namely Hölder's inequality and Minkowski's inequality, before taking up the case of pth power Lebesgue integrable functions defined on a measurable set E.

10.6.1   Theorem (Hölder's inequality) [see 1.4.3]

If $p > 1$ and q is defined by $\frac{1}{p} + \frac{1}{q} = 1$, then the following inequalities hold true

(H 1) $\displaystyle\sum_{i=1}^n |x_i y_i| \leq \left[\sum_{i=1}^n |x_i|^p \right]^{1/p} \left[\sum_{i=1}^n |y_i|^q \right]^{1/q}$

for complex numbers $x_1, x_2, \ldots, x_n, y_1, y_2, \ldots, y_n$

(H 2)   In case $x \in l_p$, i.e., pth power summable, $y \in l_q$, where p and q are defined as above $x = \{x_i\}$, $y = \{y_i\}$, we have

$$\sum_{i=1}^\infty |x_i y_i| \leq \left[\sum_{i=1}^\infty |x_i|^p \right]^{1/p} \left[\sum_{i=1}^n |y_i|^q \right]^{1/q}$$

The inequality is known as the Hölder's inequality for sum.

(H 3) If $f(x) \in L_p(]0,1[)$, i.e., pth power integrable, $g(x) \in L_q(]0,1[)$, i.e., qth power integrable, where p and q are defined as above, then

$$\int_a^b |f(x)g(x)|dx \leq \left(\int_a^b |f(x)|^p dx \right)^{1/p} \left(\int_a^b |g(x)|^q dx \right)^{1/q}$$

The above inequality is known as Hölder's inequality for integrals.

10.6.2   Theorem (Minkowski's inequality)[see 1.4.4]

(M 1) If $p \geq 1$, then

$$\left[\sum_{i=1}^n |x_i + y_i|^p \right]^{1/p} \leq \left[\sum_{i=1}^n |x_i|^p \right]^{1/p} + \left[\sum_{i=1}^n |y_i|^p \right]^{1/p}$$

for complex numbers $x_1, x_2, \ldots, x_n, y_1, y_2, \ldots, y_n$.

(M 2) If $p \geq 1$, $x = \{x_i\} \in l_p$, the pth power summable $y = \{y_i\} \in l_q$, where p and q are conjugate to each other, then

$$\left(\sum_{i=1}^{\infty} |x_i + y_i|^p \right)^{1/p} \leq \left(\sum_{i=1}^{\infty} |x_i|^p \right)^{1/p} + \left(\sum_{i=1}^{\infty} |y_i|^p \right)^{1/p}$$

(M 3) If $f(x)$ and $g(x)$ belong to $L_p(0,1)$, then

$$\left(\int_0^1 |f(x) + g(x)|^p dx \right)^{1/p} \leq \left(\int_0^1 |f(x)|^p \right)^{1/p} + \left(\int_0^1 |g(x)|^p \right)^{1/p}$$

We next consider E to be a measurable subset of $\mathbb{R}$ and $1 \leq p < \infty$. Let f be a measurable or p-integrable function on E and be **pth power integrable** on E i.e. $|f|^p$ is integrable on E. Then the following inequalities hold true.

10.6.3 Theorem

Let f and g be measurable functions on $\mathbb{R}$.

(a) **Hölder's inequality** Let $1 < p < \infty$ and $\frac{1}{p} + \frac{1}{q} = 1$. Then

$$\int_E |fg| dm \leq \left(\int_E |f|^p dm \right)^{1/p} \left(\int_E |g|^q dm \right)^{1/q}$$

(b) **Minkowski's inequality** Let $1 \leq p < \infty$.

Assume that $m(f^{-1}(\infty)) \cap g^{-1}(-\infty)) \equiv 0 \equiv m(f^{-1}(-\infty) \cap g^{-1}(\infty))$.

Then $\left(\int_E |f + g|^p dm \right)^{1/p} \leq \left(\int_E |f|^p dm \right)^{1/p} + \left(\int_E |g|^p dm \right)^{1/p}$

Proof: (a) Since f and g are measurable functions on E, fg is measurable on E (Sec. 10.2.5) and hence $\int_E |fg| dm$ is well-defined. Let

$$a = \left(\int_E |f|^p dm \right)^{1/p} \quad \text{and} \quad b = \left(\int_E |g|^p dm \right)^{1/p}$$

If $a = 0$ or $b = 0$, then $fg = 0$ almost everywhere on E and hence $\int_E |fg| dm = 0$. Therefore the inequality holds. If $a = \infty$ or $b = \infty$, then the inequality obviously holds. Next we consider $0 < a, b < \infty$. We replace x_i by f and y_i by g and the summation from $i = 0$ to n by integral over E with respect to Lebesgue measure in (H 1) of theorem 10.6.1. Then the proof of theorem 10.6.3 is obtained by putting forward arguments exactly as in the proof of theorem 10.6.1.

(b) Since f and g are measurable functions on E, $f + g$ is a measurable function on E (sec. 10.2.5). Moreover since f and g are each p-integrable, $f + g$ is p-integrable i.e. $\int_E |f + g|^p dm$ is well-defined. The proof proceeds exactly as in (M 1) of theorem 10.6.2.

10.6.4   Definition: $f \sim g$, Metric in $L_p(E)$

$f \sim g$: For measurable functions f and g on E, a measurable set, we write $f \sim g$ if $f = g$ almost everywhere on E.

It may be noted that $\sim$ is an equivalence relation on the set of measurable functions on E.

Metric: Let $1 \le p < \infty$. For any p-integrable functions f and g on E, define,

$$\rho_p(f, g) = \left(\int_E |f - g|^p dm \right)^{1/p}$$

Note that $m(\{x : |f(x)| = \infty\}) = 0 = m(\{x : |g(x)| = \infty\})$ since $\int_E |f(x)|^p dm < \infty$ and $\int_E |g(x)|^p dm < \infty$. Hence

(i) $\rho_p(f, g)$ is well-defined, non-negative and

(ii) $\rho_p(f, g) = \rho_p(g, f)$ (symmetric)

(iii) By Minkowski's inequality

$$\left(\int_E |f - g|^p dm \right)^{1/p} \le \left(\int_E |f - h|^p dm \right)^{1/p} + \left(\int_E |h - g|^p dm \right)^{1/p}$$

where h is a p-integrable function on E.

Hence $\rho_p(f, g) \le \rho_p(f, h) + \rho_p(h, g)$.

Thus the triangle inequality is fulfilled.

$L_p(E)$: Let $L_p(E)$ denote the set of all equivalence classes of p-integrable functions. ρ_p induces a metric on $L_p(E)$.

10.6.5   Definition: essentially bounded, essential supremum

Let $p = \infty$ A measurable function f is said to be **essentially bounded** on E if there exists some $\beta > 0$ such that $m\{x \in E : |f(x)| \ge \beta\} = 0$ and β is called an essential bound for $|f|$ on E.

Essential supremum

If $\alpha = \inf\{\beta : \beta$ an essential bound for $|f|$ on $E\}$, then α is itself an essential bound for $|f|$ on E. Such an 'α' is called the **essential supremum** of $|f|$ on E and will be denoted by $\text{essup}_E |f|$.

Let us consider $f(x) = \begin{cases} n & \text{at } x = \dfrac{1}{n}, \ n = 1, 2, \ldots \\ x & \text{otherwise } 0 < x \le 1. \end{cases}$ is essentially bounded and $\text{essup}_E |f| = 1$.

If f and g are essentially bounded functions on E, then it can be seen that

$$\text{essup}_E |f + g| \leq \text{essup}_E |f| + \text{essup}_E |g|$$

$L_p(E)$: Let $L_p(E)$ denote the set of all equivalence classes of essentially bounded functions on E under the equivalence relation $\sim$.

Then $\rho_\infty(f, g) = \text{essup}_E |f - g|$, $f, g \in E$ induces a metric on $L_\infty(E)$.

10.6.6 Theorem: For $1 \leq p < \infty$, the metric space $L_p(E)$ is complete

Proof: For $1 \leq p < \infty$, let $\{f_n\}$ be a Cauchy sequence in $L_p(E)$, where E is a measurable set.

To prove the space $L_p(E)$ to be complete, it is sufficient if we can show that a subsequence of $\{f_n\}$ converges to some point in $L_p(E)$. Hence, by passing to a subsequence if necessary, we may have that

$$\rho_p(f_{n+1}, f_n) \leq \frac{1}{2^n}, \quad n = 1, 2, \ldots$$

Let $f_0 = 0$ and for $x \in E$ and $n = 1, 2, \ldots$ we denote,

$$g_n(x) = \sum_{i=0}^{n} |f_{i+1}(x) - f_i(x)| \quad \text{and} \quad g(x) = \sum_{i=0}^{\infty} |f_{i+1}(x) - f_i(x)|$$

By Minkowski's inequality

$$\left(\int_E |g_n|^p dm \right)^{1/p} = \left(\int_E \left(\sum_{i=0}^{n} |f_{i+1} - f_i| \right)^p \right)^{1/p}$$

$$\leq \sum_{i=0}^{n} \left(\int_E |f_{i+1} - f_i|^p \right)^{1/p} = \sum_{i=0}^{n} \rho_p(f_{i+1}, f_i)$$

$$= \rho_p(f_1, f_0) + \sum_{i=1}^{n} \frac{1}{2^i}$$

If we apply the monotone convergence theorem (10.2.18(a)) to the above inequality, we obtain,

$$\left(\int_E |g|^p dm \right)^{1/p} = \lim_{n \to \infty} \left(\int_E |g_n|^p dm \right)^{1/p} \leq \rho_p(f_1, f_0) + 1 < \infty$$

Hence the function g is finite almost everywhere on E. Now the series $\sum_{i=0}^{\infty} |f_{i+1}(x) - f_i(x)|$ is absolutely continuous and hence summable for almost all $x \in E$. For such $x \in E$, we let

$$f(x) = \sum_{i=0}^{\infty} [f_{i+1}(x) - f_i(x)]$$

We know that $f_n(x) = \sum_{i=0}^{n-1} [f_{i+1}(x) - f_i(x)]$ for all $x \in E$, we have

$$\lim_{n \to \infty} f_n(x) = f(x) \quad \text{and} \quad |f_n(x)| \leq \sum_{i=1}^{n} |f_{i+1}(x) - f_i(x)| \leq g(x)$$

for almost all $x \in E$. Since the function g^p is integrable, the dominated convergence theorem (10.2.20(b)) yields

$$\int_E |f|^p dm = \lim_{n \to \infty} \int_E |f_n|^p dm \leq \int_E g^p dm < \infty$$

Hence $f \in L_p(E)$. By Minkowski's inequality, $|f| + g \in L_p(E)$ and $|f - f_n|^p \leq (|f| + g)^p$ for all $n = 1, 2, \ldots$ Again the dominated convergence theorem (10.2.20(b)) yields

$$\rho_p(f_n, f) = \left(\int_E |f_n - f|^p \right)^{1/p} \longrightarrow 0 \text{ as } n \to \infty.$$

Thus, the sequence $\{f_n\}$ converges to f in $L_p(E)$ showing that the metric space is complete.

Case $p = \infty$ Let us consider a Cauchy sequence $\{f_n\}$ in $L_\infty(E)$.

Let $M_j = \{x \in E : |f_j(x)| > \text{essup}_E |f_j|\}$ Thus except for the set M_j; $|f_j(x)|$ is bounded.

Moreover, let $N_{m,n} = \{x \in E | |f_m(x) - f_n(x)| > \text{essup}_E |f_m - f_n|\}$.

Thus except for the set $N_{m,n}$, $|f_m(x) - f_n(x)|$ is bounded.

Let G be the union of all M_j and $N_{m,n}$. Then $m(G) = 0$, and the sequence of $\{f_n\}$ converges uniformly to a bounded function f on the complement of G in E. Hence, $f \in L_\infty(E)$ and $\rho_\infty(f_n, f) \longrightarrow 0$ as $n \to \infty$.

Thus the sequence $\{f_n\}$ converges in $L_\infty(E)$, showing that the metric space $L_\infty(E)$ is complete.

10.6.7   The general form of linear functionals in $L_p([a, b])$

Let us consider an arbitrary linear functional $f(x)$ defined on $L_p[0, 1]$ $(p > 1)$.

Let $u_t(\zeta) = \begin{cases} 1 & \text{for } 0 \leq \zeta < t \\ 0 & \text{for } t \leq \zeta < 1 \end{cases}$ \hfill (10.20)

Let $h(t) = f(u_t(\xi))$. We first show that $h(t)$ is an absolutely continuous function. To this end, we take $\delta_i = (s_i, t_i)$, $i = 1, 2, \ldots, n$ to be an arbitrary system of non-overlapping intervals in $[0, 1]$.

Let $\epsilon_i = \text{sign } (h(t_i) - h(s_i))$.

Then $\sum_{i=1}^{n} |h(t_i) - h(s_i)| = f \left\{ \sum_{i=1}^{n} \epsilon_i [u_{t_i}(\zeta) - u_{s_i}(\zeta)] \right\}$

$$\leq ||f|| \left\| \sum_{i=1}^{n} \epsilon_i [u_{t_i}(\zeta) - u_{s_i}(\zeta)] \right\|$$

$$= ||f|| \left(\int_0^1 \left| \sum_{i=1}^{n} \epsilon_i [u_{t_i}(\zeta) - u_{s_i}(\zeta)] \right|^p d\zeta \right)^{1/p}$$

$$\leq ||f|| \left(\sum_{i=1}^{n} \int_{\delta_i} d\zeta \right)^{1/p} = ||f|| \left(\sum_{i=1}^{n} m(\delta_i) \right)^{1/p}$$

Hence, $h(t)$ is absolutely continuous. Thus $h(t)$ has an a.e. Lebesgue integrable derivative and is equal to the Lebesgue integral of this derivative. Let $h'(t) = \alpha(t)$, so that

$$h(t) - h(0) = \int_0^t \alpha(s)ds$$

Now $h(0) = f[u_0(\zeta)] = 0$, since $u_0(\zeta) \equiv 0$ is the null element of $L_p[0, 1]$. We have,

$$h(t) = \int_0^t \alpha(s)ds$$

It follows from (10.20), that

$$f[u_t(s)] = h(t) = \int_0^t \alpha(s)ds = \int_0^t u_t(s)\alpha(s)ds + \int_t^1 u_t(s)\alpha(s)ds$$

$$= \int_0^1 u_t(s)\alpha(s)ds$$

Since f is a linear functional, we have

$$f(u_{\frac{k}{n}}(s)) - f(u_{\frac{k-1}{n}}(s)) = \int_0^1 u_{\frac{k}{n}}(s)\alpha(s)ds - \int_0^1 u_{\frac{k-1}{n}}(s)\alpha(s)ds$$

$$= \int_0^1 \left[u_{\frac{k}{n}}(s) - u_{\frac{k-1}{n}}(s) \right] \alpha(s)ds$$

If $v_n(s) = \sum_{k=1}^{n} C_k[u_{\frac{k}{n}}(s) - u_{\frac{k-1}{n}}(s)]$, then $f(v_n) = \int_0^1 v_n(s)\alpha(s)ds$.

Let $x(t)$ be an arbitrary, bounded and measurable function. Then there exists a sequence of step functions $\{v_m(t)\}$, such that $v_m(t) \longrightarrow x(t)$ a.e. as $m \to \infty$, where $\{v_m(t)\}$ can be assumed to be uniformly bounded.

By the Lebesgue dominated convergence theorem (10.2.20(b)), we get

$$\lim_m f(v_m) = \lim_m \int_0^1 v_m(t)\alpha(t)dt = \int_0^1 \lim_{m \to \infty} v_m(t)\alpha(t)dt = \int_0^1 x(t)\alpha(t)dt$$

Since, on the other hand, $v_m(t) \longrightarrow x(t)$ a.e. and $v_m(t)$ is uniformly bounded, it follows that

$$||v_m - x||_p = \left(\int_0^1 |v_m(t) - x(t)|^p dt \right)^{1/p} \longrightarrow 0 \text{ as } m \to \infty$$

Therefore, $f(v_m) \to f(x)$ and consequently,

$$f(x) = \int_0^1 x(t)\alpha(t) dt$$

Consider now the function $x_n(t)$ defined as follows

$$x_n(t) = \begin{cases} |\alpha(t)|^{q-1} \text{sgn } \alpha(t) & \text{if } |\alpha(t)| \leq n \\ 0 & \text{if } |\alpha(t)| > n \end{cases}$$

where q is conjugate to p i.e. $\frac{1}{p} + \frac{1}{q} = 1$. The function $x_n(t)$ is bounded and measurable.

Therefore, $f(x_n) = \int_0^1 x_n(t)\alpha(t) dt = \int_0^1 |\alpha(t)|^{q-1} |\alpha(t)| dt$

$$\leq ||f|| \, ||x_n||_p = ||f|| \cdot \left(\int_0^1 |x_n(t)|^p dt \right)^{1/p}$$

On the other hand,

$$|f(x_n)| = f(x_n) = \int_0^1 |x_n(t)||\alpha(t)| dt$$

$$\geq \int_0^1 |x_n(t)||x_n(t)|^{\frac{1}{q-1}} dt = \int_0^1 |x_n(t)|^{\frac{q}{q-1}} dt$$

$$= \int_0^1 |x_n(t)|^p dt$$

Hence, $\int_0^1 |x_n(t)|^p dt \leq ||f|| \, ||x_n|| = ||f|| \left(\int_0^1 |x_n(t)|^p dt \right)^{\frac{1}{p}}$

Therefore, $\left(\int_0^1 |x_n(t)|^p dt \right)^{1/q} \leq ||f||$ (10.21)

Now, $\alpha(t)$ is Lebesgue integrable and becomes infinite only on a set of measure zero. Hence,

$$x_n(t) \longrightarrow |\alpha(t)|^{q-1} \text{a.e. on } [0,1]$$

Therefore, by the dominated convergence theorem (10.2.20(b))

$$\left(\int_0^1 |x_n(t)|^{(q-p)p} dt\right)^{\frac{1}{q}} \longrightarrow \left(\int_0^1 |\alpha(t)|^{(q-1)p}\right)^{\frac{1}{q}} dt \leq ||f|| \text{ as } n \to \infty$$

or, $\left(\int_0^1 |\alpha(t)|^q\right)^{1/q} \leq ||f||$ i.e. $\alpha(t) \in L_q[0,1]$ $\qquad(10.22)$

Now, let $x(t)$ be any function in $L_p[0,1]$. Then there exists $\int_0^1 x(t)\alpha(t)dt$. Furthermore, there exists a sequence $\{x_m(t)\}$ of bounded functions, such that

$$\int_0^1 |x(t) - x_m(t)|^p dt \to 0 \text{ as } m \to \infty$$

Therefore, $\int_0^1 x_m(t)\alpha(t)dt - \int_0^1 x(t)\alpha(t)dt \leq \int_0^1 |(x_m(t) - x(t)| \, |\alpha(t)|dt$

$$\leq \left(\int_0^1 |(x_m(t) - x(t)|^p dt\right)^{1/p} \cdot \left(\int_0^1 |\alpha(t)|^q dt\right)^{\frac{1}{q}}$$

Using the fact that $x_m(t) - x(t) \in L_p[0,1]$ and $\alpha(t) \in L_q([0,1])$, the above inequality is obtained by making an appeal to Hölder's inequality.

Since the sequence $\{x_m(t)\}$ are bounded and measurable functions, in $L_p([0,1])$ and $\alpha(t) \in L_q([0,1])$,

$$\int_0^1 x_m(t)\alpha(t)dt = f(x_m)$$

Hence, $f(x_m) \longrightarrow \int_0^1 x(t)\alpha(t)dt$ as $m \to \infty$

On the other hand $f(x_m) \longrightarrow f(x)$. It then follows that

$$f(x) = \int_0^1 x(t)\alpha(t)dt \qquad(10.23)$$

Thus every functional defined on $L_p([0,1])$ can be represented in the form (10.23).

Conversely, if $\beta(t)$ be an arbitrary function belonging to $L_q([0,1])$, then $g(x) = \int_0^1 x(t)\beta(t)dt$ can be shown to be a linear functional defined on $L_p([0,1])$.

If $g(x_1) = \int_0^1 x_1(t)\beta(t)dt$ and $g(x_2) = \int_0^1 x_2(t)\beta(t)dt$ then

$$g(x_1 + x_2) = \int_0^1 (x_1(t) + x_2(t))\beta(t)dt$$

$$= \int_0^1 x_1(t)\beta(t)dt + \int_0^1 x_2(t)\beta(t)dt = g(x_1) + g(x_2)$$

Moreover, $\|g(x)\|_p \leq \left(\int_0^1 |x(t)|^p dt \right)^{1/p} \left(\int_0^1 |\beta(t)|^q dt \right)^{1/q} < \infty,$

showing that $g(x) \in L_p([0,1])$.

Thus g is additive, homogeneous, i.e., linear and bounded.

The norm of the functional f given by (10.23) can be determined in terms of $\alpha(t)$.

It follows from (10.23) with the use of Hölder's inequality (1.4.4)

$$f(x) = \left| \int_0^1 x(t)\alpha(t)dt \right| \leq \left(\int_0^1 |x(t)|^p dt \right)^{1/p} \left(\int_0^1 |\alpha(t)|^q dt \right)^{1/q}$$

$$\leq \|x\|_p \|\alpha\|_q$$

Hence $\|f\| \leq \|\alpha\|_q$ (10.24)

It follows from (10.23) and (10.24) that

$$\|f\| = \|\alpha(t)\|_q = \left(\int_0^1 |\alpha(t)|^q dt \right)^{1/q}$$

10.7 L_p Convergence of Fourier Series

In 3.7.8 we have seen that in a Hilbert space H if $\{e_i\}$ is a complete orthonormal system, then every $x \in H$ can be written as

$$x = \sum_{i=0}^{\infty} c_i e_i \tag{10.25}$$

where $c_i = \langle x, e_i \rangle$, $i = 0, 1, 2, \ldots$ (10.26)

i.e., the series (10.25) converges.

c_i given by (10.26) are called **Fourier Coefficients**.

In particular, if $H = L_2([-\pi, \pi])$ and

$$e_n(t) = \left\{ \frac{e^{int}}{\sqrt{2\pi}} \right\}, \quad n = 0, \pm 1, \pm 2 \tag{10.27}$$

then any $x(t) \in L_2[-\pi, \pi]$ can be written uniquely as

$$(c_0 - id_0) + \sum_{n=1}^{\infty} \frac{2c_n \cos nt}{\sqrt{2\pi}} + \sum_{n=0}^{\infty} \frac{2d_n \sin nt}{\sqrt{2\pi}} \tag{10.28}$$

where $c_n = \frac{1}{\sqrt{2\pi}} \int_{-\pi}^{\pi} \int_{-\pi}^{\pi} x(t) \cos nt \, dt$ (10.29)

$$d_n = \frac{1}{\sqrt{2\pi}} \int_{-\pi}^{\pi} \int_{-\pi}^{\pi} x(t) \sin nt \, dt \tag{10.30}$$

Let us consider a more general problem of representing any integrable function of period 2π on $\mathbb{R}$ in terms of the special 2π-periodic function $\dfrac{e^{int}}{\sqrt{2\pi}}$, $n = 0, \pm 1, \pm 2, \ldots$. Let $x \in L_p([-\pi, \pi])$.

For $n = 0, \pm 1, \pm 2, \ldots$ the nth **Fourier coefficients** of x is defined by

$$\hat{c}_n = \frac{1}{\sqrt{2\pi}} \int_{-\pi}^{\pi} (x(t)) e^{-int} dm(t) \tag{10.31}$$

and the formal series

$$\frac{1}{\sqrt{2\pi}} \sum_{n=-\infty}^{\infty} \hat{c}_n e^{int} \tag{10.32}$$

is called the **Fourier series** of x.

For $n = 0, 1, 2, \ldots$ consider the nth partial sum,

$$s_n(t) = \sum_{k=-n}^{n} \hat{c}_k e^{ikt}, \quad t \in [-\pi, \pi]$$

10.7.1   Remark

(i) Kolmogoroff [29] gave an example of a function x in $L_1([-\pi, \pi])$ such that the corresponding sequence $\{s_n(t)\}$ diverges for each $t \in [-\pi, \pi]$.

(ii) If $x \in L_p([-\pi, \pi])$ for some $p > 1$, then $\{s_n(t)\}$ converges for almost all $t \in [-\pi, \pi]$ (see Carleson, A [10]).

A relevant theorem in this connection is the following.

10.7.2   Theorem

In order that a sequence $\{x_n(t)\} \subset L_p([0, 1])$ converges weakly to $x(t) \in L_p([0, 1])$, it is necessary and sufficient that

(i) the sequence $\{\|x_n\|\}$ is bounded,

(ii) $\displaystyle\int_0^1 x_n(s)ds \longrightarrow \int_0^t x(s)ds$ for any $t \in [0, 1]$

Proof: The assumption (i) is the same as that of theorem 6.3.22.

Therefore, we examine assumption (ii).

For this purpose, let us define,

$$\alpha_s(t) = \begin{cases} 1 & \text{for } 0 \le t \le s \\ 0 & \text{for } s \le t \le 1 \end{cases}$$

Then the sums

$$\sum_{i=1}^{n} c_i [\alpha_{s_i}(t) - \alpha_{s_{i-1}}(t)],$$

where $0 = s_0 < s_1 < \cdots < s_{n-1} < s_n = 1$ are everywhere dense in $L_q([0,1]) = L_p^*([0,1])$.

Hence, in order that $x_n(t) \xrightarrow{w} x(t)$, it is necessary and sufficient that assumption (i) is satisfied and that

$$\int_0^1 x_n(t)\alpha_s(t)dt \longrightarrow \int_0^1 x(t)\alpha_s(t)dt$$

or, $\displaystyle\int_0^s x_n(t)dt \longrightarrow \int_0^s x(t)dt,$

as $n \to \infty$ and for every $s \in [0,1]$.

10.7.3   Remark

Therefore, if $s_n(t) \in L_p([0,1])$ and fulfils the conditions of the theorem 10.7.2, then

$$\{s_n(t)\} \xrightarrow{w} s(t) \text{ as } n \to \infty.$$

CHAPTER 11

UNBOUNDED LINEAR OPERATORS

In 4.2.3 we defined a bounded linear operator in the setting of two normed linear spaces E_x and E_y and studied several interesting properties of bounded linear operators. But if said operator ceases to be bounded, then we get an **unbounded** linear operator.

The class of unbounded linear operators include a rich class of operators, notably the class of differential operators. In 4.2.11 we gave an example of an unbounded differential operator. There are usually two different approaches to treating a differential operator in the usual function space setting. The first is to define a new topology on the space so that the differential operators are continuous on a nonnormable topological linear space. This is known as L. Schwartz's theory of distribution (Schwartz, L [52]). The other approach is to retain the Banach space structure while developing and applying the general theory of unbounded linear operators (Browder, F [9]). We will use the second approach. We have already introduced closed operators in Chapter 7. The linear differential operators are usually closed operators, or at least have closed linear extensions. Closed linear operators and continuous linear operators have some common features in that many theorems which hold true for continuous linear operators are also true for closed linear operators. In this chapter we point out some salient features of the class of unbounded linear operators.

11.1   Definition: An   Unbounded   Linear Operator

11.1.1   Let E_x and E_y be two normed linear spaces

Let a linear operator $A : \mathcal{D}(A) \subset E_x \longrightarrow R(A) \subset E_y$, where $\mathcal{D}(A)$ and $R(A)$ stand for the domain and range of A, respectively.

If it does not fulfil the condition

$$\|Ax\|_{E_y} \leq K\|x\|_{E_x}, \quad \text{for all } x \in E_x \tag{11.1}$$

where K is a constant (4.2.3), the operator becomes unbounded.

See 4.2.11 for an example of an unbounded linear operator.

11.1.2   Theorem

Let A be a linear operator with domain E_x and range in E_y. The following statements are equivalent :

(i) A is continous at a point

(ii) A is uniformly continuous on E_x

(iii) A is bounded, i.e., there exists a constant K such that (11.1) holds true for all $x \in E_x$ [see theorem 4.1.5 and theorem 4.2.4].

11.2   States of a Linear Operator

Definition: State diagram is a table for keeping track of theorems between the ranges and the inverses of linear operators A and A^*. This diagram was constructed by S. Goldberg ([21]). In what follows, $A : E_x \to E_y$, E_x and E_y being normed linear (Banach) spaces.

We can classify the range of an operator into three types :

I. $R(A) = E_y$

II. $R(A) \neq E_y$, $\overline{R(A)} = E_y$,

III. $\overline{R(A)} \neq E_y$.

Similarly, A^{-1} may be of the following types:

(a) A^{-1} exists and is continuous and hence bounded

(b) A^{-1} exists but is not continuous

(c) A has no inverse.

If $R(A) = E_y$, we say A is in state I or that A is surjective written as $A \in I$. Similarly we say A is in state b written as $A \in b$ if $R(A) \neq E_y$ but $\overline{R(A)} = E_y$. Listed below are some theorems that show the impossibility of certain states for (A, A^*). For example, if A fulfills conditions I and b, then A will be said to belong to I_b. Similar meaning for $A^* \in II_c$.

Now (A, A^*) will be said to be in state (I_b, II_c) if $A \in I_b$ then $A^* \in II_c$.

11.2.1 Theorem

If A^* has a bounded inverse, then $R(A^*)$ is closed.

Proof: Let us suppose that $A^* : f_p \in E_y^* \to g \in E_x^*$. Since A^* has a bounded inverse, there exists an $m > 0$ such that

$$\|A^* f_p - A^* f_q\| \geq m\|f_p - f_q\|.$$

Thus, $\{f_p\}$ is a Cauchy sequence which converges to some f in some Banach space E_y^*. Since A^* is closed, f is in $\mathcal{D}(A^*)$ and $A^* f - g$. Hence $R(A^*)$ is closed.

11.2.2 Remark

The above theorem shows that if $A^* \in II$ then A^* cannot belong to a i.e., $A^* \notin II_a$.

11.2.3 Definition: orthogonal complements $K^\perp, {}^\perp C$

In 5.1.1. we have defined a **conjugate E_x^*** of Banach space E_x. Let A map a Banach space E_x into a Banach space E_y. In 5–6 an **adjoint, A^*** of an operator A mapping $E_y^* \longrightarrow E_x^*$ was introduced. We next introduce the notion of an **orthogonal complement** of a set in Banach space.

The orthogonal complement of a set $K \subseteq E_x$ is denoted by $K^\perp$ and is defined as

$$K^\perp = \{f \in E_x^* : f(x) = 0, \ \forall \ x \in K\} \tag{11.2}$$

Orthogonal: A set $K \subseteq E_x$ is said to be **orthogonal** to a set $F \subseteq E_x^*$, if $f(k) = 0$ for $f \in F \subseteq E_x^*$ and $\forall \ k \in K$.

Thus, $K^\perp$ is called an orthogonal complement of K, because $K^\perp$ is orthogonal to K.

Even if K is **not closed.** $K^\perp$ is a **closed subspace of E_x^*.**

${}^\perp C$: If C is a subset of E_x^*, **the orthogonal complement** of C in E_x is denoted by ${}^\perp C$ and defined by

$${}^\perp C = \{x : x \in E_x, \ F_x(f) = 0 \ \forall \ f \in C\} \tag{11.3}$$

For notion of F_x see theorem 5.6.5.

11.2.4 Remarks

$K^\perp$ and ${}^\perp C$ are closed subspaces respectively of E_x^* and E_x. Also $K^\perp = \overline{K}^\perp$ and ${}^\perp C = {}^\perp \overline{C}$.

11.2.5 Theorem

If L is a subspace of E_x, then ${}^\perp(L^\perp) = \overline{L}$.

Proof: Let $\{x_n\} \subseteq L$ be convergent and $\lim_{n \to \infty} x_n = x \in \overline{L}$. Let $\{f_m\} \subseteq L^\perp$ be convergent and $L^\perp$ being closed $\lim_{m \to \infty} f_m = f \in L^\perp$.

Now, $\quad |f_m(x_n) - f(x)| \leq |f_m(x_n) - f(x_n)| + |f(x_n) - f(x)|.$

Now, $\quad \lim_{m \to \infty} f_m(x_n) = f(x_n).$

Since $\quad x \in E_x,\ f(x) = 0$ for $f \in L^{\perp}$. Hence $f(x_m) - f(x) = 0.$

Thus, $\quad f_m(x_n) \longrightarrow f(x) = 0$ as $m, n \longrightarrow \infty.$

Hence, $x \in {}^{\perp}(L^{\perp})$. Thus ${}^{\perp}(L)^{\perp} \subseteq \overline{L}.$

On the other hand, $\overline{L} \subseteq {}^{\perp}(L)^{\perp}$ since ${}^{\perp}(L)^{\perp}$ is a closed subspace.

Thus $\quad {}^{\perp}(L)^{\perp} = \overline{L}.$

11.2.6   Theorem

If M is a subspace of E_x^*, then $({}^{\perp}M)^{\perp} \supset \overline{M}$. If E_x is reflexive, then $({}^{\perp}M)^{\perp} = \overline{M}.$

Proof: Let $\{f_n\} \subseteq M \subseteq E_x^*$ be a convergent sequence. Let $\|f_n - f\|_{E_x^*} \to 0$ as $n \to \infty$, where $f \in \overline{M}.$

Now, $\quad {}^{\perp}M = \{x_0 : x \in E_x,\ F_x(f) = 0 \ \forall \ f \in M\}$

Since $\quad f_n \in M$ and if $x \in E_x$ we have $F_x(f_n) = 0.$

Hence, $f_n(x) = F_x(f_n) = 0$ for $x \in {}^{\perp}M$ [see 5.6.5].

Thus, $|f_n(x) - f(x)| \leq \|f_n - f\| \|x\|,\ x \in {}^{\perp}M.$

$$\to 0 \quad \text{as } n \to \infty.$$

or $\quad |0 - f(x)| \to 0$ as $n \to \infty.$

Hence $f(x) = 0$ for $x \in {}^{\perp}M$. Thus $f \in ({}^{\perp}M)^{\perp}.$

Hence $f \in \overline{M} \Rightarrow f \in ({}^{\perp}M)^{\perp}$ proving that $({}^{\perp}M)^{\perp} \supseteq \overline{M}.$

For the second part see Remark 11.2.7.

11.2.7   Remark

If E_x is reflexive, i.e., $E_x = E_x^{**}$, then $({}^{\perp}M)^{\perp} = \overline{M}.$

11.2.8   Definition: domain of A^*

Domain of A^* is defined as

$$\mathcal{D}(A^*) = \{\phi : \phi \in E_y^*,\ \phi A \text{ is continuous on } \mathcal{D}(A)\}.$$

For $\phi \in \mathcal{D}(A^*)$, let A^* be the operator which takes $\phi \in \mathcal{D}(A^*)$ to $\overline{\phi A}$, where $\overline{\phi A}$ is the unique continuous linear extension of ϕA to all of $E_x.$

11.2.9   Remark

(i) $\mathcal{D}(A^*)$ is a subspace of E_y^* and A^* is linear.

(ii) $A^*\phi$ is taken to be $\overline{\phi A}$ rather than ϕA in order that $R(A^*)$ is contained in E_x^* [see 5.6.15].

11.2.10 Theorem

(i) $\overline{R}(A)^\perp = R(A)^\perp = N(A^*)$.

(ii) $\overline{R(A)} = {}^\perp N(A^*)$.

In particular, A has a dense range if and only if A^* is one-to-one.

Proof: (i) $R(A)^\perp$ is a closed subspace of E_y^*. Hence, $\overline{R}(A)^\perp = R(A)^\perp$.

$$R(A)^\perp = \{\phi : \phi \in E_y^*, \phi(v) = 0, \ v = Au \in R(A)\}.$$

Now, $\phi(v) = \phi(Au) = A^*\phi(u)$.

Therefore if $\phi \in R(A)^\perp, \ \phi \in N(A^*)$.

Hence, $R(A)^\perp \subseteq N(A^*)$. (11.4)

On the other hand, $N(A^*) = \{\psi : \psi \in \mathcal{D}(A^*) \supseteq E_y^*, \ A^*\psi = 0\}$,

Now, $A^*\psi = 0$,

Hence, $\psi(Au) = 0, \forall \, u \in \mathcal{D}(A)$ i.e. $\psi \in R(A)^\perp$.

Thus, $N(A^*) \subseteq R(A)^\perp$. (11.5)

(11.4) and (11.5) together imply that $R(A)^\perp = N(A^*)$.

(ii) It follows from (i) of theorem 11.2.6 that ${}^\perp(R(A)^\perp) = \overline{R(A)}$.

Again (i) of this theorem yields $R(A)^\perp = N(A^*)$.

Hence, ${}^\perp N(A^*) = {}^\perp(R(A)^\perp) = \overline{R(A)}$.

If $R(A)$ is dense in E_y, we have

$$\overline{R(A)} = E_y = {}^\perp N(A^*)$$
$$= \{y : y \in E_y, \ G_y(\phi) = \phi(y) = 0 \ \text{ where } A^*\phi = 0\}$$

Thus, $A^*\phi = 0 \Rightarrow \phi = 0$ showing that A^* is one-to-one.

11.2.11 Theorem

If A and A^* each has an inverse then $(A^{-1})^* = (A^*)^{-1}$.

Proof: By theorem 11.2.10. $\mathcal{D}(A^{-1}) = R(A)$ is dense in E_y. Hence $(A^{-1})^*$ is defined. Suppose $f \in \mathcal{D}((A^*)^{-1}) = R(A^*)$. Then there exists a $\phi \in \mathcal{D}(A^*)$, such that $A^*\phi = f$. To show $\phi \in \mathcal{D}((A^{-1})^*)$ we need to prove that $f(A^{-1})$ is continuous on $R(A)$. Now,

$$f((A^{-1}A)x) = A^*\phi(x) = \phi(Ax), \ x \in \mathcal{D}(A)$$

Thus, $(A^{-1})f = \phi$ on $R(A)$, where $\phi = (A^{-1})^*f = (A^{-1})^*A^*\phi$ since $R(A)$ is dense in E_y. Hence, $(A^{-1})^* = (A^*)^{-1}$ on $\mathcal{D}((A^*)^{-1})$. It remains to prove that $\mathcal{D}((A^{-1})^*) \subseteq \mathcal{D}((A^*)^{-1})$.

Let us next suppose that $\psi \in \mathcal{D}((A^{-1})^*)$. We want to show that $\psi \in \mathcal{D}((A^*)^{-1}) = R(A^*)$. For that we show that there exists an element $v^* \in \mathcal{D}(A^*)$ such that $A^*v^* = \psi$ or equivalently $v^*A = \psi$ on $\mathcal{D}(A)$.

Keeping in mind the definition of $\mathcal{D}(A^*)$ we define v^* as the continuous linear extension of ψA^{-1} to all of E_y, thereby obtaining $A^*v^* = \psi$. Thus,

$$\mathcal{D}((A^{-1})^*) \subset \mathcal{D}((A^*)^{-1}).$$

11.3   Definition: Strictly Singular Operators

The concept of a **strictly singular operator** was first introduced by T. Kato [28] in connection with the development of perturbation theory. He has shown that there are many properties common between A and $A + B$, where B is strictly singular. In what follows we take E_x and E_y as two normed linear spaces.

Definition: strictly singular operator

Let B be a bounded linear operator with domain in E_x and range in E_y. B is called **strictly singular** if it does not have a bounded inverse on any infinite dimensional subspace contained in its domain.

11.3.1   *Example*

The most important examples of strictly singular operators are compact operators. These play a significant role in the study of differential and integral equations.

In cases where the normed linear spaces are not assumed complete, it is convenient to also consider precompact operators.

11.3.2   *Definition: precompact operator*

Let A be a linear operator mapping E_x into E_y.

If $A(B(0,1))$ is **totally bounded** in E_y, then A is called **precompact**.

11.3.3   *Theorem*

Every precompact operator is strictly singular.

Proof: Let B be a precompact operator with domain in E_x and range in E_y. B is bounded since a totally bounded set is bounded. Let us assume that B has a bounded inverse on a subspace $M \subseteq D(B)$. If $B_M(0,1)$ is a unit ball in M, then $B[B_M(0,1)]$ is totally bounded. Since B has a bounded inverse on M, it follows that $B_M(0,1)$ is totally bounded in M. Since unit ball in M is totally bounded, it has a finite ϵ-net, i.e., it has finite number of points $x_1, x_2, \ldots x_n$ in the unit ball in M such that for every $x \in B_M(0,1)$, there is an x_i such that

$$\|x - x_i\| < 1 \tag{11.6}$$

Let us assume that the finite dimensional space N spanned by $x_1, x_2, \ldots x_k$ is M.

Suppose this assertion is false. Since N is a finite dimensional proper subspace of the normed linear space M, we will show that there exists an element in the unit ball of M whose distance from N is 1. Let z be a point in M but not in N. Then there exists a sequence $\{m_k\}$ in N such that $\|z - m_k\| \to d(z, N)$.

Since N is finite dimensional and $\{m_k\}$ is bounded, N must also be compact (1.6.19). Hence, $\{m_k\}$ has a convergent subsequence $\{m_{k_p}\}$ which converges to $m \in N$ (say). Hence,

$$\|z - m\| = \lim_{p \to \infty} \|z - m_{k_p}\| = d(z, N)$$
$$= d(z - m, N)$$

Since $z - m \neq 0$,

$$1 = \left\| \frac{z - m}{z - m} \right\| = \frac{d(z - m, N)}{\|z - m\|} = d\left(\frac{z - m}{\|z - m\|}, N \right).$$

This contradicts (11.6).

Hence M is finite-dimensional. Therefore, B does not have an inverse on an infinite dimensional subspace.

11.3.4 Definition: finite deficiency

A subspace L of a vector space E is said to have **finite deficiency** in E if the dimension of E/L is **finite**. This is written as

$$\dim E/L < \infty.$$

Even though L is not contained in $\mathcal{D}(A)$, or restriction of A to L will mean a restriction of A to $L \cap \mathcal{D}(A)$.

11.3.5 Theorem

Let A be a linear operator from a subspace of E_x into E_y. Assume that A does not have a bounded inverse when restricted to any closed subspace having finite deficiency in E_x. Then, given an arbitrarily small number $\epsilon > 0$, there exists an infinite dimensional subspace $L(\epsilon)$ contained in $\mathcal{D}(A)$, such that A restricted to $L(\epsilon)$ is precompact and has norm not exceeding ϵ.

Proof: Since A does not have a bounded inverse on a closed subspace E_x having finite deficiency, we can not find a $m > 0$ such that $\|Ax\| \geq m\|x\|$ on such a subspace. Therefore, there is no loss of generality in assuming that there exists an $x_1 \in E_x$ such that $\|x_1\| = 1$ and $\|Ax_1\| < \frac{\epsilon}{3}$. There is an $f_1 \in E_x^*$ such that $\|f_1\| = 1$ and $f_1(x_2) = \|x_2\| = 1$. Since $N(f_1)$ has a deficiency 1 in E_x, there exists an element $x_2 \in N(f_1)$ such that $\|x_2\| = 1$ and $\|Ax_2\| \leq \frac{\epsilon}{3^2}$. There exists an $f_2 \in E_x^*$ such that $\|f_2\| = 1$ and $f_2(x_2) = \|x_2\| = 1$. Since, $N(f_1) \cap N(f_2)$ has finite deficiency in E_x, there exists an $x_3 \in N(f_2) \cap N(f_1)$ such that $\|x_3\| = 1$ and $\|Ax_3\| \leq \frac{\epsilon}{3^3}$.

Hence, by induction, we can construct sequences $\{x_k\}$ and $\{f_k\}$ having the following properties :

$$\|x_k\| = \|f_k\| = f_k(x_k) = 1, \quad \|Ax_k\| < \frac{\epsilon}{3^k}, \quad 1 \le k < \infty \tag{11.7}$$

$$x_k \in \cap_{i=1}^{k-1} N(f_i) \text{ or equivalently, } f_i(x_k) = 0 \quad 1 \le i < \infty \tag{11.8}$$
$$i \neq k$$

We next show that $\{x_k\}$ is linearly independent.

If that is not so, we can find $\alpha_1, \alpha_2, \dots \alpha_k$ not all zeroes, such that

$$\alpha_1 x_1 + \alpha_2 x_2 + \cdots + \alpha_k x_k + \cdots = 0.$$

or $\alpha_1 f_i(x_1) + \alpha_2 f_i(x_2) + \cdots + \alpha_k f_i(x_k) + \cdots = 0.$

Using (11.7) and (11.8) we get $\alpha_i = 0$ for $1 \le i < k = 2, 3, \dots$

Hence, $\{x_k\}$ is a linearly independent set. Let $L = \text{span } \{x_1, x_2, \dots\}$. L is an infinite dimensional subspace of $\mathcal{D}(A)$. It will now be shown that the restriction A_L of A to L has norm not exceeding ϵ.

Suppose $x = \sum_{i=1}^{l} \alpha_i x_i$. Then from (11.7) and (11.8),

$$\alpha_1 = |f_1(x_1)| \le \|f_1\| \, \|x\| = \|x\|$$

We next want to establish that

$$|\alpha_k| \le 2^{k-1} \|x\| \quad 1 \le k \le l \tag{11.9}$$

Let us suppose that (11.9) is true for $k \le j < l$.

Then we get from (11.7) and (11.8) that

$$f_{j+1}(x) = \sum_{i=1}^{j} \alpha_i f_{j+1}(x_i) + \alpha_{j+1} \tag{11.10}$$

Hence, by (11.10) and the induction hypothesis,

$$|\alpha_{j+1}| \le |f_{j+1}(x)| + \sum_{i=1}^{j} |\alpha_i| \, |f_{j+1}(x_i)|$$

$$\le \|x\| + \sum_{i=1}^{j} 2^{i-1} \|x\| = 2^j \|x\|$$

Hence, (11.9) is true by induction.

Thus, $\|Ax\| \le \sum_{i=1}^{l} |\alpha_i| \|Ax_i\| \le \sum_{i=1}^{l} 2^{i-1} 3^{-i} \epsilon \|x\| \le \epsilon \|x\|$

Hence, $\|Ax\| \le \epsilon$.

To prove that A_L is precompact, we would show that A_L is the limit in $(L \longrightarrow E_y)$ of a sequence of precompact operators [see 8.1.9]. For each positive integer n, we define $A_n^L : L \longrightarrow E_y$ to be A on span $\{x_1, \ldots x_n\}$ and 0 on span $\{x_{n+1}, x_{n+2}, \ldots\}$. It may be noted that A_n^L is linear and has finite dimensional range. Moreover, A_n^L is bounded on L for if $x = \sum_{i=1}^{n+k} \alpha_i x_i$ then by (11.7) and (11.9)

$$\|A_n^L x\| \leq \sum_{i=1}^{n} |\alpha_i| \, \|A x_i\| \leq \sum_{i=1}^{n} 2^{i-1} 3^{-i} \epsilon \|x\|.$$

Thus A_n^L is bounded and finite dimensional.

Hence, A_n^L is precompact.

Since, $\|A_L x - A_n^L x\| \leq \sum_{i=n+1}^{\infty} |\alpha_i| \, \|A x_i\| \leq \epsilon \|x\| \sum_{i=n+1}^{\infty} 2^{i-1} 3^{-i} \to 0$ as $n \to \infty$, it follows that A_n^L converges to A_L in $(L \longrightarrow E_y)$. Hence, A_L is precompact [see 8.1.9].

11.4 Relationship between Singular and Compact Operators

The following theorem reveals the connection between the class of strictly singular and the class of compact operators.

11.4.1 Theorem

Suppose $B \in (E_x \longrightarrow E_y)$. The following statements are equivalent :

(a) B is strictly singular

(b) For every infinite dimensional subspace $L \subset E_x$, there exists an infinite dimensional subspace $M \subseteq L$ such that B is precompact on M.

(c) Given $\epsilon > 0$ and given L an infinite dimensional subspace of E_x, there exists an infinite dimensional subspace $M \subseteq L$ such that B restricted to M and has norm not exceeding ϵ, an arbitrary small positive number.

Proof: We first show that (a) implies (b). Let us suppose that B is strictly singular and L is an infinite-dimensional subspace of E_x. Then B_L, the restriction of B to L, is strictly singular. Therefore, B_L does not have a bounded inverse on an infinite dimensional subspace $M \subseteq L$. Hence, by theorem 11.3.4, B is precompact on such a $M \subseteq L$.

Next, we show that (b) $\Rightarrow$ (c). If (b) is true then we assert that B does not have a bounded inverse on an infinite dimensional subspace, having finite deficiency in L. If that is not so, then B would be precompact and would have a bounded inverse at the same time. This violates the conclusion of theorem 11.3.3. Hence by applying theorem 11.3.4 to B_L (c) follows.

Finally, we show that (c) $\Rightarrow$ (a). It follows from (c) that $\|B_L\| \leq \epsilon$, i.e., $\|B_L x\| \leq \epsilon \|x\|$ for an arbitrary small $\epsilon > 0$, i.e., $\|B_L x\| \not> m\|x\|$, $m > 0$ and finite, for all x belonging to an infinite dimensional subspace $M \subseteq L$. Hence, B_L does not have a bounded inverse on $M \subseteq L$. Thus B is strictly singular.

11.5   Perturbation by Bounded Operators

Suppose we want to study the properties of a given operator A mapping a normed linear space E_x into a normed linear space E_y. But if the operator A turns out to be *involved*, then A is replaced by, say, $T + V$, where T is a relatively simple operator and V is such that the knowledge about properties of T is enough to gain information about the corresponding properties of A. For example, if A is a differential operator $\sum_{k=s}^{n} a_k D^k$, T is chosen as $a_n D^n$ and V is the remaining lower order terms. The concern of the penturbation theory is to find the conditions that V should fulfil so that the properties of T can help us determine the properties of A.

In what follows, V is a linear operator with domain a subspace of E_x and range a subspace of E_y.

11.5.1   Definition: kernel index of A, deficiency index of A and index of A

Kernel index of A: The dimension of $N(A)$ will be defined as the kernel index of A and will be denoted by $\alpha(A)$.

Deficiency index of A: The deficiency of $R(A)$ in E_y written as $\beta(A)$, will be called the deficiency index of A.

Then $\alpha(A)$ and $\beta(A)$ will be either a non-negative integer or ∞.

Index of A: If $\alpha(A)$ and $\beta(A)$ are not both infinite, we say A has an inverse. The index $\kappa(A)$ is defined by $\kappa(A) = \alpha(A) - \beta(A)$.

It is understood as in the real number system, if p is any real number,

$$\infty - p = \infty \quad \text{and} \quad p - \infty = -\infty.$$

11.5.2   Examples

1. Let $E_x = L_p([a, b])$ $E_y = L_q([a, b])$ where $1 \leq p, q < \infty$.
Let us define A as follows :
$\mathcal{D}(A) = \{u : u^{(n-1)}$ exists and is absolutely continuous on $[a, b], u^{(n)} \in E_y\}$.

$Au = u^{(n)}$, $u^{(n)}$ stands for the n-th derivative of u in $[a, b]$. It may be recalled that an absolutely continuous function is differentiable almost everywhere (10.5). Here, $N(A)$ is the space of polynomials of degree at most $(n-1)$. Hence, $\alpha(A) = n$, $\beta(A) = 0$.

2. Let $E_x = E_y = l_p$, $1 \le p \le \infty$. Let $\{\lambda_\kappa\}$ be a bounded sequence of numbers and A be defined on all of E_x by $A(\{x_\kappa\}) = \{\lambda_\kappa x_\kappa\}$.

$\alpha(A)$ are the members of λ_k which are 0. $\beta(A) = 0$ if $\{1/\lambda_\kappa\}$ is a bounded sequence. $\beta(A) = \infty$ if infinitely many of the λ_κ are 0.

11.5.3 Lemma

Let L and M be subspace of E_x with dim $L >$ dim M (thus dim $M < \infty$). Then, there exists a $l \ne 0$ in L such that

$$\|l\| = \text{ dist } (l, M).$$

Note 11.5.1 This lemma does not hold if dim $L = $ dim $M < \infty$.

For example, if $E_x = \mathbb{R}^2$ and L and M are two lines through the origin which are not perpendicular to each other.

If E_x is a Hilbert space, the lemma has the following easy proof.

Proof: First we show that dim $M = \dim(M \oplus M^\perp/M^\perp) = \dim(H/M^\perp)$ where $H/M^\perp$ stands for the quotient space.

Let $x, y \in M$. We consider the mapping $x \to x + M$ in $(H/M^\perp)$. Similarly, $y \to y + M$ in $(H/M^\perp)$. Thus, $x + y \to (x + M) + (y + M) = x + y + M$ in $(H/M^\perp)$. Similarly, for any scalar λ, $\lambda x \to \lambda(x + M) = \lambda x + M$.

Thus, there is an isomorphism between M and $(H/M^\perp)$. Let us assume that $L \cap M^\perp = \Phi$. Hence, dim $M = \dim(M + M^\perp/M^\perp) = \dim(H/M^\perp) \ge \lim(L + N^\perp/N^\perp) = $ dim L. The above contradicts the hypothesis that $L \cap M^\perp = \Phi$.

Let $x \in L \cap M^\perp$ and let $x \ne \theta$. Then

$$\|x - m\|^2 = \|x\|^2 + \|m\|^2 \ge \|x\|^2 \text{ for } m \in M.$$

Thus, $d(x, M) = \|x\|$.

11.5.4 Definition: minimum module of A

Let $N(A)$, the null manifold of A, be closed. The minimum module of A is written as $\gamma(A)$ and is defined by

$$\gamma(A) = \inf_{x \epsilon D(A)} \frac{\|Ax\|}{d(x, N(A))} \qquad (11.11)$$

11.5.5 Definition

The one-to-one operator $\hat{A}$ of A induced by A is the operator from $D(A)/N(A)$ into E_y defined by

$$\hat{A}[x] = Ax,$$

where the coset $[x]$ denotes the set of elements equivalent to x and belongs to $\mathcal{D}(A)/N(A)$.

$\hat{A}$ is one-to-one and linear with same range as that of A. We next state without proof the following theorem.

11.5.6 Theorem (Goldberg [21])

Let $N(A)$ be closed and let $\mathcal{D}(A)$ be dense in E_x. If $\gamma(A) > 0$, then $\gamma(A) = \gamma(A^*)$ and A^* has a closed range.

11.5.7 Theorem

Suppose $\gamma(A) > 0$. Let V be bounded with $\mathcal{D}(V) \supset \mathcal{D}(A)$. If $\|V\| < \gamma(A)$, then

(a) $\alpha(A + V) \leq \alpha(A)$

(b) $\dim\left(E_y/\overline{R(A+V)}\right) \leq \dim\left(E_y/\overline{R(A)}\right)$

Proof: (a) For $x \neq \theta$ in $N(A+V)$ and $\|V\| < \gamma = \gamma(A)$,

since $x \neq \theta \in N(A+V)$, $Ax + Vx = \theta$ i.e., $\|Ax\| = \|Vx\|$.

$$\gamma\|[x]\| \leq \|Ax\| = \|Vx\| \leq \|V\| \, \|x\| < \gamma\|x\| \quad \text{where } [x] \in E_x/N(A).$$

Thus, $\|x\| > \|[x]\| = d(x, N(A))$.

Therefore, by lemma 11.5.3, the dimension of $N(A+V) <$ dimension of $N(A)$, or $\alpha(A+V) \leq \alpha(A)$.

(b) Let $E_x^1 = \overline{\mathcal{D}(A)}$ and let V_1 be V restricted to $\mathcal{D}(A)$. Let us consider A and V_1 as operators with domain dense in E_x^1, since $f_1^* = A^*\phi_1^*$ where $\phi_1^* \in E_y^*$ and $f^* \in E_x^*$. Therefore, the domain of A^* is in E_y^* and range in E_x^{1*}. Therefore, by theorem 11.5.6,

$$\gamma(A^*) = \gamma(A) > \|V\| \geq \|V_1\| = \|V_1^*\|.$$

We next show that

$$\dim(E_y/\overline{R(A+V)}) = \dim(E_y/\overline{R(A+V_1)}) = \dim R(A+V)^{\perp}.$$

For $g \in (E_y/\overline{R(A+V_1)})^*$, and the map W defined by:

$$(Wg(x)) = g[x], \ g \in (E_y/\overline{R(A+V_1)})^*,$$

we observe $|(Wg(x)| = |g[x]| \leq \|g\| \, \|[x]\| \leq \|g\| \, \|x\|, \ x \in E_x$

and $\qquad Wg(m) = g[m] = 0, \ m \in \overline{R(A+V_1)}$.

Thus, Wg is in $\overline{R(A+V_1)}^{\perp}$ with

$$\|Wg\| \leq \|g\| \tag{11.12}$$

Since $\qquad |g[x]| = |Wg(y)| \leq \|Vg\| \, \|y\|, \ y \in [x]$

It follows that $|g[x]| \leq \|Vg\| \, \|[x]\|$

Thus, $\qquad \|g\| \le \|Vg\|$ (11.13)

(11.13) together with (11.12) proves that W is an isometry.

Given $f \in \overline{R(A+V)}^{\perp}$, let g be a linear functional on $E_x^1/\overline{R(A+V)}^{\perp}$ defined $g[x] = g(x)$.

Now, $\qquad |g[x]| = |g(y)| \le \|g\| \, \|y\|, \; y \in [x].$

It follows that $|g[x]| \le \|g\| \, \|[x]\|$. Hence, g is in $(E_x^1/\overline{R(A+V)})^{\perp}$.

Furthermore, $Vg = f$, proving that $R(V) = (\overline{R(A+V)})^{\perp}$.

Thus, $\qquad (E_x^1/\overline{R(A+V_1)})^* = (\overline{R(A+V_1)})^{\perp}$ (11.14)

Hence, it follows from theorem 11.2.10, definition 11.5.1 and (11.14)

$$\dim(E_x^1/\overline{R(A+V_1)})^* = \dim(\overline{R(A+V_1)}^{\perp})$$
$$= \dim(N(\overline{(A+V_1)}^*)) = \alpha(A^* + V_1^*)$$
$$\le \alpha(A^*) = \dim(E_y/\overline{R(A)}).$$

11.6 Perturbation by Strictly Singular Operators

11.6.1 *Definition: normally solvable*

A **closed** linear operator with **closed range** is called **normally solvable**.

11.6.2 *Theorem*

Let E_x and E_y be complete. If A is closed but $R(A)$ is not closed, then for each $\epsilon > 0$ there exists an infinite-dimensional closed subspace $L(\epsilon)$ contained in $\mathcal{D}(A)$, such that A restricted to $L(\epsilon)$ is compact with norm not exceeding ϵ, an arbitrarily small number.

Proof: Let U be a closed subspace having finite deficiency in E_x. Assume that A has a bounded inverse on U. Since A is closed, $Ax_n \to y, x_n \in U \Rightarrow \{x_n\}$ is a Cauchy sequence and therefore converges to x in Banach space U. Thus, AU is closed.

Moreover, A being closed, $x \in \mathcal{D}(A)$ and $Ax = y$. By hypothesis, there exists a finite dimensional subspace N of E_x such that $E_x = U + N$. Hence $AE_x = AU + AN \subseteq E_y$. Thus AU is a closed subspace and AN is a finite dimensional subspace of E_y. Define a linear map B from E_y onto E_y/AU by $By = [y]$. Since $\|By\| = \|[y]\| \le \|y\|$, B is continuous. Moreover, the linearity of B and the finite dimensionality of N imply the finite dimensionality of BN. Now a finite dimensional subspace of a normed linear space is complete and hence closed. Since B is continuous $B^{-1}BN = U + N$ is closed (B^{-1} is used in the set theoretic sense). Thus

AE_x is closed. But this contradicts the hypothesis that $R(A)$ is not closed. Therefore, A does not have a bounded inverse on U. Hence there exists an $L = L(\epsilon)$ with the properties described in theorem 11.3.4. Since E_y is complete and A is closed and bounded on L, it follows that $\overline{L}$ is contained in $D(A)$. Moreover, $\|A\overline{L}\| \leq \epsilon$ and $\overline{AB_{\overline{L}}} = \overline{AB_L}$, where $B_{\overline{L}}$ and B_L are unit balls in $\overline{L}$ and L respectively and $A_{\overline{L}}$ is the restriction of A to $\overline{L}$. The precompactness of A and the completeness of E_y imply that $\overline{AB_L}$ is compact. Thus, $A_{\overline{L}}$ is compact.

11.6.3 Theorem

Suppose that A_1 is a linear extension of A such that

$$\dim(D(A_1)/D(A)) = n < \infty,$$

(a) If A is closed then A_1 is closed

(b) If A has a closed range, then A_1 has a closed range.

Proof: (a) By hypothesis, $D(A_1) = D(A) + N$, where N is a finite dimensional subspace. Hence, $G(A_1) = G(A) + H$, where $G(A_1)$ and $G(A)$ are the graphs of A_1 and A respectively and $H = \{(n, A_1n)| : n \in N\}$.

Thus, if $G(A)$ is closed, then $G(A_1)$ is closed since H is finite dimensional.

(b) $R(A_1) = R(A) + A_1N$, A_1N is finite dimensional and hence closed. Also $R(A)$ is given to be closed. Hence, $R(A_1)$ is closed.

11.6.4 Theorem

Let E_x and E_y be complete and let A be normally solvable. If L is a subspace (not necessarily closed) of E_x such that $L + N(A)$ is closed, then AL is closed. In particular, if L is closed and $N(A)$ is finite dimensional, then AL is closed.

Proof: Let A_1 be the operator A restricted to $\mathcal{D}(A) \cap (LN(A))$. Then A_1 is closed and $N(A_1) = N(A)$.

Hence $\gamma(A_1) \geq \gamma(A) > 0$. Therefore, A_1 has a closed range, i.e., $AL = A_1(L + N(A))$ is closed.

If V is strictly singular with no restriction on its norm, then we get an important stability theorem due to Kato (Goldberg [21]).

11.6.5 Theorem

Let E_x and E_y be complete and let A be normally solvable with $\alpha(A) < \infty$.

If V is strictly singular and $\mathcal{D}(A) < \mathcal{D}(V)$, then

(a) $A + V$ is normally solvable

(b) $\kappa(A + V) = \kappa(A)$

(c) $\alpha(A + \lambda V)$ and $\beta(A + \lambda V)$ have constant values p_1 and p_2, respectively, except perhaps for isolated points. At the isolated points,

$$p_1 < \lambda(A + \lambda V) < \infty \quad \text{and} \quad \beta(A + \lambda V) > p_2.$$

Proof: (a) Since $\alpha(A) < \infty$, i.e., the null space of A is finite dimensional i.e., closed, there exists a closed subspace L of E_x such that $E_x = L \oplus N(A)$. Let A_L be the operator A restricted to $L \cap \mathcal{D}(A)$. Then A being closed, A_L is closed with $R(A_L) = R(A)$. Let us suppose that $A + V$ does not have a closed range. Now $A + V$ is an extension of $A_L + V$. Then it follows from theorem 11.6.3(b) that $A_L + V$ does not have a closed range. Moreover, 11.6.3(a) yields that $A_L + V$ is closed since A_L is closed. Thus, $A_L + V$ is a closed operator but its range is not closed. It follows from theorem 11.6.2 that there exists a closed infinite–dimensional subspace L_0 contained in $\mathcal{D}(A_L) = \mathcal{D}(A_L + V)$ such that

$$\|(A_L + V)x\| < \frac{\gamma(A_L)}{2}\|x\|, \ x \in L_0 \tag{11.15}$$

Thus, since A_L is one-to-one, it follows for all x in L_0,

$$\|Vx\| \geq \|A_L x\| - \|(A_L + V)x\|$$
$$\geq \left(\gamma(A_L) - \frac{\gamma(A_L)}{2}\right)\|x\| = \frac{\gamma(A_L)}{2}\|x\|.$$

The above shows that V has a bounded inverse on the infinite dimensional space L_0. This, however, contradicts the hypothesis that V is strictly singular. We next show that $\alpha(A + V) < \infty$. There exists a closed subspace M_1 such that

$$N(A + V) = N(A + V) \cap N(A) \oplus M_1 \tag{11.16}$$

Let A_1 be the operator A restricted to M_1. Since $N(A)$ is finite dimensional, i.e., closed, $N(A) + M_1$ is closed and A is normally solvable. Hence, by theorem 11.6.4 AM_1 is closed. Thus, $R(A_1) = AM_1$ is closed. Moreover, A_1 is one-to-one. Hence, its inverse is bounded. Since $M_1 \subseteq N(A + V)\#V = -A_1$ on M_1 and V is strictly singular, M_1 must be finite dimensional. Therefore, 11.16 implies that $N(A + V)$ is finite dimensional.

(b) We have shown above that for all scalars λ, $A + \lambda V$ is normally solvable and $\alpha(A + \lambda V) < \infty$.

Let I denote the closed interval $[0, 1]$ and let Z be the set of integers together with the 'ideal' elements ∞ and $-\infty$. Let us define $\phi : I \to Z$ by $\phi(x) = \kappa(A + \lambda V)$. Let I have the usual topology and let Z have the discrete topology, i.e., points are open sets. To prove (b) it suffices to show that ϕ is continuous. If ϕ is continuous, then $\phi(I)$ is a connected set which therefore consists of only one point. In particular,

$$\kappa(A) = \phi(0) = \phi(1) = \kappa(A + V) \tag{11.17}$$

In order to show the continuity of ϕ, we first prove that
$$\kappa(A + V) = \kappa(A) \text{ for } \|V\| \text{ sufficienty small.}$$
We refer to (a) and note that A_L is closed, one-to-one and $R(A_L) = R(A)$. Hence, A_L has a bounded inverse. Then, by theorem 11.5.7, we have

$\alpha(A_L + V) = \alpha(A_L) = 0$ (11.18) Since A_L has a bounded inverse

$\dim(E_y/R\overline{(A_L + V)}) = \dim(E_y/R\overline{(A_L)})$ provided $\|V\| < \gamma(A_L)$

Hence, $\beta(A_L + V) = \beta(A_L)$.   (11.19)

Now, $D(A + V) = D(A) = D(A) \cap L \oplus N(A)$.

$D(A_L + V) = D(A_L) = D(A) \cap L$.

Thus, $D(A) = D(A_L) \oplus N(A)$.

Thus, $\dim(D(A)/D(A_L)) = \dim(N(A)) = \alpha(A)$.

Hence, $\kappa(A) = \kappa(A_L) + \alpha(A)$

$= \kappa(A_L + V) + \alpha(A) = \kappa(A + V),$ (11.20)

where $\|V\| < \gamma(A_L)$.

Hence, the continuity of ϕ is established and (11.17) is true.

(c) For proof, see Goldberg [21].

11.7 Perturbation in a Hilbert Space and Applications

11.7.1 A linear operator A defined in a Hilbert space is said to be **symmetric** if it is contained in its **adjoint** A^* and is called self-adjoint if $A = A^*$.

11.7.2 Definition: coercive operator

A symmetric operator A with domain $\mathcal{D}(A)$ dense in a Hilbert space H is said to be coercive if there exists an $\alpha > 0$ such that

$$\forall\, u \in \mathcal{D}(A),\ \langle Au, u \rangle \geq \alpha \langle u, u \rangle \tag{11.21}$$

11.7.3 Definition: scalar product [,]

Let A be a coercive linear operator with domains $\mathcal{D}(A)$ dense in a Hilbert space H. Let A satisfy (11.21) $\forall\, u \in \mathcal{D}(A)$.

We define $[u, v] = \langle Au, v \rangle$.

Since A is self-adjoint $[u, v] = [v, u]\ \forall\, u, v \in \mathcal{D}(A)$.

$$0 \leq [u, u] = |u|^2 \tag{11.22}$$

It may be seen that $[\ ,\ ]$ defines a new inner product in $\mathcal{D}(A)$. We complete $\mathcal{D}(a)$ w.r.t. the new product and call the new Hilbert space as H_A, where $|\cdot|$ defined by (11.22) will be the norm of H_A.

11.7.4 Perturbation

Our concern is to solve a complicated differential equation

$$A_1 u = f, \ u \in \mathcal{D}(A_1) \tag{11.23}$$

A_1 is coercive in the Hilbert space H_1.

We often replace the above equation by a simpler equation of the form

$$A_2 u = f \tag{11.24}$$

The question that may arise is to what extent the replacement is justified, or in other words we are to determine how close is the solution of equation (11.24) to the solution of equation (11.23). For this, let us assume that both the operators A_1 and A_2 are symmetric and **coercive** on their respective domains $\mathcal{D}(A_1)$ and $\mathcal{D}(A_2)$ respectively. We complete $\mathcal{D}(A_1)$ and $\mathcal{D}(A_2)$ respectively w.r.t. to the products $[u, u]_{A_1}$, $u \in \mathcal{D}(A_1) \subseteq H$ and $[v, v]_{A_2}$, $v \in \mathcal{D}(A_2) \subseteq H$. We call the Hilbert spaces so generated as H_{A_1} and H_{A_2} respectively.

Let $|u|_{A_1}^2 = \langle A_1 u, u \rangle \geq \alpha_1 \langle u, u \rangle = \alpha_1 \|u\|_{A_1}^2, u \in H_{A_1}$ (11.25)

Also, let $|u|_{A_2}^2 = \langle A_2 v, v \rangle \geq \alpha_2 \langle v, v \rangle = \alpha_2 \|v\|_{A_2}^2 v \in H_{A_2}$ (11.26)

11.7.5 Theorem

Let the symmetric and coercive operators A_1 and A_2 fulfill respectively the inequalities (11.25) and (11.26). Moreover, let H_{A_1} and H_{A_2} coincide and are each seperable. If u_0 and u_1 are the solutions of equations (11.23) and (11.24), then there exists some constant η such that

$$|u_1 - u_0|_{A_2} \leq \eta |u_1|_{A_2} \tag{11.27}$$

Before we prove theorem 11.7.5, a lemma is proved.

11.7.6 Lemma

Let A_1 be a symmetric coercive operator fulfilling condition (11.25). Let ΔA_1 be a symmetric nonnegative bounded linear operator and satisfy the condition

$$0 \leq \langle \Delta A_1 u, u \rangle \leq \alpha_3 \langle u, u \rangle, \alpha_3 > 0, \ \forall u \in D(\Delta A_1) \supseteq D(A_1) \tag{11.28}$$

Let $\alpha_1 > \alpha_3$ and $\alpha_1 - \alpha_3 = \alpha_2$.

Then $A_2 = A_1 - \Delta A_1$ is a symmetric linear coercive operator satisfying the condition (11.26).

Proof: $\langle A_2 u, u \rangle = \langle (A_1 - \Delta A_1) u, u \rangle = \langle A_1 u, u \rangle - \langle \Delta A_1 u, u \rangle$

$$\geq (\alpha_1 - \alpha_3)\langle u, u \rangle = \alpha_2 \langle u, u \rangle$$
$$\forall\, u \in D(A_2).$$

Moreover, since A_1 and ΔA_1 are symmetric linear operators, $A_2 = A_1 - \Delta A_1$ is symmetric and coercive.

Proof (th. 11.7.5) H_{A_1} being a Hilbert space, we can define

$$|u|_{A_1} = [u, u]_{A_1}^{\frac{1}{2}} = \langle A_1 u, u \rangle^{1/2}, \quad u \in D(A) \subseteq H_{A_1}$$

Similarly, $|u|_{A_2} = [u, u]_{A_2}^{\frac{1}{2}} = \langle A_2 u, u \rangle^{1/2}, \quad u \in D(A_2) \subseteq H_{A_2}$

H_{A_1} and H_{A_2} being seperable are isomorphic.

It follows from (11.27) $\langle (A_1 - A_2) u, u \rangle \leq \alpha_3 |u|^2$

or $\qquad \langle A_1 u, u \rangle \leq \alpha_3 |u|^2 + |u|_{A_2}^2 \leq \left(1 + \dfrac{\alpha_3}{\alpha_2}\right) |u|_{A_2}^2$

or $\qquad |u|_{A_1}^2 \leq \left(1 + \dfrac{\alpha_3}{\alpha_2}\right) |u|_{A_2}^2.$

or $\qquad |u|_{A_1} \leq \sqrt{1 + \dfrac{\alpha_3}{\alpha_2}} |u|_{A_2}$ $\qquad\qquad$ (11.29)

Again, since ΔA_1 is non-negative, we have

$$|u|_{A_1}^2 \geq |u|_{A_2}^2 \qquad\qquad (11.30)$$

Hence, we can find positive constants β_1, β_2, such that

$$\beta_1 |u|_{A_2}^2 \leq |u|_{A_1}^2 \leq \beta_2 |u|_{A_2}^2 \; \forall\, u \in H_{A_1} = H_2 \qquad (11.31)$$

If u^0 and v^0 are the respective unique solutions of (11.23) and (11.24), then the inequality

$$|v^0 - u^0|_{A_2} \leq \eta |v^0|_{A_2} \qquad\qquad (11.32)$$

holds, in which the constant η is defined by the formula

$$\eta = \max \left[\dfrac{|\beta_1 - 1|}{\beta_1}; \; \dfrac{|\beta_2 - 1|}{\beta_2}\right] \qquad (11.33)$$

Formulas (11.32) and (11.33) solve the problem (11.23) approximately with an estimation of the error involved.

11.7.7 Example

1. For error estimate due to perturbation of a second order elliptic differential equation see Mikhlin [36].

2. In the theory of small vibrations and in many problems of quantum mechanics it is important to determine how the eigenvalues and the eigenvectors of a quadratic form $K(x, x) = \displaystyle\sum_{i=1, j=1}^{n} b_{ij}, x_i x_j$, are changed if

both the form $K(x,x)$ and the unit form $E(x,x)$ are altered. Perturbation theory is applied in this case. See Courant and Hilbert [15].

3. In what follows, we consider a differential equation where perturbation method is used. We consider the differential equation

$$\frac{d^2y}{dx^2} + (1+x^2)y + 1 = 0, \ y(\pm 1) = 0 \tag{11.34}$$

We consider the perturbed equation

$$\frac{d^2y}{dx^2} + (1+\epsilon x^2)y + 1 = 0, \ y(\pm 1) = 0 \tag{11.35}$$

For $\epsilon = 0$, the equation $\dfrac{d^2y_0}{dx^2} + y_0 + 1 = 0, y_0(\pm 1) = 0$ has the solution $y_0 = \dfrac{\cos x}{\sin x} - 1.$

Let $y(x, \epsilon)$ be the solution of the equation (11.33) and we expand $y(x, \epsilon)$ in terms of ϵ

$$y(x_1\epsilon) = y_0(x) + \epsilon y_1(x) + \epsilon^2 y_2(x) \cdots \tag{11.36}$$

Substituting the power series (11.36) for y in the differential equation, we obtain,

$$\sum_{n=0}^{\infty} \epsilon^n(y_n'' + y_n) + \sum_{n=1}^{\infty} \epsilon^n x^2 y_{n-1} + 1 = 0$$

and since the coefficients of the powers of ϵ must vanish we have,

$$y_n'' + y_n + x^2 y_{n-1} = 0 \tag{11.37}$$

with the boundary conditions

$$y_n(\pm 1) = 0 \tag{11.38}$$

Thus, we have a sequence of boundary value problems of the type (11.37) subject to (11.38) from which $y_1, y_2, y_3, \ldots$ etc., can be found [see Collatz [14]].

CHAPTER 12

THE HAHN-BANACH THEOREM AND OPTIMIZATION PROBLEMS

It was mentioned at the outset that we put emphasis both on the theory and on its application. In this chapter, we outline some of the applications of the Hahn-Banach theorem on optimization problems. The Hahn-Banach theorem is the most important theorem about the structure of linear continuous functionals on normed linear spaces. In terms of geometry, the Hahn-Banach theorem guarantees the separation of convex sets in normed linear spaces by hyperplanes. This separation theorem is crucial to the investigation into the existence of an optimum of an optimization problem.

12.1 The Separation of a Convex Set

In what follows, we state a theorem which asserts the existence of a hyperplane separating two disjoint convex sets in a normed linear space.

12.1.1 Theorem (Hahn-Banach separation theorem)

Let E be a normed linear space and X_1, X_2 be two non-empty disjoint convex sets, with X_1 being an open set. Then there exists a functional $f \in E^*$ and a real number β such that

$$X_1 \subseteq \{x \in E : \mathrm{Re} f(x) < \beta\}, \ X_2 \subseteq \{x \in E : \mathrm{Re} f(x) \geq \beta\}.$$

For proof see 5.2.10.

The following theorems are in the setting of $\mathbb{R}^n$.

12.1.2 Theorem (intersection)

In the space $\mathbb{R}^n$, let $X_1, X_2, \ldots, X_m$ be compact convex sets, whose union is a convex set. If the intersection of any $(m-1)$ of them is non-empty, then the intersection of all X_j is non-empty.

Proof: We shall first prove the theorem for $m = 2$.

Let X_1 and X_2 be non-empty compact convex sets, such that $X_1 \cup X_2$ is convex. Let X_1 and X_2 be disjoint, then there is a plane P which separates them strictly. Since there exist points of $X_1 \cup X_2$ on both sides of P, and since $X_1 \cup X_2$ is convex, there exist points of $X_1 \cup X_2$ on both sides of P. But this is impossible since P separates strictly X_1 and X_2.

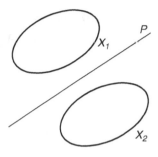

Let us next suppose that the result is true for $m = r$ convex sets. We shall prove this implies that the result holds for $m = r+1$ convex sets $X_1, X_2, \ldots, X_{r+1}$.

Fig. 12(a)

Put $X = \cap_{j=1}^{r} X_j$. Then $X \neq \Phi$ by our premise. Now, $X \neq \Phi$, $X_{r+1} \neq \Phi$. Suppose the two sets are disjoint. Then there exists a plane P' which separates them strongly. Writing $X_j' = X_j \cap P'$ we have

$$\bigcup_{j=1}^{r} X_j' = \cup(X_j \cap P') \cup (X_{r+1} \cap P')$$

$$= P' \cap \left(\bigcup_{j=1}^{r+1} X_j \right)$$

Therefore, the union of the sets $X_1', X_2', \ldots, X_r'$ is convex. Also, the intersection of any $(r-1)$ of $X_1, X_2, \ldots, X_r$ meets X and X_{r+1} and hence meets P'. Therefore, the intersection of any $(r-1)$ of $X_1', X_2', \ldots, X_r'$ is not empty.

But by hypothesis $\cap_{r=1}^{j} X_j' = X \cap P \neq \Phi$ contradicting the fact that P' is a hyperplane which separates X and X_{r+1} strictly.

It follows that $X \cap X_{r+1} \neq \Phi$ and so the result holds for $m = r+1$.

12.1.3 Theorem

A closed convex set is equal to the intersection of the half-spaces which contain it.

Proof: Let X be a closed convex set and let A be the intersection of the half-spaces which contains it. If $E_f = \{x : x \in \mathbb{R}^n, \ f(x) = \alpha\}$ is a **hyperplane** in $\mathbb{R}^n$, then $H_f = \{x : x \in \mathbb{R}^n, \ f(x) \geq \alpha\}$ is called a **half-space**.

If $x_0 \notin X$, then $\{x_0\}$ is a compact convex set not meeting X. Therefore, there exists a plane E_f separating $\{x_0\}$ and set X s.t.

$$f(x_0) < \alpha \le \inf_{x \in X} f(x).$$

We thus have $H_f \supset X$ and $x_0 \notin H_f$.

Consequently, x_0 does not belong to the intersection of the half-spaces H_f containing X, i.e., $x_0 \notin A$. Hence, $X \supset A$ and since $A \supseteq X$, $A = X$.

12.1.4   Definition: plane of support of X

Let X be any set in $\mathbb{R}^n$. A plane E_f containing at least one point of X and s.t. all points of X are on one side of E_f is called a **plane of support or supporting hyperplane of X**.

12.1.5   Observation

If X is compact, then for any linear functional f which is not identically zero, there exists a plane of support having equation $f(x) = \alpha$ (it is sufficient to take $\alpha = \min_{x \in X} f(x)$).

12.1.6   Plane of support theorem

If X is a compact non-empty convex set, it admits of an extreme point; in fact, every plane of support contains an extreme point of X.

Proof: (i) The theorem is true in $\mathbb{R}$ for, compact convex set in $\mathbb{R}$ is a closed segment $[\alpha, \beta]$ and contains two extreme points α and β; the planes of support $\{x : x \in \mathbb{R}, \ x = \alpha\}$ and $\{x : x \in \mathbb{R}, \ x = \beta\}$.

(ii) Suppose that the theorem holds for $\mathbb{R}^r$. We shall prove that it holds in $\mathbb{R}^{r+1}$. Let X be a compact convex set in $\mathbb{R}^{r+1}$ and let E_f be a plane of support. The intersection $E_f \cap X$ is a non-empty closed convex set; since $E_f \cap X$ is contained in the compact set X, it is also a compact set. The set $E_f \cap X$ can be regarded as a compact set in $\mathbb{R}^r$ and so by hypothesis, it admits of an extreme point x_0. Let $[x_1, x_2]$ be a line segment of centre x_0 with $x_1 \neq x_0$ and $x_2 \neq x_0$. Since x_0 is an extreme point of $E_f \cap X$, we have $[x_1, x_2] \not\subset E_f \cap X$. Therefore, if x_1 and $x_2 \in X$, we have $x_1, x_2 \notin E_f$ and hence x_1, x_2 are separated by E_f but this contradicts the definition of E_f as a plane of support of X. It follows that there is no segment $[x_1, x_2]$ of centre x_0 contained in X and so x_0 is an extreme point of X; by definition x_0 is in E_f.

Thus, if the theorem holds for $n = r$, it holds for $n = r + 1$. But we have seen that it holds for $n = 1$. Hence, by induction, the theorem is true for all $n \geq 1$.

12.2   Minimum Norm Problem and the Duality Theory

Let E be a real normed linear space and X be a linear subspace of E.

12.2.1   Definition: primal problem

To find $u_0 \in X$ s.t.

$$\inf_u ||u_0 - u|| = \alpha, \ u \in X \tag{12.1}$$

12.2.2   Definition: dual problem

To find $f \in X^{\perp}, \ ||f|| \leq 1$ (12.2)

s.t. $\sup f(u_0) = \beta$ (12.3)

where $X^{\perp} = \{f \in E^* : f(u) = 0 \ \ \forall \, u \in X\}$ (12.4)

12.2.3   Theorem: minimal norm problem on the normed space E

Let X be a linear subspace of the real normed space E. Let $u_0 \in E$. Then the following results are true:

(i) **Extremal values:** $\alpha = \beta$

(ii) **Dual problem:** The dual problem (12.3) has a solution u^*.

(iii) **Primal problem:** Let $\hat{f}$ be a fixed solution of the dual problem (12.3). Then the point $u_0 \in X$ is a solution of the primal problem (12.1) if and only if,

$$\hat{f}(u_0 - u) = ||u - u_0|| \tag{12.5}$$

12.2.4   Lemma

If $\dim X < \alpha$, then the primal problem (12.1) always has a solution.

Let $v \in X$ and $\phi \in X^{\perp}$ with $||\phi|| \leq 1$. Therefore, (i) we obtain the two sided error estimate for the minimum value α:

$$||v - u_0|| \geq \alpha \geq \phi(u_0).$$

Proof of theorem: (i) and (ii) Since α is the infimum in (12.1), for each $\epsilon > 0$, there is a point $u \in X$ such that,

$$||u - u_0|| \leq \alpha + \epsilon.$$

Thus, for all $f \in X^{\perp}$ with $||f|| \leq 1$,

$$f(u_0) = f(u_0 - u) \leq ||f|| \, ||u_0 - u|| \leq \alpha + \epsilon.$$

Hence $\beta \leq \alpha + \epsilon$, for all $\epsilon > 0$, that is, $\beta \leq \alpha$. Let $\alpha > 0$. Now theorem 5.1.4 yields that there is a functional $\hat{f} \in X^{\perp}$ with $||\hat{f}|| = 1$ such that

$$\hat{f}(u_0) = \alpha \tag{12.6}$$

Along with $\beta \leq \alpha$, this implies $\beta = \alpha$.

If $\alpha = 0$ then (12.6) holds with $\hat{f} = 0$ and hence we again have $\alpha = \beta$.

(iii) This follows from $\alpha = \beta$ and $\hat{f}(u) = 0 \; \forall \; u \in X$.

Proof of lemma: Since $u_0 \in X$, $\|u_0\| \le \alpha$. Thus problem (12.1) is equivalent to the finite-dimensional minimum problem

$$\min_{u \in X_0} Z = \|u - u_0\| \tag{12.7}$$

where the set $X_0 = \{u \in X : \|u\| \le \|u_0\|\}$ is compact. By the Bolzano-Weierstrass theorem (1.6.19) this problem has a solution.

12.2.5   Minimum norm problems on the dual space E^*

Let us consider the **modified primal problem:**
To find $\hat{f} \in X^{\perp}$ s.t.

$$\inf_{\hat{f}} \tilde{Z} = (\|\hat{f} - \hat{f}_0\|) = \alpha \tag{12.8}$$

along with the **dual problem.**
To find $u \in X$,

$$\sup_{\substack{u \\ \|u\|=1}} \tilde{z}(= \hat{f}_0(u)) = \beta \tag{12.9}$$

$X^{\perp} = \{\hat{f} \in E^* : \hat{f}(u) = 0 \text{ for all } u \in X\}$.

Thus, the primal problem (12.8) refers to the dual space E^*, where as the dual problem (12.9) refers to the original space E.

12.2.6   Theorem

Let X be a linear subspace of the real normed linear space E. Given $\hat{f}_0 \in E^*$, the following results hold good.

(a) **Extreme values:** $\alpha = \beta$

(b) **Primal problem:** The primal problem (12.8) has a solution $\hat{f}$.

(c) **Dual problem:** Let $\hat{f}$ be a fixed solution of the primal problem (12.8). Then, the point $u \in X$ with $\|u\| \le 1$ is a solution of the dual problem (12.9) if and only if,

$$(\hat{f}_0 - \hat{f})(u) = \|\hat{f}_0 - \hat{f}\| \tag{12.10}$$

Proof: (i) For all $\hat{f} \in X^{\perp}$,

$$\begin{aligned}
\|\hat{f} - \hat{f}_0\| &= \sup_{\|u\|\le 1} (|\hat{f}(u) - \hat{f}_0(u)|) \\
&\ge \sup_{\substack{\|u\|\le 1, \\ u \in X}} \hat{f}_0(u) = \beta
\end{aligned}$$

Since $\hat{f}(u) = 0$ for all $u \in X$. Hence, $\alpha \ge \beta$.

Let $\hat{f}_r : X \to \mathbb{R}$ be the restriction of $\hat{f}_0 : E \to \mathbb{R}$ to X.

Then $||\hat{f}_r|| = \sup\limits_{\substack{||u|| \leq 1 \\ u \in X}} \hat{f}_0(u) = \beta$.

By Hahn-Banach theorem (theorem 5.1.3) there exists an extension $\hat{F} : E \to \mathbb{R}$ of $\hat{f}_r$ with $||\hat{F}|| = ||\hat{f}_r||$.

This implies $\hat{g} := \hat{f}_0 - \hat{F} = 0$ on X, that is, $\hat{g} \in X^\perp$. Since $\alpha \geq \beta$ and

$$||\hat{g} - \hat{f}_0|| = ||\hat{F}|| = ||\hat{f}_r|| = \beta, \; \hat{g} \in X^\perp,$$

we get $\alpha = \beta$. Hence, (12.8) has a solution.

(ii) This follows from $\alpha = \beta$ with $\hat{f}(u) = 0$.

12.2.7 Definition: δ_x

Let $-\infty < a \leq x \leq b < \infty$. Set $\delta_x(u) : = u(x)$ for all $u \in C([a, b])$. Obviously, $\delta_x \in C([a, b])^*$ and $||\delta_C|| = 1$.

12.2.8 Lemma

Let $\hat{f} \in C([a, b])^*$ be such that $||\hat{f}|| \neq 0$.

Suppose that $\hat{f}(u) = ||\hat{f}|| \, ||u||$, where $||u|| = \max\limits_{a \leq x \leq b} |u(x)|$ and $u : [a, b] \to R$ is a continuous function, such that $|u(x)|$ achieves its maximum at precisely N points of $[a, b]$ denoted by $x_1, x_2, \ldots, x_N$. Then there exist real numbers $a_1, a_2, \ldots, a_N$, such that

$$\hat{f} = a_1 \delta_{x_1} + \cdots + a_N \delta_{x_N}$$

and $|a_1| + |a_2| + \cdots + |a_N| = ||\hat{f}||$.

Proof: By Riesz representation theorem (5.3.3) there exists a function $h : [a, b] \to \mathbb{R}$ of bounded variation, such that

$$\hat{f}(u) = \int_a^b u(x)h(x)dx \text{ for all } u \in C([a, b])$$

and $\mathrm{Var}(h) = ||\hat{f}||,$

where $\mathrm{Var}(h)$ stands for the total variation of h on the interval $[a, b]$. We assume that $h(a) = 0$. For simplicity, we take $N = 1$, $\pm u(x_1) = ||u||$ and $a < x_1 < b$.

Let $J = [a, b] - [x_1 - \epsilon, x_1 + \epsilon]$ for fixed $\epsilon > 0$, and let $\mathrm{Var}_J(h)$ denote the total variation of h on J. Then

$$\mathrm{Var}_J(h) + |h(x_1 + \epsilon) - h(x_1 - \epsilon)| \leq \mathrm{Var}(h) \qquad (12.11)$$

Case I Let $\mathrm{Var}_J(h) = 0$ for all $\epsilon > 0$. Then, by (12.11), h is a step function of the following form,

$$h(x) = \begin{cases} 0 & \text{if } a \leq x < x_1 \\ \pm\mathrm{Var}(h) & \text{if } x_1 < x \leq b \end{cases}$$

and by the relation

$$\hat{f}(u) = \int_a^b u(x)dh(x) = \pm u(x_1)\text{Var}(h)$$

for all $u \in C([a, b]]$.

Hence, $\hat{f} = \pm\text{Var}(h)\delta_{x_1}$.

Case II Let $\text{Var}_J(h) > 0$ for some $\epsilon > 0$. We want to show that this is impossible. By the mean value theorem, there is a point $\xi \in [x - \epsilon, x + \epsilon]$ such that

$$\hat{f}(u) = \int_J u(x)dh(x) + \int_{x_1-\epsilon}^{x_1+\epsilon} u(x)dh(x)$$
$$\leq \max_{x \in J}|u(x)|\text{Var}_J(h) + |u(\xi)||h(x+\epsilon) - h(x-\epsilon)|.$$

Since $|u(x)|$ achieves its maximum exactly at the point x_1, we get $\max_{x \in J}|u(x)| \leq ||u||$.

Thus, it follows from (12.11) that

$$\hat{f}(u) < ||u||\text{Var}(h).$$

Hence, $\hat{f}(u) < ||u||||\hat{f}||$. This is a contradiction.

For $N > 1$ we use a similar argument.

12.3   Application   to   Chebyshev Approximation

It is convenient in practice to approximate a continuous function by a polynomial for various reasons. Let u_0 be a continuous function mapping a compact interval $[a, b] \to R$. Let us consider the following **approximation** problem:

$$\max_{a \leq x \leq b}|u_0(x) - u(x)| = \text{min!}, \ u \in \mathbb{P} \qquad (12.11)$$

where $\mathbb{P}$ denotes the set of real polynomials of degree $\leq N$ for fixed $N \geq 1$. Problem (12.11) corresponds to the famous Chebyshev approximation of the function u_0 by polynomials.

12.3.1   *Theorem*

Problem (12.11) has a solution. If $p(x)$ is a solution of (12.11), then $|u_0(x) - p(x)|$ achieves its maximum at least $N + 2$ points of $[a, b]$.

Proof: Let $E = C([a, b])$ and $||v|| = \max_{a \leq x \leq b}|v(x)|$. Then, (12.11) can be written as

$$\min Z = ||u_0 - p||, \ p \in \mathbb{P} \qquad (12.12)$$

Since, $\dim X < \infty$, i.e., finite, the problem (12.12) has a solution by 12.2.4. If $u_0 \in \mathbb{P}$, then (12.12) is immediately true. Let us assume that $u_0 \notin \mathbb{P}$. Let $\hat{p}$ be a solution of (12.12). Then, since $u_0 \notin \mathbb{P}$ and $\hat{p} \in \mathbb{P}$ we have $||u_0 - p|| > 0$. By the duality theory from theorem 12.2.3 there exists a functional $f \in C([a,b])^*$ such that

$$f(u_0 - \hat{p}) = ||u_0 - p|| \qquad (12.13)$$

along with $||f|| = 1$ and

$$f(\hat{p}) = 0 \ \forall \ \hat{p} \in X \qquad (12.14)$$

Let us suppose that $|u_0(x) - p(x)|$ achieves its maximum on $[a,b]$ at precisely the points $x_1, x_2, \ldots, x_M$ where $1 \leq M < N + 2$. It follows from (12.13) and lemma 12.2.6 that there are real numbers $a_1, a_2, \ldots, a_M$ with $|a_1| + \cdots + |a_M| = 1$, such that

$$u^* = a_1 u(x_1) + a_2 u(x_2) + \cdots + a_M u(x_M)$$

where $u(x) \in C([a,b])$.

Assume that $a_M \neq 0$. Let us choose a real polynomial $\tilde{p}(x)$ of degree N, such that

$$\tilde{p}(x_1) = \tilde{p}(x_2) = \cdots = \tilde{p}(x_{M-2}) = 0 \text{ and } \tilde{p}(x_M) \neq 0.$$

This is possible since $M - 1 \leq N$. Then, $\tilde{p} \in X$ and $f(p) \neq 0$ contradicting (12.14).

12.4 Application to Optimal Control Problems

We want to study the motion of a vertically ascending rocket that reaches a given altitude H with minimum fuel expenditure [see figure 12(b)].

The mathematical model for the system is given by,

$$\frac{d^2 x}{dt^2} = u(t) - g, \qquad (12.15)$$

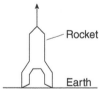

Fig. 12(b)

where x is the height of the rocket above the ground level and g is the acceleration due to gravity. $u(t)$ is the thrust exerted by the rocket.

Let h be the height attained at time T. Then the initial and boundary conditions of the equation (12.15),

$$x(0) = x'(0) = 0, \ x(T) = h \qquad (12.16)$$

We neglect the loss of mass by the burning of fuel.

Let us measure the minimal fuel expenditure during the time interval $[0, T]$ through the integral $\int_0^T |u(t)| dt$ over the rocket thrust. Let $T > 0$ be fixed. Then the minimal fuel expenditure $\alpha(T)$ during the time interval $[0, T]$ is given by a solution of the following *minimum problem*

$$\min_u \int_0^T |u(t)| dt = \alpha(T) \tag{12.17}$$

where we vary u over all the integrable functions $u : [0, T] \to \mathbb{R}$. Integrating (12.15) we get

$$x'(t) = \int_0^t u(t) dt - gt.$$

Integrating further,

$$x(t) = \int_0^t dt \int_0^t u(s) ds - \frac{1}{2} gt^2$$

$$= \int_0^t (t - s) u(s) ds - \frac{1}{2} gt^2 \quad \text{[see 4.7.16 Ex. 2]}$$

Thus, $\quad h = \int_0^T (T - s) u(s) ds - \frac{1}{2} gT^2. \tag{12.18}$

Thus for given $h > 0$, we have to determine the optimal thrust $u(\cdot)$ and the final time T as a solution of (12.17).

This formulation has the following shortcoming. If we consider only classical force functions u, then an impulse at time t of the form '$u = \delta_t$' is excluded.

However, such types of thrusts are of importance. Therefore, we consider the *generalized problem for functionals*:

(a) For a fixed altitude h and fixed final time $T > 0$, we are looking for a solution U of the following minimum problem:

$$\min \|U\| = \alpha(T), \ U \in C([0, T])^* \tag{12.19}$$

along with the side condition,

$$h = U(w) - \frac{T^2}{2} \tag{12.20}$$

where we write $w(t) = T - t$.

(b) We determine the final time T in such a way that

$$\alpha(T) = \min! \tag{12.21}$$

It may be noted that (12.19) generalizes (12.17). In fact, if the functional $U \in C([0, T])^*$ has the following special form:

$$U(v) = \int_0^T v(t) u(t) dt \text{ for all } u \in C([0, T])$$

where the fixed function $u : [0, T] \to \mathbb{R}$ is continuous, then we show that

$$||U|| = \int_0^T |u(t)| dt \qquad (12.22)$$

Let us define $h(t) = \int_0^t u(t) dt$ for all $t \in [0, T]$.

Then, $\qquad U(w) = \int_0^T w(t) dh(t)$ for all $u \in C([0, T])$

and $\qquad ||U|| = \text{Var}(h)$ by 12.2.8 $\qquad (12.23)$

For $0 = t_0 < t_1 < \cdots < t_n = T$, a partition of the interval $[0, T]$,

$$\Delta = \sum_{j=1}^n |h(t_j) - h(t_{j-1})| \leq \sum_{j=1}^n \int_{t_{j-1}}^{t_j} |u(t)| dt = \int_0^T |u(t)| dt.$$

Hence, $\text{Var}(h) \leq \int_0^T |u(t)| dt$.

By the mean value theorem,

$$\Delta = \sum_{j=1}^n |u(\xi_j)|(t_j - t_{j-1}) \quad \text{where } t_{j-1} \leq \xi_j \leq t_j.$$

Making the partition arbitrarily fine as $n \to \infty$, we get

$$\Delta \to \int_0^T |u(t)| dt \quad \text{as } n \to \infty$$

and hence

$$\text{Var}(h) = \int_0^T |u(t)| dt. \qquad (12.24)$$

Thus, (12.23) and (12.24) yield

$$||U|| = \int_0^T |u(t)| dt.$$

Thus (12.22) is true.

Theorem 12.4.1 Problem (a), (b) has the following solution:

$$U = T\delta_0 \quad \text{and} \quad T = (2h)^{\frac{1}{2}}$$

with the minimal **fuel expenditure** $||U|| = T$ [see Zeidler, E [56]].

CHAPTER 13

VARIATIONAL PROBLEMS

13.1 Minimization of Functionals in a Normed Linear Space

In this chapter we first introduce a variational problem. The purpose is to explore the conditions under which a given functional in a normed linear space admits of an optimum. Many differential equations arising out of problems of physics or of mathematical physics are difficult to solve. In such a case, a functional is built up out of the given equations and is minimized.

Let H be a Hilbert space and A be a symmetric linear operator with domain $\mathcal{D}(A)$ dense in H. Then $\langle Au, u \rangle$, for all $u \in \mathcal{D}(A)$, is a symmetric bilinear functional and is denoted by $a(u, u)$. $a(u, u)$ is a quadratic functional (9.3.1). Let $L(u)$ be a linear functional. The minimization problem can be stated as

$$\underset{u \in H}{\text{Min }} J(u) = \frac{1}{2}a(u, u) - L(u) \qquad (13.1)$$

In general, we consider a vector space E and U an open set of E. Let $J : u \in U \subset E \longrightarrow R$. Then the minimization problem is

$$\underset{u \in U}{\text{Min }} J(u) \qquad (13.2)$$

13.2 Gâteaux Derivative

Let E_1 and E_2 be normed linear spaces and $P : U \in E_1 \longrightarrow E_2$ be a mapping of an open subset U of E_1 into E_2. We shall call a vector $\phi \in E_1$,

$\phi \neq 0$ a direction in E_1.

13.2.1   Definition

The mapping P is said to be **differentiable** in the sense of **Gâteaux**, or simply **G-differentiable** at a point $u \in U$ in the direction ϕ if the difference quotient $\frac{P(u+t\phi)-P(u)}{t}$ has a **limit** $P'(u, \phi)$ in E_2 as $t \to 0$ in $\mathbb{R}$. The (unique) limit $P'(u, \phi)$ is called the **Gâteaux derivative** of P at u **in the direction of** ϕ.

P is said to be G-differentiable in a direction ϕ in a subset of U if it is G-differentiable **at every point** of the subset **in the direction** ϕ.

13.2.2   Remark

The operator $E_1 \ni \phi \longrightarrow P'(u, \phi) \in H$ is homogeneous.

For $P'(u, \alpha\phi) = \lim_{t \to 0} \dfrac{P(u + t\alpha\phi) - P(u)}{t\alpha} \cdot \alpha$

$\qquad\qquad\quad = \alpha P'(u, \phi)$ for $\alpha > 0$.

13.2.3   Remark

The operator $P'(u, \phi)$ is not, in general, **linear**.

Example 1. Let $f : \mathbb{R}^2 \longrightarrow \mathbb{R}$ be defined by

$$f(x_1, x_2) = \begin{cases} 0 & \text{if } (x_1, x_2) = (0, 0) \\ \dfrac{x_1^5}{((x_1 - x_2)^2 + x_1^4)} & \text{if } (x_1, x_2) \neq (0, 0) \end{cases}$$

If $u = (0, 0) \in \mathbb{R}^2$ and the direction $\phi = (h_1, h_2) \in \mathbb{R}^2$ $(\phi \neq 0)$, we have,

$$\frac{(f(th_1, th_2) - f(0, 0))}{t} = \frac{t^2 h_1^5}{((h_1 - h_2)^2 + t^2 h_1^4)}$$

which has a limit as $t \to 0$ and we have

$$f'(u, \phi) = f'((0, 0), (h_1, h_2)) = \begin{cases} 0 & \text{if } h_1 \neq h_2 \\ h & \text{if } h_1 = h_2 \end{cases}$$

It can be easily verified that f is G-differentiable in $\mathbb{R}^2$.

Example 2. Let Ω be an open set in $\mathbb{R}^n$ and $E = L_p(\Omega)$, $p > 1$.

Suppose $f : \mathbb{R} \ni t \longrightarrow f(t) \in \mathbb{R}$ is a continuously differentiable function, such that (i) $|f(t)| \leq K|t|^p$ and (ii) $|f'(t)| \leq K|t|^{p-1}$ for some constant $K > 0$. Then

$$J(u) = \int_\Omega f(u(x))dx \qquad\qquad (13.3)$$

defines a functional J on $L_p(\Omega) = E$ which is G-differentiable everywhere in all directions, and we have

$$J'(u, \phi) = \int_\Omega f'(u(x))\phi(x)dx \qquad\qquad (13.4)$$

We first show that the RHS of (13.4) exists.

Since $u \in L_p(\Omega)$ and since f satisfies (i) we have,

$$|J(u)| \leq \int_\Omega |f(u(x)|dx \leq K \int_\Omega |u|^p dx < \infty$$

which means that J is well-defined on $L^p(\Omega)$.

On the other hand, for any $u \in L_p(\Omega)$, since f' satisfies (ii), $f'(u) \in L_p(\Omega)$ where $\frac{1}{p} + \frac{1}{q} = 1$. This is because

$$\int_\Omega |f'(u)|^q dx \leq K \int_\Omega |u|^{(p-1)q} dx = K \int_\Omega |u|^p dx < \infty$$

Thus for any $u, \phi \in L_p(\Omega)$, we have by using Hölder's inequality (theorem 1.4.3)

$$\left| \int_\Omega f'(u)\phi dx \right| \leq ||f'(u)||_{L_p(\Omega)} ||u||_{L_p(\Omega)} \leq K ||u||_{L_p(\Omega)}^{p/q} ||\phi||_{L_p(\Omega)} < \infty$$

This proves the existence of the RHS of (13.4).

If $t \in \mathbb{R}$, we define $g : [0,1] \longrightarrow \mathbb{R}$ by setting,

$$g(\theta) = f(u + \theta t \phi).$$

Then g is continuously differentiable in $]0,1[$ and $g(1) - g(0) = \int_0^1 g'(\theta)d\theta = t\phi(x) \int_0^1 f'(u + \theta t\phi)d\theta$ $(\theta = \theta(x), |\theta(x)| \leq 1)$, so that

$$\frac{(J(u + t\phi) - J(u))}{t} = \int_\Omega \phi(x) \int_0^1 f'(u(x) + \theta t\phi(x))d\theta dx$$

Now, $\int_\Omega \phi(x) f'(u(x) + \theta t\phi(x))dx \leq K \int_\Omega |\phi(x)||u(x) + \theta t\phi(x)|^{p-1} dx$

$$\leq K \left(\int_\Omega |\phi(x)|^p dx \right)^{1/p} \left(\int_\Omega |u(x) + \theta t\phi(x)|^{(p-1)q} dx \right)^{1/q} < \infty$$

Hence, $\phi(x) f'(u(x) + \theta t\phi(x)) \in L_1(\Omega \times [0,1])$

and by Fubini's theorem (10.5.3)

$$\frac{(J(u + t\phi) - J(u))}{t} = \int_0^1 d\theta \int_\Omega \phi(x) f'(u(x) + \theta t\phi(x))dx.$$

The continuity of f' implies that $f'(u + \theta t\phi) \longrightarrow f'(u)$ as $t \to 0$ (and hence as $\theta t \to 0$) uniformly for $\theta \in]0, 1[$.

Moreover, the condition (ii) and the triangle inequality yield,

$$|\phi(x) f'(u(x) + \theta t\phi(x))| \leq K|\phi(x)|(|u(x)| + |\phi(x)|)^{p-1} \qquad (13.5)$$

The RHS of (13.5) is integrable by Hölder's inequality [see 1.4.3]. Then by dominated convergence theorem (10.2.20(b)) we conclude

$$J'(u, \phi) = \int_\Omega f'(u)\phi dx.$$

13.2.4   Definition

An operator $P : U$ $(E_1 \longrightarrow E_2)$ (U being an open set in E_1) is said to be **twice differentiable in the sense of Gâteaux** at at point $u \in U$ in the directions ϕ, ψ ($\phi, \psi \in E_1$, $\phi \neq 0$, $\psi \neq 0$ given) if the operator $u \longrightarrow P'(u, \phi) : U \subset E_1 \longrightarrow E_2$ is once G-differentiable at u in the **direction** ψ. The G-derivative of $u \longrightarrow P'(u, \phi)$ is called the **second G-derivative** of P and is denoted by $\boldsymbol{P''(u, \phi, \psi) \in E_2}$, i.e.,

$$P''(u, \phi, \psi) = \lim_{t \to 0} \frac{P'(u + t\phi, \psi) - P'(u, \psi)}{t} \tag{13.6}$$

13.2.5   Gradient

Let $J : U \subset E_1 \longrightarrow \mathbb{R}$ be a functional on an open set of a normed linear space E_1 which is once G-differentiable at a point $u \in U$. If the functional $u \longrightarrow J'(u, \phi)$ is continuous linear on E_1, then there exists a (unique) element $G(u) \in E_1^*$ (6.1), such that

$$J'(u, \phi) = G(u)(\phi) \text{ for all } \phi \in E_1.$$

Similarly, if J is twice G-differentiable at a point $u \in U$, and if the form $(\phi, \psi) \longrightarrow J''(u, \phi, \psi)$ is a bilinear (bi-) continuous form on $E_1 \times E_1$, then there exists a (unique) element $H(u) \in (E_1 \longrightarrow E_1^*)$ such that

$$J''(u, \phi, \psi) = H(u)(\phi, \psi), \quad (\phi, \psi) \in E_1 \times E_1 \tag{13.7}$$

13.2.6   Definitions: gradient, Hessian

Gradient: $G(u) \in E_1^*$ is called the gradient of J at u
Hessian: $H(u) \in (E_1 \longrightarrow E_1^*)$ is called the Hessian of J at u.

13.2.7   Mean value theorem

Let J be a functional as in 13.1. Let us assume that $[u + t\phi, t \in [0, 1]]$, is contained in U. Let the function $g : [0, 1] \to \mathbb{R}$ be defined as

$$t \longrightarrow g(t) = J(u + t\phi)$$

Let $J'(u + t\phi, \phi)$ exist. Then

$$\lim_{\theta \to 0} \frac{J(u + (\theta + t)\phi) - J(u + t\phi)}{\theta} = \lim_{\theta \to 0} \frac{g(t + \theta) - g(t)}{\theta} = g'(t)$$

showing g is once differentiable in [0,1]

Thus, $g'(t) = J'(u + t\phi, \phi)$ \hfill (13.8)

Similarly, if $J''(u + t\phi, \phi, \phi)$ exist and J is twice differentiable then

$$g''(t) = J''(u + t\phi; \phi, \phi) \tag{13.9}$$

13.2.8   Lemma

Let J be as in 13.1. Let $u \in U$, $\phi \in E_1$ be given. If $[u + t\phi : t \in [0, 1]] \in U$ and J is once G-differentiable on this set in the direction, ϕ then there exists a $t_0 \in]0, 1[$, such that

$$J(u + \phi) = J(u) + J'(u + t_0\phi, \phi) \qquad (13.10)$$

Proof: $g : [0, 1] \longrightarrow \mathbb{R}$ and g is differentiable in $]a, b[$. The classical mean value theorem yields:

$$g(1) = g(0) + 1 \cdot g'(t_0), \ t_0 \in]0, 1[\qquad (13.11)$$

Using the definition of $g(t)$ and (13.8), (13.10) follows from (13.11).

13.2.9   Lemma

Let U and J be as in lemma 13.2.8. If J is twice G-differentiable on the set $[u + t\phi; \ t \in [0, 1]]$ in the directions ϕ, ψ, then there exists a $t_0 \in]0, 1[$ such that

$$J(u + \phi) = J(u) + J'(u, \phi) + \frac{1}{2}J''(u + t_0\phi, \phi, \phi) \qquad (13.12)$$

Proof: The classical Taylor's theorem applied to g on $]0, 1[$, yields

$$g(1) = g(0) + 1 \cdot g'(0) + \frac{1}{2!} \cdot 1^2 g''(t_0) \qquad (13.13)$$

Using the definition of $g(t)$, (13.8) and (13.9), (13.12) follows from (13.13).

13.2.10   Lemma

Let E_1 and E_2 be two normed linear spaces. U an open subset of E_1 and let $\phi \in E_1$ be given. If the set $[u + t\phi; \ t \in [0, 1]] \in U$ and $P : U \subset E_1 \longrightarrow E_2$ is a mapping which is G-differentiable everywhere on the set $[u + t\phi; \ t \in [0, 1]]$ in the direction ϕ then, for any $h \in E_1^*$, there exists a $t_h \in]0, 1[$, such that

$$h(P(u + \phi)) = h(Pu) + h(P'(u + t_h\phi, \phi) \qquad (13.14)$$

Proof: Let $g : [0, 1] \longrightarrow \mathbb{R}$ be set as

$$t \longrightarrow g(t) = h(P(u + t\phi)) \qquad (13.15)$$

where $h : U \subset E_1 \longrightarrow \mathbb{R}$.

Then $g'(t)$ exists in $]0, 1[$ and

$$g'(t) = \underset{t' \to 0}{\text{Lt}} \frac{g(t + t') - g(t)}{t'} = \underset{t' \to 0}{\text{Lt}} \frac{h(P(u + (t + t')\phi) - h(P(u + t\phi))}{t'}$$

$$= h(P'(u + t\phi, \phi) \quad \text{for } t \in]0, 1[,$$

since h is a linear functional defined on E_1.

Now, (13.14) follows immediately on applying the classical mean value theorem to the function g.

13.2.11   Theorem

Let E_1, E_2, u, ϕ and U be as in above. If $P : U \in E_1 \longrightarrow E_2$ is G-differentiable in the set $[u + t\phi;\ t \in [0,1]]$ in the direction ϕ, then there exists a $t_0 \in]0,1[$, such that,

$$||P(u + \phi) - P(u)||_{E_1} \leq ||P'(u + t_0\phi, \phi)|| \qquad (13.16)$$

Proof: Let $v = P(u+\phi) - P(u)$, then $v \in E_1$. Then by th. 5.1.4, which is a consequence of the Hahn-Banach theorem, we can find a functional $h \in E_1^*$ such that

$$||h|| = 1 \quad h(v) = h(P(u + \phi) - Pu) = ||P(u + \phi) - P(u)||.$$

Since P satisfies the assumption of theorem 13.2.10, it follows that there exists $t_0 = t_h \in]0,1[$, such that

$$||P(u + \phi) - Pu|| = h(P(u + \phi) - Pu) = h(P'(u + t_0\phi, \phi))$$
$$\leq ||h||\,||P'(u + t_0\phi, \phi)|| = ||P'(u + t_0\phi, \phi)||.$$

13.2.12   Convexity and Gâteaux differentiability

Earlier we saw that a subset U of a vector space E is convex if $u, v \in E \Longrightarrow \lambda u + (1 - \lambda)v \in E, 0 \leq \lambda \leq 1$.

13.2.13   Definition

A functional $J : U \subset E_1 \longrightarrow \mathbb{R}$ on a convex set U of a vector space E_1 is said to be convex if

$$J((1 - \lambda)u + \lambda v) \leq (1 - \lambda)J(u) + \lambda J(v) \text{ for all} \qquad (13.17)$$

$u, v \in U$ and $\lambda \in [0,1]$.

J is said to be strictly convex if strict inequality holds for all $u, v \in E_1$ with $u \neq v$ and $\lambda \in]0,1[$.

13.2.14   Theorem

If a functional $J : U \subset E_1 \longrightarrow \mathbb{R}$ on a convex set is G-differentiable everywhere in U in all directions then

(i) J is convex if and only if,

$$J(v) \geq J(u) + J'(u, v - u) \text{ for all } u, v \in U \qquad (13.18)$$

(ii) J is strictly convex if and only if,

$$J(v) > J(u) + J'(u, v - u) \text{ for all } u, v \in U \qquad (13.19)$$

with $u \neq v$.

Proof: J is convex $\implies J(v) - J(u) \geq \frac{J(u+\lambda(v-u))-J(u)}{\lambda}$, $\lambda \in [0,1]$

Since $J'(u, v - u)$ exists, proceeding to the limit as $\lambda \to 0$ on the RHS, we get,

$$J(v) - J(u) \geq J'(u, v - u) \text{ which is inequality (13.18).}$$

For proving the converse, we note that (13.18) yields

$$J(v) \geq J(u + \lambda(v - u)) + J'(u + \lambda(v - u), u - (u + \lambda(v - u)))$$

or, $J(v) \geq J(u + \lambda(v - u)) - \lambda J'(u + \lambda(v - u), v - u)$ (13.20)

by the homogeneity of the mapping $\phi \longrightarrow J'(u, \phi)$.

Similarly we can write,

$$J(v) \geq J(u + \lambda(v - u)) + J'(u + \lambda(v - u)), v - (u + \lambda(v - u))$$

or, $J(v) \geq J(u + \lambda(v - u)) + (1 - \lambda)J'(u + \lambda(v - u), v - u)$ (13.21)

Multiplying (13.20) by $(1 - \lambda)$ and (13.21) by λ and adding we get back (13.17) for any $u, v \in U \subset E_1$.

If J is strictly convex, we can write,

$$J(v) - J(u) > \lambda^{-1}[J(u + \lambda(v - u)) - J(u)]$$ (13.22)

On the other hand, the mean value theorem yields

$$J(u + \lambda(v - u)) = J(u) + J'(u, \lambda_0(v - u)), \ 0 < \lambda_0 < \lambda \in]0,1[$$

or, $J(u + \lambda(v - u)) - J(u) = \lambda_0 J'(u, v - u)$ (13.23)

Using (13.23), (13.22) reduces to (13.19).

The converse can be proved exactly in the same way as in the first part.

13.2.15   Weak lower semicontinuity

Definition: weak lower semicontinuity

Let E be a normed linear space.

A functional $J : E \longrightarrow \mathbb{R}$ is said to be **weakly lower semicontinuous** if, for every sequence $v_n \overset{w}{\longrightarrow} u$ in E_1 (6.3.9), we have

$$\lim_{n \to \infty} \inf J(v_n) \geq J(u)$$ (13.24)

13.2.16   Theorem

If a functional $J : E \longrightarrow \mathbb{R}$ is convex and admits of a gradient $G(u) \in E^*$ at every $u \in E$, then J is **weakly lower semicontinuous**.

Proof: Let v_n be a sequence in E, such $v_n - u$ is in E.   Then $G(u)(v_n - u) \longrightarrow 0$ as $n \to \infty$, since v_n is weakly convergent in u. On the other hand, since J is convex, we have by theorem 13.2.14,

$$J(v_n) \geq J(u) + G(u)(v_n - u).$$

On taking limits we have

$$\lim_{n \to \infty} \inf J(v_n) \geq J(u).$$

13.2.17   Theorem

If a functional $J : U \subset E \longrightarrow \mathbb{R}$ on the open convex set of a normed linear space, E is twice G-differentiable everywhere in U in all directions, and if the form $(\phi, \psi) \longrightarrow J''(u, \phi, \psi)$ is non-negative, i.e., if $J''(u, \phi, \phi) \geq 0$ for all $u \in U$ and $\phi \in E$ with $\phi \neq 0$, then J is convex.

If the form $(\phi, \psi) \longrightarrow J''(u, \phi, \psi)$ is positive, i.e., if $J''(u, \phi, \phi) > 0$ for all $u \in U$ and $\phi \in E$ with $\phi \neq 0$, then J is strictly convex.

Proof: Since U is convex, the set $[u + \lambda(v - u), \lambda \in [0,1]]$ is contained in U whenever $u, v \in U$. Then, by Taylor's theorem [see Cea [11]], we have, with $\phi = v - u$,

$$J(v) = J(u) + J'(u, v - u) + \frac{1}{2}J''(u + \lambda_0(v - u), v - u, v - u) \quad (13.25)$$

for some $\lambda_0 \in]0, 1[$. Then the non-negativity of J'' implies

$$J(v) \geq J(u) + J'(u, v - u)$$

from which convexity of J follows from 12.2.14. Similarly, the strict convexity of J follows from positive-property of $J''(u, \phi, \phi)$.

13.2.18   Theorem

If a functional $J : E \longrightarrow \mathbb{R}$ is twice G-differentiable everywhere in E in all directions and satisfies

(a) J has a gradient $G(u) \in E^*$ at all points $u \in E$,

(b) $(\phi, \psi) \longrightarrow J''(u, \phi, \psi)$ is non-negative, i.e., $J''(u; \phi, \phi) \geq 0$ for all $u, \phi \in E$ with $\phi \neq \theta$

then J is weakly lower semicontinuous.

Proof: By theorem 13.2.14, the condition (b) implies that J is convex. Then the conditions of theorem 13.2.16 being fulfilled J is weakly lower semicontinuous.

13.3   Fréchet Derivative

Let E_1 and E_2 be two normed linear space.

13.3.1   Definition: Fréchet derivative

A mapping $P : U \subset E_1 \longrightarrow E_2$ from an open set U in E_1 to E_2 is said to be **Fréchet differentiable**, or simply F-differentiable, at a point $u \in U$ if there exists a **bounded linear operator** $P'(u) : E_1 \longrightarrow E_2$, i.e., $P'(u) \in (E_1 \longrightarrow E_2)$ such that

$$\lim_{\phi \to \theta} \frac{||P(u + \phi) - P(u) - P'(u)\phi||}{||\phi||} = 0 \qquad (13.26)$$

Clearly, $P'(u)$, if it exists, is unique and is called the **Fréchet derivative** of P at u.

13.3.2   Examples

1. f is a function defined on an open set $U \subset R^2$ and $f : U \longrightarrow \mathbb{R}$. Then f is F-differentiable if it is once differentiable in the usual sense.

   Let $u = (u_1, u_2)^T$, $\phi = (\phi_1, \phi_2)^T$.

   Then, $f(u + \phi) - f(u) = f(u_1 + \phi_1, u_2 + \phi_2) - f(u_1, u_2)$

   $$= \frac{\partial f}{\partial u_1}\phi_1 + \frac{\partial f}{\partial u_2}\phi_2 + 0(||\phi||^2)$$

   where $0(||\phi||^2)$ denotes terms of order in $||\phi||^2$ and of higher orders.

   Therefore, $\displaystyle\lim_{||\phi|| \to 0} \frac{||f(u + \phi) - f(u) - f'(u)\phi||}{||\phi||}$

   $$= \lim_{||\phi|| \to 0} \frac{\left|\left| \frac{\partial f}{\partial u_1}\phi_1 + \frac{\partial f}{\partial u_2}\phi_2 - f'(u)\phi \right|\right|}{||\phi||} = 0.$$

   Hence, $f'(u) = \text{grad } f^T = \left(\dfrac{\partial f}{\partial u_1}, \dfrac{\partial f}{\partial u_2} \right).$

2. Let $(u, v) \longrightarrow a(u, v)$ be a symmetric bi-linear form on a Hilbert space H and $v \to L(v)$ a linear form on H. Let us take $J : U \longrightarrow \mathbb{R}$ (sec 13.1) by

   $$J(v) = \frac{1}{2}a(v, v) - L(v) \text{ for } \phi \in H, \phi \neq \theta.$$

   $$J(v + \phi) - J(v) = \frac{1}{2}a(v + \phi, v + \phi) - L(v + \phi)$$

   $$-\frac{1}{2}a(v, v) + L(v)$$

   $$= \frac{1}{2}a(v, \phi) + \frac{1}{2}a(\phi, v) + \frac{1}{2}a(\phi, \phi) - L(\phi)$$

   $$= a(v, \phi) - L(\phi) + \frac{1}{2}a(\phi, \phi).$$

$$\underset{||\phi||\to 0}{\text{Lt}} \frac{||J(v+\phi) - J(v) - J'(v)\phi||}{||\phi||}$$

$$= \underset{||\phi||\to 0}{\text{Lt}} \left\| (a(v,\phi) - L(\phi) - J'(v)\phi) + \frac{1}{2}a(\phi,\phi) \right\| \quad (13.27)$$

Let us suppose that

(i) $a(\cdot,\cdot)$ is bi-continuous: there exists a constant $K > 0$ such that

$$a(u,v) \leq K||u||\,||v|| \text{ for all } u,v \in H.$$

(ii) $a(\cdot,\cdot)$ is coercive (11.7.2), i.e.,

$$a(u,v) \geq \alpha||v||_H^2 \text{ for all } v \in H.$$

(iii) L is bounded, i.e., there exists a constant M, such that

$$L(v) \leq M||v||, \text{ for all } v \in H.$$

Using condition (i) it follows from (13.27)

$$\underset{||\phi||\to 0}{\text{Lt}} \frac{||J(v+\phi) - J(v) - J'(v)\phi||}{||\phi||}$$

$$\leq \underset{||\phi||\to 0}{\text{Lt}} \left(||a(v,\phi) - L(\phi) - J'(v)\phi|| + \frac{K \cdot 0(||v||^2)}{||v||} \right).$$

The limit will be zero if $J'(v)\phi = a(v,\phi) - L(\phi)$.

13.3.3 *Remark*

If an operator $P : U \subset E_1 \longrightarrow E_2$, where E_1 and E_2 are normed linear spaces, is F-differentiable then it is also G-differentiable and its G-derivative coincides with the F-derivative.

Proof: If P has a F-derivative $P'(u)$ at $u \in U$, then

$$\lim_{||\phi||\to 0} \frac{||P(u+\phi) - P(u) - P'(u)\phi||}{||\phi||} = 0.$$

Since $\phi \neq \theta$, we put $\phi = te$ where $t = ||\phi||$ and e is an unit vector in the direction of ϕ.

The above limit yields

$$\lim_{||te||\to 0} \frac{||P(u+te) - P(u) - tP'(u)e||}{t} = 0.$$

The above shows that P is G-differential at u and $P'(u)$ is also the G-derivative.

13.3.4 Remark

The **converse** is not always true.

Example one in 13.2.3 has a G-derivative at $(0,0)$, but is not F-differentiable.

13.4 Equivalence of the Minimizing Problem for Solving Variational Inequality

13.4.1 Definition

A functional $J : U \subset E \longrightarrow \mathbb{R}$, where U is an open set in a normed linear space, is said to have a **local minimum** at a point $u \in U$ if there is a **neighbourhood** N_u of u in E such that

$$J(u) \le J(v) \text{ for all } v \in U \cap N_u \tag{13.28}$$

13.4.2 Definition

A functional J on U is said to have a global minimum in U if there exist a $u \in U$, such that

$$J(u) \le J(v) \text{ for all } v \in U \tag{13.29}$$

13.4.3 Theorem

Suppose E, U and $J : U \longrightarrow \mathbb{R}$ fulfil the following conditions:

1. E is a reflexive Banach space
2. U is weakly closed
3. U is weakly bounded and
4. $J : U \subset E \longrightarrow \mathbb{R}$ is weakly lower semicontinuous.

Then J has a global minimum in U.

Proof: Let m denote $\inf\limits_{v \in U} J(v)$. If v_n is a minimizing sequence for J, i.e.,

$$m = \inf\limits_{v \in U} J(v) = \lim\limits_{n \to \infty} J(v_n)$$

then, by the boundedness of U, (from (3)), v_n is a bounded sequence in E, i.e., there exists a constant $k > 0$, such that $\|u_n\| < k$ for all n. By the reflexivity of E, this bounded sequence is weakly compact [see theorem 6.4.4]. Hence, $\{v_n\}$ contains a weakly convergent subsequence, i.e., a sequence v_{n_p}, such that $v_{n_p} \xrightarrow{w} u \in E$ as $p \to \infty$. U being weakly closed $u \in U$. Finally, since $v_{n_p} \longrightarrow u$ and J is weakly lower semicontinuous,

$$J(u) \le \lim\limits_{p \to \infty} \inf J(v_{n_p})$$

which implies that

$$J(u) \leq \lim_{p \to \infty} J(v_{n_p}) = l \leq J(v)$$

for all $v \in E$.

13.4.4 Theorem

If E, U and J satisfy the conditions (1), (2), (4) and J satisfy the condition (5)

$$\lim_{\|v\| \to \infty} J(v) = +\infty,$$

then J admits of a global minimum in U.

Proof: Let $z \in U$ be arbitrarily fixed.

Let us consider the subset U° of U as follows,

$$U^\circ = \{v : v \in U \text{ and } J(v) \leq J(z)\}.$$

Thus, the existence of a minimum in U° ensures the existence of a minimum in U.

We would show that U° satisfies the conditions (2) and (3). If U_0 is not bounded we can find a sequence $v_n \in U^0$, such that $\|v_n\| \longrightarrow +\infty$. Then condition (5) yields $J(v_n) \longrightarrow +\infty$ which is impossible since $v_n \in U^\circ \Longrightarrow J(v_n) \leq J(z)$. Hence U^0 is bounded. To, show that U° is weakly closed, let $u_n \in U^0$ be a sequence such that $u_n \xrightarrow{w} u$ in E. Since U is weakly closed, $u \in U$. Since $u_n \in U^0$, $J(u_n) \leq J(z)$ and since $|J(u_n) - J(u)| < \epsilon$ for $n \geq n_0(\epsilon)$, it follows that $J(u) \leq J(z)$ showing that U° is weakly closed.

On the other hand, since J is weakly lower semicontinuous $u_n \xrightarrow{w} u$ in E implies that

$$J(u) \leq \lim\inf J(u_n) \leq J(z)$$

proving that $u \in U^\circ$. Now, U° and J satisfy all the conditions of theorem 13.4.3, hence J has a global minimum in U^0 and hence in U.

13.4.5 Theorem

Let $J : E \longrightarrow \mathbb{R}$ be a functional on E, U a subset of E satisfying the following conditions:

1. E is a reflexive Banach space,

2. J has a gradient $G(u) \in E^*$ everywhere in U,

3. J is twice G-differentiable in all directions $\phi, \psi \in E$ and satisfies the condition

$$J''(u, \phi, \phi) \geq \|\phi\| \mathcal{E}(\|\phi\|) \text{ for all } \phi \in \mathcal{E}$$

where $\mathcal{E}(t)$ is a function on $[t \in \mathbb{R}; t \geq 0]$ such that

$$\mathcal{E}(t) \geq 0 \quad \text{and} \quad \lim_{t \to \infty} \mathcal{E}(t) = +\infty.$$

4. U is a closed convex set

Then there exists at least one minimum $u \in U$ of J. Furthermore, if in condition (3),

5. $\mathcal{E}(t) > 0$ for $t > 0$ is satisfied by E then there exists a unique minimum of J in U

Proof: First of all, by condition (3), $J''(u, \phi, \phi) \geq 0$ and hence by Taylor's formula (13.25),

$$J(v) = J(u) + J'(u, v - u) + \frac{1}{2}J''(u + \lambda_0(v - u), (v - u), (v - w)), \ 0 < \lambda < 1.$$

We have $J(v) \geq J(u) + J'(u, v - u)$ (13.30)

Application of theorem 13.2.17 asserts the convexity of J. Similarly, condition (5) implies that J is strictly convex by theorem 13.2.17. Then by conditions (2) and (3), we conclude from theorem 13.2.16 and keeping in mind that

$$J''(u + \lambda_0(v - u), (v - u), (v - w)) \geq 0 \text{ for } 0 < \lambda_0 < 1,$$

J is weakly lower semicontinuous.

We next show that $J(v) \longrightarrow +\infty$ as $||v|| \longrightarrow +\infty$.

For this, let $z \in U$ be arbitrarily fixed. Then, because of conditions (2) and (3), we can apply Taylor's formula (13.25) to get, for $v \in E$,

$$J(v) = J(z) + G(z)(v - z) + \frac{1}{2}J''(z + \lambda_0(v - z), (v - z), (v - z))$$

$$\text{for some } \lambda_0 \in]0, 1[\quad (13.31)$$

Now, $|G(z)(v - z)| \leq ||G(z)|| \, ||v - z||,$ (13.32)

Condition (3) yields,

$$J''(z + \lambda_0(v - z), (v - z), (v - z)) \geq ||v - z||\mathcal{E}(||v - z||) \quad (13.33)$$

Using (13.32) and (13.33), (13.31) reduces to

$$J(v) \geq J(z) + ||v - z|| \left[\frac{1}{2}\mathcal{E}(||v - z||) - ||G(z)|| \right].$$

Here, since $z \in U$ is fixed, as $||v|| \longrightarrow +\infty$

$$||v - z|| \longrightarrow +\infty,$$

$J(z)$ and $||G(z)||$ are constants and $\mathcal{E}(||v - z||) \longrightarrow +\infty$ by condition (3).

Thus, $J(v) \longrightarrow +\infty$ as $||v|| \to \infty$. The theorem thus follows by virtue of theorem 13.4.5.

13.4.6 Theorem

Suppose U is a convex subset of a Banach space and $J : U \subset E \longrightarrow \mathbb{R}$ is a G-differentiable (in all directions) convex functional.

Then, $u \in U$ is a minimum for J, (i.e., $J(u) \le J(v)$ for all $v \in E$) if, and only if $u \in U$ and $J'(u, v - u) \ge 0$ for all $v \in U$.

Proof: Let $u \in U$ be a minimum for J. Then, since U is convex, $u, v \in U \implies u + \epsilon_n(v - u) \in U$ as $\epsilon_n \to 0$ for each n. Hence $J(u) \le J(u + \epsilon_n(v - u))$

Therefore, $\lim\limits_{\epsilon_n \to 0+} \dfrac{J(u + \epsilon_n(v - u)) - J(u)}{\epsilon_n} \ge 0$,

i.e., $J'(u, v - u) > 0$ for any $v \in E$.

Conversely, since J is convex and G-differentiable by condition (i) of theorem 13.2.14, we have

$$J(v) \ge J(u) + J'(u, v - u) \text{ for any } v \in E.$$

Now, using the assumption that $J'(u, v - u) \ge 0$, we get

$$J(v) \ge J(u) \text{ for all } u \in U.$$

In what follows we refer to the problem posed in (13.1).

13.4.7 Problem (PI) and problem (PII) and their equivalence

Let K be a closed convex set of **normed linear space** E.

Problem (PI): To find $u \in K$ such that

$$J(u) \le J(v) \text{ for all } v \in K \tag{13.33}$$

$J(v) = \dfrac{1}{2}a(v, v) - L(v)$ implies

(i) $J'(v, \phi) = a(v, \phi) - L(\phi)$, and

(ii) $J''(v; \phi, \phi) = a(\phi, \phi)$.

The coercivity of $a(\cdot, \cdot)$ implies that

$$J''(v; \phi, \phi) = a(\phi, \phi) \ge \alpha ||\phi||^2.$$

If we choose $\mathcal{E}(t) = \alpha t$, then all the assumptions of theorem 13.4.5 are fulfilled by E, J and K so that problem (PI) has a unique solution.

Also, by theorem 13.4.6 the problem (PI) is equivalent to

(PII): To find

$$u \in K; \ a(u, v - u) \ge L(v - u) \text{ for all } v \in K \tag{13.34}$$

We thus obtain the following theorem:

13.4.8   Theorem

(1) There exists a **unique solution** $u \in K$ of the problem (PI).

(2) **Problem (PI)** is **equivalent** to **problem (PII)**. The problem (PII) is called a variational inequality associated to the closed, convex set and the bilinear form $a(\cdot, \cdot)$.

The theorem 13.4.7 was generalized by G-stampaccia (Cea [11]) to the non-symmetric case. This generalizes and uses the classical Lax-Miligram theorem [see Reddy [45]]. We state without proof the theorem due to Stampacchia.

13.4.9   Theorem (Stampacchia)

Let K be a closed convex subset of a Hilbert space H and $a(\cdot, \cdot)$ be a bilinear bi-continuous coercive form (sec 11.7.2) on H. Then, for any given $L \in H$, the variational inequality (13.34) has a unique solution $u \in K$.

For proof see Cea [11].

13.5   Distributions

13.5.1   Definition: Support

The **support** of a function $f(x)$, $x \in \Omega \subset \mathbb{R}^n$ is defined as the closure of the set of points in $\mathbb{R}^n$ at which f is non-zero.

13.5.2   Definition: smooth function

A function $\phi : \mathbb{R}^n \longrightarrow \mathbb{R}$ is said to be *smooth* or infinitely differentiable if its derivatives of all order exist and are continuous.

13.5.3   Definition: $C_0^\infty(\Omega)$

The set of all smooth functions with *compact support* in $\Omega \subset \mathbb{R}^n$ is denoted by $C_0^\infty(\Omega)$.

13.5.4   Definition: test function

A **test function** ϕ is a smooth function with **compact support**, $\phi \in C_0^\infty(\Omega)$.

13.5.5   Definition: generalized derivative

A function $u \in C^\infty(\Omega)$ is said to have the αth **generalized derivative** $D^\alpha u$, $1 \le |\alpha| \le m$, if the following relation (**generalized Green's formula**) holds:

$$\int_\Omega D^\alpha u \phi dx = (-1)^{|\alpha|} \int_\Omega u D^\alpha \phi dx \text{ for every } \phi \in C_0^\alpha(\Omega) \qquad (13.35)$$

For $u \in C_0^\infty(\Omega)$, the generalized derivatives are derivatives in the ordinary (classical) sense.

13.5.6 Definition: distribution

A set of test functions $\{\phi_n\}$ is said to converge to a test function ϕ_0 in $C_0^\infty(\Omega)$ if there is a bounded set $\Omega_0 \subset \Omega$ containing the supports of $\phi_0, \phi_1, \phi_2, \ldots$ and if ϕ_n and all its generalized derivatives converge to ϕ_0 and its derivatives respectively. A functional f on $C_0^\infty(\Omega)$ is continuous if it maps every convergent sequence in $C_0^\infty(\Omega)$ into a convergent sequence in $\mathbb{R}$, i.e., if $f(\phi_n) \longrightarrow f(\phi_0)$ whenever $\phi_n \longrightarrow \phi$ in $C_0^\infty(\Omega)$.

A continuous linear functional on $C_0^\infty(\Omega)$ is called a *distribution* or *generalized function*.

Example An example of the distribution is provided by the delta distribution, defined by

$$\int_{-\infty}^{\infty} \delta(x)\phi(x)dx = \phi(0) \text{ for all } \phi \in C_0^\infty(\Omega) \tag{13.36}$$

Addition and scalar multiplication of distributions:

If f and g are distributions, then the distributions of $\alpha f + \beta g$, α, β, being scalars, is the sum of $\alpha \times$ distribution of f and $\beta \times$ distribution of g.

13.6   Sobolev Space

$C^\infty(\Omega)$ is an inner product space with respect to the $L_2(\Omega)$-inner product. But it is not complete with respect to the norm generated by the inner product

$$\langle u, v \rangle_p = \int_\Omega \sum_{|\alpha|=p} D^\alpha u D^\alpha v dm \tag{13.37}$$

where u and v along with their derivatives upto m, are square integrable in the Lebesgue sense [see chapter 10].

$$\int_\Omega |D^\alpha u|^2 dm < \infty \text{ for all } |\alpha| \le p.$$

The space $C^\infty(\Omega)$ can be completed by adding the limit points of all Cauchy sequences in $C^\infty(\Omega)$. It turns out that the distributions are those limits points.

We can thus introduce the Sobolev space $H^1(\Omega)$ as follows:

$$H^1(\Omega) = \left\{ v : v \in L_2(\Omega), \frac{\partial v}{\partial x_j} \in L^2(\Omega), \; j = 1, 2, \ldots, p \right\}, \tag{13.38}$$

where $\quad D^j v = \dfrac{\partial v}{\partial x_j}$ are taken in the sense of distributions,

i.e., $\quad \displaystyle\int_\Omega D^j v \phi dx = - \int_\Omega v D^j \phi dx$ for all $\phi \in \mathbb{D}(\Omega) \tag{13.39}$

where $\mathbb{D}(\Omega)$ denotes the space of all C^∞-functions with compact support on Ω. $H^1(\Omega)$ is provided with the inner product

$$\langle\langle u, v \rangle\rangle = \langle u, v \rangle_{L_2(\Omega)} + \sum_{j=1}^{n} \langle D^j u, D^j v \rangle_{L_2(\Omega)} \tag{13.40}$$

$$= \int_\Omega \left\{ uv + \sum_{j=1}^{n} (D^j u)(D^j v) \right\} dx \tag{13.41}$$

for which it becomes a Hilbert space.

13.6.1 Remark

$\mathbb{D}(\Omega) \subset C^1(\Omega) \subset H^1(\Omega)$.

We introduce the space

$$H_0^1(\Omega) = \text{the closure of } \mathbb{D}(\Omega) \text{ in } H^1(\Omega), \tag{13.42}$$

We state without proof some well-known theorems.

13.6.2 Theorem of density

If Γ, the boundary of Ω is regular (for instance, Γ is a C^1 function of dimension $n - 1$), then $C^1(\Omega)$ (or C^∞)-manifold (respectively $C^\infty(\overline{\Omega})$) is dense in $H^1(\Omega)$.

13.6.3 Theorem of trace

If Γ is regular then the linear mapping $v \mapsto v|_\Gamma$ of $C^1(\overline{\Omega}) \longrightarrow C^1(\Gamma)$ (respectively of $C^\infty(\overline{\Omega}) \longrightarrow C^\infty(\Gamma)$) extends to a continuous linear map of $H^1(\Omega)$ into $L_2(\Omega)$ denoted by γ and for any $v \in H^1(\Omega)$, γ_v is called the **trace of v in Γ**.

Moreover, $H_0^1(\Omega) = \{v : v \in H^1(\Omega), \gamma_v = 0\}$.

13.6.4 Green's formula for Sobolev spaces

Let Ω be a bounded open set with sufficiently regular boundary Γ, then there exists a unique outer normal vector $\overline{n}(x)$. We define the operator of exterior normal derivation formally

$$\frac{\partial}{\partial \overline{n}} = \sum_{j=1}^{n} n_j(x) D^j \tag{13.43}$$

Now, if $u, v \in C^1(\Omega)$ then, by the classical Green's formula [see Mikhlin [36]], we have

$$\int_\Omega (D^j u) v \, dx = -\int_\Omega u (D^j v) \, dx + \int_\Gamma uv \overline{n_j} \, d\sigma$$

where $d\sigma$ is the area element on Γ.

This formula remains valid also if $u, v \in H^1(\Omega)$ in view of the trace theorem and density theorem .

Next, if $u, v \in C^2(\overline{\Omega})$ then applying the above formula to $D^j u$, $D^j v$ and summing over $j = 1, 2, \ldots, n$, we get,

$$\sum_{j=1}^{n} \langle D^j u, D^j v \rangle_{L_2(\Omega)} = -\sum_{j=1}^{n} \int_{\Omega} ((D^j)^2 u) v dx + \int_{\Omega} \frac{\partial u}{\partial \overline{n}} \cdot v d\sigma, \quad (13.44)$$

i.e., $$\sum_{j=1}^{n} \langle D^j u, D^j v \rangle = \int_{\Omega} (\Delta u) v dx + \int_{\Gamma} \frac{\partial u}{\partial \overline{n}} \cdot v d\sigma \quad (13.45)$$

13.6.5   Remark

(i) $u \in H^2(\Omega) \implies \Delta u \in L_2(\Omega)$

(ii) Since $D^j u \in H^1(\Omega)$ by trace theorem (13.6.3) $\gamma(D^j u)$ exists and belongs to $L_2(\Omega)$ so that

$$\frac{\partial u}{\partial \overline{n}} = \sum_{j=1}^{n} n_j \gamma(D^j u) \in L_2(\Gamma).$$

Hence, by using the density and trace theorems (13.6.2 and 13.6.3), the formula (13.4.3) is valid.

13.6.6   Weak (or variational formulation of BVPs)

Example 1. Let $\Gamma = \overline{\Gamma}_1 \cup \overline{\Gamma}_2$, where Γ_j are open subsets of Γ such that $\Gamma_1 \cap \Gamma_2 = \Phi$

Consider the space

$$E = \{v : v \in H^1(\Omega); \ \gamma_v = 0 \text{ on } \Gamma_1\} \quad (13.46)$$

E is clearly a closed subspace of $H^1(\Omega)$ and is provided with the inner product induced from that in $H^1(\Omega)$ and hence it is a Hilbert space. Moreover,

$$H_0^1(\Omega) \subset E \subset H^1(\Omega) \quad (13.47)$$

and the inclusions are continuous linear. If $f \in L_2(\Omega)$ we consider the functional

$$J(v) = \frac{1}{2} \langle\langle v, v \rangle\rangle - \langle f, v \rangle_{L_2(\Omega)} \quad (13.48)$$

i.e., $a(u, v) = \langle\langle u, v \rangle\rangle$ and $L(v) = \langle f, v \rangle$.

Then $a(\cdot, \cdot)$ is bilinear, bicontinuous and coercive,

$|a(u, v)| \leq ||u||_E ||v||_E = ||u||_{H^1(\Omega)} ||v||_{H^1(\Omega)}$ for $u, v \in E$.

$a(v, v) = ||v||_{H^1(\Omega)}^2$ for $v \in E$.

$$|L(v)| \leq ||f||_{L_2(\Omega)} ||v||_{L_2(\Omega)} \leq ||f||_{L_2(\Omega)} ||v||_{H^1(\Omega)} \text{ for } v \in E.$$

Then the problems (PI) and (PII) respectively become

(PIII) To find $u \in E$, $J(u) \leq J(v)$ for all $v \in E$ $\hspace{2cm}$ (13.49)

(PIV) To find $u \in E$, $\langle\langle u, \phi \rangle\rangle = \langle f, \phi \rangle_{L_2(\Omega)}$ for all $v \in E$ $\hspace{1cm}$ (13.50)

Theorem 13.4.8 asserts that these two equivalent problems have unique solutions.

The problem (PIV) is the **Weak (or Variational)** formulation of the

(i) **Dirichlet problem** (if $\Gamma_2 = \Phi$)

(ii) **Newmann problem** (if $\Gamma_1 = \Phi$)

(iii) **Mixed boundary problem** in the **general** case.

13.6.7 Equivalence of problem (PIV) to the corresponding classical problems

Suppose, $u \in C^2(\overline{\Omega}) \cap E$ and $v \in C^1(\overline{\Omega}) \cap E$.

Using Green's formula (13.45),

$$a(u, v) = \langle\langle u, v \rangle\rangle = \int_{\Omega} (-\Delta u + u)v dx + \int_{\Gamma} \frac{\partial u}{\partial n} \cdot v d\sigma = \int_{\Omega} f v dx$$

i.e., $\hspace{0.5cm} \int_{\Omega} (-\Delta u + u - f)v dx + \int_{\Gamma} \frac{\partial u}{\partial n} \cdot v d\sigma = 0$ $\hspace{2cm}$ (13.51)

We note that this formula remains valid if $u \in H^2(\Omega) \cap E$ for any $v \in E$.

We choose $v \in \mathbb{D}(\Omega) \subset E$ then the boundary integral vanishes so that we get,

$$\int_{\Omega} (-\Delta u + u - f)v dx = 0 \quad \forall\, v \in \mathbb{D}(\Omega).$$

Since $\mathbb{D}(\Omega)$ is **dense** in $L_2(\Omega)$, this implies that (if $u \in H^2(\Omega)$) u is a solution of the differential equation,

$$-\Delta u + u - f = 0 \text{ in } \Omega \text{ (in the sense of } L_2(\Omega)).$$

More generally, without the strong regularity assumption as above, u is a solution of the differential equation.

$$-\Delta u + u - f = 0 \text{ in the sense of distribution in } \Omega \hspace{1cm} (13.52)$$

Next we choose $v \in E$ arbitrary. Since u satisfies (13.50) in Ω, we find from (13.51) that

$$\int_{\Gamma} \frac{\partial u}{\partial n} v d\sigma = 0 \,\forall\, v \in E$$

which means that $\frac{\partial u}{\partial n} = 0$ on Γ in some generalised sense. In fact by trace theorem $\gamma|v \in H^{\frac{1}{2}}(\Gamma)$ and hence $\frac{\partial u}{\partial n} = 0$ in $H^{-\frac{1}{2}}(\Gamma)$ [ee Liones and Magenes

[34]]. Thus, if the problem (IV) has a regular solution then it is the solution of the classical problem

$$\begin{cases} -\Delta u + u - f = 0 & \text{on } \Omega \\ u = 0 & \text{on } \Gamma_1 \\ \dfrac{\partial u}{\partial n} = 0 & \text{on } \Gamma_2 \end{cases} \tag{13.53}$$

13.6.8 Remark

The variational formulation (PIV) is very much used in the Finite Elements Method.

CHAPTER 14

THE WAVELET ANALYSIS

14.1 An Introduction to Wavelet Analysis

The concept of Wavelet was first introduced around 1980. It came out as a synthesis of ideas borrowed from disciplines including mathematics (Calderón Zygmund operators and Littlewood-Paley theory), physics (coherent states formalism in quantum mechanism and renormalizing group) and engineering (quadratic mirror filters, sidebend coding in signal processing and pyramidal algorithms in image processing) (Debnath [17]).

Wavelet analysis provides a systematic new way to represent and analyze **multiscale structures**. The special feature of Wavelet analysis is to generalize and expand the representations of functions by **orthogonal basis to infinite domains**. For this purpose, **compactly supported** [see 13.5] basis functions are used and this linear combination represents the function. These are the kinds of functions that are realized by physical devices.

There are many areas in which wavelets play an important role, for example

(i) Efficient algorithms for representing functions in terms of a wavelet basis,

(ii) Compression algorithms based on the wavelet expansion representation that concentrate most of the energy of a signal in a few coefficients (Resnikoff and Wells [46]).

14.2   The Scalable Structure of Information

14.2.1   Good approximations

Every **measurement**, be it with a naked eye or by a sophisticated instrument, is at best very accurate or in other words **approximate**. Even a computer can measure the finite number of decimal places which is a rational number. Even Heisenberg's uncertainty principle corroborates this type of limitations.

The onus on the technologist is thus to make a measurement or a representation as accurate as possible. In the case of speech transmission, codes are used to transmit and at the destination the codes are decoded. Any transmitted signal is sampled at a number of uniformly spaced types. The sample measurements can be used to construct a Fourier series expansion of the signal. It will also interpolate values for unmeasured instants. But the drawback of the Fourier series is that it cannot take care of local phenomenon, for example, abrupt transitions. To overcome this difficulty, compactly supported [see 13.5] wavelets are used. A simple example is to consider a time series that describes a quantity that is zero for a long time, ramps up linearly to a maximum value and falls instantly to zero where it remains thereafter (fig. 14.1).

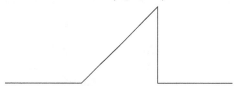

Fig. 14.1 Continuous ramp transient

Here, wavelet series approximate abrupt transitions much more accurately than Fourier series [see Resnikoff and Wells [46]]. Wavelet series expansion is less expansive too.

14.2.2   Special Features of wavelet series

That wavelet analysis provides good approximation for transient or localized phenomenon is due to following:

(a) Compact support

(b) Orthogonality of the basis functions

(c) Multiresolution representation

14.2.3   Compact support

Each term in a wavelet series has a **compact support** [13.5]. As a result, however short an interval is, there is a basis function whose support is contained within that interval. Hence, compactly supported wavelet basis function can capture local phenomenon and is not bothered by properties of the data far away from the area of interest.

14.2.4   Orthogonality

The terms in a wavelet series are **orthogonal** to one another, just like the terms in a Fourier series. This means that the information carried by one term is independent of the information carried by any other.

14.2.5   Multiresolution representation

Multiresolution representation describes what is called a *hierarchical structure.*   Hierarchical structures classify information into several categories called *levels* or *scales* so that, higher in the hierarchy a level is, the fewer the number of members it has. This hierarchy is prevalent in the social and political organization of the country. Biological sensory system, such as visions, also have this hierarchy built in.   The *human* vision system provides *wide aperture* detection (so events can be detected early) and high-resolution detection (so that the detailed structure of the visual event can be seen). Thus, a multiresolution or scalable mathematical representation provides a simpler or more efficient representation than the usual mathematical representation.

14.2.6   Functions and their representations

Representation of continuous functions

Suppose we consider a function of a real variable, namely,

$$f(x) = \cos x, \; x \in \mathbb{R}$$

where $\mathbb{R}$ denotes the continuum of real numbers.   For each $x \in \mathbb{R}$, we have a definite value for $\cos x$. Since $\cos x$ is periodic, all of its values are determined by its values on $[0, 2\pi[$. The question arises as to how best we can represent the value of the function $\cos x$ at any point $x \in [0, 2\pi[$. There is an uncountable number of points in $[0, 2\pi[$ and an uncountable number of values of $\cos x$, as x varies. If we represent

$$\cos x = \sum_{n=0}^{\infty} (-1)^n \frac{x^{2n}}{2n!} \tag{14.1}$$

for any $x \in \mathbb{R}$, we see that the sequence of numbers,

$$1, 0, -\frac{1}{2!}, 0, \frac{1}{4!}, 0, -\frac{1}{6!}, 0, \ldots, \tag{14.2}$$

which is countable, along with the sequence of power functions

$$x^0 = 1, x, x^2, x^3, \ldots, x^n$$

is sufficient information to determine $\cos x$ at any point x.   Thus we represent the *uncountable* number of values of $\cos x$ in terms of the *countable* discrete sequence (14.2), which are the coefficient of a power series representation for $\cos x$. This is the basic technique in representing a function or a class of functions in terms of more elementary or more easily computed functions.

14.2.7 Fourier series and the Fourier transform

The concept of a Fourier series was introduced in 3.7.8.

14.2.8 Definition: discrete Fourier transform

We note that if f is an integrable periodic function of period 1, then the Fourier series of f is given by

$$f(x) = \sum_{n \in \mathbb{C}} c_n e^{2\pi i n x} \tag{14.3}$$

with *Fourier coefficients* $\{c_n\}$ [see 3.7.8] given by

$$c_n = \int_0^1 f(x) e^{-2\pi i n x} dx. \tag{14.4}$$

Suppose that $\{c_n\}$ is a given discrete sequence of complex numbers in $l_2(\mathbb{C})$, that is $\sum |c_n|^2 < \infty$, then we define the *Fourier transform* of the sequence $f = \{c_n\}$ to be the Fourier series [see 14.3],

$$\hat{f}(\xi) = \sum_n c_n e^{2\pi i n \xi}$$

which is a periodic function.

14.2.9 Inverse Fourier transform of a discrete function

The inverse of the Fourier transform is the mapping from a periodic function (14.3) to its Fourier coefficients (14.4).

14.2.10 Continuous Fourier transform

If $f \in L_2(\mathbb{R})$, thus the Fourier transform of f is given by

$$\hat{f}(\xi) = \int_{-\infty}^{\infty} f(x) e^{2\pi i \xi x} dx \tag{14.5}$$

with the inverse Fourier transform given by

$$f(x) = \int_{-\infty}^{\infty} \hat{f}(\xi) e^{-2\pi i x \xi} d\xi \tag{14.6}$$

where both formulas have to be taken in a suitable limiting sense, but for nicely behaved functions that decrease sufficiently rapidly, the formulas hold (Resnikoff and Wells [46]).

14.3 Algebra and Geometry of Wavelet Matrices

A *wavelet matrix* is a generalisation of *unitary matrices* [see 9.4.1] to a larger class of rectangular matrices. Each wavelet matrix contains *the basic information* to define an associated wavelet system. Let $\mathbb{F}$ be a subfield of the field $\mathbb{C}$ of complex numbers. $\mathbb{F}$ could be the rational numbers $\mathbb{Q}$, the real numbers $\mathbb{R}$ or the field $\mathbb{C}$ itself.

Consider an array $A = (a_r^s)$, consisting of m rows of presumably infinite vectors of the form

$$\begin{pmatrix} \cdots & a_{-1}^0 & a_0^0 & a_1^0 & a_2^0 & \cdots \\ \cdots & a_{-1}^1 & a_0^1 & a_1^1 & a_2^1 & \cdots \\ \cdots & \cdots & \cdots & \cdots & \cdots & \cdots \\ \cdots & a_{-1}^{m-1} & a_0^{m-1} & a_1^{m-1} & a_2^{m-1} & \cdots \end{pmatrix} \tag{14.7}$$

In the above, a_r^s is an element of $\mathbb{F} \subseteq \mathbb{C}$ and $m \geq 2$. We call such an array A as a matrix even though the number of columns (rows) may not be finite.

Define submatrices A_p of A of size $m \times m$ in the following manner,

$$A_p = (a^s pm + q), \quad q = 0, 1, \ldots, m-1, \ s = 0, 1, \ldots, m-1 \tag{14.8}$$

for p an integer. Thus, A can be expressed in terms of submatrices in the form,

$$A = (\cdots, A_{-1}, A_0, A_1, \cdots), \tag{14.9}$$

where

$$A_p = \begin{pmatrix} a_p^0 & a_{p+1}^0 & \cdots & a_{p+m-1}^0 \\ \vdots & & & \\ a_p^{m-1} & a_{p+1}^{m-1} & \cdots & a_{p+m-1}^{m-1} \end{pmatrix}$$

From the matrix, a power series of the following form is constructed

$$A(z) = \sum_{p=-\infty}^{\infty} A_p z^p. \tag{14.10}$$

We call the above series *the Laurent series of the matrix A.*

Thus, $A(z)$ is the Laurent series with matrix coefficients. We can write $A(z)$ as a $m \times m$ matrix with Laurent series coefficients:

$$A(z) = \begin{pmatrix} \sum_j a_{mj}^0 z^j & \sum_j a_{mj+1}^0 z^j & \cdots & \sum_j a_{mj+m-1}^0 z^j \\ \vdots & & & \\ \sum_j a_{mj}^{m-1} z^j & \sum_j a_{mj+1}^{m-1} z^j & \cdots & \sum_j a_{mj+m-1}^{m-1} z^j \end{pmatrix} \tag{14.11}$$

(14.10) and (14.11) will both be referred to as the *Laurent series representation* of $A(z)$ of the *matrix A.*

14.3.1   Definition: genus of the Laurent series $A(z)$

Suppose $A(z)$ has a finite number of non-zero matrices, i.e.,

$$A(z) = \sum_{p=n_1}^{n_2} A_p z^p \tag{14.12}$$

where we assume that A_{n_1} and A_{n_2} are both non-zero matrices.

$$g = n_2 - n_1 + 1 \tag{14.13}$$

i.e., the number of terms in the series (14.12) is called the *genus* of the Laurent series $A(z)$ and the matrix A.

14.3.2   Definition: adjoint $\overline{A}(z)$ of the Laurent series $A(z)$

Let,

$$\overline{A}(z) = A^*(z^{-1})$$

$$= \sum_p A_p^* z^{-p} \tag{14.14}$$

$\overline{A}(z)$ is called the *adjoint* of the Laurent matrix $A(z)$.

In the above, $A_p^* = \overline{A}_p^T$ is the hermitian conjugate of the $m \times m$ matrix A_p.

14.3.3   Definition: the wavelet matrix

The matrix A, as defined in (14.7), is said to be a **wavelet matrix of rank m** if

(1) $A(z)\overline{A}(z) = mI$ \hfill (14.15)

(2) $\displaystyle\sum_{j=-\infty}^{\infty} a_j^s = m\delta^{s,0} \quad 0 \le s \le m-1.$ \hfill (14.16)

where $\delta^{s,0} = 1$ for $s = 0$ and is zero otherwise.

14.3.4   Lemma: a wavelet matrix with m rows has rank m

Let A be a wavelet matrix with m rows and an infinite number of columns.

Let $A(1) = \begin{pmatrix} a^0 \\ a^1 \\ \vdots \\ a^m \end{pmatrix}$ where a^i stands for the rows of A.

If the second and third rows are multiples of each other, then we can write $a^2 = \lambda a^1$ where λ is a scalar. In that case the first two rows of $A(1)$ will be multiples of each other. Therefore, the determinant of $A(1)$ would be zero.

This contradicts (14.15).

14.3.5 Definition: wavelet space $WM(m, g : \mathbb{F})$

The **wavelet space** $WM(m, g : \mathbb{F})$ denotes the set of all wavelet matrices of rank m and genus g with coefficients in the field $\mathbb{F}$.

Quadratic orthogonality relations for the rows of A

Comparison of coefficients of corresponding powers of z in (14.15) yields:

$$\sum_j a^{s'}_{j+mp'} \overline{a}^s_{j+mp} = m \delta^{s's} \delta_{p'p} \tag{14.17}$$

We will refer to (14.15) and (14.16) or equivalently (14.17) and (14.16), as the quadratic and linear conditions defining a wavelet matrix respectively.

Scaling vector: The vector a^0 is called the *scaling vector.*

Wavelet vector: a^s for $0 < s < m$ is called a *wavelet vector.*

14.3.6 Remark

1. The quadratic condition asserts that the rows of a wavelet matrix have length equal to $\sqrt{m}$ and they are pairwise disjoint when shifted by an arbitrary multiple of m.

2. The linear condition (14.16) implies that the sum of the components of the scaling vector is equal to the **rank** of A.

3. The sum of the components of each of the wavelet vector is **zero**.

14.3.7 Examples

1. *Haar matrix of rank 2*

   Let $A_1 = \begin{pmatrix} 1 & 1 \\ 1 & -1 \end{pmatrix}$. Here $A_1 A_1^T = 2I$.

   Sum of the elements of the first row $= 2 =$ rank of A.

   Sum of the elements of the second row $= 0$.

   Hence, A_1 is a wavelet matrix of rank 2.

   Similarly, $A_2 = \begin{pmatrix} 1 & 1 \\ -1 & 1 \end{pmatrix}$, $A_2 A_2^T = 2I$ and fulfils conditions (14.15) and (14.16).

   Hence, A_2 is a wavelet matrix of rank 2 too.

   The general complex Haar wavelet matrix of rank 2 has the form

   $$\begin{pmatrix} 1 & 1 \\ -e^{i\theta} & e^{i\theta} \end{pmatrix}, \ \theta \in \mathbb{R}.$$

2. *Daubechies' wavelet matrix of rank 2 and genus 2*

   Let

   $$D_2 = \frac{1}{4} \begin{pmatrix} 1+\sqrt{3} & 3+\sqrt{3} & 3-\sqrt{3} & 1-\sqrt{3} \\ -1+\sqrt{3} & 3-\sqrt{3} & -3-\sqrt{3} & 1+\sqrt{3} \end{pmatrix} \tag{14.18}$$

$$D_2 D_2^T = 2I.$$

Sum of the elements of the first row $= 2 =$ rank of D_2.

Sum of the elements of other wavelet vectors $= 0$.

14.3.8 Definition: Haar wavelet matrices

Haar wavelet matrix of rank m is denoted by $H(m; \mathbb{F})$ and is defined by

$$H(m; \mathbb{F}) = WM(m \cdot 1; \mathbb{F}) \tag{14.19}$$

Thus, Haar wavelet matrix is a wavelet matrix of genus 1.

14.3.9 The Canonical Haar matrix

In what follows, we provide a characterization of the Haar wavelet matrix.

14.3.10 Theorem

An $m \times m$ complex matrix H is a Haar wavelet matrix if and only if

$$H = \begin{pmatrix} 1 & 0 \\ 0 & U \end{pmatrix} \mathbb{H} \tag{14.20}$$

where $\mathbf{U} \in U(m-1)$ is a unitary matrix and $\mathbb{H}$ is the canonical Haar matrix of rank m which is defined by

$$\mathbb{H} = \begin{pmatrix} 1 & 1 & \cdots & \cdots & \cdots & \cdots & 1 \\ -(m-1)\sqrt{\frac{1}{m-1}} & \sqrt{\frac{1}{m-1}} & \cdots & \cdots & \cdots & \cdots & \sqrt{\frac{1}{m-1}} \\ \vdots & \vdots & & & & & \\ 0 & 0 & \cdots & -s\sqrt{\frac{m}{s^2+s}} & \sqrt{\frac{m}{s^2+s}} & \cdots & \sqrt{\frac{m}{s^2+s}} \\ \vdots & & & & & & \\ 0 & \cdots & \cdots & \cdots & 0 & -\sqrt{\frac{m}{2}} & \sqrt{\frac{m}{2}} \end{pmatrix} \tag{14.21}$$

where $s = (m - j)$ and $j = 0, 1, \ldots, (m + 1)$ are the row numbers of the matrix.

In the above, $\mathbf{U}(m - 1)$ is the group of $(m - 2) \times (m - 1)$ complex matrices U, such that $U^T U = 1$.

Before we prove the theorem, we prove the following lemma.

14.3.11 Lemma

If $H = (h_r^s)$ is a *Haar wavelet matrix*, then

$$h_r := h_r^0 = 1 \text{ for } 0 \le r < m \tag{14.22}$$

Proof: From (14.16) and (14.15), we have $\sum_{j=0}^{m-1} h_j \bar{h}_j = m$ and $\sum_{j=0}^{m-1} h_j = m$.

It follows that $\sum_{j=0}^{m-1} \bar{h}_j = m$.

Now,
$$\sum_{j=0}^{m-1} |h_j - 1|^2 = \sum_{j=0}^{m-1} (h_j \bar{h}_j - h_j - \bar{h}_j + 1)$$
$$= \sum_{j=0}^{m-1} h_j \bar{h}_j - \sum_{j=0}^{m-1} h_j - \sum_{j=0}^{m-1} \bar{h}_j + \sum_{j=0}^{m-1} 1$$
$$= m - m - m + m$$
$$= 0$$

which implies that $h_j = 1$ for $j = 0, \ldots, m-1$.

Proof of the theorem 14.3.9

We have seen that the elements of the first row of a Haar matrix are all equal to 1.

For the remaining $m-1$ rows, we proceed as follows. $\begin{pmatrix} 1 & 0 \\ 0 & U \end{pmatrix} H$ is a Haar matrix whenever H is a Haar matrix and $U \in \mathbf{U}(m-1)$. Hence the operation of U can be employed to develop a canonical form for H. The first step rotates the last row of H so that its first $(m-2)$ entries are zero. Since the rows of a Haar matrix are pairwise orthogonal and of length equal to $\sqrt{m}$, the orthogonality of the first and last rows implies that the last row can be normalized to have the form

$$\left(0, 0, 0, \ldots, -\sqrt{\frac{m}{2}}, \sqrt{\frac{m}{2}} \right)$$

Using the same argument for the proceeding rows the result can be obtained.

14.3.12   Remarks

I. If $H_1, H_2 \in \mathbb{H}(m; \mathbb{C})$ are two Haar matrices, then there exists a unitary matrix $U \in \mathbf{U}(m-1)$, such that

$$H_1 = \begin{pmatrix} 1 & 0 \\ 0 & U \end{pmatrix} H_2.$$

II. If A is a real wavelet matrix, that is, if $a_j^s \in \mathbb{R}$, then A is a Haar matrix if and only if,

$$A = \begin{pmatrix} 1 & 0 \\ 0 & O \end{pmatrix} \mathbb{H} \qquad (14.23)$$

where $O \in \mathbf{O}(m-1)$ is an orthogonal matrix and $\mathbb{H}$ is the canonical Haar matrix of rank m.

14.4 One-Dimensional Wavelet Systems

In this section, we introduce the basic **scaling** and **wavelet functions** of wavelet analysis. The principle result is that for any wavelet matrix $A \in WM(m, g; \mathbb{C})$, there is a scaling function $\phi(x)$ and $(m-1)$ wavelet functions $\psi^1(x), \ldots, \psi^{(m-1)}(x)$ which satisfy specific scaling relations defined in terms of the wavelet matrix A. These functions are all compactly supported and square-integrable.

14.4.1 The scaling equation

Let $A \in WM(m, g; \mathbb{C})$ be a wavelet matrix and consider the functional difference equation

$$\phi(x) = \sum_{j=0}^{mg-1} a_j^0 \phi(mx - j) \qquad (14.24)$$

This equation is called the **scaling equation** associated with the wavelet matrix $A = (a_j^s)$.

14.4.2 The scaling function

If $\phi \in L_2(\mathbb{R})$ is a solution of the equation (14.24), then ϕ is called a **scaling function**. It may be noted that $\begin{pmatrix} I & 0 \\ 0 & U \end{pmatrix} H$ is a Haar matrix whenever H is a Haar matrix and $\mathbf{U}$ is a group of $(m-1) \times (m-1)$ complex matrices U such that $U^*U = I$. Hence the action of U can be employed to develop a canonical form for H. Let the first step be to rotate the last row of H so that the first $(m-2)$ elements of the row are zeroes. Let the last two elements be α and β respectively. Then $\alpha + \beta = 0$ by (14.16) and $\alpha^2 + \beta^2 = m$ by (14.15). Hence, $\alpha = -\beta$ and $\alpha^2 = \frac{m}{2}$. Therefore, the last row is

$$\left(0, 0, \ldots, 0, -\sqrt{\frac{m}{2}}, \sqrt{\frac{m}{2}}\right).$$

Then next step would be to rotate the matrix so that the last three elements in $(m-1)$-th row are only non-zeroes. If these elements are α, β, γ then

$$\alpha + \beta + \gamma = 0$$
$$\alpha^2 + \beta^2 + \gamma^2 = m.$$

If we take $\beta = \gamma$ then $\alpha + 2\beta = 0$

$$\alpha^2 + 2\beta^2 = 6\beta^2 = m$$

i.e., $\beta = \sqrt{\dfrac{m}{6}}, \ \alpha = -2\sqrt{\dfrac{m}{6}}$

Hence, the last but one row is

$$\left(0, 0, \ldots, 0, -2\sqrt{\frac{m}{6}}, \sqrt{\frac{m}{6}}, \sqrt{\frac{m}{6}}\right).$$

Similarly, let the rotation yield for the $(m - s)$-th row all the first $(m - s - 1)$ elements as zeroes and the last $(s + 1)$ elements as non-zeroes.

Let these be $\alpha_1, \alpha_2, \ldots, \alpha_s, \alpha_{s+1}$.

Then $\qquad \displaystyle\sum_i \alpha_i = 0, \ \sum_i \alpha_i^2 = m.$

Taking $\qquad \alpha_2 = \alpha_3 = \cdots = \alpha_s = \alpha_{s+1}$

we have $\qquad \alpha_1 = -s\alpha_2$

$$\sum_i \alpha_i^2 = m \quad \text{or,} \quad s^2\alpha_2^2 + s\alpha_2^2 = m \quad \text{or,} \quad \alpha_2 = \sqrt{\frac{m}{s^2 + s}}.$$

Hence, the $(m - s)$-th row is

$$\left(0, 0, 0, \ldots, -s\sqrt{\frac{m}{s^2 + s}}, \sqrt{\frac{m}{s^2 + s}}, \ldots, \sqrt{\frac{m}{s^2 + s}}\right).$$

Thus we get the expression for $\mathbb{H}$ [see 14.21].

14.4.3   The wavelet function

If ϕ is a scaling function for the wavelet matrix A, then the **wavelet functions** $\{\psi^1, \psi^2, \ldots, \psi^{m-1}\}$ associated with matrix A and the scaling function ϕ are defined by the formula

$$\psi^s(x) = \sum_{j=0}^{mg-1} a_j^s \phi(mx - j) \tag{14.25}$$

14.4.4   Theorem

Let $A \in WM(m, g; \mathbb{C})$ be a wavelet matrix. Then, there exists a unique $\phi \in L_2(\mathbb{R})$, such that

(i) ϕ satisfies (14.24)

(ii) $\displaystyle\int_{\mathbb{R}} \phi(x)\,dx = 1$

(iii) $\operatorname{supp} \phi \subset \left[0, (g - 1)\left(\dfrac{m}{m - 1}\right) + 1\right].$

For proof see Resnikoff and Wells [46].

In what follows we give some examples and develop the notion of the wavelet system associated with the wavelet matrix A.

14.4.5   Examples

1. *The Haar functions* If $\mathbb{H} = \begin{pmatrix} 1 & 1 \\ -1 & 1 \end{pmatrix}$ is the canonical Haar wavelet matrix of rank 2, then the scaling function ϕ satisfies the equation

$$\phi(x) = \phi(2x) + \phi(2x - 1) \tag{14.26}$$

Hence, $\phi(x) = \chi[0, 1[$ where χ_K, the characteristic function of a subset K [see 10.2.10] is a solution of the equation (14.26) [see 14.2(b)].

The wavelet function $\psi = -\phi(2x) + \phi(2x - 1)$, where $\phi(x) = \chi[0, 1[$. For graph, see figure 14.2(c).

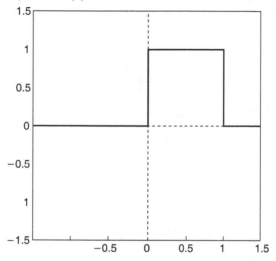

Fig. 14.2(b) The Haar scaling function for rank 2

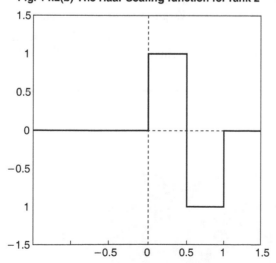

Fig. 14.2(c) The Haar scaling function for rank 2

2. *Daubechies wavelets for rank 2 and genus 2.*

$$\text{Let } D_2 = \frac{1}{4} \begin{pmatrix} 1+\sqrt{3} & 3+\sqrt{3} & 3-\sqrt{3} & 1-\sqrt{3} \\ -1+\sqrt{3} & 3-\sqrt{3} & -3-\sqrt{3} & 1+\sqrt{3} \end{pmatrix}$$

This is a wavelet matrix of rank 2 and genus 2 and discovered by Deubechies. For graphs of the corresponding scaling and wavelet functions, see Resnikoff and Wells [46]. The common support of ϕ and ψ is [0.3].

We end by stating a theorem due to Lawton, (Lawton [46]).

14.4.6   Theorem

Let $A \in WM(m, g; \mathbb{C})$. Let $W(A)$ be the wavelet system associated with A and let $f \in L_2(\mathbb{R})$ (3.1.3). Then, there exists an L_2-convergent expansion

$$f(x) = \sum_{j=-\infty}^{\infty} c_j \phi_j(x) + \sum_{s=1}^{m-1} \sum_{i=0}^{\infty} \sum_{j=-\infty}^{\infty} d_{ij}^s \psi_{ij}^s(x) \tag{14.27}$$

where the coefficients are given by

$$c_j = \int_{-\infty}^{\infty} f(x) \phi_j(x) dx \tag{14.28}$$

$$d_{ij}^s = \int_{-\infty}^{\infty} f(x) \psi_{ij}^s dx. \tag{14.29}$$

For proof, see Resnikoff and Wells [46].

14.4.7   Remark

1. For most wavelet matrices, the wavelet system $W[A]$ will be a complete orthonormal system and an orthonormal basis for $L_2(\mathbb{R})$.

2. However, for some wavelet matrices the system $W[A]$ is not orthonormal and yet the theorem 14.4.6 is still true.

CHAPTER 15

DYNAMICAL SYSTEMS

15.1   A Dynamical System and Its Properties

Let us consider a first order o.d.e. (ordinary differential equation) of the form

$$\dot{x} = \frac{dx}{dt} = \sin x \qquad (15.1)$$

The solution of the above equation is,

$$t = \log\left[\frac{\operatorname{cosec} x_0 + \cot x_0}{\operatorname{cosec} x + \cot x}\right], \qquad (15.2)$$

where $x(t)|_{t=0} = x_0$.

From (15.2), we can find the values of x for different values of t. But the determination of the values of x for different values of t is quite difficult.

On the other hand, we can get a lot of information about the solution by putting x against $\dot{x}$ of the graph $\dot{x} = \sin x$.

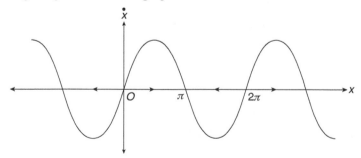

Fig. 15.1

Let us suppose $x_0 = \pi/4$.

Then the graph 15.1 describes the qualitative features of the solution $x(t)$ for all $t > 0$. We think of t as time, x as the position of an imaginary particle moving along the real line, and $\dot{x}$ as the velocity of that particle.

Then the differential equation $\dot{x} = \sin x$ represents a **vector field** on the line. The arrows indicate the directions of the corresponding velocity vector at each x. The arrows point to the right when $\dot{x} > 0$ and to the left when $\dot{x} < 0$. $\dot{x} > 0$ for $\pi/2 > x > 0$ and $\dot{x} < 0$ for $x < 0$. At $x > \pi$, $\dot{x} < 0$ and for $\pi/2 < x < \pi$, $\dot{x} > 0$. At points where $\dot{x} = 0$, there is no flow. Such points are therefore called **fixed points**. This type of study of o.d.e.-s to obtain the qualitative properties of the solution was pioneered by H. Poincaré in the late 19th century [6]. What emerged was the theory of dynamical systems. This may be treated as a special topic of the theory of o.d.e.-s. Poincaré followed by I. Benedixon [Bhatia and Sjego [6]] studied topological properties of solutions of autonomous o.d.e.-s in the plane. Almost simultaneously with Poincaré, A.M. Lyapunov [32] developed his theory of stability of a motion (solution) for a system of n first order o.d.e.-s. He defined, in a precise form, the notion of stability; asymptotic stability and instability, and developed a method for the analysis of the stability properties of a given solution of an o.d.e.. But all his analysis was strictly in a local setting. On the other hand, Poincaré studied the global properties of differential equations in a plane.

As pointed out in the example above, Poincaré introduced the concept of a trajectory, i.e., a curve in the $x, \dot{x}$ plane, parametrized by the time variable t, which can be found by eliminating the variable t from the given equations, thus reducing these to first order differential equations connecting x and $\dot{x}$. In this way, Poincaré set up a convenient geometric framework in which to study the qualitative behaviour of planar differential equations. He was not interested in the integration of particular types of equations, but in classifying all possible behaviours of the class of all second order differential equations. Great impetus to the theory of dynamical systems came from the work of G.D. Birkoff [7]. There are many other authors who contributed to a large extent to this qualitative theory of differential equations. In this chapter we give exposure to the basic elements in a dynamical system.

15.1.1   *Definition: dynamical system*

Let X denote a metric space with metric ρ. A dynamical system on X is the triplet $(X, \mathbb{R}, \pi)$, where π is a map from the product space $X \times \mathbb{R}$ into the space X satisfying the following axioms:

  (i) $\pi(x, 0) = \pi(x)$ for every $x \in X$ (identity axiom)

 (ii) $\pi(\pi(x, t_1), t_2) = \pi(x, t_1 + t_2)$ for every $x \in X$ and t_1, t_2 in $\mathbb{R}$ (group axiom)

(iii) π is continuous (continuity axiom).

Given a dynamical system on X, the space X and the map are respectively called the **phase space** and the **phase map** (of the dynamical system).

Unless otherwise stated, X is assumed to be given.

15.1.2 Example: ordinary autonomous differential systems

The differential system

$$\frac{dx}{dt} = \dot{x} = f(x, t) \tag{15.3}$$

is called an autonomous system if the RHS in (15.3) does not contain t explicitly.

We consider the equation

$$\frac{dx}{dt} = \dot{x} = f(x) \tag{15.4}$$

where $f : \mathbb{R}^n \to \mathbb{R}^n$ is continuous and, moreover, let us assume that for each $x \in \mathbb{R}^n$ a unique solution $\psi(t, x)$ which is defined on $\mathbb{R}$ and satisfies $\psi(0, x) = x$. Then following Coddington and Levinson ([13], chapters 1 and 2) it can be said that the uniqueness of the solution implies that

$$\psi(t_1, \psi(t_2, x)) = \psi(t_1 + t_2, x) \tag{15.5}$$

and considered as a function from $\mathbb{R} \times \mathbb{R}^n$ into $\mathbb{R}^n$. Moreover, ψ is continuous in its arguments [see section 4, chapter II, Coddington and Levinson [13]].

We assume that f satisfies the global Lipschitz condition, i.e.,

$$\|f(x_1) - f(x_2)\| \le M\|x_1 - x_2\|, \tag{15.5}$$

for all $x_1, x_2 \in \mathbb{R}^n$ and a given positive number M, so that the conditions on solutions of (15.4) are obtained. We next want to show that the map $\pi : \mathbb{R}^n \times \mathbb{R} \to \mathbb{R}^n$ such that $\pi(x, t) = \psi(t, x)$ defines a dynamical system on $\mathbb{R}^n$.

For that, we note that

$$\pi(x, 0) = \psi(0, x) = x$$
$$\pi(\pi(x, t_1), t_2) = \psi(t_2, \psi(t_1, x)) = \psi(t_1 + t_2, x)$$
$$= \pi(x, t_1 + t_2)$$

Moreover, $\pi(x, t)$ is continuous in its arguments. Thus all the axioms (i), (ii) and (iii) of 15.1.1 are fulfilled. Hence $\pi(x, t)$ is a phase map.

Example 2: ordinary autonomous differential systems

Let us consider the system

$$\frac{dx}{dt} = \dot{x} = F(x) \tag{15.6}$$

where $F : D \to \mathbb{R}$ is a continuous function on an open set $D \subset \mathbb{R}^n$ and for each $x \in D$ (15.6) has a unique solution $\psi(t, x), \psi(0, x) = x$ defined on a maximal interval $(a(x), b(x))$, $-\infty \leq a(x) < 0 < b(x) \leq +\infty$. For each $x \in D$, define $\Gamma^+(x) = \{\psi(t, x) : 0 \leq t < b(x)\}$ and $\Gamma^-(x) = \{\psi(t, x) : a(x) \leq t \leq 0\}$, $\Gamma^+(x)$ and $\Gamma^-(x)$ are respectively called the positive and the negative trajectories respectively through the point $x \in D$.

We will show that to each system (15.6), there corresponds a system

$$\frac{dx}{dt} = \dot{x} = F(x), \ x \in \mathbb{R}^n \tag{15.7}$$

where $F : D \to \mathbb{R}^n$ such that (15.7) defines a dynamical system on D with the property that for each $x \in D$ the systems (15.6) and (15.7) have the same positive and the same negative trajectories.

If $D = \mathbb{R}^n$, then given the equation (15.4), we set,

$$\frac{dx}{dt} = \dot{x}(t) = F_1(x) = \frac{F(x)}{1 + ||F(x)||} \tag{15.8}$$

where $|| \cdot ||$ is the Euclidean norm. If $D \neq \mathbb{R}^n$, then the boundary ∂D of $D \neq \Phi$ and is closed.

We next consider the system

$$\frac{dx}{dt} = \dot{x}(t) = F_1(x) = \frac{F(x)}{1 + ||F(x)||} \cdot \frac{\rho(x, \partial D)}{1 + \rho(x, \partial D)} \tag{15.9}$$

where $\rho(x, \partial D) = \inf\{||x - y|| : y \in \partial D\}$.

In other words, $\rho(x, \partial D)$ is the distance of x from ∂D.

Since f satisfies Lipschitz condition, equation (15.4) has a unique solution.

Now, $||F_1(x) - F_1(y)|| = \left|\left|\dfrac{F(x)}{1 + ||F(x)||} - \dfrac{F(y)}{1 + ||F(y)||}\right|\right|$

$$\leq \frac{1}{(1 + ||F(x)||)(1 + ||F(y)||)} \left[||F(x) - F(y)|| + ||F(x)|| \, ||F(y)|| \times \right.$$

$$\left. \left|\left|\left(\frac{F(x)}{||F(x)||} - \frac{F(y)}{||F(y)||}\right)\right|\right|\right]$$

$$< K||x - y|| + ||e(x) - e(y)||$$

where $e(x)$ is a vector of unit norm in the direction of $F(x)$.

Assuming $||F(x)|| > m > 0$ we have

$$||F_1(x) - F_1(y)|| < (K + k)||x - y||$$

where $||e(x) - e(y)|| \leq k||x - y||$.

Thus, F_1 satisfies the global Lipschitz condition. Thus, (15.8) defines a dynamical system. Similarly, (15.9) also defines a dynamical system. (15.8) and (15.9) have the same positive and negative trajectories as (15.4).

15.2 Homeomorphism, Diffeomorphism, Riemannian Manifold

15.2.1 Definition: homeomorphism

See 1.6.5.

15.2.2 Definition: manifold

A topological space Y [see 1.5.1] is called an **n-dimensional manifold** or (n-manifold) if every point of Y has a neighbourhood homeomorphic to an open subset of $\mathbb{R}^n$. Since one can clearly take this subset to be an open ball, and since an open ball in $\mathbb{R}^n$ is homeomorphic to $\mathbb{R}^n$ itself, the condition for a space to be a **manifold** can also be expressed by saying that every point has a **neighbourhood homeomorphic to $\mathbb{R}^n$**.

15.2.3 Remark

A homeomorphism from an open subset V of Y to an open subset of $\mathbb{R}^n$ allows one to transfer the cartesian coordinate system of $\mathbb{R}^n$ to V. This gives a *local coordinate system* or *chart* on Y [see Schwartz [51]].

15.2.4 Differentiability, C^r-,C^0-maps

Let U be an open subset of $\mathbb{R}^n$.

For definition of differentiability of $J : L \subset \mathbb{R}^n \to \mathbb{R}$ see 13.2 and 13.3.

C^r- J is said to belong to class C^r if J is r times continuously differentiable on an open set $U \subseteq \mathbb{R}^n$.

C^r-map Let $F : U \subseteq \mathbb{R}^n \to V \subseteq \mathbb{R}^m$. For definition of differentiability of F and the concrete form of the derivative see Ortega and Rheinboldt [42].

If $(x_1, \ldots, x_n)^T \in U$ and $(y_1, y_2, \ldots, y_n)^T \subset V$ then

$$y_i = F_i(x_1, x_2, \ldots, x_n), \ i = 1, 2, \ldots, m \tag{15.10}$$

The map F is called a **C^r-map** if F_i is continuously r differentiable for some $1 \leq r \leq \infty$.

Smooth: $F : U \subset \mathbb{R}^n \to \mathbb{R}^m$ is said to be **smooth** if it is a **C^∞-map**.

C^0-map: Maps that are continuous but not differentiable will be referred to C^0-maps.

15.2.5 Definition: diffeomorphism

$F : U \subset \mathbb{R}^n \to \mathbb{R}^m$ is said to be a diffeomorphism if it fulfils the following conditions:

(i) F is a **bijection** (one-to-one and onto) [see 1.2.3].
(ii) Both F and F^{-1} are **differentiable** mappings.

F is said to be **C^k-diffeomorphism** if both F and F^{-1} are **C^k-maps**.

15.2.6 Remark

Note that $G : U \to V$ is a diffeomorphism if and only if, $m = n$ and the matrix of partial derivatives,

$$G'(x_1, \ldots, x_n) = \left(\frac{\partial G_i}{\partial x_j}\right), \quad i, j = 1, \ldots, n$$

is non-singular at every $x \in U$.

Example: Let $G(x, y) = \left(\begin{array}{c} \exp y \\ \exp x \end{array}\right)^T$ with $U = \mathbb{R}^2$ and $V = \{(x, y) : x > 0, \ y > 0\}$.

$$G'(x, y) = \left(\begin{array}{cc} 0 & \exp y \\ \exp x & 0 \end{array}\right), \quad \det G'(x, y) = -\exp(x + y) \neq 0 \text{ for each}$$

$(x, y) \in \mathbb{R}^2$.

Thus, G is a **diffeomorphism**.

15.2.7 Two types of dynamical systems

We note that in a Dynamical system the **state** changes with time $\{t\}$. The two types of dynamical system encountered in practices are as follows:

(i) $x_{t+1} = G(x_t)$, $t \in \mathbb{Z}$ or $\mathbb{N}$ (15.11)

Such a system is called a **discrete system**.

(ii) When t is continuous, the dynamics are usually described by a differential equation,

$$\frac{dx}{dt} = \dot{x} = F(x) \tag{15.12}$$

In (15.11), x represents the state of the system and takes values in the state space or phase space X [see 15.1.1]. Sometimes the phase space is the Euclidean space or a subspace of it. But it can also be a non-Euclidean structure such as a circle, sphere, a torus or some other **differential manifold**.

15.2.8 Advantages of taking the phase space as a differential manifold

If the phase space X is Euclidean, then it is easy to analyse. But if the phase space X is non-Euclidean but a differential manifold there is also an advantage. This is because a differential manifold is 'locally Euclidean' and this allows us to extend the idea of differentiability to functions defined on them. If Y is a manifold of dimension n, then for any $x \in Y$ we can find a neighbourhood $N_x \subseteq Y$ containing x and a homomorphism $h : N_x \to \mathbb{R}^n$ which maps N_x onto a neighbourhood of $h(x) \in \mathbb{R}^n$. Since we can define coordinates in $U = h(N_x) \subseteq \mathbb{R}^n$ (the coordinate curves of which can be mapped back onto N_x) we can think of h as defining local coordinates on the patch N_x of Y [see figure 15.2].

The pair (U, h) is called a **chart** and we can use it to give differentiability on N_x. Let us assume that $G : N_x \to N_x$ then G induces a mapping $\hat{G} = h \cdot G \cdot h^{-1} : U \to U$ [see figure 15.3]. We say G is a C^k-map on N_x if $\hat{G}$ is a C^k-map on U.

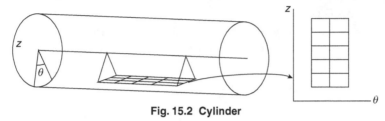

Fig. 15.2 Cylinder

Example of a differential manifold

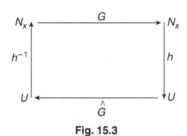

Fig. 15.3

15.3 Stable Points, Periodic Points and Critical Points

Let Y be a differential manifold and $G : Y \to Y$ be a diffeomorphism. For $x \in Y$, the iteration (15.11) generates a sequence $\{G^k_{(x_t)}\}$. The distinct points of the sequence define the orbit or trajectory of x under G. More generally, the orbit of x under G is $\{G^m(x) : m \in \mathbb{Z}\}$. For $m \in \mathbb{Z}^+$, G^m is the composition of G with itself m times. Since G is a diffeomorphism, G^{-1} exists and $G^{-m} = (G^{-1})^m \cdot G^0 = Idy$, the identity map on Y. Thus, the orbit of x is an infinite (on both sides) sequence of distinct points of Y.

15.3.1 Definition: fixed point

A point $x^* \in Y$ is called a **fixed point** of G if $G^{(m)}(x^*) = x^*$ for all $m \in \mathbb{Z}$.

15.3.2 Example 1

To find the fixed points for $\dot{x} = F(x)$, where $F(x) = x^2 - 1$.

Now, $x(t+1) - x(t) \approx F(x(t))$.

If x^* is a fixed point of G, then

$$x^* - x^* = F(x^*), \text{ i.e., } F(x^*) = 0, \text{ i.e., } x^* = \pm 1.$$

It may be noted that at the fixed points, $x^* = \pm 1$, there is no **flow**.

15.3.3   Definition: periodic points

A point $x^* \in Y$ is said to be a **periodic point** of Y if $G^p(x^*) = x^*$ for some integer $p \geq 1$.

The least value of p for which definition of a periodic point is true is called the period of the point x^* and the orbit of x^* is

$$\{x^*, G(x^*), G^2(x^*), \ldots, G^{p-1}(x^*)\} \tag{15.13}$$

is said to be a periodic orbit of period p or a p-cycle of G.

15.3.4   Remark

1. A fixed point is a periodic point of period one.

2. Since $(G^p)^q(x^*) = G^p(x^*)$, $q \in \mathbb{Z}$ for a periodic point x^* of G with period p, x^* is a fixed point of $G^p(x^*)$.

3. If x^* is a periodic point of period p for G, then all other points in the orbit of x^* are periodic points of period p of G. For if $G^p(x^*) = x^*$, then $G^p(G^i(x^*)) = G^i(G^p(x^*)) = G^i(x^*)$, $i = 0, 1, 2, \ldots, q - 1$.

15.3.5   Definition: stability according to Lyapunov stable point

A fixed point x^* is said to be stable if, for every neighbourhood N_{x^*} of x^*, there is a neighbourhood $N'_{x^*} \subseteq N_{x^*}$ of x^* such that if $x \in N'_{x^*}$ then $G^m(x^*) \subseteq N_{x^*}$ for all $m > 0$. The above definition implies that iterates of points *near to* a *stable fixed point* remain *near to* it for $m \in \mathbb{Z}^+$. This is in conformity with the definition of a **stable equilibrium point** for a **moving particle**.

15.3.6   Remark

1. If a fixed point x^* is stable and $\lim_{m \to \infty} G^m(x^*) = x^*$, for all x in some neighbourhood of x^*, then the fixed point is said to be **asymptotically stable**.

2. **Unstable point.** A fixed point x^* is said to be **unstable** if for every neighbourhood N_{x^*} of x^* $G^m(x) \notin N_{x^*}$ for all $x \in N_{x^*}$.

15.3.7   Example

We refer to the example in 15.3.2. To determine stability, we plot $x^2 - 1$ and then sketch the vector field [see figure 15.4]. The flow is to the right where $x^2 - 1 > 0$ and to the left where $x^2 - 1 < 0$. Thus, $x^* = -1$ is **stable** and $x^* = 1$ is **unstable**.

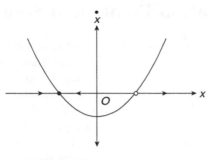

Fig. 15.4

We end this section with an important theorem.

15.4 Existence, Uniqueness and Topological Consequences

15.4.1 Theorem: existence and uniqueness

Consider $\dot{x} = F(x)$, where $x \in \mathbb{R}^n$ and $F : U \subset \mathbb{R}^n \to C^1(\mathbb{R}^n)$, where U is an open connected set in $\mathbb{R}^n$.

Then, for $x_0 \in U$, the initial value problem has a solution $x(t)$ on some time interval about $t = 0$ and the solution is **unique**. (Strogatz [54])

15.4.2 Remark

1. These theorems have deep implications. **Different trajectories do not intersect.**

For if two trajectories intersect at some point in the phase space, then starting from the crossing point we get **two solutions along the two trajectories.** This contradicts the **uniqueness** of the solution.

2. In two dimensional phase spaces let us consider a closed orbit C in the phase plane. Then any trajectory starting inside C will always lie within C. If there are fixed points inside C, then the trajectory may approach one of them.

But if there are no fixed points inside C, then by intuition we can say that the trajectory can not move inside the orbit endlessly. This is supported by the following famous theorem.

Poincaré-Bendixon theorem:

If a trajectory is confined to a closed, bounded region and there are no fixed points in the region, then the trajectory must eventually approach a closed circuit (Arrowsmith and Pace [3]).

15.5   Bifurcation Points and Some Results

Bifurcation phenomenon is the outcome of the presence of a parameter in a dynamical system. A physical example may stimulate the study of the bifurcation theory. Suppose a body is resting on a vertical iron pillar. If the weight is gradually increased, a stage may come when the pillar may become unstable and buckle. Here the weight plays the role of a control parameter and the deflection of the pillar from vertical plays the role of the dynamical variable x. The bifurcation of fixed points for flows on the line occurs in several physical phenomenon, such as the onset of coherent radiation in a laser and the outbreak of an insect population, etc. [see Strogatz [54]]. Against the above backdrop, we can have the formal definition of bifurcation as follows. Let $F : \mathbb{R}^m \times \mathbb{R}^n \to \mathbb{R}^n (G : \mathbb{R}^m \times \mathbb{R}^n \to \mathbb{R}^n)$ to be an m-parameter, C^r-family of vector fields (diffeomorphisms) on $\mathbb{R}^n$, i.e., $(\mu, x) \longmapsto F(\mu, x)(G(\mu, x))$, $\mu \in \mathbb{R}^m$, $x \in \mathbb{R}^n$. The family $F(G)$ is said to have a **bifurcation** point at $\mu = \mu^*$ if, in every neighbourhood of μ^*, there exists values of μ such that corresponding vector fields $F(\mu, \cdot) = F_\mu(\cdot)$ (diffeomorphisms $G(\mu, \cdot) = G_\mu(\cdot)$) show topologically distinct behaviour. For details see Arrowsmith and Place [3]. We provide here some examples.

Example 1. $F(x, \mu) = \mu - x^2$ (15.14)

Let us sketch the phase portraits for $\dot{x} = F_\mu(x)$, with (μ, x) near $(0,0)$. Since our study is confined to a neioghbourhood of $(0,0)$, the bifurcation study is local in nature (15.5).

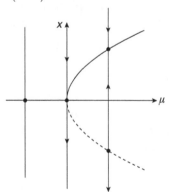

Fig. 15.5

For $\mu < 0$, the equation (15.14) yields $\dot{x} < 0$ for all $x \in \mathbb{R}$. When $\mu = 0$ there is a non-hyperbolic fixed point at $x = 0$, but $\dot{x} < 0$ for all $x \neq 0$ (Arrowsmith and Place [3]). For $\mu > 0$, $\mu - x^2 = 0 \Rightarrow x = \pm \mu^{1/2}$ for $x > \mu^{1/2}$, $\dot{x} < 0$ and for $x < \mu^{1/2}$, $\dot{x} > 0$. Hence, $x = \mu^{1/2}$ is a **stable** fixed point. On the other hand, $x = -\mu^{1/2}$ is an **unstable** fixed point.

Example 2. $F(\mu, x) = \mu x - x^2$ (15.15)

If $\mu > 0$, $x = \mu$ is stable and $x = 0$ is unstable. The stabilities are reversed when $\mu < 0$. At $\mu = 0$, there is one singularity at $x = 0$ and $\dot{x} < 0$

for all $x \neq 0$. This leads to a bifurcation as depicted below [see figure 15.6].

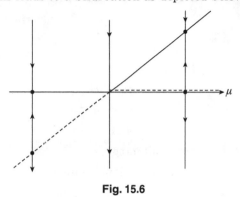

Fig. 15.6

List of Symbols

Φ (phi)	Null set
$\in$	Belongs to
$A \subseteq B$	A is a subset of B
$B \supseteq A$	B is a superset of A
$P(A)$	Power set of A
J	Set of all integers
$\mathbb{Q}$	Set of all rational numbers
$\mathbb{N} = \{1, 2, \ldots, n, \ldots\}$	The set of natural numbers
P	Set of all polynomials
$\mathbb{R}^n$	n-dimensional real Euclidean space
$]a, b[$	An open interval containing a and b
$[a, b]$	A closed interval containing a and b
$\aleph_0$	Class of all denumerable sets
$\mathbb{R}$	Set of real numbers
$\mathbb{C}$	Set of complex numbers
$A \cup B$	Union of sets A and B
$A + B$	The sum of sets A and B
$A \cap B$	The intersection of A and B
AB	The product of A and B
$A - B$	The set of elements in A which are not elements of B
A^c	The complement of the set A
g.l.b (infimum)	Greatest lower bound
l.u.b (supremum)	Least upper bound
$X \times Y$	Cartesian product of sets X and Y
(x, y)	Ordered pair of elements x and y
$\mathbb{R}^{m \times n}$	Space of $m \times n$ matrices
$A = \{a_{ij}\}$	A, a matrix
l_∞	Sequence space
$C([a, b])$	Space of functions continuous in $[a, b]$
l_p	p^{th} power summable space
l_2	Hilbert sequence space
$L_p([a, b])$	Lebesgue p^{th} integrable functions

$P_n([a,b])$	Space of real polynomials of order n defined on $[a,b]$
$X_1 \oplus X_2$	Direct sum of subspaces
X/Y	Quotient space of a linear space X by a subspace Y
Codimension (codim) of Y in X	Dimension of Y in X
$\rho(x,y)$	Distance between x and y
(X,ρ)	Metric space where X is a set and ρ a metric
$M([a,b])$	Space of bounded real functions
c	Space of convergent numerical sequences
m	Space of bounded numerical sequences
s	Space of not necessarily bounded sequences
$\subset$	Set inclusion
$D(A)$	Diameter of a set A
$D(A,B)$	Distance between two sets A and B
$D(x,A)$	Distance of a point x from a set A
$B(x_0,r)$	Open ball with centre x_0 and radius r
$\overline{B}(x_0,r)$	Closed ball with centre x_0 and radius r
$S(x_0,r)$	Sphere with centre x_0 and radius r
K^c	The complement of the set K
N_{x_0}	Neighbourhood of x_0
$\overline{A}$	Closure of the set A
$\mathbb{C}^n$	n-dimensional complex space
$S([a,b])$	Space of continuous real valued functions on $[a,b]$ with the metric $\rho(x,y) = \int_a^b |x(t)-y(t)|dt$
B_x	Basic $\mathcal{F}$-neighbourhood of x
$D(A)$	Derived set of A
Int A	Interior of A
$\bigcap\limits_{i=1}^{n} F_i$	Intersection of all sets for $i = 1, 2, \ldots n$
Inf	Infimum
Sup	Supremum
$\|\cdot\|$	Norm

$x_m \to x$	$\{x_m\}$ tends to x
lim sup	Limit supremum
lim inf	Limit infimum
$\|\cdot\|_1$	l_1—norm
$\|\cdot\|_2$	l_2—norm
$\|\cdot\|_\infty$	l_∞—norm
$\sum\limits_{n=1}^{\infty} x_n$	Summation of x_n for $n = 1, \ldots \infty$
$\sum\limits_{n=1}^{m} x_n$	Summation of x_n for $n = 1, 2, \ldots m$ (finite)
$l_p^{(n)}$	n-dimensional p^{th} summable space
X^0	The interior of the set X
c_0	Space of all sequences converging to 0
$\|x + L\|_q$	Quotient norm of $x + L$
Span L	Set of linear combinations of elements in L
$\sigma_a(A)$	Approximate eigenspectrum of an operator A
$r_\sigma(A)$	Spectral radius of an operator A
A^T	Transpose of a matrix A
$\mathcal{D}(A)$	Domain of an operator A
$R(A)$	Range of an operator A
$N(A)$	Null space of an operator A
$BV([a,b])$	Space of scalar-valued functions of bounded variation on $[a, b]$
$NBV([a,b])$	Space of scalar-valued normalized functions of bounded variation on $[a, b]$
Δ	Forward difference operator
$x_n \xrightarrow{w} x$	$\{x_n\}$ is weak convergent to x
$\langle x, y \rangle$	Inner product of x and y
$x \perp y$	x is orthogonal to y
$E \perp F$	E is orthogonal to F
$\overline{M}^T$	Conjugate transpose of a matrix M
$A \geq 0$	A is a non-negative operator
$w(A)$	Numerical range of an operator A

$m(E)$	Lebesgue measure of a subset E of $\mathbb{R}$
$\displaystyle\int_E x\, dm$	Lebesgue integral of a function x over a set E
Var (x)	Total variation of a scalar valued function x
essup$_E|x|$	Essential supremum of a function $|x|$ over a set E
$L_\infty(E)$	The set of all equivalent classes of essentially bounded functions on E
$\overline{\text{Span } L}$	Closure of the span of L
$M^\perp$	Set orthogonal to M
$M^{\perp\perp}$	Set orthogonal to $M^{\perp\perp}$
$(E_x \to E_y)$	Space of all bounded linear operators mapping n.l.s. $E_x \to$ n.l.s. E_y
H	Hilbert space
A^{-1}	Inverse of an operator A
A^*	Adjoint of an operator A
$\overline{A}$	Closure of A
A_λ	Operator of the form $A - \lambda I$
$\|A\|$	Norm of A
$A(x,x)$	Quadratic Hermitian form
$A(x,y)$	Bilinear Hermitian form
$C^k([0,1])$	Space of continuous functions $x(t)$ on $[0,1]$ and having derivatives to within k-th order
E_x^*	Conjugate (dual) of E_x
E_x^{**}	Conjugate (dual) of E_x^*
$\Pi_{E_x}(x)$	Canonical embedding of E_x into E_x^{**}
χ_E	Characteristic function of a set E
δ_{ij}	Kronecker delta
sgn z	Signum of $z \in \mathbb{C}$
$\{e_1, e_2, \ldots\}$	Standard Schauder basis for $\mathbb{R}^n$ or $\mathbb{C}^n$ or $l^p, 1 \le p < \infty$
$Gr(F)$	Graph of a map F
I	Identity operator
$\rho(A)$	Resolvent set of an operator A
$\sigma(A)$	Spectrum of an operator A

$\sigma_e(A)$ Eigenspectrum of an operator A

Abbreviations

BVP   Boundary value problems
LHS   Left hand side
ODE   Ordinary differential equations
RHS   Right hand side
s.t.   Such that
WRT   With regards to
a.e.   Almost everywhere

Bibliography

[1] Aliprantis, C.D. and Burkinshaw, O. (2000) : Principles of Real Analysis, Harcourt Asia Pte Ltd, Englewood Cliffs, N.J.

[2] Anselone, P.M. (1971) : Collectively Compact Operator Approximation Theory, Prentice-Hall.

[3] Arrowsmith, D.K. and Place, C.M. (1994) : An Introduction to Dynamical Systems, Cambridge University Press, Cambridge.

[4] Bachman, G. and Narici, L. (1966) : Functional Analysis, Academic Press, New York.

[5] Banach, S. (1932) : Théories des opérations linéaires, Monografje Matematyczne, Warsaw.

[6] Bhatia, N.P. and Szegö, G.P. (1970) : Stability Theory of Dynamical Systems, Springer-Verlag, New York.

[7] Birkoff, G.D. (1927) : Dynamical Systems (Amer. Math. Soc. Colloquium Publications Vol. 9), New York.

[8] Bohnenblust, H.F. and Sobczyk, A. (1938) : Extension of Functionals on Complex Linear Spaces, Bull. Amer. Math. Soc. **44**, 91–3.

[9] Browder, F.E. (1961) : On the Spectral Theory of Elliptic Differential Operators I, Math. Ann., Vol. **142**, 22–130.

[10] Carleson, A. (1966) : On the Convergence and Growth of Partial Sums of Fourier series, Acta Math. **116**, 135–7.

[11] Cea, J. (1978) : Lectures on Optimization—Theory and Algorithms, Tata Institute of Fundamental Reseach, Narosa Publishing House, New Delhi.

[12] Clarkson, J.A. (1936) : Uniformly Convex Spaces, Trans. Amer. Math. Soc. **40**, 396–414.

[13] Coddington, E.A., Levinson, N. (1955) : Theory of Ordinary Differential Equations, McGraw-Hill Book Company, New York.

[14] Collatz, L. (1966) : Functional Analysis and Numerical Mathematices, Academic Press, New York.

[15] Courant, R. and Hilbert, D. (1953) : Methods of Mathematical Physics, Interscience, New York.

[16] Daubechies, I. (1988) : Orthonormal Bases of Compactly Supported Wavelets. Commun. Pure App. Math., **41**; 909–96.

[17] Debnath, L. (1998) : Wavelet Transforms and their Applications, Pinsa-A, **64**, A, 6.

[18] Dieudońne, J. (1969) : Foundations of Modern Analysis, Academic Press, New York.

[19] Gelfand, I. (1941) : Normiert Ringe, Mat. Sbornik, N.S. 9 (51), 3–24.

[20] Goffman, C and Pedrick, G. (1974) : First Course in Functional Analysis, Prentice-Hall of India Private Ltd., New Delhi.

[21] Goldberg, S. (1966) : Unbounded Linear Operators with Applications, Mc-Graw Hill Book Company, New York.

[22] Haar, L. (1910) : Zur Theorie der Orthogonalen Functionensysteme, Math. Ann. **69**, 331–371.

[23] Hahn, H. (1927) : Über Lineare Gheichungssysteme in Linearen Raümen, Journal Reine Angew Math. **157**, 214–29.

[24] Hilbert, D. (1912) : Grundzüge Einerallgemeinen Theorie der Lineraren Integralgleichungen, Repr. 1953, New York.

[25] Jain, P.K., Ahuja, O.P. and Ahmed, K. (1997) : Functional Analysis, New Age International (P) Limited, New Delhi.

[26] James, R.C. (1950) : Bases and Reflexively in Banach spaces, Ann. Math., **52**, 518–27.

[27] Kantorovich, L.V. (1948) : Functional Analysis and Applied Mathematics, (Russian) Uspekhi Matem. Nauk, **3**, 6, 89–185.

[28] Kato, T. (1958) : Perturbation Theory for Nullity, Deficiency and Other Quantities of Linear Operators, J. Analyse Math. vol. 6, pp. 273–322.

[29] Kolmogoroff, A. and Fomin, S. (1954) : Elements of the Theory of Functions and Functional Analysis, Izdatb Moscow Univ., Moscow; transl. by L. Boron. Grayrock Press, Rochester, New York, 1957.

[30] Kreyszig, E. (1978) : Introductory Functional Analysis with Applications, John Wiley & Sons. New York.

[31] Lahiri, B.K. (1982) : Elements of Functional Analysis, World Press, Kolkata.

[32] Liapunov, A.M. (1966) : Stability of Motion (English translation), Academic Press, New York.

[33] Limaye, B.V. (1996) : Functional Analysis, New Age International Ltd., New Delhi.

[34] Lions, J.L. and Magenes, E. (1972) : Non-homogeneous Boundary Value Problems, vol. I, Springer-Verlag, Berlin.

[35] Lusternik, L.A. and Sobolev, V.J. (1985) : Elements of Functional Analysis, Hindusthan Publishing Corporation, New Delhi.

[36] Mikhlin, S. (1964) : Variational Methods in Mathematical Physics, Pergamon Press, New York.

[37] Mikhlin, S. (1965) : The Problem of the Minimum of a Quadratic Functional, Holden-day, San Francisco.

[38] Mansfield, M.J. (1963) : Introduction to Topology, Litton Educational Publishing Inc., New York.

[39] Nair, M.T. (2002) : Functional Analysis, Prentice-Hall of India Private Limited, New Delhi.

[40] Natanson, I.P. (1955) : Konstruktive Funktionentheories (translated from Russian), Akademic Verlag, Berlin.

[41] Neumann, J. Von. (1927) : Mathematische Begründung der Quantenmechanik. Nachr. Ges.Wiss. Götingen. Math. Phys. Kl 1–37.

[42] Ortega, J.M. and Rheinboldt, W.C. (1970) : Iterative Solution of Nonlinear Equations in Several Variables, Academic Press, New York.

[43] Ralston, A. (1965) : A First Course in Numerical Analysis, McGraw-Hill Book Company, New York.

[44] Rall, L.B. (1962) : Computational Solution of Nonlinear Operator Equations, John Wiley & Sons, New York.

[45] Reddy, J.N. (1986) : Applied Functional Analysis and Variational Methods in Engineering, McGraw-Hill Book Company, New York.

[46] Resnikoff, H.L. and Wells, O Jr. (1998) : Wavelet Analysis, Springer, New York.

[47] Royden, H.L. (1988) : Real Analysis, Macmillan, New York.

[48] Rudin, W. (1976) : Principles of Mathematical of Analysis, 3rd. ed., McGraw-Hill Book Company, New York.

[49] Schauder, J (1930) : Über Lineare, Vollstetige Funktionaloperationen, Studia Math. **2**, 1–6.

[50] Schmidt, E. (1908) : Über die Auflösung Linearer Gleichungen mit Unendlich vielen Unbekannten, Rendi. Circ. Math. Palermo **25**, 53–77.

[51] Schwartz, A.S. (1996) : Topology for Physicists, Springer-Verlag, Berlin.

[52] Schwartz, L (1951) : Théorie des Distributions, vols. I and II. Hermann & Cie, Paris.

[53] Simmons, G.F. (1963) : Introduction to Topology and Modern Analysis, Mc-Graw Hill Book Company, Tokyō.

[54] Strogatz, S.H. (2007) : Nonlinear Dynamics and Chaos, Levant Books, Kolkata.

[55] Taylor, A.E. (1958) : Introduction to Functional Analysis, John Wiley & Sons, New York.

[56] Zeidler, E. (1995) : Applied Functional Analysis: Main Principles and their Applications, Springer-Verlag, Berlin.

Index